Introduction to Unified Mechanics Theory with Applications

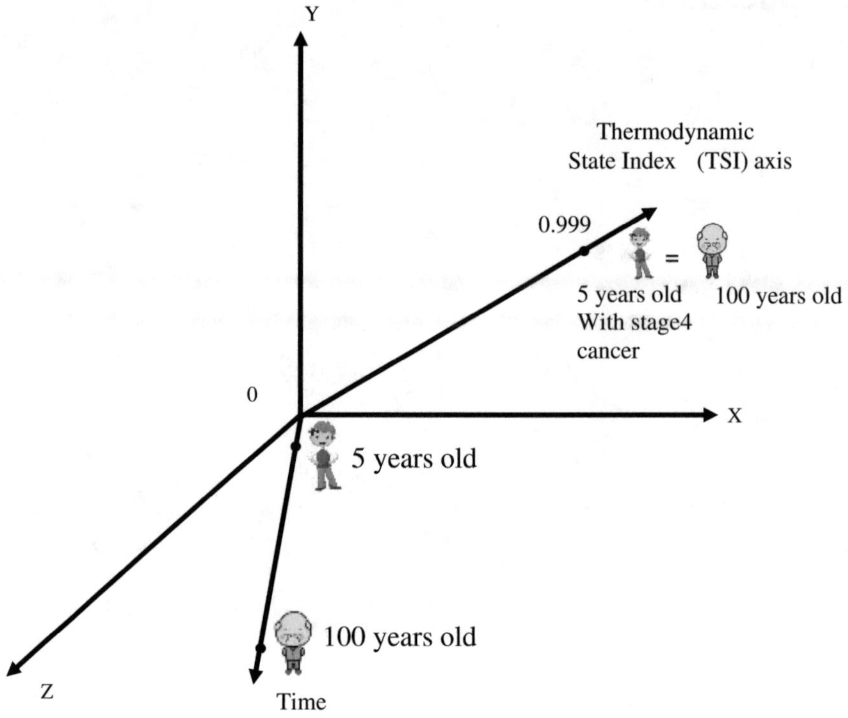

Y
Thermodynamic
State Index (TSI) axis
0.999
=
5 years old
With stage4
cancer
100 years old
0
X
5 years old
100 years old
Z
Time

Cemal Basaran

Introduction to Unified Mechanics Theory with Applications

Springer

Cemal Basaran
Department of Civil, Structural, and Environmental Engineering
University at Buffalo, SUNY
Buffalo, NY, USA

ISBN 978-3-030-57774-2 ISBN 978-3-030-57772-8 (eBook)
https://doi.org/10.1007/978-3-030-57772-8

This Springer imprint is published by the registered company Springer Nature Switzerland AG
The registered company address is: Gewerbestrasse 11, 6330 Cham, Switzerland

Foreword

"A new scientific truth does not triumph by convincing opponents and making them see the light, but rather because its opponents eventually die, and a new generation grows up that is familiar with it."

Prof. Max Planck,

1918 Nobel Prize in Physics for his work on quantum theory.

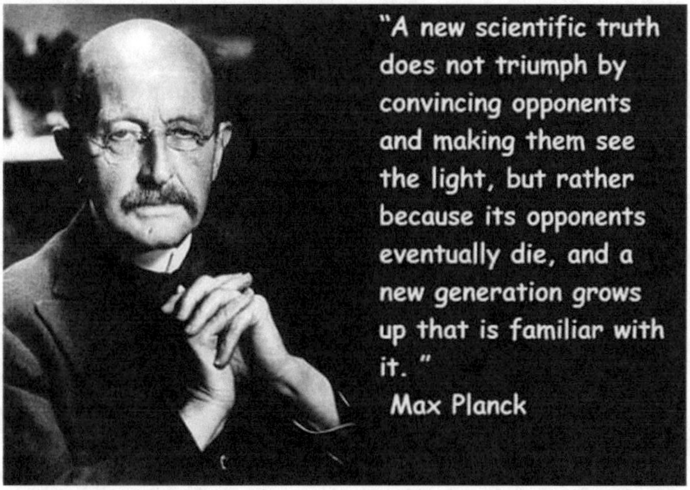
"A new scientific truth does not triumph by convincing opponents and making them see the light, but rather because its opponents eventually die, and a new generation grows up that is familiar with it."
Max Planck

Preface

This book is culmination of my research in the last 26 years at the University at Buffalo, The State University of New York. Most of the book heavily relies on the work of my PhD students and our joint publications.

Traditional (Newtonian) continuum mechanics portion of this book is primarily based on the seminal textbook by Malvern (1969). Since I consider Prof. Lawrence Malvern's textbook one of the best ever written on continuum mechanics, I did not find any reason to change the formulation or his presentation of the basics. Yet, when I myself, was as a student, I found his book hard to follow because it is written at a high level. Therefore, I have tried to simplify many of the formulae derivation steps and made it easier for students to understand. While I did change the order of the topics that I copied from Malvern (1969), I tried to stay true to the message Prof. Malvern had in his formulation.

I tried to explain Unified Mechanics Theory in the simplest terms possible. In order to accomplish this objective, I found it necessary to include the entire paper by Sharp and Matschisky (2015), which is the English translation of Boltzmann's (1877) original paper. While I acknowledge that including papers from a journal in its entirety is a very unorthodox practice, it was necessary because Boltzmann's original paper is misunderstood. Therefore, I had to present his entire derivation annotated with my explanations. To this day, many physics books refer to the equation of Boltzmann's second law as being valid for gasses only, with no reference to his derivation. Boltzmann did not make any such assumption in his formulation that would invalidate the applicability of his formulation to solids. For the past quarter century, my PhD students and I have provided in the literature the experimental proof of Boltzmann's second law equation for solids; Sharp and Matschisky (2015) have also addressed this validity of major misunderstanding in their paper. I quoted them directly. The fact that Boltzmann did not consider any interaction between the particles only increases the possible scenarios of combination. Interactions, like friction or molecular bonds, would only reduce the number of possible combinations. Therefore, Boltzmann's formulation can be considered an upper bound of possible permutations. However, Boltzmann's mathematical derivation is valid for solids, gasses, liquids, or any other materials. It was not possible to include Boltzmann's paper without Max Planck (1901) paper. Because they complement each other.

In introducing basics of thermodynamics, I relied heavily on Callen (1985). While there are many books on thermodynamics in the literature, I found the work of Callen (1985) to be most comprehensive and most understandable from a graduate student's perspective. I also kept in mind that unfortunately most mechanics students never take a basic course on thermodynamics at the undergraduate level. Therefore, the chapter on thermodynamics is based on primarily on the work of Callen (1985).

The rest of the chapters are all taken from my work with my former and two current PhD students whom I am eternally grateful. My former PhD students who contributed to the development of this theory over the years are Drs. ChengYong Yan, Rumpa Chandaroy, Hong Tang, Ying Zhao, Yujun Wen, Shihua Nie, Juan Gomez, Eray Mustafa Gunel, Shidong Li, Wei Yao, Hua Ye, Ming Hui Lin, Michael Sellers, Mohammad Fouad Abdul Hamid, YongChang Lee, and soon to be Drs. Noushad Bin Jamal and Hsiao Wei Lee.

I am especially grateful to the US Navy, Office of Naval Research for believing in me and sponsoring most of the work reported in this book. Their continous support, [starting in 1997 with ONR Young Investigator Award program under program director Dr. Roshdy Barsoum] has been an immense help which without it, this idea would have never come to realization.

Buffalo, NY, USA Cemal Basaran

Contents

1 Introduction 1
1.1 What Is Mechanics of Continuous Medium? 1
References 3

2 Stress and Strain in Continuum 5
2.1 Newton's Universal Laws of Motion 5
2.1.1 First Universal Law of Motion 5
2.1.2 Second Universal Law of Motion 7
2.1.3 Third Universal Law of Motion 9
2.1.4 Range of Validity of Newton's Universal Laws of Motion 11
2.1.5 Relation to the Thermodynamics and Conservation Laws 12
2.2 Stress 12
2.2.1 Definitions of Stress and Traction 12
2.2.2 Stress Vector on an Arbitrary Plane 15
2.3 Symmetry of Stress Tensor 19
2.4 Couple Stresses 20
2.4.1 Example 2.1 20
2.5 Principal Stresses and Principal Axes 21
2.6 Stress Tensor Invariants 24
2.6.1 Stress Invariants in Principal Axes 25
2.6.2 Representation of Stress Tensor in Spherical and Deviatoric Components 26
2.6.3 Invariants of the Deviatoric Stress Tensor 27
2.7 Deformation and Strain 28
2.7.1 Small Strain Definition 29
2.7.2 Small Strain and Small Rotation Formulation 31
2.7.3 Small Strain and Rotation in 3-D 40
2.8 Kinematics of Continuous Medium 41
2.8.1 Material (Local) Description 42

2.8.2 Referential Description (Lagrangian Description) 42
2.8.3 Spatial Description (Eulerian Description) 42
2.9 Rate of Deformation and Rate of Spin Formulation 43
2.9.1 Comparison of Rate of Deformation Tensor, D, and Time Derivative of the Strain Tensor, $\dot{\varepsilon}$ 44
2.9.2 True Strain (Natural Strain) (Logarithmic Strain) 45
2.10 Finite Strain and Deformation 46
2.10.1 Green Deformation Tensor, $\boldsymbol{C}$, and Cauchy Deformation Tensor, $\boldsymbol{B}^{-1}$ 49
2.10.2 Relation Between Deformation, Strain, and Deformation Gradient Tensors 50
2.10.3 Comparing Small Strain and Large (Finite) Strain 52
2.10.4 Strain Rate and Rate of Deformation Relation 54
2.10.5 Relation Between the Spatial Gradient of Velocity Tensor, L,and the Deformation Gradient Tensor, F 56
2.11 Rotation and Stretch Tensors in Finite Strain 57
2.12 Compatibility Conditions in Continuum Mechanics 57
2.13 Piola-Kirchhoff Stress Tensors 60
2.13.1 First Piola-Kirchhoff Stress Tensor σ^0 60
2.13.2 Second Piola-Kirchhoff Stress Tensor $\widetilde{\sigma}$ 61
2.14 Direct Relation Between Cauchy Stress Tensor and Piola-Kirchhoff Stress Tensors 62
2.15 Conservation of Mass Principle 63
2.16 The Incompressible Materials 65
2.17 Conservation of Momentum Principle 65
2.18 Conservation of Moment of Momentum Principle 68
References 70

3 Thermodynamics 73
3.1 Thermodynamic Equilibrium 74
3.2 First Law of Thermodynamics 75
3.2.1 Work Done on the System (Power Input) 76
3.2.2 Heat Input 78
3.3 Second Law of Thermodynamics 80
3.3.1 Entropy 81
3.3.2 Quantification of Entropy in Thermodynamics 85
References 112

4 Unified Mechanics Theory 115
4.1 Literature Review of Use of Thermodynamics in Continuum Mechanics 115
4.2 Laws of Unified Mechanics Theory 132
4.2.1 Second Law of Unified Mechanics Theory 133
4.2.2 Third Law of Unified Mechanics Theory 134
4.3 Evolution of Thermodynamic State Index (Φ) 135

4.3.1 On the Relationship Between the Second Fundamental Theorem of the Mechanical Theory of Heat and Probability Calculations Regarding the Conditions for Thermal Equilibrium, by Ludwig Boltzmann (1877) 137
4.3.2 Kinetic Energy Has Discrete Values 139
4.4 Critic of Boltzmann's Mathematical Derivation 180
4.5 On the Law of Distribution of Energy in the Normal Spectrum, By Max Planck, Annalen Der Physik, Vol. 4, P. 553 Ff (1901) 181
4.5.1 Calculations of the Entropy of a Resonator as a Function of Its Energy 183
4.5.2 Introduction of Wien's Displacement Law 186
4.5.3 Numerical Values 189
4.6 Thermodynamic State Index (TSI) in Unified Mechanics Theory 190
4.6.1 Experimental Verification Example 194
References 197

5 Unified Mechanics of Thermo-mechanical Analysis 203
5.1 Introduction 203
5.2 Unified Mechanics Theory-Based Constitutive Model 203
5.2.1 Flow Theory 203
5.2.2 Effective Stress Concept and Strain Equivalence Principle 207
5.3 Return Mapping Algorithm 209
5.3.1 Linearization (Consistent Jacobian) 211
5.4 Thermodynamic Fundamental Equation in Thermo-mechanical Problems 213
5.4.1 Conservation Laws 216
5.5 Numerical Validation of the Thermo-mechanical Constitutive Model 233
5.5.1 Thin Layer Solder Joint-Monotonic and Fatigue Shear Simulations 233
5.6 Thermo-mechanical Analysis of Cosserat Continuum: Length-Scale Effects 234
5.6.1 Introduction 234
5.6.2 Cosserat Couple Stress Theory 237
5.6.3 Toupin-Mindlin Higher-Order Stress Theory 242
5.6.4 Equilibrium Equations and Problem Formulation 245
5.7 Finite Element Method Implementation 248
5.7.1 General Couple Stress Theory: Variational Formulation 250
5.7.2 Reduced Couple Stress Theory: Variational Formulation 251
5.7.3 Reduced Couple Stress Theory: Mixed Variational Principle 251

5.7.4 General Couple Stress Theory Implementation 252
5.8 Cosserat Continuum Implementation in Unified Mechanics Theory 256
5.8.1 Rate-Independent Material Without Degradation 256
5.8.2 Rate-Dependent Material Without Degradation 262
5.8.3 Thermodynamic State Index Coupling 263
5.8.4 Entropy Generation Rate in Cosserat Continuum 266
5.8.5 Integration Algorithms 266
References 274

6 Unified Micromechanics of Particulate Composites 277
6.1 Introduction 277
6.2 Ensemble-Volume Averaged Micromechanical Field Equations . . . 280
6.3 Noninteracting Solution for Two-Phase Composites 284
6.3.1 Average Stress Norm in Matrix 285
6.3.2 Average Stress in Particles 290
6.4 Pairwise Interacting Solution for Two-Phase Composites 292
6.4.1 Approximate Solution of Two-Phase Interaction 292
6.4.2 Ensemble-Average Stress Norm in the Matrix 295
6.4.3 Ensemble-Average Stress in the Filler Particles 298
6.5 Noninteracting Solution for Three-Phase Composites 300
6.5.1 Effective Elastic Modulus of Multiphase Composites . . . 301
6.5.2 Ensemble-Average Stress Norm in the Matrix 302
6.5.3 Ensemble-Average Stress in Filler Particles 306
6.6 Effective Thermo-Mechanical Properties 309
6.6.1 Effective Bulk Modulus 310
6.6.2 Effective Coefficient of Thermal Expansion (ECTE) 310
6.6.3 Effective Shear Modulus 310
6.6.4 Effective Young's Modulus and Effective Poisson's Ratio 312
6.6.5 Numerical Examples 312
6.7 Micromechanical Constitutive Model of the Particulate Composite 316
6.7.1 Modeling Procedures for Particulate Composites 316
6.7.2 Elastic Properties of Particulate Composites 317
6.7.3 A Viscoplasticity Model 322
6.7.4 Thermodynamic State Index 324
6.7.5 Solution Algorithm 324
6.7.6 Consistent Elastic-Viscoplastic Tangent Modulus 329
6.8 Verification Examples 332
6.8.1 Material Properties of ATH 332
6.8.2 Properties of PMMA 332
6.8.3 Properties of Matrix-Filler Interphase 333
6.8.4 Cyclic Stress-Strain Response 337
References 341

7 Unified Micromechanics of Finite Deformations 343
7.1 Introduction to Finite Deformations 343
7.2 Frame of Reference Indifference 347
7.3 Unified Mechanics Theory Formulation for Finite Strain 351
7.3.1 Thermodynamic Restrictions 352
7.3.2 Constitutive Relations 360
7.4 Thermodynamic State Index 371
7.5 Definition of Material Properties 373
7.6 Applications of Finite Deformation Models 376
7.6.1 Material Properties 377
7.7 Numerical Implementation of Dual-Mechanism Viscoplastic Model 381
7.7.1 Simulating Isothermal Stretching of PMMA 384
7.7.2 Simulating Non-isothermal Stretching of PMMA 390
References 393

8 Unified Mechanics of Metals under High Electrical Current Density: Electromigration and Thermomigration 395
8.1 Introduction 395
8.2 Physics of Electromigration Process 395
8.2.1 Driving Forces of Electromigration Process 396
8.2.2 Laws Governing Electromigration and Thermomigration Process 397
8.2.3 Electromigration Electron Wind Force 400
8.2.4 Temperature Gradient Diffusion Driving Force 402
8.2.5 Stress Gradient Diffusion Driving Force 404
8.3 Laws of Conservation 406
8.3.1 Vacancy Conservation 407
8.4 Newtonian Mechanics Force Equilibrium 411
8.5 Heat Transfer 412
8.6 Electrical Conduction Equations 415
8.7 Fundamental Equation for Electromigration and Thermomigration 416
8.7.1 Entropy Balance Equations 418
8.8 Example 421
References 423

Index 427

Abbreviations[1]

γ	Shear strain
δ_{ij}	Kronecker delta
$\boldsymbol{\varepsilon}$	Total strain tensor
ϵ^p	Plastic strain rate
ρ	Mass density
$\boldsymbol{\sigma}$	Total stress tensor
$\boldsymbol{\sigma}^0$	First Piola-Kirchhoff stress tensor
$\widetilde{\boldsymbol{\sigma}}$	Second Piola-Kirchhoff tensor
φ	Dissipation potential
Φ	Thermodynamic state index
Ψ	Helmholtz-free energy
A	Area
b	Body force per unit mass
B^{-1}	Cauchy deformation tensor
C	Green deformation tensor
D	Rate of deformation tensor
E	Lagrange strain tensor
E_{total}	Total energy
F	Deformation gradient tensor
F	Force
F	Plasticity yield surface
$\widetilde{F}$	Second Piola-Kirchhoff force
g	Gibbs function
h	Enthalpy

[1]Most of the time, I tried to use the same character to define a variable. However, due to large number of characters needed to define variables, some characters were used more than once to define different variables. In order to clarify the definitions, each variable is defined in each subsection.

I	Impulse
I_1, I_2, I_3	Stress tensor invariants
k	Boltzmann's constant
K	Kinetic energy
L	Spatial gradient of velocity
m	Mass
m_{ij}	Distributed moment in couple stress theory
n_i	Direction cosine
p	Momentum
q	Heat flux vector
Q	Plastic potential
Q	Total heat input
s	Entropy
S	Deviatoric stress tensor
t	Time
T.S.I.	Thermodynamics State Index
u, v, w	Displacements along spatial coordinates, x, y, z, respectively.
U	Strain energy
v	Velocity
V	Volume
W	Spin tensor

Chapter 1
Introduction

1.1 What Is Mechanics of Continuous Medium?

Probably one of the best definitions of mechanics of a continuous medium is provided by Malvern [1969] in his seminal work. Continuum mechanics is a branch of mechanics concerned with the deformation or flow of solids, liquids, and gases. In continuum mechanics, the molecular structure and electronics structure of the material are ignored. When molecular structure must be taken into account, molecular dynamics is used, and when electronic structure must be considered, quantum mechanics is used. In continuum mechanics, it is assumed that the material is continuous without empty spaces. It is also assumed that all differential equations governing the continuous medium are continuous functions, except at boundaries between continuous regions. It is also assumed that the derivatives of the differential equations are continuous as well. A material satisfying these requirements is considered a continuous medium. It is important to point out that "empty space" does not mean there cannot be atomic vacancies. There will always be atomic vacancies. It just means that there cannot be an empty space proportional to the size of the object being studied.

The concept of a continuous medium allows us to define stress and strain at an "imaginary point," a geometric point in space assumed to be occupying no volume. When we say a point in continuum mechanics, we do not mean an atomic point. It is an imaginary point with no volume. Continuum stresses and strain are defined at this point. Atomic stresses are defined differently, which is outside the scope of this book. This approach allows us to use differential calculus to study nonuniform distributions of strain. This assumption of continuous medium allows us to study deformation in most engineering problems but not all. When continuous medium assumption is not satisfied or molecular dynamics [movement of atoms] or electronics structure influences the mechanical behavior, then continuum mechanics cannot be used. In these latter instances, a multi-scale mechanics analysis becomes essential.

C. Basaran, *Introduction to Unified Mechanics Theory with Applications*,
https://doi.org/10.1007/978-3-030-57772-8_1

Traditional definition of continuous medium also makes three assumptions (Malvern 1969):

Continuity: A material is continuous if it completely fills the space that it occupies, leaving no pores or empty spaces, and if furthermore its properties are describable by continuous functions.

Homogeneity: A homogeneous material has identical properties at all points. The size is larger than representative volume element (RVE) of the material. RVE is the smallest volume over which a property measurement can be made that will yield a value representative of the whole (Hill 1965).

Isotropy: A material is isotropic with respect to certain properties if these properties are the same in all directions in space.

Only the first assumption is needed to define concepts of stress and strain. The last two assumptions are needed when we introduce stress-strain relations (constitutive relation) into continuum mechanics equations.

Of course, anisotropic, inhomogeneous, and discontinues systems can also be analyzed by continuum mechanics by representing them as separate pieces, as it is done in computational mechanics, such as finite element method.

Chapter 2 of this book covers only the general principles of continuum mechanics, and associated definitions, such as stress and strain. The fundamentals of continuum mechanics, presented in the Chap. 2 has been well established by Isaac Newton (1687); Leonhard Euler (1736); Cauchy, Green, Truesdell, and Toupin (1960); Trusdell and Noll (1965); Truesdell (1965, 1966); and others. Maugin (2016) published a comprehensive account of the foundations of continuum mechanics as well as a complete list of references of earlier work. Therefore, there is no attempt in this book to cover earlier fundamental developments.

Continuum mechanics cannot be discussed without including material constitutive equations. While material behavior modeling is a very important topic of major current research efforts, it is not the focus of this book. It is covered in the applications of the unified mechanics theory section, Chaps. 5–8.

Chapter 3 of the book covers basics of thermodynamics. This chapter starts with a comprehensive literature survey of the use of thermodynamics in continuum mechanics in the last 150 years. While we do not claim to have included every paper published on the topic, the survey covers all significant developments. Degradation of materials and structures is discussed in the context of thermodynamics. Since not all engineering students are required to take the course in thermodynamics, the reader is assumed to be a beginner; as such, an elementary information is included.

Chapter 4 presents the formulation of the unified mechanics theory. Great effort has been made to include all the details of the formulation for mechanical, thermo-mechanical, and electro-thermo-mechanical loading and for both metal- and particle-filled composite materials. The last part of Chap. 4 is the finite element method implementation of the unified mechanics theory. Chapter 4 also discusses motivation behind the development of unified mechanics theory. Essentially, the present-day mechanics equations are all based on three laws of motion of Isaac Newton.

However, these three laws are "incomplete." The following examples can help us explain what we mean by "incomplete." For example, if a soccer ball is given an initial acceleration by a kick according to Newton's second law, that acceleration is constant, and it does not degrade. The ball will travel with that acceleration forever according to $F = ma$. However, that is obviously not true. In the same fashion, Newton's third law (also Hooke's law, which was published by Robert Hooke 10 years before Newton's *Principia*) assumes that reaction due to an applied load will be constant, without degradation of the material. Of course, materials degrade and response displacement does not remain constant. Fortunately, energy loss in the ball and material degradation can be modeled by laws of thermodynamics. Basaran and Yan published the first paper in (1998) using entropy generation rate as a metric to account for degradation in electronic solder joints. They implemented it by modifying Newton's laws with Boltzmann's second law of thermodynamics. Since then, the theory has gone through significant consolidation by experimental verifications and mathematical derivations by many researchers around the world. These developments are included in Chap. 4.

Chapter 5 presents formulation for thermo-mechanical analysis using the unified mechanics theory. The chapter includes nonlocal and local mechanics formulation. Cosserat continuum is explained in great detail and used for introducing length scale into the continuum mechanics formulation It is presented with simple steps from a beginner graduate student's perspective

Chapter 6 is about particle-filled composite materials' formulation using the unified mechanics theory. Particle-filled acrylic composite formulation and implementation is presented in the context of small strain formulation.

Chapter 7 covers finite strain formulation of unified mechanics theory. Again, particle-filled composite material application is presented in detail.

Chapter 8 presents electro-thermo-mechanical loading application of the unified mechanics theory.

I tried to provide a list of references at the end of each chapter. While they are not complete, it can help a researcher in the right direction.

References

Hill, R. (1965). Continuum micro-mechanics of elastoplastic polycrystals. *Journal of the Mechanics and Physics of Solids, 13*(2), 89–101.

Malvern, L. E. (1969). *Introduction to the Mechanics of continuous medium*. Englewood Cliffs, NJ: Prentice-Hall.

Chapter 2
Stress and Strain in Continuum

2.1 Newton's Universal Laws of Motion

It is essential to start the discussion about continuum mechanics with universal laws of motion as given by Newton. Understanding their implications in historical context is essential to understand these simple and probably most famous physics equations Of course, Newton's universal laws of motion are very well-known to most readers of this book; however, their historical context and derivations are not. Therefore, we find it necessary to quote these basics from Sir Isaac Newton's own Wikipedia web site with their sources. The information below is quoted directly with little or no editing.

Philosophiæ Naturalis Principia Mathematica (Latin for *Mathematical Principles of Natural Philosophy*) (Newton n.d.-a), often referred to as simply the *Principia*, is a work in three books by Isaac Newton, in Latin, first published on 5 July 1687. After annotating and correcting his personal copy of the first edition (Newton n.d.-a), Newton published two further editions, in 1713 and 1726 (Harvard UP 1972). The *Principia* states Newton's laws of motion, forming the foundation of classical mechanics, Newton's law of universal gravitation, and a derivation of Kepler's laws of planetary motion (which Kepler first obtained empirically) (Anon 2020; Cohen 2002; Galili and Tseitlin 2003; Kleppner and Kolenkow 1973; Plastino and Muzzio 1992; Halliday et al. 1992) (Fig. 2.1).

2.1.1 First Universal Law of Motion

Original text in Latin reads:

Newton's original Latin text in *Principia* states:

> " Lex I: Corpus omne perseverare in statu suo quiescendi vel movendi uniformiter in directum, nisi quatenus a viribus impressis cogitur statum illum mutare."

Translated to English, this reads:

C. Basaran, *Introduction to Unified Mechanics Theory with Applications*,
https://doi.org/10.1007/978-3-030-57772-8_2

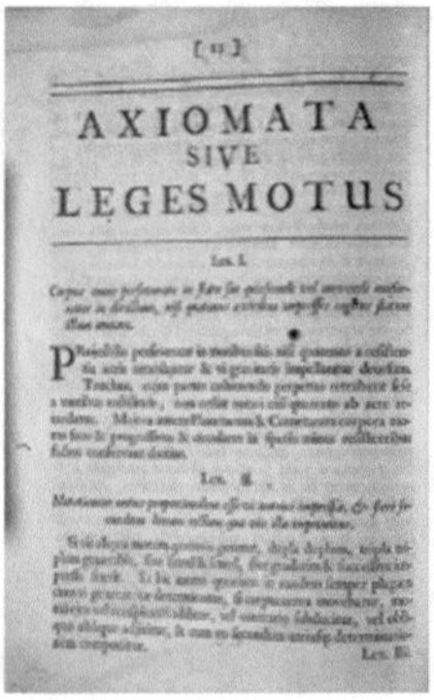

Fig. 2.1 Original 1687 *Principia Mathematica*, where Sir Isaac Newton published the three laws, in Latin

"Law I: Every body persists in its state of being at rest or of moving uniformly straight forward, except insofar as it is compelled to change its state by force impressed" (Newton 1999).

(Modern translation) first law: In an inertial frame of reference, an object either remains at rest or continues to move at a constant velocity, unless acted upon by a force.

The ancient Greek philosopher Aristotle had the view that all objects have a natural place in the universe: that heavy objects (such as rocks) wanted to be at rest on the Earth and that light objects like smoke wanted to be at rest in the sky and the stars wanted to remain in the heavens. He thought that a body was in its natural state when it was at rest, and for the body to move in a straight line at a constant speed, an external agent was needed continually to propel it; otherwise, it would stop moving. Galileo Galilei, however, realized that a force is necessary to change the velocity of a body, i.e., acceleration, but no force is needed to maintain its velocity. In other words, Galileo stated that, in the *absence* of a force, a moving object will continue moving. (The tendency of objects to resist changes in motion was what Johannes Kepler had called *inertia*.) This insight was refined by Newton, who made it into his first law, also known as the "law of inertia"—no force means no acceleration, and hence the body will maintain its velocity. As Newton's first law is a restatement of the law of inertia which Galileo had already described, Newton appropriately gave credit to Galileo.

The law of inertia apparently occurred to several different natural philosophers and scientists independently, including Thomas Hobbes in his *Leviathan* (Hobbes 1651). The seventeenth-century philosopher and mathematician René Descartes also formulated the law, although he did not perform any experiments to confirm it (Hellingman 1992; Resnick and Halliday 1977).

2.1.1.1 Formulation of the First Law

The first law states that if the net force (the vector sum of all forces acting on an object) is zero, then the velocity of the object is constant. Velocity is a vector quantity which expresses both the object's speed and the direction of its motion;

therefore, the statement that the object's velocity is constant is a statement that both its speed and the direction of its motion are constant.

The first law can be stated mathematically, when the mass is a nonzero constant, as

$$F = 0, \frac{dv}{dt} = 0$$

where F is the total net force acting on the system and $\frac{dv}{dt}$ is the time derivative of velocity. Consequently, an object that is at rest will stay at rest unless a force acts upon it. An object that is in motion will not change its velocity unless a force acts upon it. This is known as uniform motion. An object continues to do whatever it happens to be doing unless a force is exerted upon it. If it is at rest, it continues in a state of rest (demonstrated when a tablecloth is skillfully whipped from under dishes on a tabletop and the dishes remain in their initial state of rest), where the inertia of dishes balances the force exerted by the tablecloth. If an object is moving, it continues to move without turning or changing its speed. This is evident in space probes that continuously move in outer space. Changes in motion must be imposed against the tendency of an object to retain its state of motion. In the absence of net forces, a moving object tends to move along a straight-line path indefinitely.

Newton placed the first law of motion to establish frame of reference for which the other laws are applicable. The first law of motion postulates the existence of at least one frame of reference called a Newtonian or inertial reference frame, relative to which the motion of a particle not subject to forces is a straight line at a constant speed (Newton 1999; Anon 2020). Newton's first law is often referred to as the law of inertia. Thus, a condition necessary for the uniform motion of a particle relative to an inertial reference frame is that the total net force acting on it is zero. In this sense, the first law can be restated as:

> "In every material universe, the motion of a particle in a preferential reference frame X is determined by the action of forces whose total vanished for all times when and only when the velocity of the particle is constant in X. That is, a particle initially at rest or in uniform motion in the preferential frame X continues in that state unless compelled by forces to change it." [Motte's 1729 translation]
>
> "Newton's first and second laws are valid only in an inertial reference frame. Any reference frame that is in uniform motion with respect to an inertial frame is also an inertial frame, i.e. Galilean invariance or the principle of Newtonian relativity" (Resnick and Halliday 1977). If someone is kicking a ball on a moving train car, the acceleration the ball will gain will be with respect to the train not with respect to the ground.

2.1.2 *Second Universal Law of Motion*

Newton's original Latin text reads:

"Lex II: Mutationem motus proportionalem esse vi motrici impressae, et fieri secundum lineam rectam qua vis illa imprimitur. "

This was translated quite closely in Motte's 1729 translation as:

"Law II: The alteration of motion is ever proportional to the motive force impress'd; and is made in the direction of the right line in which that force is impress'd."

(Modern translation) second law: In an inertial reference frame, the vector sum of the forces "F" on an object is equal to the mass "m" of that object multiplied by the acceleration "a" of the object.

According to modern ideas of how Newton was using his terminology, this is understood, in modern terms, as an equivalent of (Anon 2020):

The change of momentum of a body is proportional to the impulse impressed on the body and happens along the straight line on which that impulse is impressed.

This may be expressed by the formula $F = p'$, where p' is the time derivative of the momentum p. This equation can be seen clearly in the Wren Library of Trinity College, Cambridge, in a glass case in which Newton's manuscript is open to the relevant page.

Motte's 1729 translation of Newton's Latin continued with Newton's commentary on the second law of motion, reading:

If a force generates a motion, a double force will generate double the motion, a triple force triple the motion, whether that force be impressed altogether and at once, or gradually and successively. And this motion (being always directed the same way with the generating force), if the body moved before, is added to or subtracted from the former motion, according as they directly conspire with or are directly contrary to each other; or obliquely joined, when they are oblique, so as to produce a new motion compounded from the determination of both.

The sense or senses in which Newton used his terminology, and how he understood the second law and intended it to be understood, have been extensively discussed by historians of science, along with the relations between Newton's formulation and modern formulations (Newton 1729).

2.1.2.1 Formulation of Newton's Second Law

The second law states that the rate of change of momentum of a body is directly proportional to the force applied, and this change in $\frac{dp}{dt}$ momentum takes place in the direction of the applied force, F:

$$F = \frac{dp}{dt} = \frac{d(mv)}{dt}$$

The second law can also be stated in terms of an object's acceleration. It is important to note that we cannot derive a general expression for Newton's second law for variable mass systems by treating the mass in $F = dP/dt = d(mv)$ as a variable. We can use $F = dP/dt$ to analyze variable mass systems only if we apply it to an entire system of constant mass having parts among which there is an interchange of mass. Since Newton's second law is valid only for constant mass systems (Beatty 2006; Thornton 2004; Hellingman 1992), m can be taken outside the differentiation operator by the constant factor rule in differentiation. Thus,

$$F = m\,\frac{dv}{dt} = m\,a$$

An impulse I occurs when a force F acts over an interval of time Δt, and it is given by

$$I = \int_{\Delta t} F\ dt$$

Since force is the time derivative of momentum, it follows that

$$I = \Delta p = m\,\Delta v$$

This relation between impulse and momentum is closer to Newton's wording of the second law. Impulse is a concept frequently used in the analysis of collisions, impacts, and dynamic loads.

2.1.3 *Third Universal Law of Motion*

Newton's original Latin text reads:

"Lex III: Actioni contrariam semper et æqualem esse reactionem: sive corporum duorum actiones in se mutuo semper esse æquales et in partes contrarias dirigi."

Directly translated to English, this reads:

"Law III: To every action there is always opposed an equal reaction: or the mutual actions of two bodies upon each other are always equal, and directed to contrary parts."

(Modern translation) third law: When one body exerts a force on a second body $F_{12} = F_{21}$, the second body simultaneously exerts a force F_{21} equal in magnitude and opposite in direction on the first body.

Newton's explanatory comment to this law:

"Whatever draws or presses another is as much drawn or pressed by that other. If you press a stone with your finger, the finger is also pressed by the stone. If a horse draws a stone tied to a rope, the horse (if I may so say) will be equally drawn back towards the stone: for the distended rope, by the same endeavor to relax or unbend itself, will draw the horse as much towards the stone, as it does the stone towards the horse, and will obstruct the progress of the one as much as it advances that of the other. If a body impinges upon another, and by its force changes the motion of the other, that body also (because of the equality of the mutual pressure) will undergo an equal change, in its own motion, toward the contrary part. The changes made by these actions are equal, not in the velocities but in the motions of the bodies; that is to say, if the bodies are not hindered by any other impediments. For, as the motions are equally changed, the changes of the velocities made toward contrary parts are reciprocally proportional to the bodies. This law takes place also in attractions, as will be proved in the next scholium" (Fairlie and Cayley 1965).

In the above, as usual, *motion* is Newton's name for momentum, hence his careful distinction between motion and velocity.

Newton used the third law to derive the law of conservation of momentum (Cohen 1995), but from a deeper perspective, however, conservation of momentum is the more fundamental idea (derived via Noether's theorem from Galilean invariance), and holds in cases where Newton's third law appears to fail, for instance, when force fields as well as particles carry momentum, and in quantum mechanics:

$$F_{12} = F_{21},$$

An illustration of Newton's third law in which two people push against each other. The first person on the left exerts a normal force F_{12} on the second person directed towards the right, and the second person exerts a normal force F_{21} on the first person directed toward the left.

The magnitudes of both forces are equal, but they have opposite directions, as dictated by Newton's third law.

The third law states that all forces between two objects exist in equal magnitude and opposite direction: if one object A exerts a force F_A on a second object B, then B simultaneously exerts a force F_B on A, and the two forces are equal in magnitude and opposite in direction: $F_A = -F_B$. The third law means that all forces are interactions between different bodies, or different regions within one body, and thus that there is no such thing as a force that is not accompanied by an equal and opposite force. In some situations, the magnitude and direction of the forces are determined entirely by one of the two bodies, say Body A; the force exerted by Body A on Body B is called the "action," and the force exerted by Body B on Body A is called the "reaction." This law is sometimes referred to as the action-reaction law, with F_A called the "action" and F_B the "reaction." In other situations, the magnitude and directions of the forces are determined jointly by both bodies, and it is not necessary to identify one force as the "action" and the other as the "reaction." The action and the reaction are simultaneous, and it does not matter which is called the action and which is called the reaction; both forces are part of a single interaction, and neither force exists without the other (Anon 2020; Bernard Cohen 1967).

The two forces in Newton's third law are of the same type (e.g., if the road exerts a forward frictional force on an accelerating car's tires, then it is also a frictional force that Newton's third law predicts for the tires pushing backward on the road).

From a conceptual standpoint, Newton's third law is seen when a person walks: they push against the floor, and the floor pushes against the person. Similarly, the tires of a car apply a shear force on the road, while the road pushes back on the tires with a reaction shear force—the tires and road simultaneously push against each other. The reaction force is responsible for the motion of the person or the car. These forces depend on friction; however, if a person or car is on ice, there will be no motion because of failure to produce the needed reaction force.

It is important to point out that Robert Hooke (1687) published the third law for deformable bodies as $F = k\,u$, where k is the stiffness of a one-dimensional spring and u is the displacement response in the direction of the force. "Hooke stated in his 1687 work that he was aware of the law in 1600, which is long before Newton published his Principia" (Robert Hooke's Wikipedia page).

> "In their original form, Newton's laws of motion are not adequate to characterize the motion of rigid bodies and deformable bodies. Leonhard Euler in 1750 introduced a generalization of Newton's laws of motion for rigid bodies called Euler's laws of motion, later applied as well for deformable bodies assumed as a continuum. If a body is represented as an assemblage of discrete particles, each governed by Newton's laws of motion, then Euler's laws can be derived from Newton's laws. Euler's laws can, however, be taken as axioms describing the laws of motion for extended bodies, independently of any particle structure (Lubliner 2008)" (Leonard Euler Wikipedia page).

2.1.4 Range of Validity of Newton's Universal Laws of Motion

"Newton's laws were verified by experiments and observations for over 200 years, and they are excellent approximations at the scales and speeds of most of the motions we observe in daily life. Newton's laws of motion, together with his law of universal gravitation and the mathematical techniques of calculus, provided for the first time a unified quantitative explanation for a wide range of physical phenomena.

These three laws hold to a good approximation for macroscopic objects under everyday conditions. However, Newton's laws are inappropriate for use in certain circumstances, most notably, at very high speeds (in special relativity, the Lorentz factor must be included in the expression for momentum along with the rest mass and velocity) or at very strong gravitational fields. Therefore, Newton's laws cannot be used to explain phenomena such as conduction of electricity in a semiconductor, optical properties of materials, errors in nonrelativistically corrected GPS systems, and superconductivity. Explanation of these phenomena requires more sophisticated physical theories, including general relativity and quantum field theory.

However, in quantum mechanics, concepts such as force, momentum, and position are defined by linear operators that operate on the quantum state; at speeds that are much lower than the speed of light, Newton's laws are just as exact for these operators as they are for classical objects" (Anon 2020).

2.1.5 Relation to the Thermodynamics and Conservation Laws

"In modern physics, the laws of conservation of momentum, energy, and angular momentum are of more general validity than Newton's laws, since they apply to both light and matter and to both classical and nonclassical physics. Because force is the time derivative of momentum, the concept of force is redundant and subordinate to the conservation of momentum and is not used in fundamental theories (e.g., quantum mechanics, quantum electrodynamics, general relativity, etc.). Other forces, such as gravity, also arise from the momentum conservation. Newton stated the third law within a worldview that assumed instantaneous action at a distance between material particles. However, he was prepared for philosophical criticism of this action at a distance, and it was in this context that he stated the famous phrase 'I feign no hypotheses.' The discovery of the second law of thermodynamics in the nineteenth century showed that not every physical quantity is conserved over time, thus disproving the validity of inducing the opposite metaphysical view from Newton's laws. Hence, a 'steady-state' worldview based solely on Newton's laws and the conservation laws do not take entropy into account" (Anon 2020).

That brings us to the main topic of this book. In unified mechanics theory, entropy generation is incorporated into Newton's universal laws of motion.

2.2 Stress

2.2.1 Definitions of Stress and Traction

The vector *t* applied on the external surface per unit area is called the traction. It is a distributed external load. The vector ΔF acting at an imaginary point in the interior surface is called the internal force vector (Fig. 2.2).

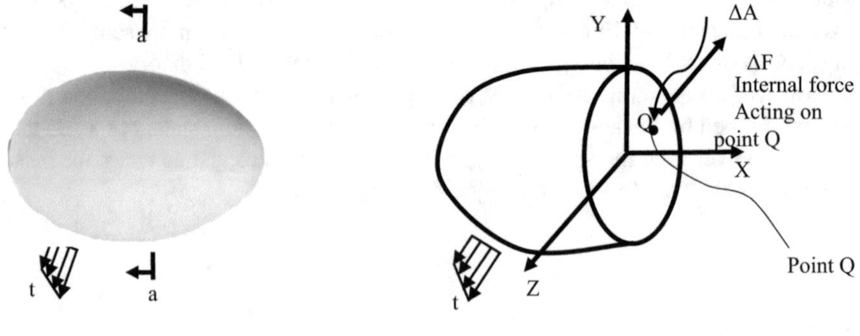

Fig. 2.2 Definition of stress at an imaginary point Q in the internal surface at cross-sectional cut a-a

ΔA is the area of the internal imaginary point Q, and ΔF is the force acting on this area as a result of external loads applied on the body.

The average stress σ acting on ΔA is the vector $\frac{\Delta F}{\Delta A}$, ΔF is the vector sum of the forces acting at point Q, and ΔA is the area of the chosen imaginary point. The average stress is also a vector having the same direction as ΔF.

Assuming that ΔA is selected so that it represents a point with no volume as such the value of ΔA approaches to zero:

$$\lim_{\Delta A \to 0}$$

Of course, ΔA can never be equal to zero. If it were zero, stress value would be infinite. We let ΔA approach zero to be able to define stress at an imaginary point:

$$\sigma = \lim_{\Delta A \to 0} \left(\frac{\Delta F}{\Delta A} \right)$$

Because the area corresponding to internal force is approaching zero, the numerator, ΔF, also approaches zero. However, the fraction, in general, approaches to a finite limit:

$$\sigma = \lim_{k \to \infty} \left(\frac{\Delta F}{\Delta A} \right)_k$$

where k represent the number of points in the body. This is a basic postulate of continuum mechanics that such a limit exists and is independent of area used (Malvern 1969).

It is important to explain what we mean by an "imaginary point" with zero volume; as the area ΔA, it is natural to assume that this point represents an atom. It cannot be smaller than an atom. However, this point does not represent an atom, because at the atomic level, fundamental definitions of continuum do not exist, and this definition of stress at the atomic level is not true. This "imaginary point" therefore represents a point in the continuum, but not an atom or any other quantum mechanics matter. Since quantum mechanics is outside the scope of this book, we will refer the reader to a textbook on quantum mechanics.

x, y, z Cartesian coordinate system stresses at point Q on the cross section that has x axis as its normal can be defined by

$$\sigma_{xx} = \lim_{\Delta A_x \to 0} \frac{\Delta F_x}{\Delta A_x} = \frac{dF_x}{dA_x}$$

$$\sigma_{xy} = \lim_{\Delta A_x \to 0} \frac{\Delta F_y}{\Delta A_x} = \frac{dF_y}{dA_x}$$

Fig. 2.3 Definition of stress at point Q

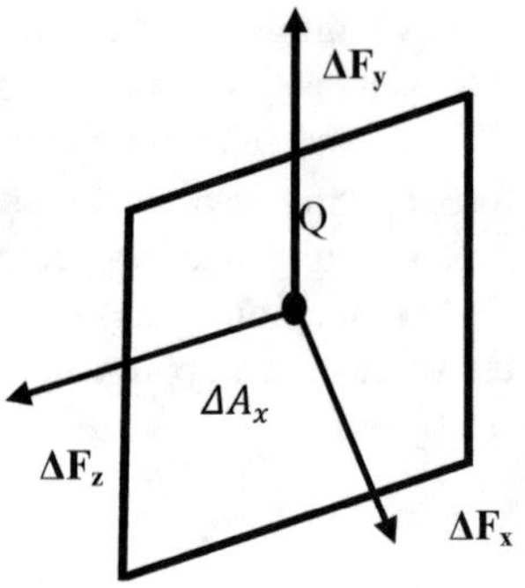

Figure 2.4 Stresses acting at a point in Cartesian coordinates

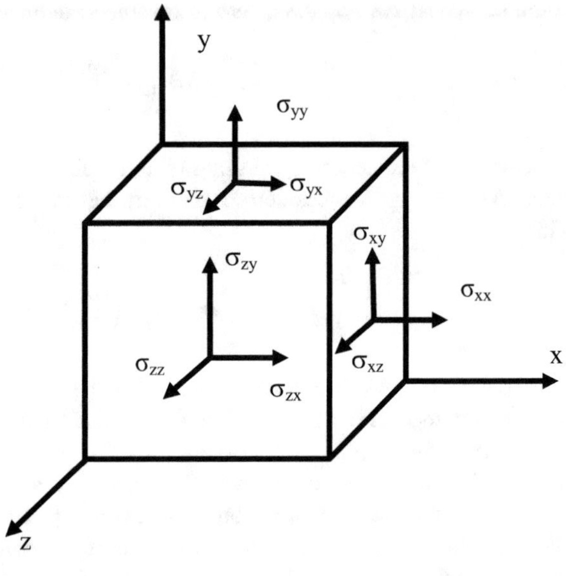

$$\sigma_{xz} = \lim_{\Delta A_x \to 0} \frac{\Delta F_z}{\Delta A_x} = \frac{dF_z}{dA_x}$$

Subscript x in dA_x indicates that x axis is the normal to area dA_x.

In the same fashion, we can define stress vectors acting on other surfaces, *y and z*. As a result, the state of stress at a point is defined by a second-order tensor with nine components (Figs. 2.3 and 2.4).

$$\boldsymbol{\sigma} = \begin{bmatrix} \sigma_{xx} & \sigma_{xy} & \sigma_{xz} \\ \sigma_{yx} & \sigma_{yy} & \sigma_{yz} \\ \sigma_{zx} & \sigma_{zy} & \sigma_{zz} \end{bmatrix}$$

2.2.2 Stress Vector on an Arbitrary Plane

Let's define a plane ABC at an arbitrary slope passing through point Q (Fig. 2.5).

In order to obtain the equations governing the state of stress at an arbitrarily oriented plane ABC, we will use conservation of momentum principle. Forces acting on the tetrahedron are five vectors representing the resultant force on each of the faces (ΔA_X, ΔA_Y, ΔA_Z, ΔA) and the resultant body force $\rho b \Delta V$, where ρ is the mass density, b is the body force per unit mass, and ΔV is the volume, the tetrahedron.

$\sigma^{(n)}$ is the average value of stress on the oblique face which has n as its normal. $\sigma^{(X)}$ is the average stress on the area ΔA_X that has X axis as its normal; similarly $\sigma^{(Y)}$ and $\sigma^{(Z)}$ are defined.

Equilibrium of the tetrahedron and stress vector components on the inclined plane can be derived from conservation of momentum principle of a collection of particles. While conservation of momentum is discussed later in the book, assuming students have a rudimentary knowledge of the principle, we will use it at this stage to derive the equilibrium equations.

Momentum principle of a collection of particles states that the vector sum of all external forces acting on the free body is equal to the rate of change of the total momentum. The total momentum of collection of particles in a given volume is given by

$$\int_{\Delta V} v\rho dV = \int_{\Delta m} v dm$$

where dm is the mass of the tetrahedron, dV is the volume, ρ is mass density, and v is the velocity of the particle.

The time rate of change of the total momentum is given by

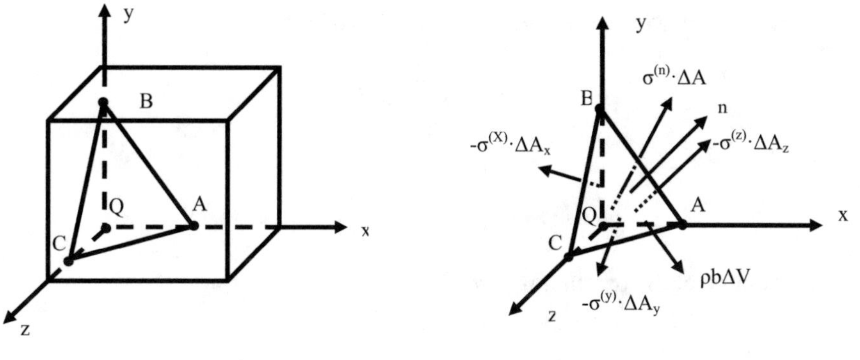

Fig. 2.5 (**a**) An arbitrary tetrahedron QABC at point Q. (**b**) Free body diagram of tetrahedron QABC

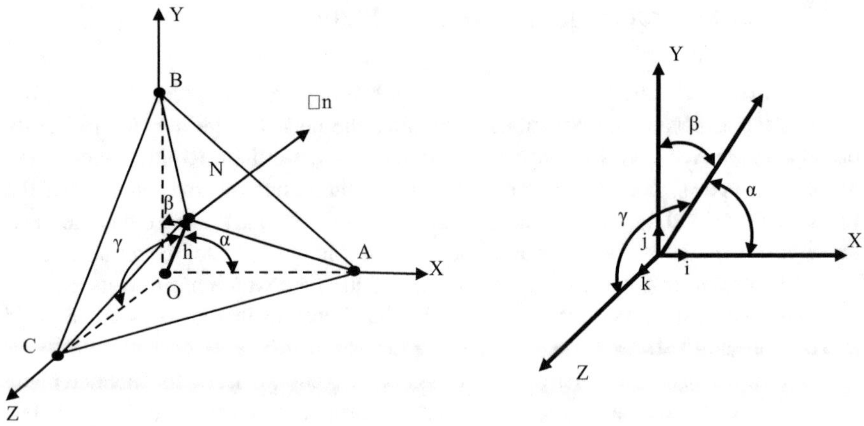

Fig. 2.6 Geometry of the tetrahedron

$$\Delta m \frac{dv}{dt} = (\rho \Delta V) \frac{dv}{dt}$$

It is assumed that Δm does not change by time. The conservation of momentum principle yields the following equilibrium equation for the free body (Fig. 2.6):

$$\sigma^{(n)} \Delta A + \rho b \Delta V - \sigma^{(X)} \Delta A_X - \sigma^{(Y)} \Delta A_Y - \sigma^{(Z)} \Delta A_Z = (\rho \Delta V) \frac{dv}{dt}$$

We need to calculate the height and volume of the arbitrary tetrahedron at point Q. Then we let the height h go to zero, to be able define an imaginary point Q:

$$\cos\alpha = \frac{h}{\mathrm{OA}}, \cos\beta = \frac{h}{\mathrm{OB}}, \cos\gamma = \frac{h}{\mathrm{OC}}$$

$$n_X = \cos\alpha, \quad n_Y = \cos\beta, \quad n_Z = \cos\gamma$$

Three direction cosines, n_X, n_Y, n_z, are also defined by

$$n_X^2 + n_y^2 + n_z^2 = 1$$

As a result,

$$h = \mathrm{OA} \cdot n_X = \mathrm{OB} \cdot n_Y = \mathrm{OC} \cdot n_Z$$

The volume of the pyramid is given by

$$\Delta V = \frac{1}{3} h \Delta A$$

Substituting h in the previous equation yield

$$\Delta V = \frac{1}{3}(\mathrm{OA} \cdot \cos\,\alpha)\Delta A = \frac{1}{3}(OB \cdot \cos\,\beta)\Delta A = \frac{1}{3}(\mathrm{OC} \cdot \cos\,\gamma)\Delta A$$

Note that we can define the direction cosines also using the relationship between the areas of the faces of the tetrahedron:

$$\cos\,\alpha = \frac{\Delta A_X}{\Delta A}$$

because ΔA_X is the projection of the inclined area of interest ΔA on $Y - Z$ plane with X axis as its normal. Hence,

$$\Delta V = \frac{1}{3} OA\ \Delta A_X$$

Similarly, we can write

$$\Delta V = \frac{1}{3}\mathrm{OB}\ \Delta A_Y \text{ and } \Delta V = \frac{1}{3} OC\ \Delta A_Z$$

Substituting ΔV, ΔA_X, ΔA_Y, and ΔA_Z in conservation of momentum equation, we obtain

$$\sigma^{(n)}(\Delta A) + \rho b\left(\frac{1}{3}h\Delta A\right) = \sigma^{(X)}\Delta A\ n_X + \sigma^{(Y)}\Delta A\ n_Y + \sigma^{(Z)}\Delta A\ n_Z + \rho\left(\frac{1}{3}h\Delta A\right)\frac{dv}{dt}$$

Eliminating ΔA from each term leads to

$$\sigma^{(n)} + \rho b\frac{1}{3}h = \sigma^{(X)}n_X + \sigma^{(Y)}n_Y + \sigma^{(Z)}n_Z + \rho\frac{1}{3}h\frac{dv}{dt}$$

Let the height of the tetrahedron h go to zero to define our imaginary point. The terms with h disappear:

$$\sigma^{(n)} = \sigma^{(X)}n_X + \sigma^{(Y)}n_Y + \sigma^{(Z)}n_Z \tag{3.1}$$

This equation defines the stress at a point on an arbitrary oblique plane passing through point Q. This equation was derived from the conservation of momentum principle of a collection of particles. Hence, it applies to solid mechanics as well as fluid mechanics. However, it important to point out that this is not an equilibrium of stress equation. There is an equilibrium of forces but not stresses. If direction cosines are replaced by their respective equations in terms of areas, this equation becomes equilibrium of forces at point Q. We should again emphasize that this equation is only in terms of stresses and direction cosine. It is not possible to write equilibrium of stresses in solid mechanics.

The resultant stress on any inclined plane can be determined from the stress tensor of the point and the direction cosines of the inclined plane.

The Eq. (3.1) is a vector equation, because stresses are vectors. The corresponding algebraic equations yield

$$\begin{aligned}\sigma_X^{(n)} &= \sigma_{XX}n_X + \sigma_{XY}n_Y + \sigma_{XZ}n_Z \\ \sigma_Y^{(n)} &= \sigma_{YX}n_X + \sigma_{YY}n_Y + \sigma_{YZ}n_Z \\ \sigma_Z^{(n)} &= \sigma_{ZX}n_X + \sigma_{ZY}n_Y + \sigma_{ZZ}n_Z\end{aligned} \tag{3.2}$$

We can write a second-order tensor, σ_{ij}, Cauchy stress tensor, from these algebraic equations. In indicial notation, algebraic equations can be written as

$$\sigma_j^{(n)} = \sigma_{ij}n_i \quad i = X, Y, Z \quad j = X, Y, Z$$

in matrix notation $\{\sigma^n\} = \{n\}[\sigma]$. Remember that { } defines a vector and [ ] defines a matrix.

Cauchy stress tensor in any other Cartesian coordinate system can easily be obtained. Assume that the new coordinate system is rotated with respect to the original coordinate system by α, β, γ. The components of the new rotated stress tensor $\overline{\sigma}_{ij}$ can be defined in terms of the three stress vectors acting across the three planes normal to the new coordinates.

Components of stresses will be

$$\overline{\sigma}_i^{(n)} = \overline{\sigma}_{ij}n_j.$$

$\overline{\sigma}_{ij}$ is a second-order tensor (linear vector function); therefore, the components of the rotated stress tensor $\overline{\sigma}_{ij}$ can be obtained by the tensor transformation equations:

$$\overline{\sigma}_{ij} = n_i^k n_j^l \sigma_{kl}$$

or in matrix notation

$$[\overline{\sigma}] = [N]^T[\sigma][N] \quad [\sigma] = [N][\overline{\sigma}][N]^T$$

where $[N]$ is the matrix of direction cosines $n_i^k = \cos\left(\overline{X}_i, X_k\right)$ of the angles between the new and old axes. $[N]^T$ is its transpose. Matrix of direction cosine is given by

$$[N] = \begin{bmatrix} \cos\left(\overline{X},X\right) & \cos\left(\overline{X},Y\right) & \cos\left(\overline{X},Z\right) \\ \cos\left(\overline{Y},X\right) & \cos\left(\overline{Y},Y\right) & \cos\left(\overline{Y},Z\right) \\ \cos\left(\overline{Z},X\right) & \cos\left(\overline{Z},Y\right) & \cos\left(\overline{Z},Z\right) \end{bmatrix}$$

2.3 Symmetry of Stress Tensor

Normal stresses act orthogonal to the surface they are acting on. However, shear stresses act parallel to the plane in each surface of the cubic control volume, which actually represents a point in space.

Shear stresses are assumed to be positive when they are on the positive face of the cube and acting in the positive coordinate axis direction. On the negative face, shear stress is positive if the direction of the stress is negative

When there are no distributed (body or surface) couples (moment) or the material has no length-scale effect, all off-diagonal terms of the stress tensor are assumed to be equal, due to moment equilibrium of the point Q:

$$\sigma_{XY} = \sigma_{YZ}, \quad \sigma_{XZ} = \sigma_{ZX}, \quad \sigma_{YZ} = \sigma_{ZY}$$

Writing equilibrium of moment about Z axis leads to $\sum M_Z = 0$:

$$(\sigma_{XY}\mathrm{dydz})dx - (\sigma_{YX}dxdz)dy = 0$$

$$\sigma_{XY} = \sigma_{YX}$$

Similarly writing $\sum M_y = 0$ and $\sum M_X = 0$ equilibrium equations leads to

$$\sigma_{XZ} = \sigma_{ZX} \quad \text{and} \quad \sigma_{ZY} = \sigma_{YZ}$$

However, we should make it clear that we obtained this result because we assumed that at point Q only linear stress vectors are acting and the are no stress couple vectors (moments) acting at point Q or the material does not exhibit length-scale effect that leads to differential shear stresses on opposite sides of the point Q. Therefore, in the absence of distributed moment acting at a point, the Cauchy stress tensor is symmetric (Fig. 2.7).

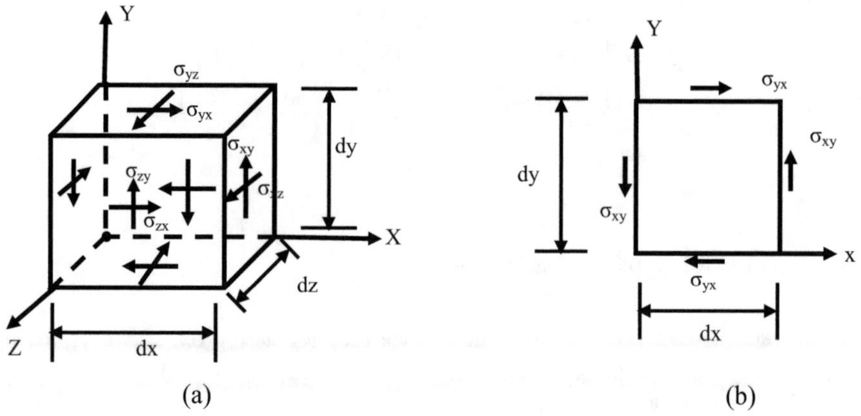

Fig. 2.7 (**a**) Linear shear stresses acting at point Q. (**b**) *x-y* plane view of point Q

2.4 Couple Stresses

Couple stress at a point can exist due to external loads, like a magnetic field, due to materials' microstructure or due to small size of the structure which can cause large deformation gradients.

If we take moment with respect to Z axis

$$(\sigma_{XY}\Delta Y\Delta Z)\Delta X + m_{XZ} - (\sigma_{YX}\Delta X\Delta Z) + m_{YZ} = 0$$

where m_{XZ}, m_{YZ} are distributed moments.

Therefore, $\sigma_{XY} \neq \sigma_{YX}$; similarly $\sigma_{XZ} \neq \sigma_{ZX}$, $\sigma_{YZ} \neq \sigma_{ZY}$.

The skew symmetric second-order couple stress tensor can be given by

$$[m] = \begin{bmatrix} 0 & \frac{1}{2}m_{XZ} & -\frac{1}{2}m_{XY} \\ -\frac{1}{2}m_{YZ} & 0 & \frac{1}{2}m_{YX} \\ \frac{1}{2}m_{ZY} & -\frac{1}{2}m_{ZX} & 0 \end{bmatrix}$$

It is important to point out that 1/2 multiplier in front of the moment is due to the relation between shear stresses and normal stresses [$2\,\varepsilon_{xy} = \gamma_{xy}$] (Fig. 2.8).

2.4.1 *Example 2.1*

From equilibrium of forces along x axis, we can write the following relations, where A represents the area along the inclined surface (Fig. 2.9):

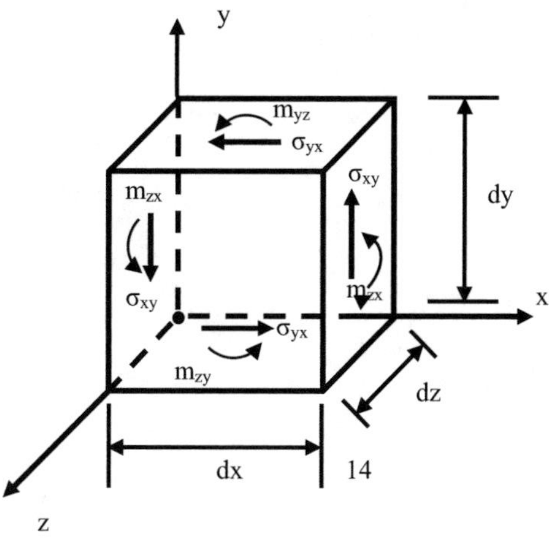

Fig. 2.8 Couple stresses acting in conjunction with linear shear stresses at a point in *x*-*y* plane only

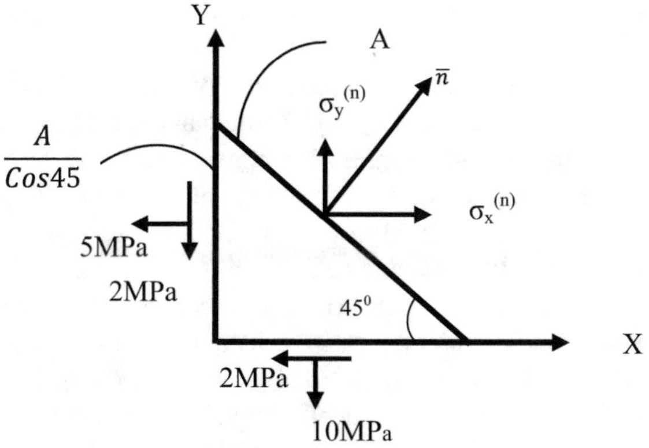

Fig. 2.9 State of stress at a point

$$A\,\sigma_X^{(n)} = \frac{A}{(\cos 45^\circ)}[-5MPa - 2MPa]$$

$$\sigma_X^{(n)} = \sigma_n^n(\cos 45^\circ)$$

2.5 Principal Stresses and Principal Axes

The stress tensor given in the original Cartesian coordinate system changes values if we change the orientation of the coordinate system, i.e., simply rotating it around the origin O. Inclined plane represents the orientation of a new coordinate system. Therefore, stress tensor becomes a function of the rotation angle with respect to

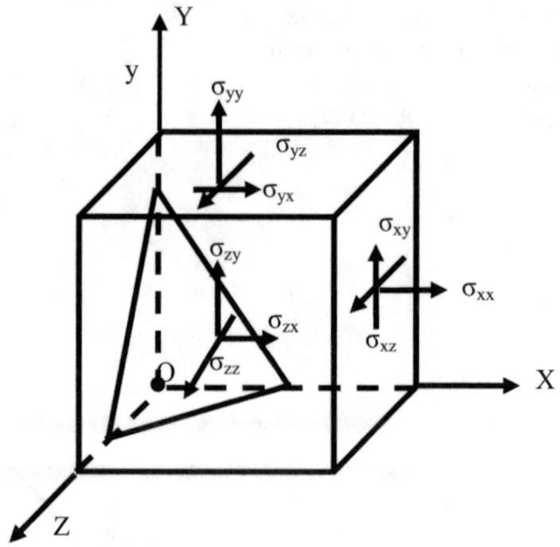

Fig. 2.10 Arbitrary tetrahedron in a stress cube

the original coordinate system. One of these orientations will give a maximum stress value.

For three-dimensional stress tensor, these principal values and principal orientation can be identified by eigenvalue analysis (Fig. 2.10).

Normal and shear stress components acting on the inclined face of the tetrahedron are a function of the orientation of the inclined plane. If the stress tensor is symmetric, we can choose an orientation of the new coordinate axes such that the shear stress components vanish in this new coordinate system. These special axes are called principal axes or principal directions. The three planes that are orthogonal to three principal axes are called principal planes. On principal planes, all shear stresses are zero. There are only normal stresses on principal planes. Maximum normal stress is defined as the algebraically largest of the three principal stresses. The minimum normal stress is defined as the algebraically smallest of the three principal stresses. There is always such a set of three mutually orthogonal directions at any point, if the state of stress at a point is a symmetric second-order tensor. This is a property common to all symmetric second-order tensors.

The principal planes and principal stresses can be found by traditional eigenvalue analysis.

Let $[I]$ be a matrix of unit vectors in one of the arbitrary orientations of the coordinate system. Let λ be the principal stress components in the new orientation, whose normals are given by $[\overline{n}]$.

Since there is no shear stress on the principal planes, the stress vector on the principal plane will be parallel to the normal:

$$\sigma_i^{(n)} = \lambda \overline{n}_i$$

where $i = X, Y, Z$.

In matrix notation

$$\left\{\sigma^{(n)}\right\} = \lambda\{\overline{n}\}$$

Since

$$\left\{\sigma^{(n)}\right\} = \{\overline{n}\}[\sigma]$$

Therefore, we can write

$$\{\overline{n}\} \cdot [\sigma] = \lambda\{\overline{n}\}$$

This can be written as

$$\{\overline{n}\}[[\sigma] - \lambda[I]] = 0 \tag{3.3}$$

where $[I]$ is an identity matrix

$$I = \begin{bmatrix} 1 & 0 & 0 \\ 0 & 1 & 0 \\ 0 & 0 & 1 \end{bmatrix}$$

$\{\overline{n}\}$ is a row matrix $\{n_1, n_2, n_3\}$ where n_i defines the direction cosine. λ is the principal stress values sought. Thus,

$$\lambda[I] = \begin{bmatrix} \lambda & 0 & 0 \\ 0 & \lambda & 0 \\ 0 & 0 & \lambda \end{bmatrix}$$

Equation (3.3) has a solution only if the determinant of the matrix is equal to zero since direction cosines' vector cannot be zero and because $n_1^2 + n_2^2 + n_3^2 = 1$ must be satisfied:

Hence,

$$[[\sigma] - \lambda[I]] = 0$$

$$\begin{vmatrix} \sigma_{XX} - \lambda & \sigma_{XY} & \sigma_{XZ} \\ \sigma_{YX} & \sigma_{YY} - \lambda & \sigma_{YZ} \\ \sigma_{ZX} & \sigma_{ZY} & \sigma_{ZZ} - \lambda \end{vmatrix} = 0$$

This is a cubic equation for the unknown λ. If the stress matrix is symmetric and has real stress values (opposed to imaginary values), the three roots of the cubic equation $\lambda^3 + a\lambda^2 + b\lambda + c = 0$ are all real numbers.

The three roots of the cubic equation λ_1, λ_2, λ_3 are the three principal stresses $\lambda_1 = \sigma_1$, $\lambda_2 = \sigma_2$, and $\lambda_3 = \sigma_3$.

In order to find the principal directions, we can use

$$\{n\}[[\sigma] - \lambda[I]] = 0$$

If we substitute σ_1 for λ in the above equation, these three equations reduce to only two linearly independent equations. These two equations can be solved with the help of the $n_1^2 + n_2^2 + n_3^2 = 1$ requirement, in order to determine the direction cosines of the normal $\{n^{(1)}\} = \left\{n_X^{(1)} n_Y^{(1)} n_X^{(1)}\right\}$ to the plane which σ_1 acts on.

When we are solving for direction cosines, since one of the equations $\left[n_1^2 + n_2^2 + n_1^3 = l\right]$ is quadratic, two solutions will be found representing two oppositely directed normals to the same plane. The choice of which one is the positive is arbitrary.

In the same manner, the process is repeated for σ_2 to find $\left\{n_X^{(2)}, n_Y^{(2)}, n_Z^{(2)}\right\}$ and σ_3 to find $\left\{n_X^{(3)}, n_Y^{(3)}, n_Z^{(3)}\right\}$.

The third principal direction can also be found by taking direction orthogonal to the first two directions.

When all the roots of the characteristic equation are different, then these are three unique orthogonal principal directions.

However, when we find the roots of the characteristic equation, if two of them are equal and one is different, then the direction of the different root is unique; however, the other two must be arbitrarily chosen as any two perpendicular axes to each other and perpendicular to the unique axis. Three axes must define a right-handed system. If all three principal stresses are equal, the state of stress is said to be hydrostatic state of stress because this is only the state of stress that can exist in fluids.

2.6 Stress Tensor Invariants

The determinant of the characteristic equation:

$$\begin{vmatrix} \sigma_{XX} - \lambda & \sigma_{XY} & \sigma_{XZ} \\ \sigma_{YX} & \sigma_{YY} - \lambda & \sigma_{YZ} \\ \sigma_{ZX} & \sigma_{ZY} & \sigma_{ZZ} - \lambda \end{vmatrix} = 0$$

$$\lambda^3 - I_1\lambda^2 - I_2\lambda - I_3 = 0 \tag{3.4}$$

where I_1, I_2, and I_3 are the first, second, and third invariants of the stress tensor. Invariants are given by the following equations:

$$I_1 = \sigma_{XX} + \sigma_{YY} + \sigma_{ZZ} = tr\sigma \tag{3.5}$$

$$I_2 = (\sigma_{XX}\sigma_{YY} + \sigma_{YY}\sigma_{ZZ} + \sigma_{ZZ}\sigma_{XX}) - \sigma_{XY}^2 - \sigma_{XZ}^2 - \tau_{YZ}^2$$

$$I_2 = \frac{1}{2}\left(\sigma_{ii}\sigma_{jj} - \tau_{ij}\tau_{ij}\right) \tag{3.6}$$

where $i = X, Y, Z$ $j = X, Y, Z$

$$I_3 = \begin{bmatrix} \sigma_{XX} & \sigma_{XY} & \sigma_{XZ} \\ \sigma_{YZ} & \sigma_{YY} & \sigma_{YZ} \\ \sigma_{ZX} & \sigma_{ZY} & \sigma_{ZZ} \end{bmatrix} = \det\ |\ \sigma\ | \tag{3.7}$$

Stress invariant is a scalar quantity, and it is an intrinsic property of a stress tensor that does not depend on the coordinate system. When a stress tensor is rotated, each vector entry in the tensor matrix changes value; however, stress invariants stay the same. This is because in eigen Eq. (3.4), three roots are independent of coordinate system.

2.6.1 Stress Invariants in Principal Axes

In principal planes, these are no shear stresses; therefore, the invariant terms become simpler:

$$I_1 = \sigma_1 + \sigma_2 + \sigma_3 \tag{3.8}$$

$$I_2 = \sigma_1\sigma_2 + \sigma_2\sigma_3 + \sigma_3\sigma_1 \tag{3.9}$$

$$I_3 = \sigma_1\sigma_2\sigma_3 \tag{3.10}$$

Stress invariants are mostly used in constitutive modeling of materials. Any function constructed with invariants is also invariant with respect to coordinate systems (Fig. 2.11).

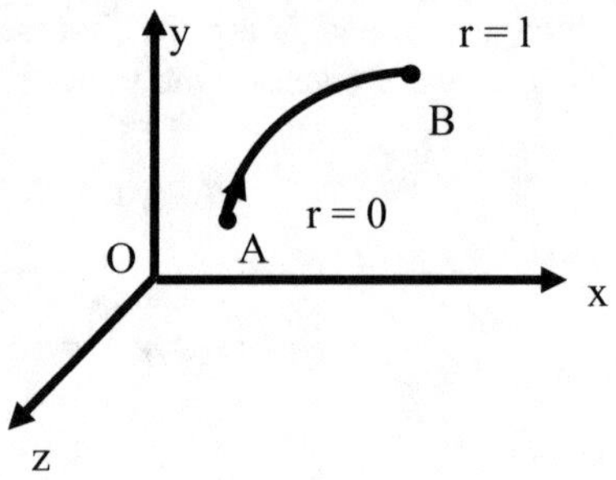

Fig. 2.11 Definition of local coordinates

2.6.2 Representation of Stress Tensor in Spherical and Deviatoric Components

Hydrostatic stress component for any stress tensor is given by

$$p = \frac{1}{3}(\sigma_{11} + \sigma_{22} + \sigma_{33}) \tag{3.11}$$

Using this definition, we can represent the stress tensor as a summation of hydrostatic stress tensor and the remainder, which is called deviatoric stress tensor:

$$[\sigma] = \begin{bmatrix} p & 0 & 0 \\ 0 & p & 0 \\ 0 & 0 & p \end{bmatrix} + \begin{bmatrix} \sigma_{xx} - p & \sigma_{xy} & \sigma_{xz} \\ \sigma_{yx} & \sigma_{yy} - p & \sigma_{yz} \\ \sigma_{zx} & \sigma_{zy} & \sigma_{zz} - p \end{bmatrix} \tag{3.12}$$

In the Eq. (3.12), the first tensor is called hydrostatic stress tensor and the second tensor is called deviatoric stress tensor. Hydrostatic tensor is also called spherical stress tensor. We will use S_{ij} to represent deviatoric stress tensor:

$$[S_{ij}] = \begin{bmatrix} S_{xx} & S_{xy} & S_{xz} \\ S_{yx} & S_{yy} & S_{yz} \\ S_{zx} & S_{zy} & S_{zz} \end{bmatrix}$$

where

$$S_{XX} = \sigma_{XX} - \frac{1}{3}(\sigma_{xx} + \sigma_{YY} + \sigma_{ZZ}) \qquad S_{XY} = \sigma_{XY}, \;\; S_{XZ} = \sigma_{XZ}, \;\; S_{YZ} = \sigma_{YZ}$$

In the same fashion S_{YY} *and* S_{ZZ} can be written in Cartesian form, deviatoric stress tensor is given by

$$S_{ij} = \sigma_{ij} - \frac{1}{3}\sigma_{kk}\delta_{ij}$$

where δ_{ij} is the Kronecker delta, when

$$\text{if } i = j \quad \delta_{ij} = 1, \qquad \text{if } i \neq j \quad \delta_{ik} = 0$$

Hydrostatic stress is the same in all three directions; therefore, in isotropic materials, hydrostatic stress causes elastic volumetric change only. However, deviatoric stress is different in all directions; as a result, it can cause shape change and inelastic (irreversible) deformation. In anisotropic materials, both hydrostatic and deviatoric stresses can lead to inelastic shape change.

2.6.3 Invariants of the Deviatoric Stress Tensor

Principal deviatoric stresses are in the same planes as the total stress tensor principal planes. Following the same procedure we used for the total stress tensor, we can obtain the characteristic equation for the deviatoric stress tensor as follows:

$$\begin{bmatrix} S_{XX} - \lambda & S_{YY} & S_{XZ} \\ S_{YX} & S_{YY} - \lambda & S_{YZ} \\ S_{ZX} & S_{ZY} & S_{ZZ} - \lambda \end{bmatrix} = 0$$

This equation can be expanded to the following form:

$$\lambda^2 - J_{2D}\lambda - J_{3D} = 0$$

where J_{2D} and J_{3D} are the second and third invariants of the deviatoric stress tensor, respectively. The first invariant of the deviatoric stress tensor is zero,$J_{1D} = 0$:

$$\begin{aligned} J_{2D} &= \frac{1}{2} S_{ij} S_{ij} \\ &= \frac{1}{6}\left[(\sigma_{XX} - \sigma_{YY})^2 + (\sigma_{YY} - \sigma_{ZZ})^2 + (\sigma_{XX} - \sigma_{ZZ})^2 + \sigma_{XY}^2 + \sigma_{XZ}^2 + \sigma_{YZ}^2\right] \end{aligned}$$

$$J_{2D} = \frac{1}{2}\left(S_{XX}^2 + S_{YY}^2 + S_{ZZ}^2\right) + \sigma_{XY}^2 + \sigma_{XZ}^2 + \sigma_{YZ}^2$$

$$J_{2D} = \frac{1}{2} S_{ij} S_{jk} S_{ki}$$

$$= \begin{vmatrix} S_{XX} & S_{XY} & S_{XZ} \\ S_{YX} & S_{YY} & S_{YZ} \\ S_{ZX} & S_{ZY} & S_{ZZ} \end{vmatrix}$$

It is important to point out that the second invariant of the deviatoric stress tensor is a representation of shear stress; however, physical interpretation of the third invariant is more complex.

2.7 Deformation and Strain

Let's assume we have a deformable solid object defined by corners A to H. Under external loading, the corner points displace to new locations given by points A' to H' (Fig. 2.12).

Using Cartesian coordinates X, Y, Z, we can easily measure the relative displacement of corner points (or any point in the object) with respect to their original location. However, just knowing the relative displacement dr of each point in the deformable body does not help us to understand the strain imposed on the body due to new configuration. Therefore, we must establish a strain definition. There are many definitions of strain. We can choose any definition of strain and calculate the strain in the deformable body. The point we are trying to emphasize is that strain is a definition (a human construct) calculated using measured displacement quantities. Therefore, strain cannot be measured directly.

Strain formulation without any constraints on the magnitude of the strain gives more accurate representation of the deformation. However, mathematical description

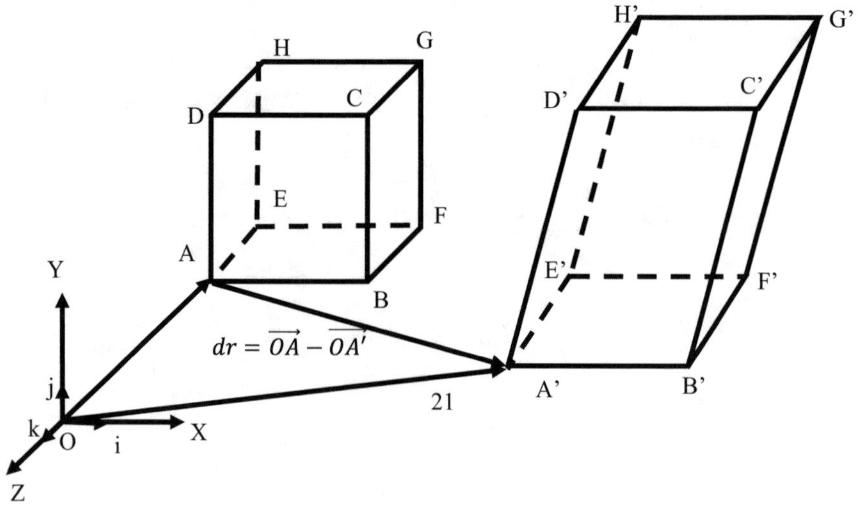

Fig. 2.12 Definition of strain deformation and rigid body motion

of finite strain (large strain) is quite complex. Therefore, for many engineering problems, small strain theory is used. For many problems, this approach is satisfactory with a reasonable degree of error.

First, we will present the small strain formulation and then more complex large strain (finite strain) formulation. Rigid body motion does not lead to any strain.

2.7.1 *Small Strain Definition*

There is no magic number that defines the boundary between small strain theory and large strain theory. However, any strain value less than 2% is usually, but not always, considered a small strain. Decision to use small or large strain formulation depends on the problem at hand and of the material properties. Now we will define strains with respect to initial (undeformed) configuration.

2.7.1.1 Elementary Definition of Pure Uniaxial Strain

At the point defined by ABCD, corners B and C are stretched to new locations B' and C' (Fig. 2.13).

As a result unit, extension ϵ_X is the change in length per unit initial length in X direction:

$$\varepsilon_X = \frac{\Delta U}{\Delta X} \quad \varepsilon_Y = \frac{\Delta V}{\Delta Y} \quad \varepsilon_Z = \frac{\Delta W}{\Delta Z}$$

We can define the unit extension per unit length in the same fashion for Y and Z directions, where ΔV *and* ΔW are extensions in Y and Z directions, respectively. We let $\Delta X, \Delta Y,$ *and* ΔZ approach zero to be able to define the strain at an imaginary point.

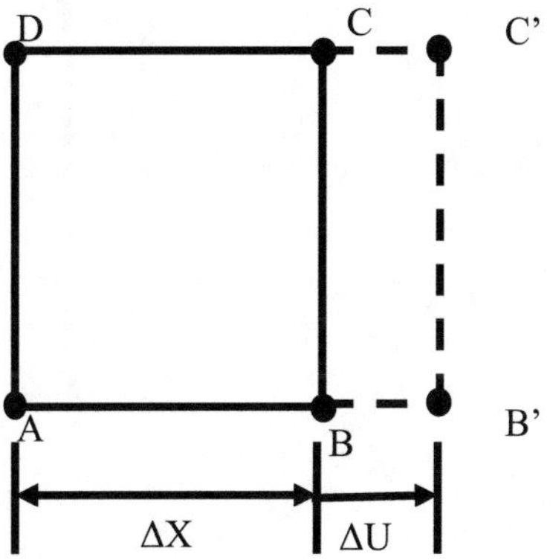

Fig. 2.13 Uniaxial extension in one dimension

2.7.1.2 Pure Shear Strain

The shear strain is defined as the change in initial right angle at point A (Fig. 2.14):

$$\gamma_{XY} = \frac{\pi}{2} - \Psi = \theta_1 + \theta_2$$

For small angles [less than 0.5 radian]

$$\tan\theta_1 = \theta_1 \text{ and } \tan\theta_2 = \theta_2$$

$$\tan\,\theta_1 = \theta_1 = \frac{\Delta V}{\Delta X} \quad \text{and} \quad \tan\,\theta_2 = \theta_2 = \frac{\Delta U}{\Delta Y}$$

As a result, we can write

$$\gamma_{\mathrm{XY}} = \theta_1 + \theta_2 = \frac{\Delta V}{\Delta X} + \frac{\Delta U}{\Delta Y}.$$

Since we define our strain at a point in the limit dimensions of the differential element, as ΔX and ΔY approach zero. Hence, we can define strain in differential form as

$$\epsilon_X = \frac{\partial U}{\partial X} \quad \epsilon_Y = \frac{\partial V}{\partial Y} \quad \gamma_{XY} = \frac{\partial U}{\partial Y} + \frac{\partial V}{\partial X}$$

where U is the displacement in X axis and V is the displacement in Y axis.

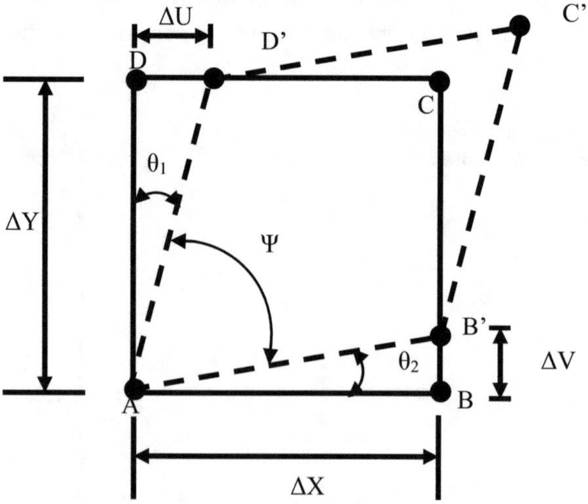

Fig. 2.14 Shear strain definition in two dimensions

It is important to point out that the initial (undeformed) position is the independent variable and as the reference state in these small strain calculations. In other words, the change in per unit length is with respect to the initial length. If we take the deformed position as the reference state and as the independent variable, we end up with a different formulation of strain. Of course, these two reference states will lead to two different numerically different strain values.

2.7.1.3 Pure Rigid Body Motion

Let's assume the cantilever beam shown in Fig. 2.15 is supported at point A and subjected to vertical and axial loads at point B. Point *C* in the beam will be subjected to rigid body displacement and rigid body rotation, but it will not experience any strain because point *C* undergoes rigid body motion only.

Point *C* undergoes displacement in *X* and *Y* axis directions as well as rotation around *Z* axis. However, these motions will not load to any strain or stress field at point *C* (Fig. 2.16).

2.7.2 Small Strain and Small Rotation Formulation

We will start with defining spatial coordinate and material coordinate systems. Let's say *f*(*u*, *t*) is a function that defines displacement of point *A* located at a particular point A' in space at time *t*.

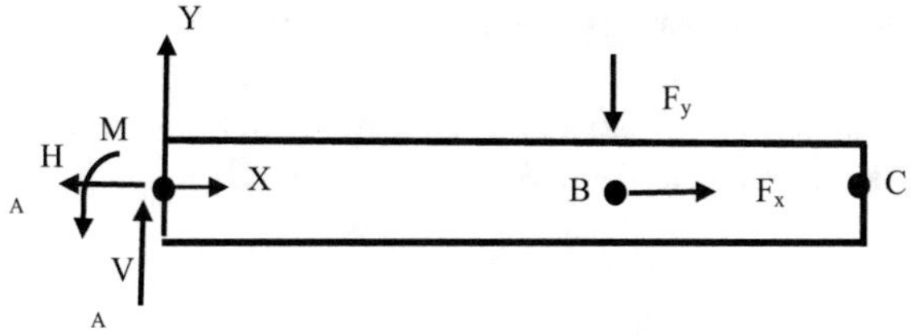

Fig. 2.15 Definition of pure rigid body motion

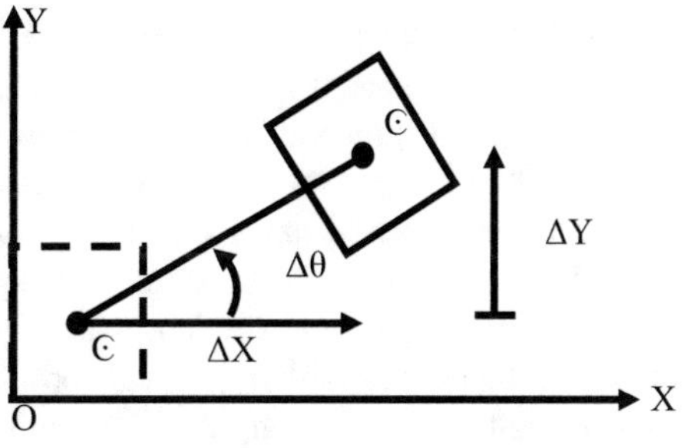

Fig. 2.16 Pure rigid body rotation

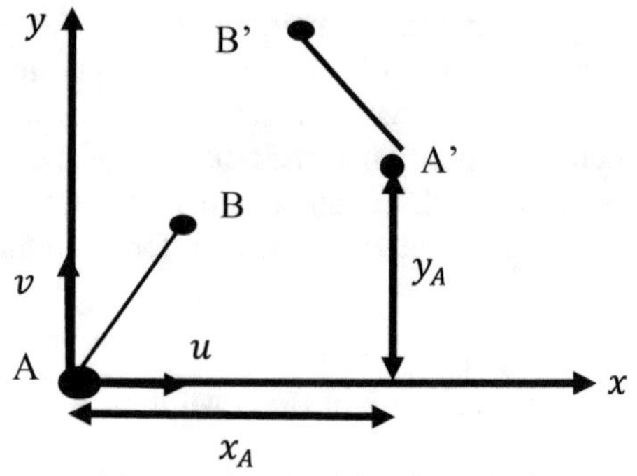

Fig. 2.17 Spatial description of the new location of point A

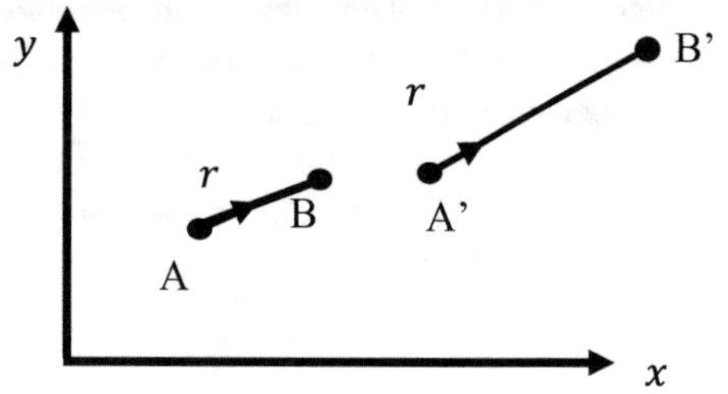

Fig. 2.18 Material (local) coordinate system description

This spatial description is called Eulerian description. Another description is material field description. In this case, function $g(r, t)$ defines displacement of point A at time t regardless of where point A is located in space. This is called material or local description (Fig. 2.17).

The coordinates of a point in local reference coordinate system (configuration) r are referred to as material coordinates.

The coordinates of a point in spatial coordinate system x, y, z are referred to as spatial coordinates. Of course, we can link the material coordinate system and spatial coordinate system using deformation mapping.

In computational solid mechanics, we prefer to use material coordinates because it is more convenient. We will explain this now.

2.7.2.1 Definition of Material (Local) Coordinates

In Fig. 2.18, we can define a coordinate system that is shaped in the original shape of the line AB. We place the origin of our local coordinate axis r at point A. At point A r $= 0$ and at point B the value of local coordinate r $=1$. By normalizing the coordinate between 0 and 1, we can define displacement of any point between A and B with respect to initial coordinates very easily. Of course, we can define displacement of point A using global Cartesian coordinate system x, y and time or we can use material (local) coordinate system (which is usually called local coordinate system in finite element method terminology). Local coordinate r is in the axis direction of the initial 1-D element shown in the figure. The origin of the local coordinate system is at point

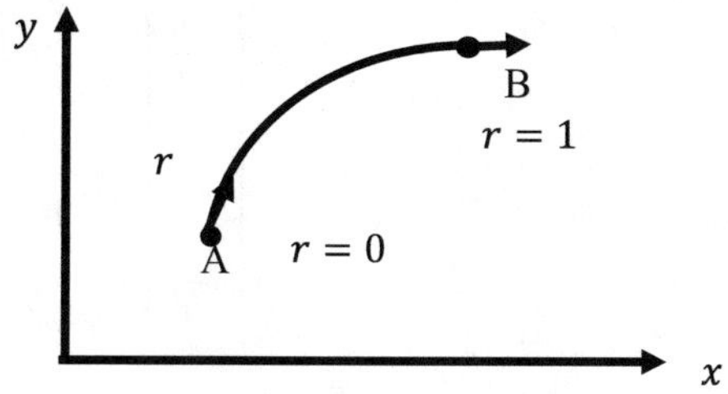

Fig. 2.19 Material (local) coordinate system

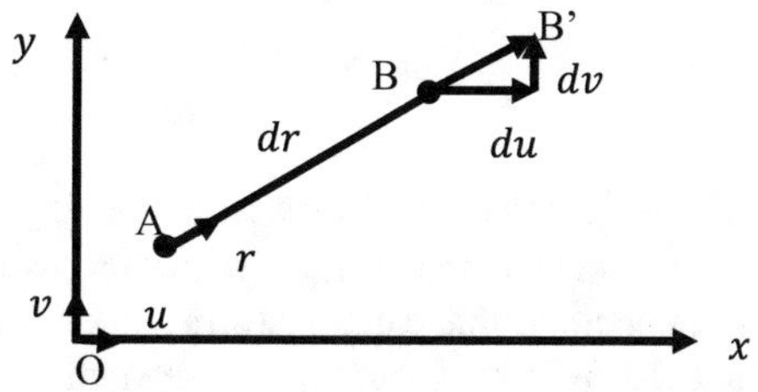

Fig. 2.20 Local (material) coordinate system

A, or it can be located at any point between *A* and *B* using a linear interpolation function. As a result, if point *A* moves with respect to initial location due to loading, origin A also moves. Therefore, coordinate *r* will define displacement of point *A* at any time regardless of where point *A* is (Fig. 2.19).

We will use the terminology local coordinate rather than material coordinates. For a 2-D case, let's *x*-*y* define our global Cartesian coordinate system and *r* define local coordinate system. Let *dr* define relative deformation of *B* with respect to *A* (Fig. 2.20), where *r* is the material (local) coordinate axis. It is always in the natural direction of the member, *u* is the displacement vector along spatial *x* axis and *v* is the displacement vector along spatial *y* axis.

The local coordinate relative displacement *dr* in spatial Cartesian coordinate system *x*, *y* can be given by

$$\begin{aligned} \frac{du}{dr} &= \frac{du}{dx}\frac{dx}{dr} + \frac{du}{dy}\frac{dy}{dr} \\ \frac{dv}{dr} &= \frac{dv}{dx}\frac{dx}{dr} + \frac{dv}{dy}\frac{dy}{dr} \end{aligned} \tag{3.13}$$

It is important to point out that the left-hand side of this equation represents strain components with respect to the initial length. Equation (3.13) can be written in matrix form:

$$\begin{bmatrix} \frac{du}{dr} \\ \\ \frac{dv}{dr} \end{bmatrix} = \begin{bmatrix} \frac{du}{dx}\frac{du}{dy} \\ \\ \frac{dv}{dx}\frac{dv}{dy} \end{bmatrix} \begin{bmatrix} \frac{dx}{dr} \\ \\ \frac{dy}{dr} \end{bmatrix}$$

or

$$\frac{\mathrm{dU}}{\boldsymbol{dr}} = \boldsymbol{J} \cdot \boldsymbol{n} = \boldsymbol{u}\boldsymbol{\nabla} \cdot \widetilde{\boldsymbol{n}}$$

where $\frac{dU}{dr}$ is the column vector containing relative displacement vector components and n is the column vector of direction cosines. $\boldsymbol{J}$ is called the Jacobian matrix, which is an operator that can transform any unit relative displacements for an infinitesimal line AB from spatial coordinate system x, y to local (material) coordinate system r.

Assume that the length of $AB \rightarrow 0$; as a result, all derivatives are evaluated at point B.

We can split the Jacobian matrix into two components. One component will include deformations around point B and the other will include rigid body relative displacements. It is assumed that rigid body relative displacements do not cause any strain in the body. This splitting process can be accomplished in the following manner:

$$\begin{bmatrix} \frac{\partial u}{\partial x} & \frac{\partial u}{\partial y} \\ \frac{\partial v}{\partial x} & \frac{\partial v}{\partial y} \end{bmatrix} = \begin{bmatrix} \frac{\partial u}{\partial x} & \frac{1}{2}\left(\frac{\partial u}{\partial y} + \frac{\partial v}{\partial x}\right) \\ \frac{1}{2}\left(\frac{\partial v}{\partial x} + \frac{\partial u}{\partial y}\right) & \frac{\partial v}{\partial y} \end{bmatrix} + \begin{bmatrix} 0 & \frac{1}{2}\left(\frac{\partial u}{\partial y} - \frac{\partial v}{\partial x}\right) \\ \frac{1}{2}\left(\frac{\partial v}{\partial x} - \frac{\partial u}{\partial y}\right) & 0 \end{bmatrix}$$

Earlier we defined small strain components as follows:

$$\epsilon_{xx} = \frac{\partial u}{\partial x}, \quad \varepsilon_{yy} = \frac{\partial v}{\partial y}, \quad \gamma_{xy} = \left(\frac{\partial u}{\partial y} + \frac{\partial v}{\partial x}\right)$$

For convenience, we will define shear strain ϵ_{xy} as half the decrease γ_{xy} in the right angle to be able to define strain vectors as components of a second-order strain tensor, $\boldsymbol{\varepsilon}$:

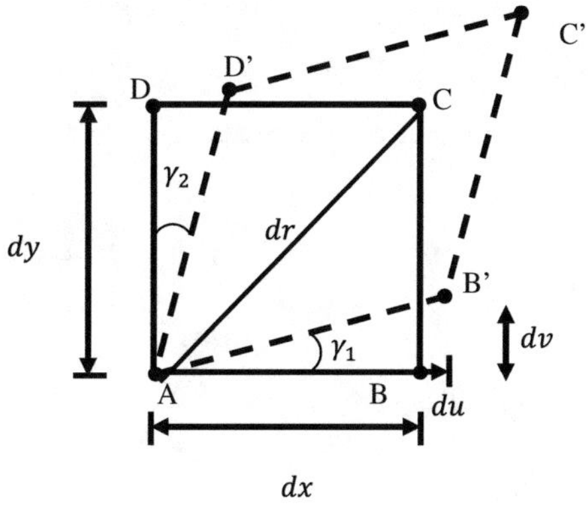

Fig. 2.21 Deformed differential element without rigid body motion

$$\epsilon_{xy} = \frac{1}{2}\gamma_{xy}$$

As a result, the first matrix is the strain matrix and the second matrix is the rotation matrix:

$$\Omega_{xy} = \frac{1}{2}\left(\frac{\partial u}{\partial y} - \frac{\partial v}{\partial x}\right) \quad \Omega_{yx} = \frac{1}{2}\left(\frac{\partial v}{\partial x} - \frac{\partial u}{\partial y}\right) = -\Omega_{xy}$$

As a result, the Jacobian matrix can be written as

$$\boldsymbol{J} = \begin{bmatrix} \epsilon_{xx} & \epsilon_{xy} \\ \epsilon_{yx} & \epsilon_{yy} \end{bmatrix} + \begin{bmatrix} 0 & \Omega_{xy} \\ \Omega_{yx} & 0 \end{bmatrix}$$

In our stress and strain definitions, the first subscript defines the plane and the second subscript defines the direction. Now we can show that the strain matrix leads to deformations (shape change), and the rotation matrix leads to rigid body motion only with no shape change (strain) (Fig. 2.21).

We can write earlier transformation equations between local (material) coordinates and global (spatial) Cartesian coordinates in order to find unit relative displacement of B with respect to A:

$$\begin{bmatrix} \frac{du}{dr} \\ \frac{dv}{dr} \end{bmatrix} = \begin{bmatrix} \frac{du}{dx} & \frac{1}{2}\left(\frac{du}{dy} + \frac{dv}{dx}\right) \\ \frac{1}{2}\left(\frac{dv}{dx} + \frac{du}{dy}\right) & \frac{dv}{dy} \end{bmatrix} \begin{bmatrix} \frac{dx}{dr} \\ \frac{dy}{dr} \end{bmatrix}$$

Assume γ_1 is very small $cos\,\gamma_1 = \frac{dx}{dr} \cdot = \cos 0 = 1$.

And

$$\frac{dy}{dr} = \cos\ \frac{\pi}{2} = 0$$

As a result, we can write

$$\begin{bmatrix} \frac{du}{\mathrm{dr}} \\ \frac{dv}{dr} \end{bmatrix} = \begin{bmatrix} \epsilon_{xx} & \epsilon_{xy} \\ \epsilon_{yx} & \epsilon_{yy} \end{bmatrix} \begin{bmatrix} 1 \\ 0 \end{bmatrix} = \begin{bmatrix} \epsilon_{xx} \\ \epsilon_{yx} \end{bmatrix}$$

Thus, for *AB*

$$\epsilon_{xx} = \frac{du}{dr}, \quad \epsilon_{yx} = \frac{dv}{dr}$$

It is important to point out that these equations are only valid for small deformations.

Similarly, for *AD* we can write

Assuming γ_2 is very small:

$$\frac{dx}{dr} = \cos\ \frac{\pi}{2} = 0, \quad \frac{dy}{dr} = \cos\ 0 = 1$$

$$\begin{bmatrix} \frac{du}{dr} \\ \frac{dv}{dr} \end{bmatrix} = \begin{bmatrix} \epsilon_{xx} & \epsilon_{xy} \\ \epsilon_{yx} & \epsilon_{yy} \end{bmatrix} \begin{bmatrix} 0 \\ 1 \end{bmatrix} = \begin{bmatrix} \epsilon_{xy} \\ \epsilon_{yy} \end{bmatrix}$$

$$\frac{du}{dr} = \epsilon_{xy} \quad \frac{dv}{dr} = \epsilon_{yy}$$

2.7.2.2 Small Strain in Local Coordinates

Displacement of corner point *C* and *C'* is plotted in detail Figs. 2.22, 2.23, and 2.24.

Strain along local axis *r* can be written as

$$\epsilon_{rr} = \frac{\left(\epsilon_{xx}dx + \gamma_{xy}dy\right)\cos\ \theta + \epsilon_{yy}dy\sin\ \theta}{dr}$$

where *dr* is the initial length of AC in local coordinates,

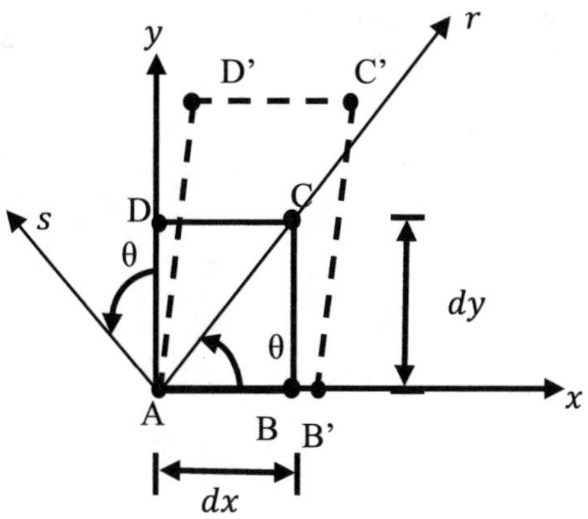

Figure 2.22 Deformation of rectangle shape ABCD into AB'C'D' in local and global coordinates

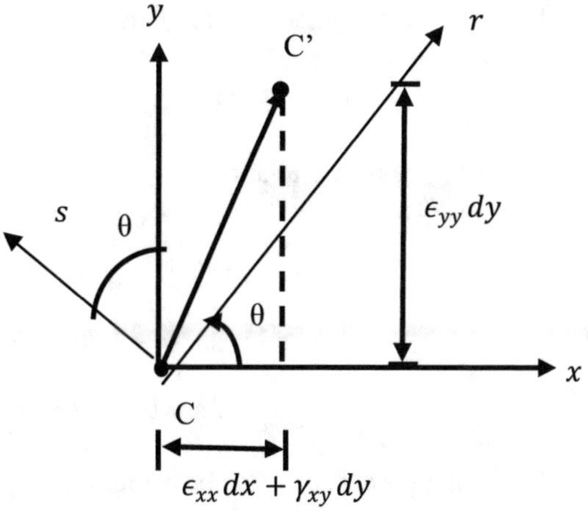

Figure 2.23 Displacement of point C to C'

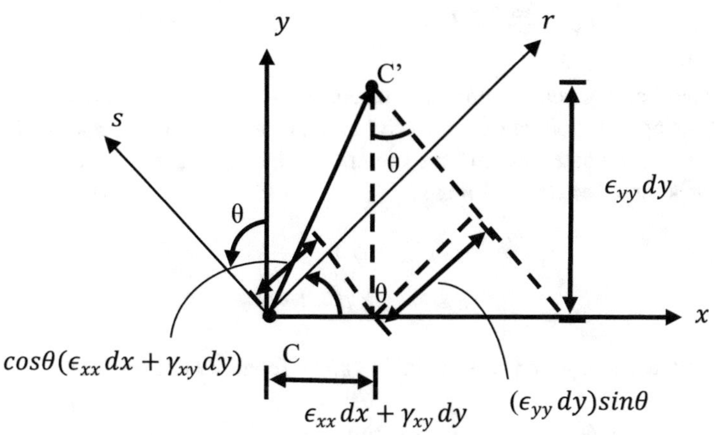

Fig. 2.24 Displacement of a point in material (local) and spatial coordinates

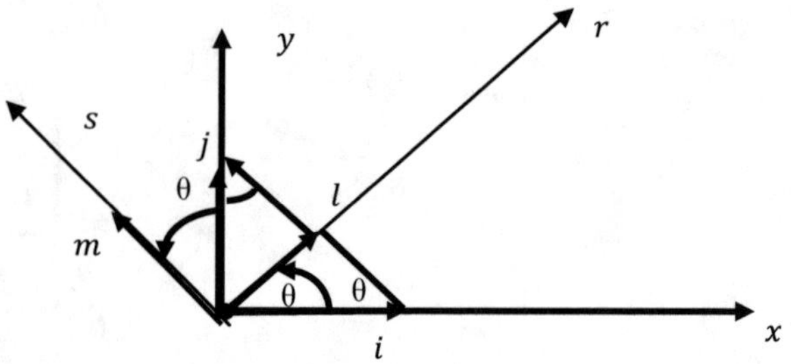

Fig. 2.25 Unit vectors in local (material) ***r-s*** coordinates and spatial coordinates ***x,y***

$$\epsilon_{rr} = \left(\epsilon_{xx}\frac{dx}{dr} + \gamma_{xy}\frac{dy}{dr}\right)\cos\,\theta + \epsilon_{yy}\frac{dy}{dr}\sin\,\theta$$

Knowing that $\cos\,\theta = \frac{dx}{dr}$, $\sin\,\theta = \frac{dy}{dr}$
Inserting these equations in ε_{rr} yields

$$\epsilon_{rr} = \epsilon_{xx}\cos^2\theta + \gamma_{xy}\sin\,\theta\cos\,\theta + \epsilon_{yy}\sin^2\theta$$

Using trigonometric relations

$$\cos^2\theta = \frac{1}{2}(1 + \cos\,2\theta)$$

$$\sin^2\theta = \frac{1}{2}(1 - \cos\,2\theta)$$

$$2\sin\,\theta\cos\,\theta = \sin\,2\theta$$

Strain along r axis can also be expressed as

$$\epsilon_{rr} = \frac{\epsilon_{xx} + \epsilon_{yy}}{2} + \frac{\epsilon_{xx} - \epsilon_{yy}}{2}\cos\,2\theta + \frac{\gamma_{xy}}{2}\sin\,2\theta$$

We could also obtain components of displacement CC' in r and s local coordinates by using dot product of vector CC' and unit vectors along r and s (Fig. 2.25).

Material coordinate system unit vectors can be represented in terms of spatial coordinates' unit vectors as follows:

$$l = i\cos\,\theta + j\sin\,\theta$$

$$m = -i\sin\,\theta + j\cos\,\theta$$

The displacement in r and s direction can be found by dot product:

$$CC' \cdot l = \left[\left(\epsilon_{xx}dx + \gamma_{xy}dy\right)i + \epsilon_y dyj\right]\left[i\cos\theta + j\sin\theta\right]$$

$$= \left[\epsilon_x dx + \gamma_{xy} dy\right]\cos\theta + \epsilon_{yy} dy\sin\theta$$

In s direction,

$$CC' \cdot m = \left[\left(\epsilon_{xx} dx + \gamma_{xy} dy\right)i + \epsilon_y dy\, j\right]\left[i\ \sin\theta + j\cos\theta\right]$$

$$= -\left(\epsilon_{xx}\, dx + \gamma_{xy}\, dy\right)\sin\theta + \epsilon_{yy}\, dy\cos\theta$$

From here, we can obtain

$$\epsilon_{rr} = \frac{CC' \cdot l}{dr}$$

and

$$\epsilon_{ss} = \frac{CC' \cdot m}{dr}$$

2.7.2.3 Shear Strain in Local (Material) Coordinates

Displacement along axis s is (Fig. 2.26)

$$\epsilon_{yy} dy\ cos\theta - \left(\epsilon_{xx} dx + \gamma_{xy}\right)\sin\theta$$

We can also obtain displacement of point C along s axis from the arc length (Fig. 2.27).

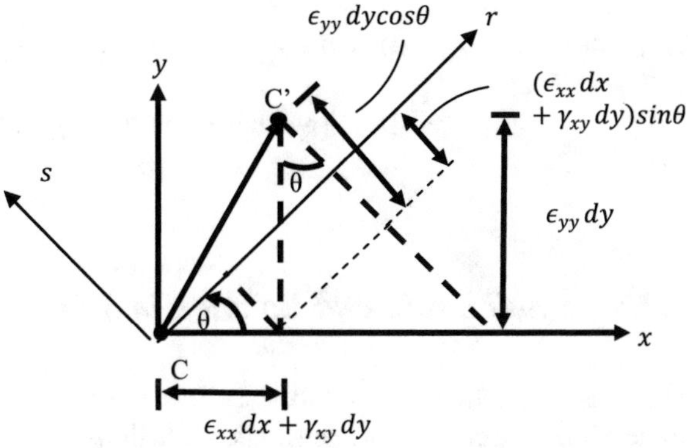

Fig. 2.26 Shear strain in local (material) coordinates

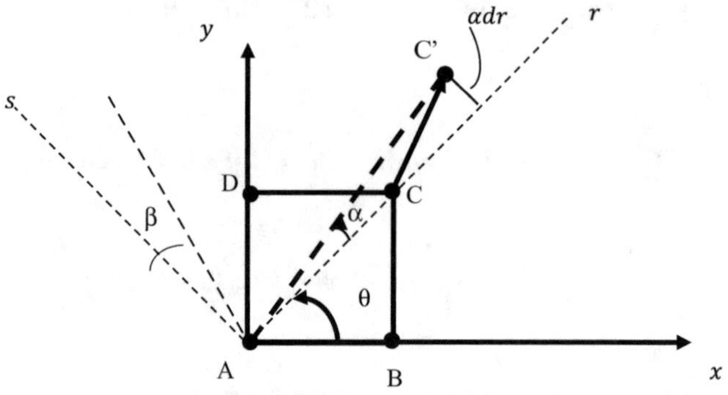

Fig. 2.27 Relations between local (material) and spatial coordinates

Assuming small deformations and small strains

$$\alpha dr = \epsilon_{yy} dy\ cos\theta - \left(\epsilon_{xx} dx + \gamma_{xy} dy\right) \sin\theta$$

$$\alpha = \epsilon_{yy} \frac{dy}{dr} cos\theta - \left(\epsilon_{xx} \frac{dx}{dr} + \gamma_{xy} \frac{dy}{dr}\right) \sin\theta$$

$$\alpha = \epsilon_{yy} \sin\theta\ \cos\theta - \left(\epsilon_{xx} \cos\theta + \gamma_{xy} \sin\theta\right) \sin\theta$$

Rotation of s axis (β) can be found by substituting $\theta + \frac{\pi}{2}$ in equation for α:

$$\beta = -\epsilon_{yy} \sin\theta\ \cos\theta + \epsilon_{xx} \sin\theta\ \cos\theta - \gamma_{xy} \cos^2\theta$$

In Fig. 2.27, it is assumed that shear strain is positive. Therefore, α is counter-clockwise and β is clockwise. The shear strain is the change in the initial right angle:

$$\gamma_{rs} = (\alpha - \beta)$$

$$= \left[\epsilon_{yy} \sin\theta \cos\theta - \epsilon_{xx} \sin\theta \cos\theta - \gamma_{xy} \sin^2\theta\right] - \left[-\epsilon_{yy} \sin\theta \cos\theta + \epsilon_{xx} \sin\theta \cos\theta - \gamma_{xy} \cos^2\theta\right]$$

$$\gamma_{rs} = -2\left(\epsilon_{xx} - \epsilon_{yy}\right) \sin\ \theta \cos\ \theta + \gamma_{XY}\left(\cos^2\theta - \sin^2\theta\right)$$

2.7.3 Small Strain and Rotation in 3-D

Using the derivation we used for 2-D case, we can derive the formulation for three-dimensional case. In the dimensions, we will use u for displacement along x, v for displacement along y, and w for displacement along z axis. r axis is again our local (material) coordinate axis:

$$\begin{Bmatrix} \frac{du}{dr} \\ \frac{dv}{dr} \\ \frac{dw}{dr} \end{Bmatrix} = \begin{bmatrix} \frac{\partial u}{\partial x} & \frac{\partial u}{\partial y} & \frac{\partial u}{\partial z} \\ \frac{\partial v}{\partial x} & \frac{\partial v}{\partial y} & \frac{\partial v}{\partial z} \\ \frac{\partial w}{\partial x} & \frac{\partial w}{\partial y} & \frac{\partial w}{\partial z} \end{bmatrix} \begin{Bmatrix} \frac{dx}{dr} \\ \frac{dy}{dr} \\ \frac{dz}{dr} \end{Bmatrix}$$

Again Jacobian matrix $\boldsymbol{J}$ can be written as

$$\boldsymbol{J} = \boldsymbol{E} + \boldsymbol{\Omega}$$

where

$$\boldsymbol{E} = \begin{bmatrix} \frac{\partial u}{\partial x} & \frac{1}{2}\left(\frac{\partial u}{\partial y} + \frac{\partial v}{\partial x}\right) & \frac{1}{2}\left(\frac{\partial u}{\partial z} + \frac{\partial w}{\partial x}\right) \\ \frac{1}{2}\left(\frac{\partial u}{\partial y} + \frac{\partial v}{\partial x}\right) & \frac{\partial v}{\partial y} & \frac{1}{2}\left(\frac{\partial v}{\partial z} + \frac{\partial w}{\partial y}\right) \\ \frac{1}{2}\left(\frac{\partial u}{\partial z} + \frac{\partial w}{\partial x}\right) & \frac{1}{2}\left(\frac{\partial w}{\partial y} + \frac{\partial v}{\partial z}\right) & \frac{\partial w}{\partial z} \end{bmatrix}$$

$$\boldsymbol{\Omega} = \begin{bmatrix} 0 & \frac{1}{2}\left(\frac{\partial u}{\partial y} - \frac{\partial v}{\partial x}\right) & \frac{1}{2}\left(\frac{\partial u}{\partial z} - \frac{\partial w}{\partial x}\right) \\ -\frac{1}{2}\left(\frac{\partial u}{\partial y} - \frac{\partial v}{\partial x}\right) & 0 & \frac{1}{2}\left(\frac{\partial v}{\partial z} - \frac{\partial w}{\partial y}\right) \\ -\frac{1}{2}\left(\frac{\partial u}{\partial z} - \frac{\partial w}{\partial x}\right) & -\frac{1}{2}\left(\frac{\partial v}{\partial z} - \frac{\partial w}{\partial y}\right) & 0 \end{bmatrix}$$

When components of $\boldsymbol{\Omega}$ are small compared to one radian, it represents rigid body rotation.

2.8 Kinematics of Continuous Medium

There are three distinct descriptions of the displacements in a continuum.

2.8.1 Material (Local) Description

The independent variables are the local (material) coordinates of a particle at time t, regardless of where the point is in spatial coordinates.

2.8.2 Referential Description (Lagrangian Description)

This is usually the reference configuration description. The independent variables are the x, y, z coordinates and time with respect to initial configuration and initial state. The chosen reference configuration is the initial (unstressed) state at $t = 0$.

2.8.3 Spatial Description (Eulerian Description)

This is the most common coordinate description used in fluid mechanics. The coordinate system is defined with respect to current location of the object at time t. As a result, the origin of the coordinate system is located at the new (deformed) location of the object.

The origin of the material description and Eulerian description of coordinate systems are usually the same, meaning that they use the same origin and the same coordinate axes, although they could be different.

Of course transformation of coordinates between these descriptions can easily be accomplished (Fig. 2.28). where r and s, are local coordinate system axes that move with the point. O is the origin for referential (Lagrangian) coordinate system. O' is the origin for the material (local) and Eulerian coordinate system.

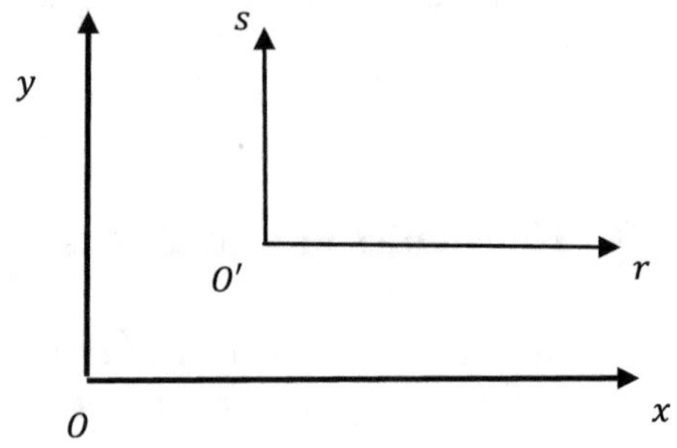

Fig. 2.28 Material (local), spatial, and referential coordinate system description

2.9 Rate of Deformation and Rate of Spin Formulation

Rate of deformation tensor is also called the stretching tensor or velocity of strain (strain rate) tensor. The spin tensor is also called the vorticity tensor.

We realize that the rate of deformation tensor is defined differently by different authors. We will subscribe to Malvern's (1969) definition and use his formulation and derivation with few modifications (Fig. 2.29).

The relative velocity components of dv_i of point A' relative to point B' can be given by

$$d\widetilde{v}_i = \frac{\partial \widetilde{v}_i}{\partial x_m} dx_m$$

or in matrix form

$$[d\widetilde{v}_i] = [\widetilde{v}_{i,m}][dx_m] \tag{3.14}$$

or in tensorial notation

$$d\widetilde{\boldsymbol{v}} = \boldsymbol{L} \cdot d\boldsymbol{x} = d\boldsymbol{x} \cdot \boldsymbol{L}^T$$

where $\boldsymbol{L}$ is $\qquad L_{im} = \widetilde{v}_{i,m},\ (L^T)_{im} = \widetilde{v}_{m,i}$

The components of tensor $\boldsymbol{L}$ are spatial gradients of the velocity. $L_{im} = \widetilde{v}_{i,m}$ can be written as the sum of a symmetric tensor $\boldsymbol{D}$, which we will call the rate-of-deformation tensor (also called stretching tensor) and a skew-symmetric tensor **W** called the spin tensor (also called vorticity tensor).

Hence, we can write

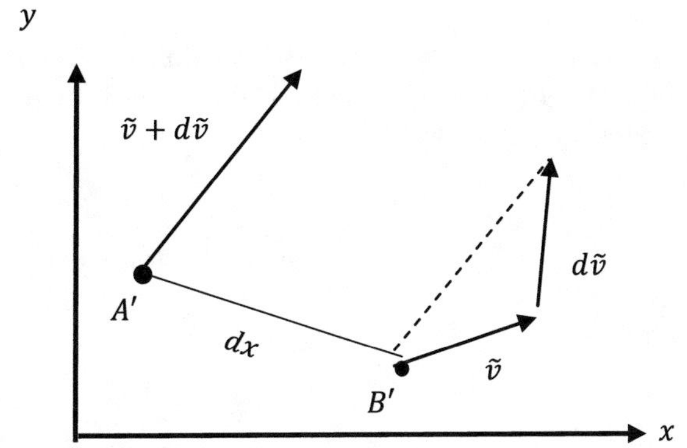

Figure 2.29 Relative velocity of $d\widetilde{v}_i$ of point A' relative to particle B'

$$\boldsymbol{L} = \boldsymbol{D} + \boldsymbol{W}$$

$$\boldsymbol{D} = \frac{\mathbf{1}}{\mathbf{2}}\left(\boldsymbol{L} + \boldsymbol{L}^{\boldsymbol{T}}\right) \quad \text{and} \quad \boldsymbol{W} = \frac{\mathbf{1}}{\mathbf{2}}\left(\boldsymbol{L} - \boldsymbol{L}^{\boldsymbol{T}}\right)$$

or in Cartesian coordinates

$$D_{im} = \frac{1}{2}\left(\widetilde{v}_{i,m} + \widetilde{v}_{m,i}\right) \quad \text{and} \quad W_{im} = \frac{1}{2}\left(\widetilde{v}_{i,m} - \widetilde{v}_{m,i}\right)$$

$\boldsymbol{W}$ spin matrix is different from $\boldsymbol{\Omega}$ rotation matrix. They should not be confused with each other. The same thing is true, for the rate of deformation tensor $\boldsymbol{D}$ which is not a strain matrix. The rate of deformation tensor, $\boldsymbol{D}$, allows us defining the relative velocity of point A' with respect to point B'. As a result, Equation (3.12) the relative velocity can be given by

$$d\widetilde{v}_i = D_{im}dx_m + W_{im}dx_m$$

2.9.1 Comparison of Rate of Deformation Tensor, D, and Time Derivative of the Strain Tensor, $\dot{\boldsymbol{\varepsilon}}$

The small strain tensor $\boldsymbol{\varepsilon}$ is defined in terms of local coordinates (material coordinates) or Lagrangian description (initial coordinates). Therefore, it is given by

$$\epsilon_{ij} = \frac{1}{2}\left(\frac{\partial u_i}{\partial r_j} + \frac{\partial u_j}{\partial r_i}\right)$$

where r represents local (material) coordinates or where Lagrangian coordinates, u, represents the amount of deformation in each axis. Therefore, time derivative of small strain tensor would be given by

$$\frac{d\epsilon_{ij}}{dt} = \frac{1}{2}\left(\frac{\partial \widetilde{v}_i}{\partial r_j} + \frac{\partial \widetilde{v}_j}{\partial r_i}\right)$$

where

$$\frac{\partial \widetilde{v}_i}{\partial x_j} = \frac{d}{dt}\left(\frac{\partial u_i}{\partial r_j}\right)$$

On the other hand, the rate of deformation tensor, **D**, is given by

$$D_{ij} = \frac{1}{2}\left(\frac{\partial \widetilde{v}_i}{\partial x_j} + \frac{\partial \widetilde{v}_j}{\partial x_i}\right)$$

The time derivative of the strain tensor is with respect to local coordinates (material coordinates), while the rate of deformation tensor is defined with respect to spatial coordinates x, y, z (Eulerian description). For small displacement and small strain problems, both tensors are the same. However, $\boldsymbol{D}$, the rate of deformation tensor, is necessary for large displacement and large strain problems, where $i = 1, 2, 3$, $j = 1, 2, 3$. It is important to point out that indicial notation axes are represented by x_i. However, when convenient, x, y, z spatial coordinates and r, s, t local coordinates are also used throughout the book.

2.9.2 True Strain (Natural Strain) (Logarithmic Strain)

Engineering strain is defined by elongation divided by initial length:

$$de = \frac{L - L_0}{L_0}$$

and in incremental form as

$$de = \frac{dL}{L_0} \quad e = \frac{1}{L_0}\int_{L_0}^{L} dL$$

However, if we use the instantaneous length (new length after deformation) in the denominator, we obtain the true strain (natural or logarithmic strain):

$$d\epsilon = \frac{dL}{L}$$

Here the increment of true strain is defined by change in length per unit of instantaneous (new) length. In order to find the total true strain, we can integrate the increment:

$$\int_{L_0}^{L} d\epsilon = \epsilon = \int_{L_0}^{L} \frac{dL}{L} = \ln \frac{L}{L_0} = \ln\left(\frac{L_0 + \Delta L}{L_0}\right) = \ln\,(1 + e)$$

where ln is the natural logarithm. Relation between true strain and engineering strain can also be written as

$$e^{\varepsilon} = (1 + e) \quad \text{or} \quad e = e^{\varepsilon} - 1$$

From Taylor series, we can write the following relation:

$$e^x = \sum_{k=0}^{\infty} \frac{x^k}{k!} = 1 + x + \frac{x^2}{2!} + \frac{x^3}{3!} + \ldots\ldots$$

As a result, we can write the following relation between true strain and engineering strain:

$$\varepsilon + \frac{\epsilon^2}{2!} + \frac{\varepsilon^3}{3!} + \ldots\ldots = e$$

In three dimensions for small strain, this relationship can be generalized as follows:

$$d\epsilon_{ij} = D_{ij}dt$$

where D_{ij} is the rate-of-deformation tensor and dt is the time increment.

It is important to point out that increments of natural strain, $d\epsilon_{ij}$, are components of a Cartesian tensor; as a result, the transformation formulas and principal axis theory all apply. The quantities ϵ_{ij} defined by integration are not components of a Cartesian tensor, because during the deformation, principal axes continuously rotate at each increment (Malvern 1969).

2.10 Finite Strain and Deformation

There are many definitions of finite (large) strain. They can be categorized into two classes.

1. Defining strain with respect to undeformed original configuration and geometry. This approach is called Lagrangian formulation.
2. Defining strain with respect to deformed configuration and geometry. This approach is called Eulerian formulation.

Large (finite) strain formulation is the easiest to define in terms of a deformation gradient tensor. However, deformation gradient tensor includes the strain tensor and the rotation tensor. As a result, this can make it tricky to be used in material modeling. On the other hand, strain tensor only includes strains, where x, y is the global Cartesian coordinate (referential description) system and r is the local coordinate (material description) system (Figs. 2.30 and 2.31).

Deformation equation for point B can be defined in local (material) coordinate system and can be mapped onto referential coordinate system. For example,

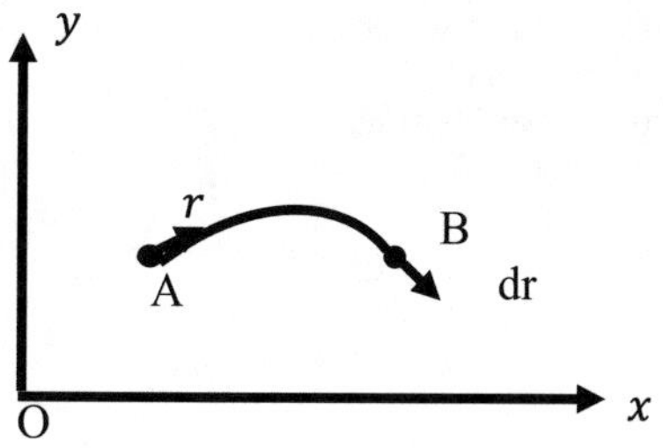

Fig. 2.30 Relative displacement and rotation of point B with respect to A

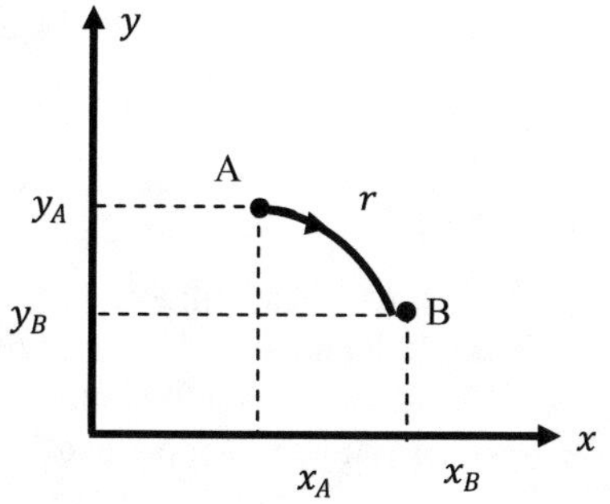

Fig. 2.31 Normalized local (material) coordinate system and spatial coordinates

$$x = x_A + (x_B - x_A)r$$

$$y = y_A + (y_B - y_A)r$$

We will derive the finite strain formulation using Malvern (1969) formulation and notation with few changes. We will define deformation gradient tensor F as a tensor that its components are the partial derivatives between the referential (Lagrangian) coordinate axes and local (material) coordinates. F, the deformation gradient tensor, will be defined in terms of the undeformed configuration. x, y, z referential coordinates are represented by bold $\boldsymbol{X}$. Local (material) coordinates are represented by bold $\boldsymbol{r}$.

The relation between the referential coordinates and local (material) coordinates are given by:

$$d\boldsymbol{X} = \boldsymbol{F} \cdot d\boldsymbol{r} \quad \text{or} \quad d\boldsymbol{r} = d\boldsymbol{X} \cdot \boldsymbol{F}^T$$

or in indicial matrix notation

$$\{dx_i\} = \left[\frac{\partial x_i}{\partial r_j}\right]\{dr_j\}$$

Therefore,

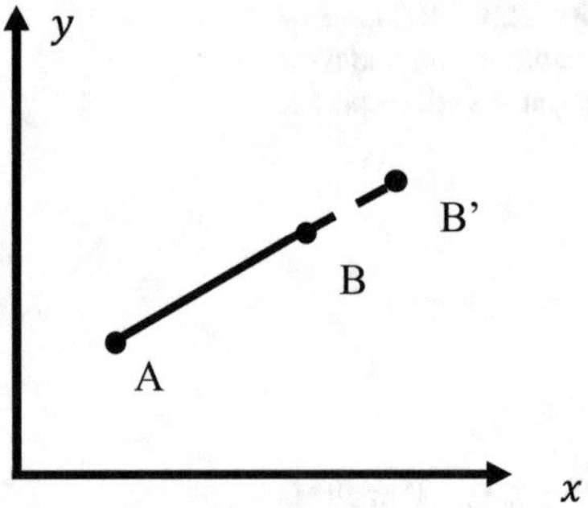

Fig. 2.32 Axial strain definition in one-dimensional line element

$$[F_{ij}] = \left[\frac{\partial x_i}{\partial r_j}\right]$$

The first index identifies the row and the second index identifies the column.

When Eulerian description is used and deformation gradient is defined with respect to deformed configuration (the spatial deformation gradient $\boldsymbol{F}^{-1}$), relations between spatial coordinate $\boldsymbol{x}$ and local (material) coordinates $\boldsymbol{r}$ is given by

$$d\boldsymbol{r} = \boldsymbol{F}^{-1} \cdot d\boldsymbol{x}$$

or in indicial matrix notation,

$$\{dr_i\} = \left[\frac{\partial r_i}{\partial x_j}\right] dx_j$$

From the last equation, we can infer that the spatial deformation gradient, $\boldsymbol{F}^{-1}$, at point $\boldsymbol{x}$ is inverse of the deformation gradient tensor $\boldsymbol{F}$

Therefore,

$$\boldsymbol{F} \cdot \boldsymbol{F}^{-1} = \boldsymbol{I}$$

can be written between the deformation gradient tensor and the spatial deformation gradient tensor, where $\boldsymbol{I}$ is the unit matrix.

Strain is a human construct. It is not a physical quantity that can be measured directly like displacement or time. It is defined to be able to formulate material behavior using mathematics. Because strain is not a direct physical quantity, it can be defined in many ways. For large strain formulation, the strain will be defined as the one half the change in squared length of the local (material) vector AB as follows (Fig. 2.32):

$$\left[\frac{1}{2}\frac{|AB'|^2 - |AB|^2}{|AB|^2}\right] = \frac{1}{2}\left[\frac{|AB + dr|^2 - |AB|^2}{|AB|^2}\right] = \frac{1}{2}\left[\frac{|2drAB + dr^2|}{|AB|^2}\right]$$

$$= \left|\frac{dr}{\boldsymbol{AB}}\right| + \frac{1}{2}\left|\frac{dr}{AB}\right|^2$$

$$\text{Axial strain} = \epsilon_r + \frac{1}{2}\epsilon_r^2$$

For Lagrangian formulation (referential description) in three dimensions, the strain will be defined as

$$(ds)^2 - (dS)^2 = 2d\boldsymbol{r} \cdot \boldsymbol{E} \cdot d\boldsymbol{r}$$

where $dS = \text{AB}$ *and* $ds = \text{AB}'$. And in indicial notation, the last equation can be given by

$$(ds)^2 - (dS)^2 = dr_i \cdot E_{ij} \cdot dr_j$$

For Eulerian formulation (spatial description),

$$(dS)^2 - (dS)^2 = 2d\boldsymbol{x} \cdot \boldsymbol{E}^* \cdot d\boldsymbol{x}$$

or in indicial notation,

$$(ds)^2 - (dS)^2 = 2dx_i E_{ij}^\star dx_j$$

where s is the new deformed length and S is the original undeformed length.

2.10.1 *Green Deformation Tensor,* $\boldsymbol{C}$, and Cauchy Deformation Tensor, $\boldsymbol{B}^{-1}$

Deformation tensors $\boldsymbol{C}$ and $\boldsymbol{B}^{-1}$ are related to the strain tensors. Instead of giving the one half the change in squared length per unit squared initial length, Green deformation tensor, $\boldsymbol{C}$, refers to the undeformed configuration, and its tensor entries provide the new squared length $(ds)^2$ into which the given vector $d\boldsymbol{r}$ is deformed into Cauchy deformation tensor, $\boldsymbol{B}^{-1}$, which gives the initial squared length $(dS)^2$ of a vector $d\boldsymbol{x}$ defined in the deformed configuration.

As a result, we can write Green deformation as follows:

$$(ds)^2 = \boldsymbol{dr} \cdot \boldsymbol{C} \cdot \boldsymbol{dr}$$

or in indicial notation as

$$(ds)^2 = dr_i C_{ij} dr_j$$

In the same manner, Cauchy deformation tensor $\boldsymbol{B}^{-1}$ can be written as

$$(dS)^2 = \boldsymbol{dx} \cdot \boldsymbol{B}^{-1} \cdot \boldsymbol{dx}$$

or in indicial notation as

$$(dS)^2 = dx_i \left(B^{-1}\right)_{ij} dx_j$$

Comparing Lagrangian strain tensor, $\boldsymbol{E}$, and Green deformation tensor, $\boldsymbol{C}$, we observe that

$$2\boldsymbol{E} = \boldsymbol{C} - \mathbf{I} \quad \text{or} \quad 2E_{ij} = C_{ij} - \delta_{ij}$$

In the same manner, comparing Eulerian strain tensor $\boldsymbol{E}^*$ and Cauchy deformation tensor $\boldsymbol{B}^{-1}$, we observe that

$$2\boldsymbol{E}^* = \mathbf{I} - \boldsymbol{B}^{-1} \quad \text{or} \quad 2E^*_{ij} = \delta_{ij} - \left(B^{-1}\right)_{ij}$$

There is no special reason for defining Cauchy deformation tensor as $\boldsymbol{B}^{-1}$. It could be named just about any character. However, we are following the notation used by Cauchy in 1827.

In the formulation given above, both Green deformation tensor, $\boldsymbol{C}$, and Cauchy deformation tensor, $\boldsymbol{B}^{-1}$, reduce to unit tensor when the strain is zero (Malvern 1969).

2.10.2 Relation Between Deformation, Strain, and Deformation Gradient Tensors

The aquare of the new length $(ds)^2$ can be written as

$$(ds)^2 = \boldsymbol{dx} \cdot \boldsymbol{dx}$$

where $\boldsymbol{dx}$ is the vector in the deformed configuration. On the other hand, the relation between $\boldsymbol{dx}$ and the undeformed vector $\boldsymbol{dr}$ in local (material) coordinates is given by

$$d\boldsymbol{x} = \boldsymbol{F} \cdot d\boldsymbol{r}$$

Hence, we can write

$$(ds)^2 = \left(d\boldsymbol{r} \cdot \boldsymbol{F}^T\right)(\boldsymbol{F} \cdot d\boldsymbol{r}) = d\boldsymbol{r} \cdot \left[\boldsymbol{F}^T \cdot \boldsymbol{F}\right] \cdot d\boldsymbol{r}$$

Similarly, the square of the original length $(dS)^2$ can be written as

$$(dS)^2 = d\boldsymbol{r} \cdot d\boldsymbol{r}$$

Because $d\boldsymbol{r} = \boldsymbol{F}^{-1} \cdot d\boldsymbol{x}$, we can write

$$(dS)^2 = \left[d\boldsymbol{x} \cdot \left(\boldsymbol{F}^{-1}\right)^T\right] \cdot \left[\boldsymbol{F}^{-1} \cdot d\boldsymbol{x}\right] = d\boldsymbol{x} \cdot \left[\left(\boldsymbol{F}^{-1}\right)^T\left(\boldsymbol{F}^{-1}\right)\right] d\boldsymbol{x}$$

Let's compare these last equations with our strain definition equations where Green deformation tensor, $\boldsymbol{C}$ is given by

$$(ds)^2 = d\boldsymbol{r} \cdot \boldsymbol{C} \cdot d\boldsymbol{r}$$

Therefore, Green deformation tensor is also defined by

$$\boldsymbol{C} = \boldsymbol{F}^T \cdot \boldsymbol{F}$$

and Cauchy deformation tensor was defined by

$$(ds)^2 = d\boldsymbol{x}\boldsymbol{B}^{-1}d\boldsymbol{x}$$

Therefore, we can write

$$\boldsymbol{B}^{-1} = \left[\left(\boldsymbol{F}^{-1}\right)^T \cdot \left(\boldsymbol{F}^{-1}\right)\right]$$

in indicial notation, we can write the following relation:

$$C_{ij} = \frac{\partial x_k}{\partial r_i}\frac{\partial x_k}{r_j}$$

and

$$B_{ij}^{-1} = \frac{\partial r_k}{\partial x_i}\frac{\partial r_k}{\partial x_j}$$

Summation on repeated indices applies in both of these equations.

The Lagrangian strain tensor can then be obtained as follows:

$$\boldsymbol{2E} = \boldsymbol{C} - \mathbf{I}$$

since $\boldsymbol{C} = \boldsymbol{F}^{\boldsymbol{T}} \cdot \boldsymbol{F}$

$$\boldsymbol{E} = \frac{\mathbf{1}}{\mathbf{2}}\left[\boldsymbol{F}^{\boldsymbol{T}} \cdot \boldsymbol{F} - \mathbf{I}\right]$$

or the indicial notation

$$E_{ij} = \frac{1}{2}\left[\frac{\partial x_k}{\partial r_i}\frac{\partial x_k}{\partial r_j} - \delta_{ij}\right]$$

and Eulerian strain tensor can be given by

$$2\boldsymbol{E}^{*} = \mathbf{I} - \boldsymbol{B}^{-1}$$

since $\boldsymbol{B}^{-\mathbf{1}} = [(\boldsymbol{F}^{-\mathbf{1}}\boldsymbol{T} \cdot \boldsymbol{F}^{-\mathbf{1}})]$

$$E_{ij}^{\star} = \frac{1}{2}\left[\delta_{ij} - \frac{\partial r_k}{\partial x_i}\frac{\partial r_k}{\partial x_j}\right]$$

Both Green deformation tensor $\boldsymbol{C}$ and Lagrangian strain tensor $\boldsymbol{E}$ are symmetric tensors. Therefore, they both have three eigenvalues (principal values) in eigen directions (principal directions). Also principal directions of $\boldsymbol{C}$ and $\boldsymbol{E}$ coincide, for obvious reasons, because in principal directions, off-diagonal terms are zero in deformation tensor $\boldsymbol{C}$. Then they have to be zero in strain tensor $\boldsymbol{E}$. Of course, the same arguments can be made between Cauchy deformation tensor $\boldsymbol{B}^{-\mathbf{1}}$ and Euler strain tensor $\boldsymbol{E}^{*}$.

However, principal (eigen) directions of [$\boldsymbol{B}^{-\mathbf{1}}$ and $\boldsymbol{E}^{*}$] and [$\boldsymbol{C}$ and $\boldsymbol{E}$] will not coincide.

2.10.3 Comparing Small Strain and Large (Finite) Strain

For the sake of simplicity, we will assume that material (local) coordinate axes and referential Cartesian coordinates are parallel (Fig. 2.33).

We can write the Lagrangian strain terms as

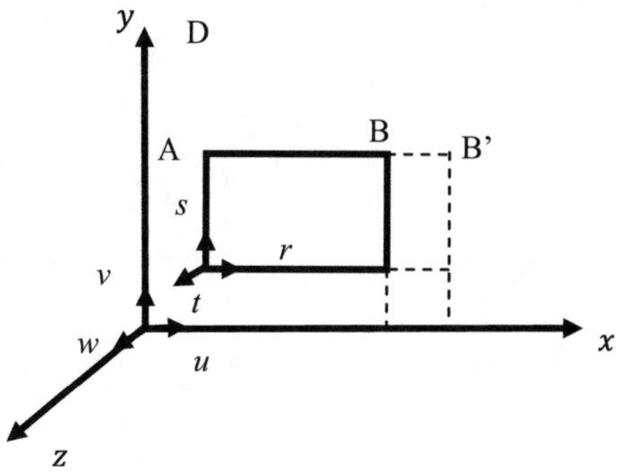

Fig. 2.33 Spatial and local (material) coordinates. x_i = spatial coordinate, r_i = local (material) coordinate. displacement u_i is expressed in local (material) coordinates, $u_i = u_i(r, s, t, \text{time})$

$$E_{ij} = \frac{1}{2}\left[\frac{\partial x_k}{\partial r_i}\frac{\partial x_k}{\partial r_j} - \delta_{ij}\right]$$

$$E_{11} = \frac{1}{2}\left[\left(1+\frac{\partial u}{\partial r}\right)\left(1+\frac{\partial u}{\partial r}\right) + \frac{\partial v}{\partial r}\frac{\partial v}{\partial r} + \frac{\partial w}{\partial r}\frac{\partial w}{\partial r} - 1\right]$$

or

$$E_{11} = \frac{\partial u}{\partial r} + \frac{1}{2}\left[\left(\frac{\partial u}{\partial r}\right)^2 + \left(\frac{\partial v^2}{\partial r}\right) + \left(\frac{\partial w}{\partial r}\right)^2\right]$$

Similarly

$$E_{22} = \frac{\partial v}{\partial s} + \frac{1}{2}\left[\left(\frac{\partial u}{\partial s}\right)^2 + \left(\frac{\partial v}{\partial s}\right)^2 + \left(\frac{\partial w}{\partial s}\right)^2\right]$$

$$E_{33} = \frac{\partial w}{\partial t} + \frac{1}{2}\left[\left(\frac{\partial u}{\partial t}\right)^2 + \left(\frac{\partial v}{\partial t}\right)^2 + \left(\frac{\partial w}{\partial t}\right)^2\right]$$

off-diagonal (shear strain terms)

$$E_{12} = \frac{1}{2}\left[\left(1+\frac{\partial u}{\partial r}\right)\frac{\partial u}{\partial s} + \frac{\partial v}{\partial r}\left(1+\frac{\partial v}{\partial s}\right) + \frac{\partial w}{\partial r}\frac{\partial w}{\partial s} - 0\right]$$

or

$$E_{12} = \frac{1}{2}\left(\frac{\partial u}{\partial s} + \frac{\partial v}{\partial r}\right) + \frac{1}{2}\left[\frac{\partial u}{\partial r}\frac{\partial u}{\partial s} + \frac{\partial v}{\partial r}\frac{\partial v}{\partial s} + \frac{\partial w}{\partial r}\frac{\partial w}{\partial s}\right]$$

$$E_{13} = \frac{1}{2}\left(\frac{\partial u}{\partial t} + \frac{\partial w}{\partial r}\right) + \frac{1}{2}\left[\frac{\partial u}{\partial r}\frac{\partial u}{\partial t} + \frac{\partial v}{\partial r}\frac{\partial v}{\partial t} + \frac{\partial w}{\partial r}\frac{\partial w}{\partial t}\right]$$

$$E_{23} = \frac{1}{2}\left(\frac{\partial v}{\partial t} + \frac{\partial w}{\partial s}\right) + \frac{1}{2}\left[\frac{\partial u}{\partial s}\frac{\partial u}{\partial t} + \frac{\partial v}{\partial s}\frac{\partial v}{\partial t} + \frac{\partial w}{\partial s}\frac{\partial w}{\partial t}\right]$$

or in indicial notation,

$$E_{ij} = \frac{1}{2}\left[\frac{\partial u_i}{\partial r_j} + \frac{\partial u_j}{\partial r_i} + \frac{\partial u_k}{\partial r_i}\frac{\partial u_k}{\partial r_j}\right]$$

In indicial notation, r_j refers to r, s, t local coordinates and u_i refers to u, v, w displacement vectors in x, y, z spatial coordinates. It is obvious that the first term in large (finite) strain formulation gives the small strain formulation.

In the same way, the Euler strain formulation can be given by

$$E_{ij}^{*} = \frac{1}{2}\left[\frac{\partial u_i}{\partial x_j} + \frac{\partial u_j}{\partial x_i} - \frac{\partial u_k \partial u_k}{\partial x_i \partial x_j}\right]$$

Again u_i represents u, v, w for subscripts 1, 2, 3 and x_i represents x, y, z spatial coordinates for 1, 2, 3.

The only difference between Lagrangian strain tensor and Eulerian strain tensor is the fact that in Lagrangian strain tensor, all derivatives are with respect to local (material) coordinates, while in Eulerian strain tensor, components are with respect to spatial (deformed) coordinates.

For small displacement and small strain cases, the difference between Lagrangian strain and Eulerian strain is small.

2.10.4 Strain Rate and Rate of Deformation Relation

Strain rate $\dot{\boldsymbol{E}}$ is given by

$$\frac{d}{dt}(\mathrm{ds})^2 - \frac{d}{dt}(dS)^2 = \frac{d}{dt}(2d\boldsymbol{r} \cdot \boldsymbol{E} \cdot d\boldsymbol{r})$$

$d\boldsymbol{r}$ and dS are constant with respect to time. Therefore, we can write

$$\frac{d}{dt}(ds)^2 = 2d\boldsymbol{r} \cdot \frac{d\boldsymbol{E}}{dt} \cdot d\boldsymbol{r}$$

On the other hand, the rate of deformation $\boldsymbol{D}$ is given by

$$\frac{d}{dt}(ds)^2 = 2d\boldsymbol{x} \cdot \boldsymbol{D} \cdot d\boldsymbol{x}$$

Also, we have shown that $d\boldsymbol{x} = \boldsymbol{F} \cdot d\boldsymbol{r}$. Hence, we write the following relation:

$$\frac{d}{dt}(ds)^2 = 2\left(d\boldsymbol{r}\boldsymbol{F}^T\right) \cdot \boldsymbol{D} \cdot (\boldsymbol{F} \cdot d\boldsymbol{r})$$

$$= 2d\boldsymbol{r} \cdot \left(\boldsymbol{F}^T \cdot \boldsymbol{D} \cdot \boldsymbol{F}\right) \cdot d\boldsymbol{r}$$

Subtracting strain rate equation from rate of deformation,

$$2d\boldsymbol{r} \cdot \frac{d\boldsymbol{E}}{dt} \cdot d\boldsymbol{r} - 2d\boldsymbol{r} \cdot \left(\boldsymbol{F}^T \cdot \boldsymbol{D} \cdot \boldsymbol{F}\right) \cdot d\boldsymbol{r} = 0$$

$$d\boldsymbol{r} \cdot \left[\frac{d\boldsymbol{E}}{dt} - \left(\boldsymbol{F}^T \cdot \boldsymbol{D} \cdot \boldsymbol{F}\right)\right] \cdot d\boldsymbol{r} = 0$$

Of course $d\boldsymbol{r} = 0$ is the trivial solution. Hence, the following relation must be satisfied:

$$\frac{d\boldsymbol{E}}{dt} = \boldsymbol{F}^T \cdot \boldsymbol{D} \cdot \boldsymbol{F}$$

or in indicial notation,

$$\frac{dE_{ij}}{dt} = \frac{\partial x_m}{\partial r_i} D_{mn} \frac{\partial x_n}{\partial r_j}$$

The relationship between Green deformation rate $\boldsymbol{C}$ and strain rate tensor $\dot{E}$ can be given by

$$\frac{d\boldsymbol{C}}{dt} = 2\frac{d\boldsymbol{E}}{dt}$$

When displacement gradient components are small compared to unity, strain rate is equal to rate of deformation:

$$\frac{d\boldsymbol{E}}{dt} = \boldsymbol{D}$$

or in indicial notation,

$$\frac{dE_{ij}}{dt} = D_{ij}$$

2.10.5 Relation Between the Spatial Gradient of Velocity Tensor, L,and the Deformation Gradient Tensor, F

The definition of deformation gradient tensor $\boldsymbol{F}$ is given by

$$d\boldsymbol{x} = \boldsymbol{F} \cdot d\boldsymbol{r}$$

or

$$dx_i = \frac{\partial x_i}{\partial r_j} \partial r_j$$

The rate of change of the deformation gradient is $\dot{\boldsymbol{F}}$. The spatial gradient of velocity $\boldsymbol{L}$ is given by

$$d\boldsymbol{v} = \boldsymbol{L} \cdot d\boldsymbol{x}$$

Hence,

$$L_{ij} = v_{i,j} = \frac{\partial v_i}{\partial x_j}$$

If we write time derivative of $\boldsymbol{F} = \frac{\partial x_i}{\partial r_j}$

$$\dot{\boldsymbol{F}} = \frac{d}{dt}\left(\frac{\partial x_i}{\partial r_j}\right) = \frac{\partial \dot{x}_i}{\partial r_j} = \frac{\partial \dot{x}_i}{\partial x_m}\frac{\partial x_m}{\partial r_j}$$

$$\dot{\boldsymbol{F}} = \boldsymbol{L} \cdot \boldsymbol{F} \quad \text{and} \quad \boldsymbol{L} = \dot{\boldsymbol{F}} \cdot \boldsymbol{F}^{-1}$$

$$\frac{d}{dt}\left(\frac{\partial x_i}{\partial r_j}\right) = \frac{\partial v_i}{\partial x_m}\frac{\partial x_m}{\partial r_j} \qquad \frac{\partial v_i}{\partial x_m} = \frac{d}{dt}\left(\frac{\partial x_i}{\partial r_j}\right)\left(\frac{\partial r_j}{\partial x_m}\right)$$

Relations between Euler strain rate E^*, the rate of deformation tensor D, and the spatial gradients of velocity L tensors can be given by

$$\dot{\boldsymbol{E}}^{*} = \boldsymbol{D} - \left(\boldsymbol{E}^{*} \cdot \boldsymbol{L} + \boldsymbol{L}^{T} \cdot \boldsymbol{E}^{*}\right)$$

Proof of this equation is provided by Malvern (1969).

2.11 Rotation and Stretch Tensors in Finite Strain

When strain is finite (large), the symmetric and skew-symmetric parts of the displacement gradient matrix, $\boldsymbol{J}$, cannot be represented by additive decomposition of a pure strain matrix and a pure rotation matrix. However, other types of multiplication decompositions are possible, where one of the two tensors will represent a rigid body rotation and the second tensor will be a symmetric positive-definite.

2.12 Compatibility Conditions in Continuum Mechanics

When the displacements are known, strain field can easily be calculated. For the case of small strain, relations are given by

$$\epsilon_{ij} = \frac{1}{2}\left(\frac{\partial u_i}{\partial r_j} + \frac{\partial u_j}{\partial r_i}\right)$$

where u_i represent displacement vectors u, v, w and r_i represents local (material) coordinate axes r, s, t. There are nine strain equations but because of symmetry only six of these are linearly independent.

For any given strain field to be admissible, certain compatibility conditions that guarantee continuum character of the medium must be satisfied. This requirement is also due to mathematics. If there are six known strain equations, it is not possible to obtain three unknown displacement components as unique values. The number of unknowns and the number of linearly independent equations must be the same. St. Venant's compatibility equations must be satisfied by the six strain equations in order to find unique displacement values.

In a three-dimensional solid mechanics boundary value problem St. Venant's compatibility equation's provide us with six equations. There are also three force equilibrium equations.

In addition, there are six stress-strain constitutive relations. As a result in total there are 15 equations and 12 unknowns (6 stresses and 6 strains). However, only three of the compatibility equations are linearly independent. Hence, we have 12 equations and 12 unknowns.

However, if the displacements are unknown, then compatibility equations are not needed, because there are 15 equations (6 strain-displacement relations, 6 stress-strain constitutive relations, and 3 equilibrium equations). On the unknown side,

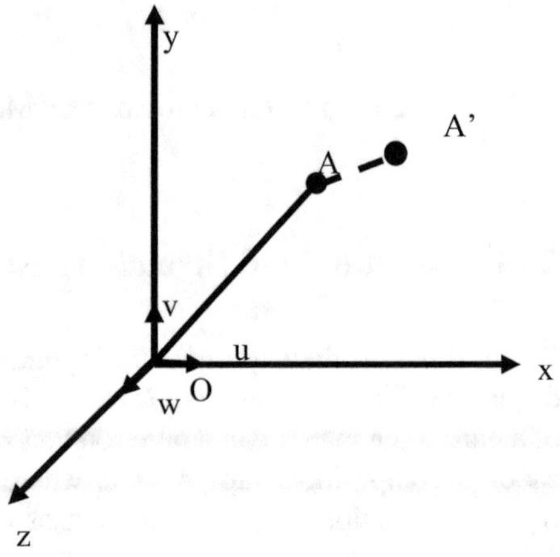

Fig. 2.34 Strain-displacement relation

there are three unknown displacements, six unknown stresses, and six unknown strains. Therefore, the number of equations is equal to the number of unknowns.

In most computational mechanics boundary value problems, usually the displacements are included as unknowns, as in displacement-based finite element method. As a result we do not need the compatibility equations. However, in force-based analysis methods, compatibility functions become necessary.

In summary, when displacements are not explicitly retained as unknowns, the compatibility conditions (this should actually be called the compatibility of the displacement field) are needed to make sure that strain field leads to a continuous displacement field that has a single value at any point. Strain-displacement relations are given by (Fig. 2.34). Assuming small deformations, we will define small strain in x, y, z coordinate system:

$$\epsilon_{xx} = \frac{\partial u}{\partial x}$$

$$\epsilon_{yy} = \frac{\partial v}{\partial y}$$

$$\epsilon_{zz} = \frac{\partial w}{\partial x}$$

$$\epsilon_{xy} = \frac{1}{2}\left(\frac{\partial u}{\partial y} + \frac{\partial v}{\partial x}\right)$$

$$\epsilon_{xz} = \frac{1}{2}\left(\frac{\partial u}{\partial z} + \frac{\partial w}{\partial x}\right)$$

$$\epsilon_{yz} = \frac{1}{2}\left(\frac{\partial v}{\partial z} + \frac{\partial w}{\partial y}\right)$$

We assume $\frac{\partial^2 \epsilon_{xy}}{\partial x \partial y}$ exists:

$$\frac{\partial^2 \epsilon_{xy}}{\partial x \partial y} = \frac{1}{2}\left[\frac{\partial^3 u}{\partial^2 y \partial x} + \frac{\partial^3 v}{\partial x^2 \partial y}\right]$$

We can also write second derivatives of ϵ_{xx} and ϵ_{yy} and then sum them up:

$$\frac{\partial^2 \epsilon_{xx}}{\partial y^2} + \frac{\partial^2 \epsilon_{yy}}{\partial x^2} = \frac{\partial^3 u}{\partial y^2 \partial x} + \frac{\partial^2 v}{\partial x^2 \partial y}$$

Hence,

$$\frac{\partial^2 \epsilon_{xx}}{\partial y^2} + \frac{\partial^2 \epsilon_{yy}}{\partial x^2} = 2\frac{\partial^2 \epsilon_{xy}}{\partial x \partial y}$$

We can repeat this process for ϵ_{xz} and ϵ_{yz}. Then we have

$$\left[\frac{\partial^2 \epsilon_{xy}}{\partial x \partial z} + \frac{\partial^2 \epsilon_{zx}}{\partial y \partial x}\right] = \frac{1}{2}\left[\frac{\partial^2}{\partial x \partial z}\left(\frac{\partial u}{\partial y} + \frac{\partial v}{\partial x}\right) + \frac{\partial^2}{\partial y \partial x}\left(\frac{\partial w}{\partial x} + \frac{\partial u}{\partial z}\right)\right]$$

$$= \frac{1}{2}\left[\frac{\partial^3 u}{\partial x \partial y \partial z} + \frac{\partial^3 v}{\partial x^2 \partial y} + \frac{\partial^3 w}{\partial x^2 \partial y} + \frac{\partial^3 u}{\partial x \partial y \partial z}\right]$$

$$= \frac{1}{2}\left[2\frac{\partial^2}{\partial y \partial z}\left(\frac{\partial u}{\partial x}\right) + \frac{\partial^2}{\partial x^2}\left(\frac{\partial v}{\partial y} + \frac{\partial w}{\partial y}\right)\right]$$

$$= \frac{1}{2}\left[2\frac{\partial^2 \epsilon_{xx}}{\partial y \partial z} + \frac{\partial^2 \epsilon_{yz}}{\partial x^2}\right]$$

$$\frac{\partial^2 \epsilon_{xx}}{\partial y \partial z} = \frac{\partial}{\partial x}\left(\frac{\partial \epsilon_{xy}}{\partial z} + \frac{\partial \epsilon_{zx}}{\partial y} - \frac{1}{2}\frac{\partial \epsilon_{yz}}{\partial x}\right)$$

Similar process can be repeated for other shear strain pairs ϵ_{yz} and ϵ_{xy} and, ϵ_{yz} and ϵ_{xz}.

These equations prove integrability of strain field. Based on this derivation, the following six St. Venant's compatibility equations can be given:

$$\frac{\partial^2 \epsilon_{xx}}{\partial y^2} + \frac{\partial^2 \epsilon_{yy}}{\partial x^2} - 2\frac{\partial^2 \epsilon_{xy}}{\partial x \partial y} = 0$$

$$\frac{\partial^2 \epsilon_{yy}}{\partial z^2} + \frac{\partial^2 \epsilon_{zz}}{\partial y^2} - 2\frac{\partial^2 \epsilon_{yz}}{\partial y \partial z} = 0$$

$$\frac{\partial^2 \epsilon_{zz}}{\partial x^2} + \frac{\partial^2 \epsilon_{xx}}{\partial z^2} - 2\frac{\partial^2 \epsilon_{zx}}{\partial z \partial x} = 0$$

$$-\frac{\partial^2 \epsilon_{xx}}{\partial y \partial z} + \frac{\partial}{\partial x}\left(-\frac{\partial \epsilon_{yz}}{\partial x} + \frac{\partial \epsilon_{zx}}{\partial y} + \frac{\partial \epsilon_{xy}}{\partial z}\right) = 0$$

$$-\frac{\partial^2 \epsilon_{yy}}{\partial z \partial x} + \frac{\partial}{\partial y}\left(\frac{\partial \epsilon_{yz}}{\partial x} - \frac{\partial \epsilon_{zx}}{\partial y} + \frac{\partial \epsilon_{xy}}{\partial z}\right) = 0$$

$$-\frac{\partial^2 \epsilon_{zz}}{\partial x \partial y} + \frac{\partial}{\partial z}\left(\frac{\partial \epsilon_{yz}}{\partial x} + \frac{\partial \epsilon_{zx}}{\partial y} - \frac{\partial \epsilon_{xy}}{\partial z}\right) = 0$$

These six compatibility equations must be satisfied to ensure a compatible strain field and to ensure that only a single-valued displacement exist at any point. Therefore, compatibility conditions and sufficient conditions for unique displacement values are necessary.

While there are six compatibility equations, only three of them can be linearly independent because they are obtained from three independent displacements u, v, w.

2.13 Piola-Kirchhoff Stress Tensors

Cauchy stress tensor is defined in the spatial coordinates (in the deformed configuration). Euler strain is also defined in spatial position in the deformed configuration. Therefore, using Euler strain definition with Cauchy stress tensor is appropriate.

On the other hand, Lagrangian and material (local) formulations are based on undeformed configuration or any other reference state configuration that does not change. In this case, it is also more appropriate to define the Piola-Kirchhoff stress tensor in local (material) reference coordinates. There are two Piola-Kirchhoff stress definitions using the undeformed (or any other stationary reference) state as the basis for formulation.

2.13.1 First Piola-Kirchhoff Stress Tensor $\boldsymbol{\sigma}^0$

The first Piola-Kirchhoff stress tensor is also called the Lagrangian stress tensor. The derivatives are defined with respect to material (local) coordinates. However, this

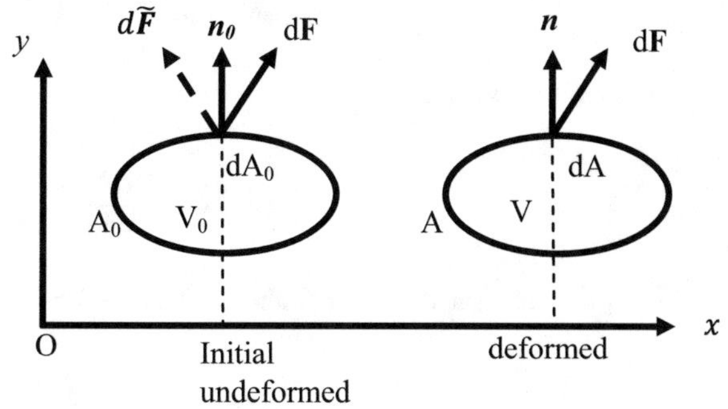

Fig. 2.35 Piola-Kirchhoff stress definition

stress tensor is not symmetric, even when there is no distributed body moment (nonpolar case) on the system.

Stress is defined as a force per unit undeformed area. Keep in mind that stress is a human construct not a physical measurable quantity; thus, it is possible to define it as we please. The second Piola-Kirchhoff stress tensor is defined by the following equation (Fig. 2.35):

$$\left(\boldsymbol{n}_0 \cdot \boldsymbol{\sigma}^0\right) = \frac{\boldsymbol{dF}}{\boldsymbol{dA}_0}$$

where $\boldsymbol{\sigma}^{\mathbf{0}}$ is the first Piola-Kirchhoff tensor and $\boldsymbol{dF}$ is the actual force acting on the deformed configuration dA; however, it is formulated based on the initial undeformed area dA_0.

$\boldsymbol{dF}$ can also be written using Cauchy stress tensor $\boldsymbol{\sigma}$ as $\boldsymbol{dF} = (\boldsymbol{n} \cdot \boldsymbol{\sigma})dA$. Hence, we can write the following equation between the first Piola-Kirchhoff stress tensor and Cauchy stress tensor:

$$(\boldsymbol{n_0}\boldsymbol{\sigma_0})dA_0 = \boldsymbol{dF} = (\boldsymbol{n} \cdot \boldsymbol{\sigma})dA$$

2.13.2 Second Piola-Kirchhoff Stress Tensor $\widetilde{\sigma}$

The second Piola-Kirchhoff stress tensor, $\widetilde{\sigma}$, uses a force $d\widetilde{\boldsymbol{F}}$ that is related to the actual force $d\mathbf{F}$ by the inverse of deformation gradient F^{-1}:

$$d\widetilde{\boldsymbol{F}} = \boldsymbol{F}^{-1} \cdot \boldsymbol{dF}$$

Remember that material (local) coordinates and spatial coordinates are also related by

$$\boldsymbol{dr} = \boldsymbol{F}^{-1} \cdot \boldsymbol{dx}$$

Then we can write

$$(\boldsymbol{n_0} \cdot \widetilde{\boldsymbol{\sigma}}) = \frac{d\widetilde{\boldsymbol{F}}}{dA_0}$$

Using the relation we defined above, we can write

$$(\boldsymbol{n_0} \cdot \widetilde{\boldsymbol{\sigma}}) = \frac{\boldsymbol{F}^{-1} \cdot d\boldsymbol{F}}{dA_0}$$

At the same time, dF can be also defined in terms of Cauchy stress tensor as

$$(\boldsymbol{n_0} \cdot \widetilde{\boldsymbol{\sigma}}) = \frac{\boldsymbol{F}^{-1} \cdot (\boldsymbol{n} \cdot \boldsymbol{\sigma})dA}{dA_0} = \frac{(\boldsymbol{n} \cdot \boldsymbol{\sigma}) \cdot \left(\boldsymbol{F}^{-1}\right)^T dA}{dA_0}$$

2.14 Direct Relation Between Cauchy Stress Tensor and Piola-Kirchhoff Stress Tensors

The following relation can be given:

$$\widetilde{\boldsymbol{\sigma}} = \frac{\boldsymbol{\rho_0}}{\boldsymbol{\rho}} \boldsymbol{F}^{-1} \cdot \boldsymbol{\sigma} \cdot \left(\boldsymbol{F}^{-1}\right)^T \quad \text{and} \quad \widetilde{\boldsymbol{\sigma}} = \boldsymbol{\sigma}^{\mathbf{0}} \cdot \left(\boldsymbol{F}^{-1}\right)^T$$

Details of the long derivations for these relations are provided by Malvern (1969).

The second Piola-Kirchhoff stress tensor is usually preferred in finite strain elasticity problems.

Equations of motions in undeformed (referenced) state first Piola-Kirchhoff stress tensor is given by

$$\int_{A_0} n_0 \cdot \sigma^0 dA_0 + \int_{V} \rho_0 b_0 dV_0 = \int_{V_0} \rho \frac{d^2 r}{dt^2} dV_0$$

Transforming the surface integral to volume integral by Gauss's theorem (divergence theorem) in indicial notation, we can write

$$\frac{\partial \sigma_{ij}^{0}}{\partial r_i} + \rho_0 b_{0j} = \rho_0 \frac{d^2 x_i}{dt^2}$$

where r_i is the reference coordinate system axis. Since

$$\boldsymbol{\sigma_0} = \widetilde{\boldsymbol{\sigma}} \cdot \boldsymbol{F^T}$$

or in indicial notation

$$\boldsymbol{\sigma_{ij}^{0}} = \widetilde{\boldsymbol{\sigma}}_{\boldsymbol{ik}} \boldsymbol{x_{i,k}}$$

by substituting these equations in the equilibrium equations above, we get the equation of motion in terms of the second Piola-Kirchhoff stress tensor

$$\left[\widetilde{\sigma}_{\mathrm{ij}} x_{k,j}\right]_{,i} + \rho_0 b_{0_k} = \rho_0 \frac{d^2 x_k}{dt^2}$$

2.15 Conservation of Mass Principle

The rate of change in mass is given by

$$\frac{\partial m}{\partial t} = \int_V \frac{\partial \rho}{\partial t} dV$$

where ρ is the material density, which can be a function of x, y, z coordinates and time. The flux, which is the rate of mass flowing through an area dA with a velocity of

$\widetilde{v}_n = \widetilde{v} \cdot n,$ can be given by

$$\int_A \rho \widetilde{v}_n \, dA = \int_A \rho \widetilde{v} \cdot n \, dA$$

Using the divergence theorem (Gauss's theorem), the integral over a closed surface can be converted to an integral over a volume bounded by the closed surface by the following equation:

$$\int_A n \times \tilde{v}\, dA = \int_V \nabla \times \tilde{v} dV$$

where n is the vector normal to the surface, $\tilde{v}$ is velocity vector, and $\nabla \times \tilde{v}$ is given by

$$\nabla \times \tilde{v} = \begin{vmatrix} i & j & k \\ \dfrac{\partial}{\partial x} & \dfrac{\partial}{\partial y} & \dfrac{\partial}{\partial z} \\ \tilde{v}_x & \tilde{v}_y & \tilde{v}_z \end{vmatrix}$$

where i, j, k are the unit vectors in Cartesian coordinates x, y, z Hence,

$$\int_A \rho\tilde{v} \cdot ndA = \int_V \nabla \cdot (\rho\tilde{v}) dV$$

Time rate of change in mass was defined by

$$\int_V \frac{\partial \rho}{\partial t} dV$$

Equating time rate of change in mass to rate of inflow through area A yields

$$\int_V \frac{\partial \rho}{\partial t} dV = -\int_V \nabla \cdot (\rho\tilde{v}) dV$$

The negative on the right-hand side is due to the fact that the normal of the surface is defined positive outward. Therefore, inflow is in negative normal direction.

Hence, we can write

$$\int_V \left[\frac{\partial \rho}{\partial t} + \nabla \cdot (\rho\tilde{v})\right] dV = 0$$

Here, the integral must be zero for an arbitrary volume. Therefore, the integrand must be equal to zero:

$$\frac{\partial \rho}{\partial t} + \nabla \cdot (\rho\tilde{v}) = 0$$

This equation is called continuity equation, as a result of the conservation of mass principle. Or in indicial notation, it can be given as

$$\frac{\partial \rho}{\partial t} + \rho \left[\frac{\partial \widetilde{v}_x}{\partial x} + \frac{\partial \widetilde{v}_y}{\partial y} + \frac{\partial \widetilde{v}_z}{\partial z} \right] = 0$$

2.16 The Incompressible Materials

When the material is incompressible,

$$\frac{d\rho}{dt} = 0$$

Therefore,

$$\frac{\partial \widetilde{v}_x}{\partial x} + \frac{\partial \widetilde{v}_x}{\partial y} + \frac{\partial \widetilde{v}_z}{\partial z} = 0$$

must be satisfied. It is important to point out that in the theory of plasticity during plastic deformation, materials, like metals, are considered to be incompressible.

2.17 Conservation of Momentum Principle

"The expression for equilibrium of forces is obtained by using the conservation of momentum principle of a collection of particles. This principle applies to a collection of particles as well as the continuous medium. This principle states that the vector sum of all the external forces acting on the free body is equal to the rate of change of the total momentum."

Normal stresses, σ_{ii}, are positive when the vector is in the positive axis direction. Shear stresses σ_{ij} are positive when the vector component is in the positive direction on the positive face of the block and in the negative direction on the negative face of the block. Shear stresses σ_{ij} are negative when the vector component is in the negative direction on the positive face of the block and in the positive direction on the negative face of the block.

Tensile stress is defined to be positive and compression is defined as negative. However, we realize that this definition is not universal. For example, in geo-mechanics, tension is negative and compression is positive.

While tension and compression lead to completely different results when applied on any object. Negative shear stress and positive shear stress are considered to be the same kind of loading in the opposite directions from a physical point of view. There is no compression shear and tension shear.

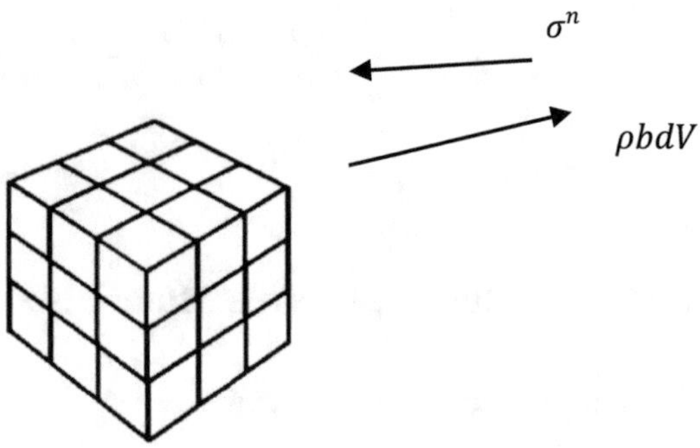

Figure 2.36 Conservation of momentum

Force is the time derivative of momentum; therefore, the concept of force is redundant and subordinate to the conservation of momentum. In this section, we will derive the force from conservation of momentum. According to the conservation of momentum principle for a collection of particles, the time rate of change of the total momentum of a given set of particles equals the vector sum of all the external forces acting on the particles. It is assumed that Newton's third law of action and reaction governs the internal forces.

The continuum form of conservation of momentum principle is the basis for the Newtonian continuum mechanics. In this book, the term "Newtonian mechanics" is for all mechanics formulations based on Newton's laws (Fig. 2.36).

Consider the cube shown in Fig. 2.36, occupying volume V and bounded by surface A. There is an external surface loading vector $\sigma^{(n)}$, per unit area on any arbitrary plane, and b is the body force per unit mass.

Momentum, p, is defined by

$$p = \int \rho\, \tilde{v}\, dV$$

where ρ is the density, v is the velocity, and dV is the volume. The time rate of change of the total momentum of a portion of the particles can be given by the material derivative of the integral, which is defined as the time rate of change of any quantity for a portion of a material. Hence, the time rate of change of the total momentum is

$$\frac{dp}{dt} = \frac{d}{dt}\int \rho \tilde{v}\, dV$$

External forces acting on the particles

$$\int_A \sigma^{(n)}\, dA + \int_V \rho b\, dV$$

According to the conservation of momentum principle, the time rate of change of the total momentum equals the vector sum of all the external forces:

$$\int_A \sigma^{(n)}\, dA + \int_V \rho b\, dV = \frac{d}{dt}\int \rho\widetilde{v}\, dV$$

Using Cartesian coordinate system where *x*, *y*, *and* *z* are represented by indices i yields

$$\int_A \sigma_i^{(n)}\, dA + \int_V \rho b_i\, dV = \frac{d}{dt}\int \rho\, \widetilde{v}_i\, dV$$

Components of external surface loading vector $\sigma^{(n)}$ can be given by

$$\sigma_i^{(n)} = \sigma_{ij\, n_j}$$

or in Cartesian coordinates

$$\sigma_x^{(n)} = \sigma_{xx} n_x + \sigma_{xy} n_y + \sigma_{xz} n_z$$

$$\sigma_y^{(n)} = \sigma_{yx} n_x + \sigma_{yy} n_y + \sigma_{yz} n_z$$

$$\sigma_z^{(n)} = \sigma_{zx} n_x + \sigma_{zy} n_z + \sigma_{zz} n_z$$

where $n_{x,}$ $n_{x,}$ $n_{x,}$ are direction cosines between the normal to the arbitrary plane and spatial coordinates. In matrix notation, this can be written as

$$\left\{\boldsymbol{\sigma}^{(n)}\right\} = \{\boldsymbol{n}\}\, [\boldsymbol{\sigma}]$$

$$\{\boldsymbol{\sigma}^{(n)}\} = \left\{\sigma_x^{(n)}\ \sigma_y^{(n)}\ \sigma_z^{(n)}\right\},\ \{\boldsymbol{n}\} = \{n_x\ \ n_y\ \ n_z\},\ [\boldsymbol{\sigma}] = \begin{bmatrix} \sigma_{xx} & \sigma_{xy} & \sigma_{xz} \\ \sigma_{yx} & \sigma_{yy} & \sigma_{yz} \\ \sigma_{zx} & \sigma_{zy} & \sigma_{zz} \end{bmatrix}$$

$[\boldsymbol{\sigma}]$ is a second-order Cauchy stress tensor, which is a linear vector function. Using the divergence theorem, we can write the following conversion:

$$\int_A \sigma_i^{(n)}\, dA = \int_V \frac{\partial \sigma_{ji}}{\partial x_j}\, dV$$

Meanwhile, due to conservation of mass, material time derivative of volume integral can be written as (Malvern 1969)

$$\frac{\partial}{\partial t}\int_V \rho\, \widetilde{v}_i\, dV = \frac{\partial}{\partial t}\int_V \rho\, \widetilde{v}_i\, dV$$

Substituting the last two relations in the conservation of momentum equation yields

$$\int_V \left(\frac{\partial \sigma_{ji}}{\partial x_j} + \rho\, b_i\right) dV = \int_V \rho\, \frac{d\widetilde{v}_i}{dt}\, dV$$

Hence, the conservation of momentum principles takes the final form in the following equation:

$$\int_V \left(\frac{\partial \sigma_{ji}}{\partial x_j} + \rho\, b_i - \rho\, \frac{d\widetilde{v}_i}{dt}\right) dV =$$

For any arbitrary volume, we can write Cauchy's equation of motion as follows:

$$\frac{\partial \sigma_{ji}}{\partial x_j} + \rho\, b_i = \rho\, \frac{d\widetilde{v}_i}{dt}$$

Of course, for static equilibrium, the right-hand side of this equation is equal to zero. Hence,

$$\frac{\partial \sigma_{ji}}{\partial x_j} + \rho\, b_i = 0$$

In any other Cartesian coordinate, the system rotated with respect to the original coordinate system stress tensor can be defined using the tensor transformation equations:

$$[\overline{\sigma}] = [N]^T [\sigma]\, [N]$$

where [N] is the matrix of direction cosines $n_k^i = \cos\left(\overline{x}_r, x_i\right)$ of the angles between the axes of the original and rotated coordinate systems. Superscript T is used to denote transpose.

2.18 Conservation of Moment of Momentum Principle

The time rate of change of the total moment of momentum for a collection of masses is equal to the vector sum of the moments of the external forces acting on these masses. Assuming there are no distributed couples, we can write

$$\frac{d}{dt}\int_{v}(r\times\rho\widetilde{v})dV=\int_{A}\left(r\times\sigma^{(n)}\right)dA+\int_{V}(r\times\rho b)dV$$

Using vector product definition

$$a\times b=e_{ijr}a_jb_r=e_{jri}a_jb_r$$

we can rewrite the conservation of momentum of moment principle in indicial notation as follows:

$$\frac{d}{dt}\int_{V}e_{ijr}x_j\rho\widetilde{v}_rdV=\int_{A}e_{ijr}x_j\sigma_r^{(n)}dA+\int_{V}e_{ijr}x_jb_r\rho dV$$

We have defined $\sigma_r^{(n)}$ previously for an arbitrary direction as

$$\sigma_i^{(n)}=\sigma_{ij}n_j$$

We can also transform the surface integral to a volume integral using the relation (divergence (Gauss's) theorem):

$$\int_{A}e_{ijk}n_j\widetilde{v}_kdA=\int_{V}e_{ijk}\frac{\partial}{\partial x_j}(\widetilde{v}_k)dV$$

where ρ is the density of the material per unit mass, and we can transform the moment of momentum principle to the following form:

$$\int_{V}e_{ijr}\frac{d}{\mathrm{dt}}(x_j\widetilde{v}_r)\rho dV=\int_{V}e_{ijr}\left[\frac{\partial(x_j\sigma_{kr})}{\partial x_k}+x_j\rho b_r\right]dV$$

Time derivative of displacement is velocity; hence, we can substitute $\widetilde{v}_i=\frac{dx_i}{dt}$:

$$\int_{V}e_{ijr}\left(\widetilde{v}_jv_r+x_j\frac{d\widetilde{v}_r}{dt}\right)\rho dV=\int_{V}e_{ijr}\left[x_j\left(\frac{\partial\sigma_{kr}}{\partial x_k}+\rho b_r\right)+\sigma_{kr}\delta_{jk}\right]dV$$

$e_{ijr}\widetilde{v}_j\widetilde{v}_r=0$ because $\widetilde{v}_j\widetilde{v}_r$ is symmetric for indices jr, while e_{ijr} is antisymmetric. Knowing that Cauchy's equation of motion is given by

$$\frac{\partial\sigma_{\mathrm{kr}}}{\partial x_k} + \rho b_r = \rho\frac{d\tilde{v}_r}{dt}$$

Substituting this equilibrium equation of motion above yields

$$\int_V e_{ijr}\sigma_{kr}\delta_{jk}dV = 0$$

which is equal to

$$\int_V e_{\mathrm{ijr}}\sigma_{jr}dV = 0$$

However, the last equation is equal to zero. Because for an arbitrary volume

$$e_{ijr}\sigma_{jr}$$

for

- $i = 1: \quad \sigma_{23} = \sigma_{32}$
- $i = 2: \quad \sigma_{31} = \sigma_{13}$
- $i = 3: \quad \sigma_{12} = \sigma_{21}$

which proves that the stress tensor is symmetric when there are no couple (moment) stresses. This proof also establishes the symmetry of the stress tensor in general without any assumption of equilibrium or uniformity of the stress distribution.

References

Anon. (2020). Wikipedia Sir Isaac Newton webpage, https://en.wikipedia.org/wiki/Isaac_Newton.

Beatty, M. F. (2006). Principles of engineering mechanics Volume 2 of Principles of Engineering Mechanics: Dynamics-The Analysis of Motion,. Springer. p. 24. ISBN 0-387-23704-6.

Bernard Cohen, I. (1967). Newton's second law and the concept of force in the principia. In *The Annus Mirabilis of Sir Isaac Newton* (pp. 1666–1966). Cambridge, MA: The MIT Press.

Bernard Cohen, I. (2002). In P. M. Harman & A. E. Shapiro (Eds.), *The investigation of difficult things: essays on Newton and the history of the exact sciences in honour of D.T. Whiteside* (p. 353). Cambridge UK: Cambridge University Press. ISBN 0-521-89266-X.

Cohen, I. B. (1995). *Science and the founding fathers: Science in the Political Thought of Jefferson, Franklin, Adams and Madison* (p. 117). New York: W.W. Norton. ISBN 978-0-393-24715-2.

Fairlie, G., & Cayley, E. (1965). *The life of a genius* (p. 163). Hodder and Stoughton.

Galili, I., & Tseitlin, M. (2003). Newton's first law: Text, translations, interpretations and physics education. *Science & Education, 12*(1), 45–73. Bibcode:2003Sc&Ed..12...45G. https://doi.org/10.1023/A:1022632600805.

Halliday, D., Resnick, R., & Krane, K. S. (1992). *Physics*, John Wiley & Sons.

Harvard UP (1972) [In Latin] Isaac Newton's Philosophiae Naturalis Principia Mathematica: the Third edition (1726) with variant readings, assembled and ed. by Alexandre Koyré and I Bernard Cohen with the assistance of Anne Whitman. Cambridge, MA: Harvard UP.
Hellingman. (1992). Newton's third law revisited. *Phys. Educ., 27*(2), 112–115. https://doi.org/10.1088/0031-9120/27/2/011. Bibcode:1992PhyEd..27..112H.
Hobbes, T. (1651). Leviathan – Oxford University Press.
Kleppner, D., & Kolenkow, R. (1973). *An introduction to mechanics* (pp. 133–134). McGraw-Hill. ISBN 0-07-035048-5.
Lubliner, J. (2008). Plasticity theory (Revised Edition) (PDF). Dover Publications. ISBN 0-486-46290-0. Archived from the original (PDF) on 31 March 2010.
Malvern, L. (1969). *E introduction to the mechanics of continuous medium*. Prentice-Hall.
Newton. (1729). *Principia*, Corollary III to the laws of motion.
Newton, I. (1999). *The Principia: Mathematical Principles of Natural Philosophy*. A new translation by I.B. Cohen and A. Whitman. Berkeley: University of California press.
Newton, I. (n.d.-a). Philosophiæ Naturalis Principia Mathematica (Newton's personally annotated 1st edition).
Plastino, A. R., & Muzzio, J. C. (1992). On the use and abuse of Newton's second law for variable mass problems. *Celestial Mechanics and Dynamical Astronomy, 53*(3), 227–232. Netherlands: Kluwer Academic Publishers. Bibcode:1992CeMDA..53..227P.
Resnick, R., & Halliday. D. (1977). *Physics* (3rd ed., pp. 78–79). John Wiley & Sons. Any single force is only one aspect of a mutual interaction between two bodies.
Thornton, M. (2004). *Classical dynamics of particles and systems* (5th ed., p. 53). Brooks/Cole. ISBN 0-534-40896-6.

Chapter 3
Thermodynamics

Newtonian mechanics is a study of determination of location of points in space-time (x, y, z and time) coordinate system after being subjected to external forces. Since Newton's 1687 formulations, many other mechanics theories were proposed. However, they are all based on Newton's laws of motion and are referred to as Newtonian mechanics. In this book, the term "Newtonian mechanics" is used to refer to all mechanics theories that use Newton's universal laws of motion.

In Newtonian mechanics formulation, the object being studied is assumed to be ageless, and energy loss (entropy generation) is not considered.

Thermodynamics is about past, present, and future of objects, under given set of conditions. Directly quoting Callen's (1985) excellent description, "Thermodynamics is concerned with the macroscopic consequences of the myriads of atomic coordinates that, by virtue of the coarseness of macroscopic observations, do not appear explicitly in a macroscopic description of the system." Therefore, thermodynamics is actually the study of nature at macroscopic scale that reflects what happens at the atomic scale. However, "Among the many consequences of the 'hidden' atomic modes of motion, the most evident is the ability of these modes to act as a repository for energy" Callen (1985).

Thermodynamics is a universal subject. As such, it is assumed that laws of thermodynamics are universally valid for all systems at macro level under equilibrium conditions.

Therefore, we can treat any structure or system or any piece of material as a thermodynamic system. The size of the system studied is very important for the laws of thermodynamics. In this book, we are only concerned with continuum mechanics where Newton's laws are applicable. The boundary between quantum mechanics and continuum mechanics is not an easily quantifiable one. There is a gray area. However, continuum mechanics is the primary topic in this book.

When we treat a continuum system (solid or fluid), we will assume that it is a closed system. This means that for a given time period (time increment), there is no exchange of matter or energy with its surroundings. This can be considered as a fixed

C. Basaran, *Introduction to Unified Mechanics Theory with Applications*,
https://doi.org/10.1007/978-3-030-57772-8_3

control box in space and time. The control box is insulated, to be able to satisfy thermodynamic conservation laws.

3.1 Thermodynamic Equilibrium

All macroscopic systems have "memory." This memory is always with respect to their stress-free state, which is the minimum energy state. All macroscopic systems tend to evolve toward a minimum energy point, which is defined by intrinsic properties. Evolution toward these minimum energy points can be slow or fast, depending on external factors.

These minimum energy points of the system are called thermodynamic equilibrium states. These states usually are asymptotic states. Hence, they are static. Callen (1985) provides a good description of thermodynamic equilibrium from the atomic point of view: "The macroscopic thermodynamic equilibrium state is associated with incessant and rapid transitions among all the atomic states consistent with the given boundary conditions. If the transition mechanism among the atomic states is sufficiently effective, the system passes rapidly through all representative atomic states in the course of a macroscopic observation; such a system is in thermodynamic equilibrium. In actuality, few systems are in absolute and true thermodynamic equilibrium. In absolute thermodynamic equilibrium, from the atomic point of view, all radioactive materials would have decayed completely and nuclear reactions would have transmuted all nuclei to the most stable of isotopes. Such processes, would take cosmic times to complete, generally can be ignored. A system that has completed the relevant [significant] processes of spontaneous evolution and that can be described by a reasonable small number of parameters can be considered to be in metastable thermodynamic equilibrium. Such a limited thermodynamic equilibrium is sufficient for the application of thermodynamics." This is the definition of "thermodynamic equilibrium" we use in the rest of the book.

From a practical point of view, a system is considered to be in equilibrium if it satisfies the laws of thermodynamics. However, this justification is actually circular. That is to say, if a system satisfies the laws of thermodynamics, it is said to be in equilibrium.

A succinct definition of thermodynamics with an example is given by Callen (1985): "Thermodynamics is the determination of the equilibrium state that eventually results in a closed system." (Fig. 3.1).

> "Let us assume we have a container separated into two sections with a moveable rigid wall. Container wall is assumed to be impermeable to matter, and adiabatic (heat does not enter or leave the container). This container is the definition of closed system in thermodynamics. Initially the separation wall is fixed. If we release the separation wall, it will move to a new location due to gradient of pressure on each sides of the wall.
>
> Then if we remove the adiabatic coating from the separation wall, the heat can freely flow between two sections of the container. Now we drill holes in the separation wall, the matter can flow freely between two sections depending on the concentration gradient. Every time a constraint is removed, a spontaneous process will take place that will result in a new

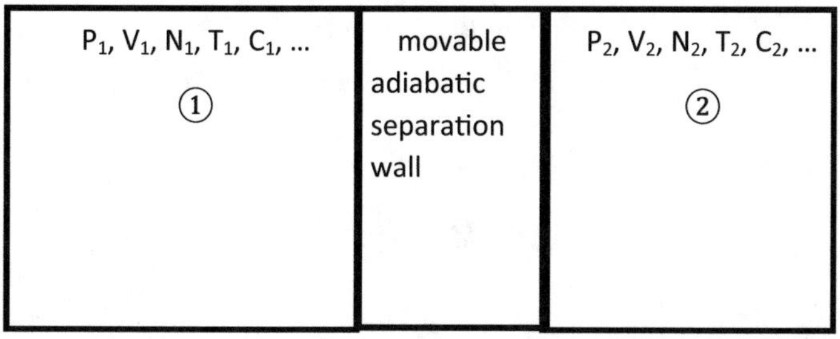

Fig. 3.1 Thermodynamic chambers, P = pressure, V = volume, N = number of particles, T = temperature, C = concentration

equilibrium state. Initial values of pressure, volume, temperature concentration and all other parameters in each chamber will take on new thermodynamic equilibrium values. The primary problem in thermodynamics is the computation of these parameters at the thermodynamic equilibrium.

It is important to point out again, by what we mean by a closed system. Because our thermodynamic equilibrium is defined for a closed system. According to thermodynamics a system is considered to be closed if it cannot exchange any energy, matter or anything whatsoever with its surrounding. This assumption does not prevent us from solving any engineering problem with fluctuating loads or masses. Because problem is always solved in incremental time steps. At each step equilibrium must be satisfied" Callen (1985).

3.2 First Law of Thermodynamics

This law of thermodynamics states that the total energy is conserved in every process. Therefore, it is sometimes referred to as the law of conservation of energy. The law is independent of path taken between the initial and final states of the system.

This law is generalized from experimental observations. Therefore, it is an empirical conclusion. While mathematically, it is very difficult to prove this law, we assume its validity because there is no way to disprove it mathematically or experimentally [at least until now]. Now we can formulate the first law of thermodynamics for application in continuum mechanics. The conservation of energy can be written as

$$\frac{d}{dt}(u+k) = W_{input} + Q_{input} \tag{3.1}$$

where u is the internal potential energy; k is the macro level kinetic energy; W_{input} is the work done by the system, due to external loads; and Q_{input} is the energy added to the system by heat transfer. Both W_{input} and Q_{input} do not have exact differentials. Therefore,

$$\oint W_{input} dt \neq 0 \tag{3.2}$$

and

$$\oint Q_{input} dt \neq 0 \tag{3.3}$$

However,

$$\oint \left[W_{input} + Q_{input}\right] dt = 0 \tag{3.4}$$

Here, exact differential means that there exists a continuously differentiable function called the potential function. In physical terms, if a system is loaded and plastically deformed and then returned to its initial state, W_{input} and Q_{input} will not be recovered completely. $\oint dt$ denotes the integral through the loading cycle. However, the total energy of the system is an exact differential:

$$\oint (du + dk) dt = 0 \tag{3.5}$$

3.2.1 Work Done on the System (Power Input)

The work done on the system by external (forces) surface tractions and body forces can be given by

$$W_{input} = \int_A \boldsymbol{\sigma}^{(n)} \cdot \boldsymbol{v} \, dA + \int_V \rho \boldsymbol{b} \cdot \boldsymbol{v} \, dV \tag{3.6}$$

where $\boldsymbol{\sigma}^{(n)}$ is the external surface traction per unit area, $\widetilde{v}$ is the velocity field in the body, $\boldsymbol{b}$ is the per unit mass body forces, and ρ is the unit weight. Concentrated point loads are also represented by surface tractions. We can write this equation in indicial notation as

$$W_{input} = \int_A \sigma_j^{(n)} \widetilde{v}_j \, dA + \int_V \rho b_j \widetilde{v}_j dV \tag{3.7}$$

The relation between external surface traction $\sigma_j^{(n)}$ and internal stresses σ_{ij} is given by

$$\sigma_j^{(n)} = \sigma_{ij} n_i \tag{3.8}$$

The surface integral can be transformed to volume integral by the divergence theorem:

$$\int_A \sigma_j^{(n)} \widetilde{v}_j dA = \int_V \left(\widetilde{v}_j \frac{\partial \sigma_{ij}}{\partial x_i} + \sigma_{ij} \frac{\partial \widetilde{v}_i}{\partial x_i} \right) dV \tag{3.9}$$

Hence, W_{input} can be written as

$$W_{input} = \int_V \left[\widetilde{v}_j \left(\frac{\partial \sigma_{ij}}{\partial x_i} + \rho b_j \right) + \sigma_{ij} \frac{\partial \widetilde{v}_j}{\partial x_i} \right] dV \tag{3.10}$$

where x_j denotes Cartesian coordinate system axes. Cauchy's equation of motion is given by

$$\frac{\partial \sigma_{ij}}{\partial x_i} + \rho b_j = \rho \frac{d\widetilde{v}_j}{dt} \tag{3.11}$$

We can substitute this equilibrium equation in W_{input}, which yields

$$W_{input} = \int_V \left[\widetilde{v}_j \left(\rho \frac{d\widetilde{v}_j}{dt} \right) + \sigma_{ij} \left(\frac{\partial \widetilde{v}_j}{\partial x_i} \right) \right] dV \tag{3.12}$$

$v_j \left(\rho \frac{d\widetilde{v}_j}{dt} \right)$ can be written as

$$\widetilde{v}_j \rho \left(\frac{d\widetilde{v}_j}{dt} \right) = \frac{d}{dt} \left(\frac{1}{2} \rho \widetilde{v}_j \widetilde{v}_j \right) \tag{3.13}$$

Hence,

$$W_{input} = \frac{d}{dt} \int_V \left[\frac{1}{2} \rho \widetilde{v}_j \widetilde{v}_j dV \right] + \int_V \sigma_{ij} \frac{\partial \widetilde{v}_j}{\partial x_i} dV \tag{3.14}$$

$\frac{\partial v_j}{\partial x_i}$ is the spatial gradient of velocity.

If we assume that there are no distributed moment (couple forces) and/or polarized fields acting on the system, then the stress tensor σ_{ij} is symmetric. Earlier spatial gradient of velocity tensor $\boldsymbol{L}$ was given by

$$\frac{\partial \widetilde{v}_i}{\partial x_j} = \boldsymbol{L} = \boldsymbol{D} + \boldsymbol{W} \tag{3.15}$$

where $\boldsymbol{D}$ is the rate of deformation tensor and $\boldsymbol{W}$ is the rate of spin tensor.

We have shown that for nonpolar case, the stress matrix σ_{ij} is symmetric; on the other hand, the rate of spin tensor W_{ij} is skew-symmetric. Therefore, we can write the following relations:

$$\sigma_{ij} W_{ij} = 0 \quad and \quad \sigma_{ij} D_{ij} \neq 0 \tag{3.16}$$

Substituting these facts into W_{input} leads to

$$W_{input} = \frac{d}{dt} \int_V \frac{1}{2} \rho \widetilde{v}_j \widetilde{v}_j dV + \int_V \sigma_{ij} D_{ij} dV \tag{3.17}$$

The first term represents the kinetic energy and the second term represents the internal (strain) energy of the system.

Here, we should point out that this equation is based on Cauchy's equilibrium equation, which is based on Newtonian mechanics laws.

3.2.2 Heat Input

The heat input, Q, has two parts. One is heat conduction through the surface coming from outside, and the other part is the distributed internal heat generation with a source strength of r per unit mass.

Internal heat generation is due to internal scattering, friction, or chemical reactions. Obviously, these two components of the heat travel in opposite directions. One is coming from outside through the surface and the other part is directly generated by the material:

$$Q_{input} = \int_A q \cdot n \, dA - \int_V \rho r dV \tag{3.18}$$

where q is the heat flux vector and n is the surface normal vector. The negative sign due to internal heat generation is in opposite direction of the heat coming from outside and is outward. More importantly, r is generated by the system just like mechanical work.

We can substitute all these terms in the first law of thermodynamics equation. For the general case, the total energy of the system will be summation of the kinetic energy and internal potential energy. Kinetic energy here is associated with macroscopic level velocity of the matter. It does not refer to kinetic energy of the atoms.

That is included in the internal potential energy, which refers to many forms of the energy stored in the lattice.

For arbitrary values of W_{input} and Q_{input}, we can write the following relations:

$$dE = dU + dK = \left(Q_{input} + W_{input}\right)dt \tag{3.19}$$

Total energy rate $d\dot{E}_{\text{total}}$ can be given by

$$\frac{dE}{dt} = \frac{dU}{dt} + \frac{dK}{dt} = \frac{d}{dt}\left\{\left[\rho u + \frac{1}{2}\rho\tilde{v}\cdot\tilde{v}\right]\right\}dV \tag{3.20}$$

where u is the internal strain energy per unit mass. We can write the first law of thermodynamics remembering that Q_{input} is negative when heat enters the system from outside and W_{input} is positive when the work is done by the system.

Since

$$d\dot{U} = Q_{input} + W_{input} - \frac{dK}{dt} \tag{3.21}$$

$$\frac{d}{dt}\int_V \rho u dV = \left[\int_A -q\cdot ndA + \int_V \rho r dV\right] + \left[\frac{d}{dt}\int_V \frac{1}{2}\rho\tilde{v}\cdot\tilde{v}dV + \int_V \boldsymbol{\sigma} : \mathbf{D}dV\right] - \frac{d}{dt}\int_V \frac{1}{2}\rho\,\tilde{v}\cdot\tilde{v}dV \tag{3.22}$$

Kinetic energy terms cancel each other. Using divergence theorem, we can transform surface integral to volume integral:

$$\int_A q\cdot ndA = \int_V \nabla\cdot qdV \tag{3.23}$$

We finally collect all terms on the one side and set it to equal to zero:

$$\int_V \left[\rho\frac{du}{dt} + \nabla\cdot q - \rho r - \boldsymbol{\sigma} : \mathbf{D}\right]dV = 0 \tag{3.24}$$

In indicial notation, energy conservation law can be written as

$$\rho\frac{du}{dt} = \sigma_{ij}D_{ij} + \rho r - \frac{\partial qj}{x_j} \tag{3.25}$$

where u is the internal potential energy per unit mass. The left side of the equation is the rate of increase in the internal potential energy. The right side of the equation represents the input power that is converted into internal work as mechanical work and as heat. However, excluding the kinetic energy at macro level, the last term is negative because it represents heat inflow per unit volume through the boundaries.

We should point out that this equation is very a simple form of the first law of thermodynamics because it assumes that only W_{input} is due to surface traction (distributed loads). If we have distributed fields such as electromagnetic and chemical loads, they must be included in this equation. These load cases will be covered in Chap. 8. Moreover, the first law of thermodynamics given above does not include energy loss terms because mechanical work terms is defined according to Newtonian mechanics. In the next chapter, the same equation will be derived for unified mechanics theory where energy loss will be included.

3.3 Second Law of Thermodynamics

There are several approaches to define the second law; however, we believe Kelvin-Planck's statement is the most concise and classical description. According to this statement, it is impossible to construct an engine that will produce no other effect than the extraction of heat from a single heat reservoir and the performance of an equivalent amount of work. In other terms, we can state that it is not possible to construct an engine that has an efficiency of 100%, meaning that the input energy and output energy for the intended work cannot be equal. There will always be energy loss for unintended work. This energy loss will only occur in positive direction, meaning that we can only dissipate energy, but cannot gain energy in a closed system. This law of nature is expressed mathematically as an inequality stating that the internal entropy production is always nonnegative and is positive for an irreversible process. This inequality is called Clausius-Duhem inequality.

Simplest definition of entropy is that entropy quantifies how much energy is unavailable for work.

Another way of stating the second law of thermodynamics is that the entropy of all natural processes increases. The concept of entropy will allow us to tie probability and statistics to continuum mechanics.

In the first law of thermodynamics, we saw interconvertibility of energy between different forms such as heat and mechanical work. In Newtonian mechanics, kinetic energy and potential energy may be converted from one to another with no energy loss. The transformation can proceed in either direction, like a pendulum with no friction that swings to eternity.

On the other hand, the second law of thermodynamics states that such a pendulum is not possible. Each time the pendulum swings, it loses some of its kinetic energy to friction; as a result, there is less total energy to continue swinging. Complete reversal is not possible. The frictional dissipation is an irreversible process.

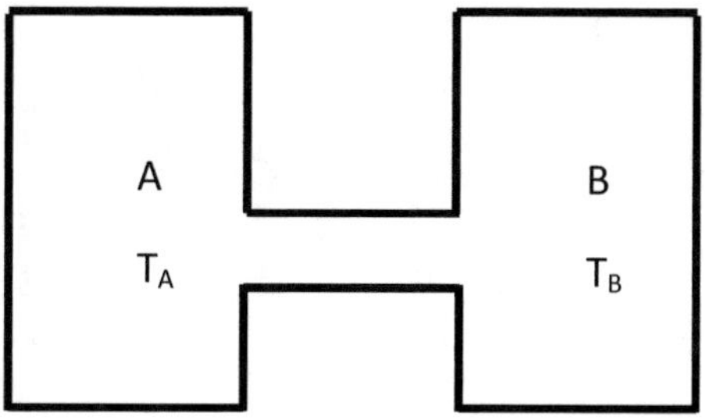

Fig. 3.2 Twin chambers

The second law of thermodynamics also defines a preferred direction in any natural process. This is clearly stated by the definition of the second law given by Clausius (1850). He stated that the heat never flows on its own from the colder system to the warmer system. This defines a preferred direction in the flow of heat between two bodies at different temperatures (Fig. 3.2). At this point, it is important to define what temperature means. While at the macroscale, in continuum mechanics, we consider objects as stationary, at the atomic scale, this is not true. Even if an object is not subjected to any dynamic loading, if the temperature of the object is above zero Kelvin, all its atoms are vibrating continuously. This phenomenon is called atomic vibrations. At any given point in time, not all atoms vibrate at the same frequency and amplitude, nor with the same energy. Vibrational modes of atoms are called phonons. All modes can be determined by solving the eigenvalue equation, just like in continuum mechanics. Atoms will be vibrating at different levels of energies; however, the average vibrational energy of all atoms in the object will be fluctuating around a constant energy level.

This constant average energy level will increase when the temperature of the object increases. Therefore, temperature is just a measure of average vibrational energy of all atoms in a body.

Going back to the second law of thermodynamics, consider two containers filled with the same gas. Assume the temperature in A is lower and B is higher. It is not possible for low-energy atoms (cold) to move toward B. Because atoms in B have higher energy levels, as such their movement toward B is possible. This is a directional preference in another statement of the second law of thermodynamics.

3.3.1 *Entropy*

The term "entropy" was first used by Clausius (1856). The second law of thermodynamics is related to a variable called entropy, s, which is a mathematical quantity. Therefore, entropy allows us to quantity the second law of thermodynamics. While the concept of entropy is not easily understood from continuum mechanics point of view, it can be explained by means of molecular behavior.

First, we should clarify that the second law of thermodynamics cannot be proven mathematically, but it is a generalization from observations (inductive-reasoning).

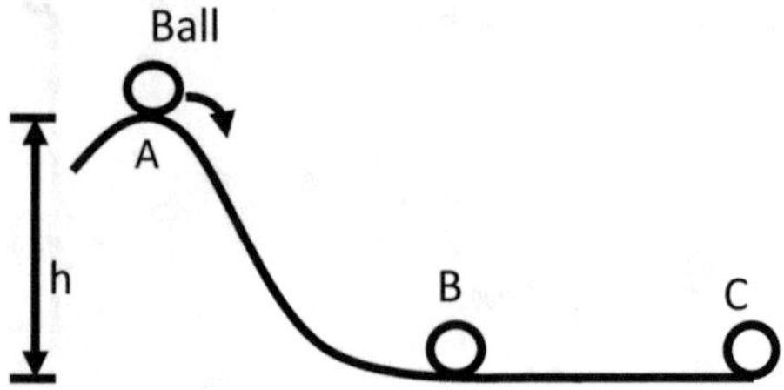

Fig. 3.3 Different states of energy

There are several simplified descriptions of entropy while they are not mathematical; however, they help in explaining an abstract quantity (Fig. 3.3).

Assume we have a ball at point *A*. It has a potential energy of *mgh*. Once the ball is slightly pushed, it rolls down and travels along the flat surface. At point *B*, the ball will have no potential energy but just kinetic energy. Finally, at point *C*, the ball will come to a stop. It will have zero energy.

As the ball is traveling from point *A* to point *C*, the amount of energy available for work continually decreases. Hence, the amount of energy unavailable for work (entropy) increases. This latter one is called entropy. In other words "production of entropy is associated with dissipation (or consumption) of the capacity for spontaneous change when any process occurs" DeHoff (1993). Entropy of the closed system increases or remains constant in all processes. Therefore, entropy always increases. Energy and entropy are governed by different laws. Energy can enter or exit a body, but it cannot be created inside a body. However, entropy can be created inside the body.

Callen (1985) formulates the interpretation of entropy in a set of postulates depending upon a posteriori justifications. The postulates are based on the basic principle that the minimization of energy function happens at equilibrium. In the following section, we will quote these postulates as defined by Callen (1985).

Postulate I

There exists particular states (called equilibrium states) of simple systems that macroscopically, are characterized completely by the internal energy, and mole numbers of the chemical components.

Postulate II

There exists a function (called the entropy *s*) of the extensive parameters of any composite system, defined for all equilibrium states and having the following property. The values assumed by the extensive parameters in the absence of an internal constraint are those that maximize the entropy over the manifold of constrained equilibrium states.

Extensive parameter in a homogeneous system is proportional to the total mass of the system.

An **intensive variable** is defined as a variable has the same value everywhere in a homogeneous system. The densities of extensive variables (e.g., *u*, *s*, *v*) are intensive variables.

This postulate assumes the existence of the entropy only for equilibrium states. In the absence of constraints, the system is free to select anyone number of states; however, the state of maximum entropy is always selected by the system.

As stated earlier, the basic problem of thermodynamics is "The single, all-encompassing problem of thermodynamic is the determination of the equilibrium state that eventually results after the removal of internal constraints in a closed composite system" Callen (1985). This basic problem can be solved if the entropy of the system is known as a function of the intrinsic parameters of the system. The relation that gives the entropy as a function of the system parameters is known as a **fundamental relation**. It, therefore, follows that if the fundamental relation of a particular system is known, all conceivable thermodynamic information about the system can be obtained from it. Essentially, if the fundamental relation of a system is known, every thermodynamic attribute is completely and precisely determined. Callen (1985) does an outstanding job in establishing postulator basis for thermodynamics.

Postulate III

The entropy of a composite system is additive over the constituent subsystems. The entropy is continuous and differentiable and is a monotonically increasing function of the energy.

The additive property of entropy is just summation of the entropies of the constituent subsystems:

$$s = \sum_{i=1}^{n} s_i = \sum_{i=1}^{n} s_i(U_i, V_i, N_i, \ldots N_r) \tag{3.26}$$

where n is the number of subsystems.

The entropy of each subsystem is a function of extensive parameters of that subsystem alone when the additive property is used for spatially separate subsystems. The following property must be satisfied. "The entropy of a simple system is a homogeneous first-order function of the extensive parameters." In other words, if all the extensive parameters are multiplied by a constant, say λ, the entropy is multiplied by the same constant:

$$s(\lambda U, \lambda V, \lambda N_1 \ldots \lambda N_r) = \lambda s(U, V, N_1 \ldots N_r) \tag{3.27}$$

Postulate II implies that the partial derivative of entropy with respect to internal energy is a positive quantity:

$$\left(\frac{\partial s}{\partial U}\right)_{V,N_i} > 0 \tag{3.28}$$

The possibility of negative values of this derivative is a topic of research. It was first mentioned by Ramsey (1956). However, such instances are nonequilibrium states. They can only be produced in unique systems and most importantly, they decay away very quickly. Therefore, these nonequilibrium transient points do not invalidate the positive value of the partial derivative for steady-state equilibrium states. Since entropy is a continuous, differentiable, and monotonic function of energy, the opposite is also true. We can express energy as a single-valued, continuous, and differentiable function of entropy and other system parameters as follows:

$$U = U(s, V, N_1, \ldots N_r) \tag{3.29}$$

where V is the volume and $N_1, \ldots N_r$ are the entire system parameters.

Postulate IV

At zero Kelvin temperature, the entropy of any system vanishes in the state for which

$$\frac{\partial U}{\partial s} = 0 \tag{3.30}$$

Essentially, this postulate states that when atoms have no vibrational energy, it is not possible to generate entropy. Therefore, at zero Kelvin temperature, systems have zero entropy generation.

Postulate IV is an extension of the Nernst postulate or also known as the **third law of thermodynamics.**

These four postulates are the rational basis for development of thermodynamics. Based on these postulates, we can solve any thermodynamics problem. Since everything organic or inorganic is a thermodynamic system, these postulates are applicable. Using these postulates, we can solve any problem that we are able to derive the fundamental relation, in the following procedure.

Let us assume the fundamental equations governing each of the constituent systems (mechanisms) are known. These fundamental governing equations allow us to calculate the entropy generation in each subsystem (mechanisms) when these mechanisms are in equilibrium. If the entire system is in a constrained equilibrium state, the total entropy is obtained by addition of the individual entropies for each subsystem (mechanism). Entropy generation in each subsystem (micro mechanism) is calculated. Therefore, the total entropy is a function of the various parameters of the subsystems (micro mechanisms). Taking straightforward differentiation of the entropy function, we can compute the extrema of the function. Of course, extrema can be a minimum, maximum, or horizontal inflection point. Then by taking the second derivative, we classify these equilibrium states as stable equilibrium, unstable equilibrium, or metastable equilibrium.

When a system is in stable equilibrium point for energy, it is also in stable equilibrium state for entropy function (fundamental relation). Minimization of energy function corresponds to maximization of entropy function, because energy is a function of the system parameters and entropy.

3.3.2 *Quantification of Entropy in Thermodynamics*

Carnot in 1824 defined entropy for a hypothetical reversible engine, where he shows the mechanical equivalence of heat energy. Entropy is a variable defined by

$$ds = -\left(\frac{dQ}{T}\right) \quad (\textit{for a reversible process}) \tag{3.31}$$

where Q is the heat energy in Joule or Calorie and T is temperature in Kelvin. In 1851, William Thomson (Baron Kelvin) and Clausius published Carnot's work posthumously, where it is shown that entropy is a state variable. "If the integral of a quantity around any closed path is zero, that quantity is called a state variable, that it has a value that is characteristic only of the state of the system regardless of how that state was arrived at" Halliday and Resnick (1966):

$$\oint ds = 0 \quad (\textit{reversible process}) \tag{3.32}$$

We should point out that neither heat energy Q nor temperature T is a perfect differential of any function. However, $ds = \frac{dQ}{T}$ is a perfect differential of a function. The fact that entropy is a state function can easily be proven for an ideal gas. Entropy will return to its initial value whenever the temperature returns to its initial value, for any reversible process. Of course, reversible process does not exist in real life. It is just an imaginary idealization created for expedience in formulation.

3.3.2.1 Gibbs-Duhem Relation

The fundamental equation for total energy U is a function of entropy, volume, and extensive parameters given by

$$U = U(S, V, N_1, \ldots N_r) \tag{3.33}$$

Taking the first differential of the total energy, we can write

$$dU = \left(\frac{\partial U}{\partial S}\right)dS + \left(\frac{\partial U}{\partial V}\right)dV + \left(\frac{\partial U}{\partial N_1}\right)dN_1 + \cdots + \left(\frac{\partial U}{\partial N_r}\right)dN_r \tag{3.34}$$

Partial derivatives for individual terms are defined as follows:

$$\left(\frac{\partial U}{\partial S}\right) = T, (temperature) \tag{3.35}$$

$$-\left(\frac{\partial U}{\partial V}\right) = P, (pressure) \tag{3.36}$$

$$\left(\frac{\partial U}{\partial N_i}\right) = \mu_i, \left(the\ \ electrochemical\ \ potential\ \ of\ \ the\ \ i^{th}\ \ component\right) \tag{3.37}$$

Temperature *T*, pressure *P*, and electrochemical potential of any subcomponent, μ_i, are called **intensive parameters**. Because they have the same value at all points in a homogeneous system, of course, this does not mean that we cannot use them for heterogeneous systems. In computational mechanics, we discretize a heterogeneous system in small pieces of homogeneous elements.

Based on these definitions, differential of total energy can be written as

$$dU = TdS - PdV + \mu_1 dN_1 + \cdots + \mu_r dN_r \tag{3.38}$$

for a homogeneous system in equilibrium.

Formal thermodynamics definitions of temperature (at macro level) agrees with the intuitive definition given by ($\frac{\partial U}{\partial S} = T$). We should point out that this is a macro definition of temperature, since at the atomic level, temperature is just a vibrational energy. In addition, the definition of pressure given above agrees with the mechanics definition of pressure. [$-PdV$] is called quasi-static work:

$$dW_M = -PdV \tag{3.39}$$

The term quasi-static used here contrasts to a dynamic load where a load is applied very fast; as a result, there is inertia effect.

It is necessary to discuss the negative sign in quasi-static work. The quasi-static work is assumed positive if it increases the total energy of the system.

If change in volume *dV* is negative, work done on the system is positive, increasing its energy, because of the negative sign in the equation:

$$dW_M = -P(-dV) = +PdV \tag{3.40}$$

Heat flux can be defined quantitatively with the help of quasi-static work.

For a quasi-static process at constant number of moles, heat energy *dQ* can be defined by

$$dQ = dU - dW_M \tag{3.40a}$$

Or

$$dQ = dU + PdV \tag{3.40b}$$

In thermodynamics terms, heat and heat flux are used interchangeably. In thermodynamics terminology, "heat" like "work" is only a form of energy transfer. It is assumed that once the energy is transferred to a system either in the form of heat or as mechanical work, it is indistinguishable from any energy that may have been transferred directly to the system. It is important to remind that Joule is the unit for energy and work. Therefore, there cannot be distinction between variables with the same units.

The total energy U of a state is not just a sum of work (dW_M) and heat (dQ), because dW_M and dQ are imperfect differentials, which means that the differential is path dependent. [exact differential is path independent].

The integrals of dW_M and dQ are the work and heat for that particular path of the process. Their sum is the total energy difference ΔU. However, ΔU is independent of the path taken.

We should point out that when the fundamental equation is written as $U = U(S, V, N_1, \ldots N_r)$, the total energy U is a dependent variable and entropy S is an independent variable. This is the fundamental equation unified mechanics theory is based on. Since entropy is an independent variable, it can only be defined in space on its own axis. The relation $U = U(S, V, N_1, \ldots N_r)$ is also named the **energetic fundamental relation**.

However, $S = S(U, V, N_1, \ldots N_r)$ is said to be an **entropic fundamental relation**.

The remaining terms in dU equation represent an increase of internal energy associated with the addition of matter to a system. These terms are called the quasi-static electrochemical work. The term matter here does not exclude electron flow as a matter due to an electrical bias. The quasi-static electrochemical work is given by

$$dW_c = \sum_{i=1}^{r} \mu_i N_i \tag{3.41}$$

Therefore, energy equilibrium energy becomes a summation of heat, mechanical work, and electrochemical work:

$$dU = dQ + dW_M + dW_C \tag{3.42}$$

3.3.2.2 Euler Equation

Quoting Callen (1985) formulation, the homogeneous first-order property of the energetic fundamental relation permits that equation to be written in a convenient form called the Euler form:

$$U(\lambda s, \lambda X_1, \ldots, \lambda X_r) = \lambda U(s, X_1, \ldots X_r) \tag{3.43}$$

Following the same differentiation process and using the same definitions we used earlier in this section for the energy equilibrium equation and differentiating with respect λ yield

$$\frac{\partial U}{\partial(\lambda S)}\frac{\partial(\lambda S)}{\partial\lambda} + \frac{\partial U}{\partial(\lambda X_i)}\left(\frac{\partial\lambda X_i}{\partial\lambda}\right) + \cdots = U\,(S, X_1, \ldots X_r) \tag{3.44}$$

Assuming $\lambda = 1$, then

$$\frac{\partial U}{\partial S}s + \sum_{i=1}^{t}\frac{\partial U}{\partial X_i}X_i = U \tag{3.45}$$

$$U = Ts + \sum_{i=1}^{r} P_i X_i \tag{3.46}$$

For a simple system, this equation becomes

$$U = TS - PV + \mu_1 N_1 + \cdots + \mu_r N_r \tag{3.47}$$

The last two equations are called Euler relations.

A differential form of the relation among the intensive parameters can be obtained directly from the Euler relation and is known as the Gibbs-Duhem relation.

Euler (1736) relation is given by

$$U = TS + \sum_{i=1}^{r} P_i X_i \tag{3.48}$$

Taking differentiation of this equation leads to

$$dU = TdS + SdT + \sum_{i=1}^{r} P_i dX_i + \sum_{i=1}^{r} X_i dP_i \tag{3.49}$$

Earlier we derived that

$$dU = TdS + \sum_{i=1}^{r} P_i dX_i \tag{3.50}$$

We should note that in derivation of the last equation, we separated quasi-static mechanical work and quasi-static electrochemical work. Here, they are both represented in one term:

$$dU = dQ + dW_M + dW_C \tag{3.51}$$

$$\sum_{i=1}^{r} P_i dX_i \tag{3.52}$$

Substituting this last equation in the differential above, we end up with

$$SdT + \sum_{i=1}^{r} X_i dP_i = 0 \tag{3.53}$$

This equation is referred to as Gibbs-Duhem relation. For a single component system, this equation can be written as

$$SdT - VdP + Nd\mu = 0 \tag{3.54}$$

The Gibbs-Duhem relation presents the relationship among the intensive parameters in differential form. The number of linearly independent intensive parameters that have independent variations is called the number of thermodynamic degrees of freedom of any system, which are mapped onto thermodynamic state index axis in unified mechanics theory, which is discussed in the next chapter.

At this point, it is necessary to restate the thermodynamic formalism in energy representation. This fundamental equation is

$$U = U(S, V, N_1, \ldots N_r) \tag{3.55}$$

which contains all thermodynamic information of any system. Relying on our definitions, the equations of state can be given by

$$\begin{aligned} T &= T(S, V, N_1, \ldots N_r) \\ P &= P(S, V, N_1, \ldots N_r) \\ \mu_1 &= \mu_1(S, V, N_1, \ldots N_r) \\ \mu_r &= \mu_r(S, V, N_1, \ldots N_r) \end{aligned} \tag{3.56}$$

When all equations of state are combined, it is equivalent to the **fundamental equation**.

Assuming a simple single component system of two equations of the state is known, the Gibbs-Duhem relation can be integrated to obtain the third equation of state.

As pointed out by Callen (1985), it is possible to express internal energy U without the term entropy S. However, such a total energy equation is not a fundamental relation and does not contain all thermodynamic information about the system.

3.3.2.3 Entropy Production in Irreversible Process

The cantilever beam shown in Fig. 3.4 is subjected to linear elastic static cyclic loading. Let us assume the force is small enough to produce stress level well below the yield stress. A student observing this beam will see that after every load removal, the beam returns to the original shape. To the observer this gives the impression that deformation under elastic loading is a perpetual reversible process. However, after certain number of cycles, the cantilever beam will fatigue and fail. It will probably break away from the support where stresses are maximum. This simple experiment shows that while we can observe macro behavior to be reversible, our observation is not accurate. If the linear elastic loading-unloading cycles were reversible, the beam would never fail and fatigue, and deflections from the cycle would not change.

According to the second law of thermodynamics, reversible processes are imaginary. They cannot happen in real domain. Actually, what happens in the lattice is when the beam is mechanically loaded, there is strain in the lattice. However, when the load is removed, atoms do not go back to exactly their original lattice site. They go to a lattice site with a lower energy level, in order to minimize their energy level. In the process, they leave behind a vacancy, of course. So as a result, the process is not reversible. Keep in mind that thermodynamic reversible process is where the initial and the final states are the same. If an atom moves to a new lattice site with lower energy state and left behind, a vacancy that is a different thermodynamic state than the initial one state. There is a positive entropy generation, which is a quantitative measure of dissipation in the system. DeHoff (1993) states that "changes in the real world are always accompanied by friction and something [energy] that is dissipated. When the pebble is dropped into pond its kinetic energy at impact is dispersed by the wave motion throughout the pond and the [initial kinetic energy] is dissipated in the absorbing water of the pond." In this irreversible process, the energy

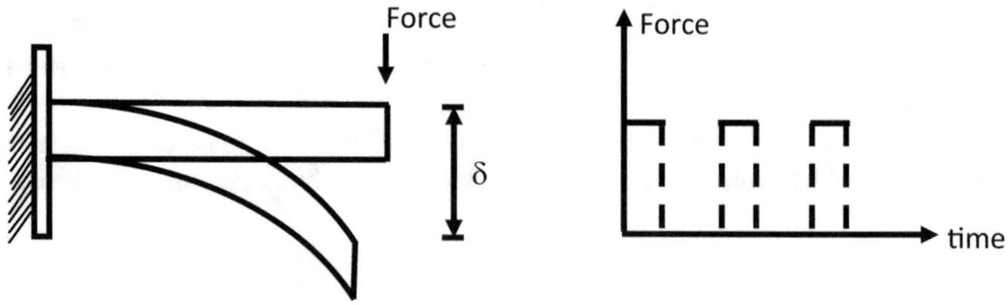

Fig. 3.4 A cantilever beam subjected to cycling shear loading

cannot be recovered. "The entropy production in a system is a quantitative measure of this energy dissipation [mechanism]."

Eddington (1958) referred to entropy as "entropy is time's arrow." The total entropy of any system plus its surroundings always increases with increasing time. However, their relationship is not linear and each one is linearly independent. Time can change without change in entropy. If we had a process where entropy is destroyed, then the direction of the time arrow must be reversed for that process to happen. However, this is not possible in a real system. A reader may question accuracy of theory of elasticity where all mechanisms are assumed reversible. Because of the second law of thermodynamics, the theory of elasticity provides an approximate solution to an imaginary system that cannot exist in reality. In the words of DeHoff (1993), "All real processes have a finite rate in response to finite influences; such real processes are called irreversible to emphasize the contrast with imaginary reversible processes. Real processes are irreversible and suffer dissipations that result in the production of entropy and thus a permanent (irreversible) change in the system." Therefore, for an irreversible process, we can write

$$\Delta S = S_2 - S_1 = \int_1^2 \left(\frac{dQ}{T}\right)_{irreversible} > 0 \tag{3.57}$$

Of course, here, it is assumed that it is a closed system and it is an adiabatic process.

While we acknowledge that all real processes are irreversible, sometimes for the sake of simplicity, a system is modeled as an imaginary reversible process, to be able to get an understanding of the system near time=0.

3.3.2.4 Clausius-Duhem Inequality

There are different approaches to derive this inequality. We will quote the derivation given by Malvern (1969). This inequality is considered a quantitative explanation of the second law of thermodynamics. The theory explains the relationship between heat energy flow in a system and the entropy generation in the system and its surroundings (Fig. 3.5).

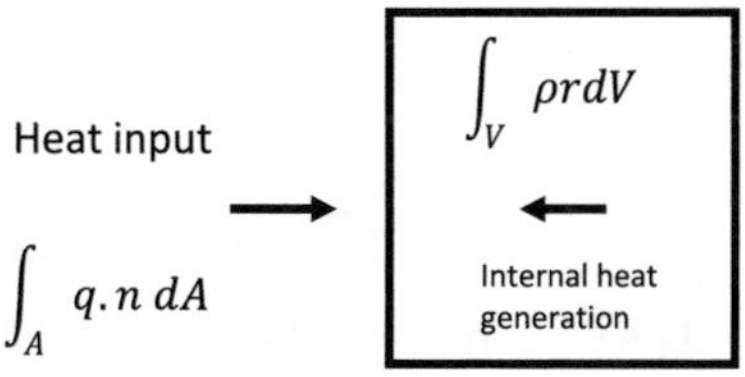

Fig. 3.5 Heat exchange in a chamber

Heat input in a system is given by difference between the heat conduction coming from outside through the surface A and distributed internal heat generation r per unit mass:

$$Q_{input} = \int_A -q \cdot n\,dA + \int_V \rho r dV \tag{3.58}$$

Entropy was defined by $dS = \frac{dQ}{T}$. Hence, entropy input rate can be given by

$$\int_A -\frac{q}{T} \cdot n\,dA + \int_V \frac{\rho_r}{T} dV \tag{3.59}$$

We are assuming that the system is closed. If the system is open in a control surface, [fixed in space], additional entropy input due to mass flux would be accounted for by

$$\int_A -\rho s V \cdot n\,dA \tag{3.60}$$

where s is entropy per unit mass. According to the second law of thermodynamics, entropy increase rate in the system $\geq$ entropy Input rate.or

$$\frac{d}{dt}\int_V \rho s dV \geq \int_A -\frac{q}{T} \cdot n dA + \int_V \frac{r}{T} \rho dV \tag{3.61}$$

This is the integral form of the Clausius-Duhem inequality. This equation means that in an irreversible process, internal entropy production is always greater than input. In this equation, outward normal is a positive sign. Equal sign ensures that the equation holds for reversible [imaginary] process.

We can transform the surface integral to a volume integral. Since volume can be an arbitrary value, we can write the local version of the Clausius-Duhem inequality that must be satisfied at each point in any given volume as follows:

$$\frac{ds}{dt} \geq \frac{r}{T} - \frac{1}{\rho} div\left(\frac{q}{T}\right) \tag{3.62}$$

or

$$\gamma \equiv \frac{ds}{dt} - \frac{r}{T} + \frac{1}{\rho T} divq - \frac{q}{\rho T^2} \cdot grad\,T \geq 0 \tag{3.63}$$

where γ is the internal entropy production rate per unit mass.

A stronger assumption of the inequality was proposed by Truesdell and Noll (1985):

$$\frac{ds}{dt} - \frac{r}{T} + \frac{1}{\rho T} div\, q \geq 0 \qquad -\frac{1}{\rho T^2} q \cdot grad\, T \geq 0 \tag{3.64}$$

where the first inequality represents the local entropy production by the system and the second inequality represents the entropy production by heat conduction.

The Truesdell and Noll (1985) interpretation of the second law of thermodynamics is important for unified mechanics theory, because we will see that the disorder in the system increases only due to internal entropy production.

Entropy production by heat conduction in a free (unconstrained) system cannot lead to irreversible disorder because the system travels between stress-free states. The interatomic equilibrium distance is a function of temperature; therefore, due to heat conduction, the temperature of the system increases (atoms have higher vibrational energy), and the equilibrium interatomic distance also increases. However, there will be no strain (or stress) exerted on the atoms. Therefore, internal work will be zero; as a result, internal entropy production will be zero. If the atoms move from one thermal equilibrium to another thermal equilibrium, it is assumed that there will be no internal entropy generation. Of course, this assumption is not 100% true. When there is a heat transport in a lattice, there is always scattering of phonons. We are assuming this scattering-induced internal entropy generation is so small that we can ignore it.

When there are no external constraint blocking atoms moving freely at each temperature, they have a stress-free equilibrium interatomic separation distance r_0. When the interatomic distance is r_0, the attraction potential energy is at maximum and the force acting on the atoms is zero (Fig. 3.7). Therefore, internal work is not possible, because internal work and internal entropy production are proportional to dislocation of an atom from equilibrium, lattice site, and force acting on the atom. Of course, when a system is at equilibrium and the force acting on the atom is zero, the work done must be zero. However, this argument assumes that the system is free to move and the boundary conditions do not impose any constraint on the system that prevents it from moving freely. If the cantilever shown in Fig. 3.6 is fixed on both sides, of course, there will be internal work and internal entropy generation.

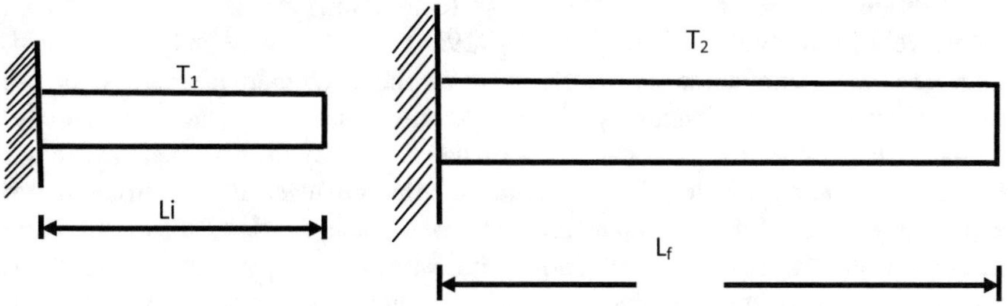

Fig. 3.6 Expansion of a cantilever beam due to increasing temperature

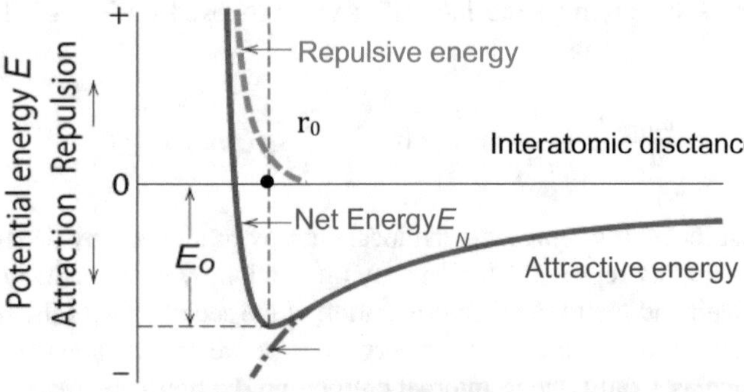

Fig. 3.7 Repulsive and attractive bonding energy between two isolated atoms as a function of interatomic distance. (After Callister Jr. and Rethwisch (2010)

Because Clausius-Duhem inequality is just a quantitative statement of the second law of thermodynamics, it must be satisfied by every process in organic or inorganic systems. The coordinate system assumed in derivation of the formulation in previous pages is arbitrary; however, positive definite property of the inequality is not arbitrary. It must be satisfied for all and any coordinate system.

3.3.2.5 Traditional Use of Entropy as a Functional in Continuum Mechanics

The unified mechanics theory uses entropy as a linearly independent state variable. Furthermore, entropy (thermodynamics state index) defines a new additional axis in addition to Newtonian space-time axes. This is needed to be able to define location of a system in space-time-thermodynamic state index coordinate system. While there are similarities between traditional interpretation of entropy in Newtonian (continuum) mechanics and unified mechanics, there are also major differences. In Newtonian (continuum) mechanics, derivative with respect to entropy is taken as zero. In the unified mechanics theory, derivatives with respect to entropy are not zero. When derivatives with respect entropy are taken as zero, the unified mechanics theory collapses to Newtonian mechanics.

Traditional use of entropy in Newtonian (continuum) mechanics is considered extensively by Malvern (1969), Coleman (1964), and Coleman and Mizel (1964, 1967). Here, we summarize their work. In Newtonian continuum mechanics, the use of the Clausius-Duhem inequality differs from the usual second law of thermodynamics, where the entropy is a state function determined by the instantaneous values of the other state variables. In Newtonian continuum mechanics, entropy is not explicitly postulated to be a state function. As a result, in Newtonian continuum mechanics, entropy maximization (and minimization of entropy generation rate) has no effect on system state variables or system properties. In simple terms, in Newtonian continuum mechanics, degradation of the system is not possible. Traditionally, many researchers and authors of thermodynamics have postulated that entropy is a state function of all the state variables including possibly some variable not observable by macroscale specimen testing.

It is important that we clarify what we mean by "state variables." For a simple ideal gas, pressure, temperature, and volume are thermodynamic variables. However, the internal energy and the entropy are state variables (also called state functions). State variables are expressed by means of thermodynamic variables. For a variable to be classified as a state variable (also called state function), it must have an exact differential, based on a vanishing integral in an arbitrary cycle. This definition assumes that the material is capable of going through an arbitrary cycle and be restored to its initial state. As an example, we can plastically deform a steel beam, then melt it or anneal it, and return it to its original state. Therefore, for the total energy of the system is a state variable,

$$\oint \left[P_{input} + Q_{input}\right] dt = 0 \tag{3.65}$$

Malvern (1969) states that if the "state" of a continuum is taken to be defined by a limited number of explicitly enumerated macroscopic state variables, observable at least in principle, then the entropy must in general depend on the history of this limited number of state variables and not merely on their current values. This is of course the case for any solid that experiences inelastic deformations. Because plastic work is path dependent, therefore entropy is also of course path dependent. Coleman (1964) and Malvern (1969) define entropy as a *functional* of history of deformation gradient F, temperature and by the instantaneous value of temperature gradient.

Malvern (1969) makes the distinction between function and functional by defining that a function depends only on the instantaneous values of its variables. Functional is a function of the history of the variables. Malvern (1969) citing a derivation by Coleman and Mizel (1964) defines caloric equation of state in the following form:

$$s = f(u, F) \tag{3.66}$$

where entropy is a function of internal energy and deformation gradient. However, this definition ignores other electro-thermo-chemical-radiation mechanisms that could contribute to entropy generation. Earlier Truesdell and Toupin (1960) have postulated a caloric equation of state as follows:

$$u = u(s, V_i, \ldots, V_n) \tag{3.67}$$

Here s is entropy per unit mass and V_i is a thermodynamic substate variable accounting for all thermo-electro-mechanical-chemical-radiation processes. Of course, the inverse of the last equation also holds true:

$$s = s(u, V_1, \ldots, V_n) \tag{3.68}$$

which is a more general **caloric equation of the state**, which is called thermodynamic fundamental relation by Callen (1985).

3.3.2.6 Entropy as a Measure of Disorder

Order and disorder are relative terms, meaning that they are defined with respect to an initial reference state. However, in general, the initial or strain-free state of a system is considered as ordered. Of course, this does not mean that the strain-free state has no disorder at the atomic scale. It is just a benchmark state. Any deviation from the initial reference state is an increase in disorder. For a disorder to happen in a system, there must be some inhomogeneity. For example, if we have a box with four identical baseballs, regardless how much we shake the box, the final "disordered" state will be identical to the initial ordered state.

There are 24 distinct configurations possible for these balls in the container (Fig. 3.8). However, the positions the balls could take are not observable at the macroscale, because all the balls are identical. Fortunately, most systems and materials at the microaggregate scale are not made up of identical crystals. Now assume that we mark the balls as *A*, *B*, *C*, *D* and initially we place them as in the box in alphabetical order. We assume that the initial alphabetical order configuration is a reference "ordered" state. In this case, after we shake the box among $4\,! = 24$ possible configurations, the initial ordered state is only possible 1/24 of the times. That means 23/24 times it will be a disordered state. Possibility of getting a disordered state is five times more likely. If we have 26 baseballs marked *A* to *Z*, the possibility of getting the ordered state is one in millions. All organic and inorganic systems are made up of large number of atoms (or molecules). When subjected to any external load (disturbance), they move from their initial ordered state to a new configuration that we call "disordered state"; of course, there is energy cost associated with changing the ordered state. It does not happen on its own with no external energy input. "In statistical mechanics the entropy of a state is related to the probability of the occurrence of that state among all the possible states that could occur" Malvern (1969). This is only rational because entropy generation requires work. As more entropy is generated, other configurations that are possible are achieved. Malvern (1969) states: "Thus, increasing entropy is associated with increasing disorder. The second law of thermodynamics seems to imply an almost metaphysical principle of preference for disorder." In nature, it is observed that changes of state are more likely to occur in the direction of greater disorder. However, the second law of thermodynamics also implies that when the disorder is maximum, entropy is also at maximum and entropy generation rate is zero. The balls in a box example are an illustration of this argument.

Callen (1985) explains eloquently why entropy is a measure of disorder, quantitatively. We will summarize his approach. Callen (1985) states that, in statistical

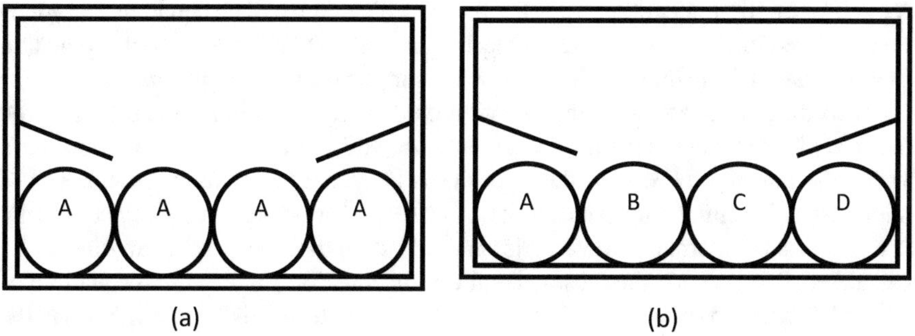

Fig. 3.8 Description of (**a**) order and (**b**) disorder

mechanics, the number of microstates [arrangement of *A*, *B*, *C*, *D* balls in our example, Fig. 3.8] among which the system undergoes transitions and which thereby share uniform probability of occupation increases to the maximum permitted by the constraints. This statement is strikingly reminiscent of the entropy postulate of thermodynamics, according to which the entropy increases to the maximum permitted by the imposed constraints, for a closed system. It is possible to conclude that entropy can be identified with the number of microstates in a closed system.

However, Callen (1985) points out that entropy is additive, but the number of microstates is multiplicative. The number of microstates available to two systems is the product of the number of microstates available to each system. If we have two dices, each has six microstates. But two dices rolled together have $6 \times 6 = 36$ microstates. Callen (1985) points out that in order to interpret the entropy, then we require an additive quantity that measures the number of microstates available to a closed [isolated] system. Boltzmann (1877) [English translation by Sharp and Matschinsky (2015)] and Max Planck (1900a, b, c, d) suggested that this problem could be solved by identifying the entropy with the logarithm of the number of available microstates, because the logarithm of a product is the sum of the logarithms:

$$ln\ (a \cdot b) = ln\ a + ln\ b \tag{3.69}$$

Thus, the following equation resulted:

$$S = kln\ w \tag{3.70}$$

where *w* is the number of microstates consistent with the macroscopic constraints of the closed [isolated] system. The confusion about *w* being the number of microstates or probability of a state was covered earlier. The Boltzmann constant

$$k = R/N_A = 1.3807 \times 10^{-23} J/K \tag{3.71}$$

is used to define temperature in Kelvin scale and to ensure consistency of units on both sides of the equation. $R = 8.31 \frac{\text{Joule}}{\text{mol} \cdot \text{K}}$ is the gas constant and

$N_A = 6.022x\ 10^{23}$molecules/mol is the Avogadro's number. The equation $S = k \ln w$ is considered the basis for statistical mechanics. Callen (1985) refers to this equation as a postulate that is dramatic in its brevity, simplicity, and completeness.

According to Boltzmann, the equation calculating the natural logarithm of the number of microstates available to the system, and multiplying with a constant k, yields the entropy. However, the opposite is also true to calculate the number of microstates. In unified mechanics theory, this second approach is used; since entropy is a function of internal state variables and all active mechanisms, it can directly be calculated, because it is the most general caloric equation [fundamental equation] of state. Callen (1985) refers to Boltzmann's equation as the statistical mechanics in the micro canonical formation.

Entropy calculations in continuum mechanics are not readily available for all mechanisms. Of course, very often we do not know what these mechanisms are. This is expected to be a research field in the near future. Entropy in thermodynamics and statistical physics is the same thing. Statistical mechanics interpretation of entropy where natural progress toward more disorder establishes a concrete and understandable concept of entropy. However, in statistical mechanics to be able to calculate the entropy, all active mechanisms and processes taking place in a material must be accounted for. Then entropy generation for each process must be calculated. Unfortunately, this is a very primitive field in mechanics. There are no explicit formulas for entropy generation due to thermo-mechanical-electrical-chemical-radiation loads. This is a wide-open research topic and very challenging one to say the least, because material modeling field has always been based on empirical observations. We crash a concrete cylinder and model it with macro measurement variables, such as stress or strain and curve fitting. However, macroscale testing does not give us much information about actual mechanisms that are responsible for degradation and final failure. These are the mechanisms that we need to compute entropy generation for, because it is always much easier to curve fit to a test data and use it then to understand the actual mechanisms and processes leading to failure.

Theoretical physicist and mathematician Freeman Dyson in *Scientific American's* September (1954) issue suggested that heat is a disordered energy. The author gave an example of a flying rifle bullet. The bullet has kinetic energy but no disorder. After the bullet hits a steel plate target, its kinetic energy is transferred to random motions of the atoms in the plate and bullet. This disordered energy makes itself felt in the form of heat. However, the author ignores other mechanisms. Because heat is not the only product of the initial kinetic energy, there are plastic deformations, melting, possible phase change in the material, and other mechanisms that could result from an impact.

Freeman Dyson (1954) also adds that the quantity of disorder is measured in terms of the mathematical concept called entropy; of course, entropy is a functional of energy.

Therefore, we can compute a disorder in terms of its energy using Boltzmann's equation. It is important to point out that a disorder is the perfect way to express degradation of all organic and inorganic systems, because as the entropy increases, the amount of disorder from the initial "ordered" state increases. Realignment of

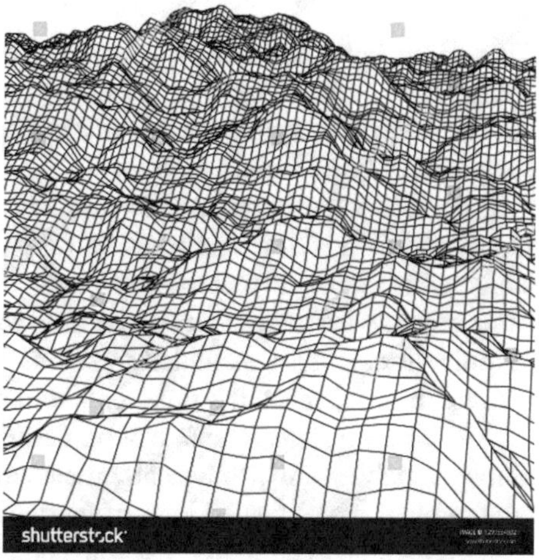

Fig. 3.9 Energy terrain along different paths

atoms in any system just takes one of the possible disordered states. Of course, because of very large number of possible disordered states, the system cannot get into every possible disordered configuration before entropy becomes maximum, because of the closed-isolated system requirement. This is easier to explain with an example. Assume we have a box with 26 baseballs marked A to Z. Initially, they are all in alphabetical order. Say the box is 10 kg. We are allowed to spend 100 Joules of energy to shake the box. Assume we have probably ten chances to shake the box. Of course, for our closed system when 100 Joule is consumed, the maximum entropy is reached. However, we could not discover all possible disorder states that 26 baseballs could take. In the same fashion, when a material fails in a fatigue loading or monotonic loading, the failure is always through a different path. Yet the maximum entropy value to reach the failure is always the same (Naderi et al. 2009 Yun and Modarres 2019 Tu and Gusak 2019). Failure or increase in disorder in a system can be considered a travel over an energy terrain. A soccer ball sitting in a valley after a kick will travel. However, there are many paths it can take. Each one of them is possible. Each one represents a disorder path. However, the ball will finally come to a stop at a valley. We can write the fundamental relation. Of course, we have to know what path the ball will take to use the properties relevant to that path. Then we can exactly predict in which valley the ball will stop at (Fig. 3.9).

3.3.2.7 Thermodynamic Potential

The choice of thermodynamic potential in continuum mechanics is subjective. One can use different thermodynamic potentials depending on the problem. However, there are some mathematical conditions that all thermodynamic potentials must satisfy. To be able to define thermodynamic potentials, first we must define thermodynamic variables that are also called state variables or independent variables.

Thermodynamic Variables

Thermodynamic local state of a system at any given time and space in a point can be completely defined by the thermodynamic variables using the fundamental relations. Evolution of thermodynamic state under external loading is defined by a succession of equilibrium states.

Each microstate in energy landscape is an equilibrium point. However, that does not mean that it is at a maximum entropy point. As Lemaitre and Chaboche (1990) suggest, the ultrarapid phenomenon for which the time scale of the evolution is in the same order as the atomic relaxation time for a return to thermodynamic equilibrium (atomic vibrations) is excluded from this theory's field of applications, because in continuum mechanics, the time scale to reach equilibrium is much longer.

Any physical phenomena can be defined with appropriate choice of thermodynamic variables. Any physical process defined by thermodynamic variables is admissible if it satisfies all laws of thermodynamics.

It is essential to point out that when degradation is introduced into Newton's laws of motion, Lemaitre and Chaboche (1990) opine that there is no objective way of choosing the internal variables best suited to study a phenomena. This is an opinion that the choice is dictated by empirical experience, "physical feeling," and very often by the type of application. We believe researchers' latter opinion is based on their phenomenological curve fitting approach for Kachanov-type damage mechanics models. However, in unified mechanics theory, all thermodynamic variables must be chosen objectively to be able to assemble the fundamental equation. They must represent the active micro mechanisms responsible for entropy production. In Lemaitre and Chaboche (1990) approach, which is based on Newtonian mechanics, thermodynamic variables are up to the scientist/engineer to decide so that model and experimental data can be fit to the same curve. In unified mechanics theory, the actual micro mechanisms that are responsible for entropy generation must be identified objectively, and they must be physically formulated in the fundamental equations without curve fitting.

Observable Thermodynamic Variables

1. Temperature.
2. Space coordinates.

While some authors define total strain and stress as observable variables, we do not subscribe to this school of thought, because strain and stress are human, construct definitions, not physical quantities. We can only measure displacement and then calculate strain and stress.

Independent Internal Thermodynamic Variables

1. Entropy.
2. Internal energy (this is not limited to mechanical energy).
3. Density of dislocations.
4. Crystal structure (phase).
5. Etc.

Formulation of Thermodynamic Potential

Postulate There exists a thermodynamic potential function defined by thermodynamic variables. The thermodynamic state laws are derived from this potential. Thermodynamic potential must be concave [$f''(X) < 0$] with respect to temperature and convex [$f''(X) > 0$] with respect to other variables. These convex and concave requirements ensure the thermodynamic stability requirement imposed by the Clausius-Duhem inequality. Malvern (1969) proposes the following thermodynamic potentials (Table 3.1):

where v_i is any independent thermodynamic state variable. The Helmholtz free specific energy potential Ψ is the portion of the internal energy available for doing work at constant temperature. In continuum mechanics, free energy Ψ is a function of thermodynamic variable defined earlier:

$$\Psi = \Psi(T, D, V_1, \ldots V_r) \tag{3.72}$$

Enthalpy, h, is the portion of the internal energy, u, that can be released as heat when the thermodynamic tensions are held constant. Enthalpy can also be given by

$$h = u - \tau_i v_i \tag{3.73}$$

where τ_i is a thermodynamic tension

Table 3.1 Thermodynamic potentials (Malvern 1969)

Potential		Relation to u	Thermodynamic-independent variables
Internal energy	u	u	s, v_j
Helmholtz free energy	Ψ	$\Psi = u - s\theta$	θ, v_j
Enthalpy	h	$h = u - \tau_j v_j$	s, τ_j
Free enthalpy or Gibbs function	g	$g = u - s\theta - \tau_j v_j$ $= h - s\theta$	θ, τ_j

$$\tau_i = \left(\frac{\partial u}{\partial v_i}\right)_S \tag{3.74}$$

In differential form, thermodynamic potentials can be given by

$$du = Tds + \tau_i dv_i \tag{3.75}$$

$$d\Psi = -sdT + \tau_i dv_i \tag{3.76}$$

$$dh = Tds - v_i dT_i \tag{3.77}$$

$$dg = -sdT - v_i dT_i \tag{3.78}$$

Assuming state variables defined by subscripts are held constant, the following relationships can be written:

$$T = \left(\frac{\partial u}{\partial s}\right)_{v_i} \tag{3.79}$$

$$\tau_i = \left(\frac{\partial u}{\partial v_i}\right)_s \tag{3.80}$$

$$s = -\left(\frac{\partial \Psi}{\partial T}\right)_{v_i} \tag{3.81}$$

$$\tau_i = \left(\frac{\partial \Psi}{\partial v_i}\right)_T \tag{3.82}$$

$$T = \left(\frac{\partial h}{\partial s}\right)_{\tau_i} \tag{3.83}$$

$$v_i = -\left(\frac{\partial h}{\partial \tau_i}\right)_S \tag{3.84}$$

$$s = -\left(\frac{\partial g}{\partial T}\right)_{\tau_i} \tag{3.85}$$

$$v_j = -\left(\frac{\partial g}{\partial \tau_j}\right)_\theta \tag{3.86}$$

It is assumed that thermodynamics tensions can be identified with the Piola-Kirchhoff stress tensors and with recoverable work assumption. Malvern (1969) states that "The concept of recoverable work is bound to the existence of a caloric equation [fundamental equation] of state and only thermodynamic tensions derived from a potential defined by such an equation do work not contributing to entropy production. The assumed existence of such a caloric equation of state does not imply our knowledge of such a formula for it."

Here, it is assumed that internal energy u is a potential for the thermodynamic tensions when entropy is constant. Of course, this assumption is eliminated using internal energy as a thermodynamic potential in any real process. However, Helmholtz free energy density is a thermodynamic potential in an isothermal process. Since most engineering problems are solved in incremental format, Helmholtz free energy is more appropriate as a thermodynamic potential. It should be clarified that the term thermodynamic tension is not the same as universal thermodynamic force, or just any stress tensor.

Helmholtz free energy can be implemented as a thermodynamic potential, for an elastic-plastic solid under thermo-mechanical loads. Helmholtz free energy for an elastic-plastic solid is given by

$$\Psi = \Psi([D - D^p], T) = \Psi(D^e, T) \tag{3.87}$$

Then we can write the following relation based on deformation gradient tensor which is a summation of elastic and plastic parts:

$$\frac{\partial \Psi}{\partial D^e} = \frac{\partial \Psi}{\partial D} - \frac{\partial \Psi}{\partial D^p} \tag{3.88}$$

We can also write

$$\dot{\Psi} = \frac{\partial \Psi}{\partial D^e} : D^e + \frac{\partial \Psi}{\partial T} \dot{T} \tag{3.89}$$

Thermodynamic potential must satisfy Clausius-Duhem inequality, which is given by

$$\frac{ds}{dt} - \frac{r}{T} + \frac{1}{\rho T} div\, q - \frac{q}{\rho} .grad\, T \geq 0 \tag{3.90}$$

If we substitute r from the conservation of energy equation, we obtain the fundamental inequality that contains both the first and second laws of thermodynamics:

$$\frac{ds}{dt} + \frac{1}{\rho} div \left(\frac{q}{T}\right) - \frac{1}{T}\left(\frac{du}{dt} - \boldsymbol{\sigma} \cdot \boldsymbol{D} + div\, q\right) \geq 0 \tag{3.91}$$

Since

$$div \left(\frac{q}{T}\right) = \frac{div\, q}{T} - \frac{q \cdot grad\, T}{T^2} \tag{3.92}$$

Multiplying each term with T,

$$\rho\left(T\frac{ds}{dt} - \frac{du}{dt}\right) + \boldsymbol{\sigma} : \boldsymbol{D} - q \cdot \frac{grad\, T}{T} \geq 0 \tag{3.93}$$

Helmholtz specific free energy is given by

$$\Psi = u - Ts \tag{3.94}$$

Differentiating this, we obtain

$$\frac{d\Psi}{dt} = \frac{du}{dt} - T\frac{ds}{dt} - s\frac{dT}{dt} \tag{3.95}$$

We can write

$$-\left(\frac{d\Psi}{dt} + s\frac{dT}{dt}\right) = T\frac{ds}{dt} - \frac{du}{dt} \tag{3.96}$$

If we substitute this new relationship in fundamental inequality, we get

$$\boldsymbol{\sigma} : \boldsymbol{D} - \rho\left(\frac{\partial \Psi}{dt} + s\frac{dT}{dt}\right) - q \cdot \frac{grad\, T}{T} \geq 0 \tag{3.97}$$

We should point out that in conservation of energy, time derivative of internal energy is multiplied by ρ density. It was dropped in this derivation for consistency of units.

Now we can substitute $\frac{d\Psi}{dt}$ in the fundamental equation.

Meanwhile, we can split D, the deformation gradient tensor, into elastic and plastic components:

$$\boldsymbol{D} = \boldsymbol{D}^e + \boldsymbol{D}^p \tag{3.98}$$

Hence, we can write

$$\boldsymbol{\sigma} : (\boldsymbol{D}^p + \boldsymbol{D}^e) - \rho\left(\frac{\partial \boldsymbol{\Psi}}{\partial \boldsymbol{D}^e} : \boldsymbol{D}^e + \frac{\partial \Psi}{\partial T}\dot{T} + s\dot{T}\right) - q\frac{grad\ T}{T} \geq 0 \tag{3.99}$$

$$\left(\boldsymbol{\sigma} - \rho\frac{\partial \boldsymbol{\Psi}}{\partial \boldsymbol{D}^p}\right) : \boldsymbol{D}^e + \boldsymbol{\sigma} : \boldsymbol{D}^p - \rho\left(s + \frac{\partial \Psi}{\partial T}\right)\dot{T} - q\frac{grad\ T}{T} \geq 0 \tag{3.100}$$

If we assume small strain formulation under pseudo-static loading, the deformation gradient tensor is equal to small strain tensor:

$$\boldsymbol{D}^e + \boldsymbol{D}^P = \boldsymbol{\varepsilon}^e + \boldsymbol{\varepsilon}^p \tag{3.101}$$

In many continuum mechanics textbooks for thermoelastic materials, the following assumption is made for $\dot{T} = 0$:

$$s + \frac{\partial \Psi}{\partial T} = 0 \tag{3.102}$$

However, this is not true. Because any reversible process is imaginary, it cannot happen in real life, without violating the second law of thermodynamics. There is always entropy generation. Thermoelastic laws are defined for an imaginary process where $\dot{\varepsilon}^p = 0, grad\ T = 0,$ and T is arbitrary. Then the following relations define thermoelastic laws:

$$\boldsymbol{\sigma} = \rho\frac{\partial \Psi}{\partial \boldsymbol{\varepsilon}^e} \quad and\ s = -\frac{\partial \Psi}{\partial T} \tag{3.103}$$

Thermodynamic Forces

Thermodynamic potential $\Psi(D, T, V_i, \ldots V_r)$ is a function of thermodynamic state variables. In constitutive modeling, the concept of thermodynamic force is very convenient for material modeling. It is defined by

$$A_i = \rho\left(\frac{\partial \Psi}{\partial V_i}\right) \tag{3.104}$$

where Ψ is the thermodynamic specific free energy potential

$$\Psi = \Psi(\varepsilon, T, V_k) \tag{3.105}$$

Ψ is a function of observable state variables and internal variables.

Of course, the vector of this force is normal to the thermodynamic potential Ψ surface.

Dissipation Potential

In Newtonian mechanics, based on constitutive modeling, a phenomenological dissipation potential is defined (Malvern 1969). This is also the basis for yield surface or plastic potential used in the theory of plasticity. Lemaitre and Chaboche (1990) indicate correctly that the whole problem of modeling a phenomenon [material] lies in the determination of the analytical expressions for the thermodynamic potential Ψ and dissipation potential φ or its dual φ^* and their identification in characteristic experiments. Therefore, constitutive modeling of Newtonian mechanics of materials is a phenomenological process.

In Newtonian mechanics, it is postulated that there exists a dissipation potential (also called pseudo potential in some books). This dissipation potential is expressed as a continuous and convex scalar-valued function of the state variables $\dot{\epsilon}^P$, $-\dot{V}_k$, $-\frac{\vec{q}}{T}$. Of course, this definition limits the potential to be valid for thermo-mechanical loading only. Hence, it can be given by

$$\varphi\left(\dot{\epsilon}, \dot{V}_k, \vec{q}/T\right) \tag{3.106}$$

The state variable V_k is used to define any internal state variable. Dissipation potential is assumed to have a zero value at the origin of the state variables. However, this is not a mathematical requirement. It is possible to define a dissipation function with nonzero value at the origin. The complementary relationships between internal state variables and associated variables can be defined by means of dissipation potential. Normality rule requires that the first derivative of the dissipation potential with respect to state variables is normal to the potential surface and directed outward:

$$\sigma = \frac{\partial \varphi}{\partial \dot{\epsilon}} \quad A_k = -\frac{\partial \varphi}{\partial V_k} \quad \vec{g} = -\frac{\partial \varphi}{\partial\left(\vec{q}/T\right)} \tag{3.107}$$

We should point out that in most continuum mechanics textbooks, dissipation potential is a function of plastic strain not the elastic strain. However, this approach assumes that elastic strain does not cause dissipation and is completely recoverable. However, this approach is not correct, because there is dissipation during elastic response. If there were no dissipation during elastic response, there would not be fatigue under elastic loading.

According to Lemaitre and Chaboche (1990), the thermodynamic forces are the components of the vector $\overrightarrow{\text{grad}\varphi}$, which are normal to the φ surface in the state variable space.

The term "thermodynamic force" is an abstract force concept that is supposed to satisfy laws of thermodynamics. However, because of their empirical nature and simplifications like ignoring ignoring derivative of entropy with respect to

Table 3.2 Dissipation flux variables and dual variables (Lemaitre and Chaboche 1990)

Flux of state variables	Dual variables
$\dot{\epsilon}$	σ
$-V_k$	A_k
$-\frac{\vec{q}}{T}$	$\vec{g} = \overrightarrow{grad}\, T$

displacement, elastic deformations and ignoring many entropy-generating mechanisms, they do not satisfy laws of thermodynamics, in the strict sense. Therefore, it is more appropriate to call these forces pseudo-thermodynamic forces or pseudo-dissipative forces.

It is easier to formulate material model constitutive laws in state variables that can easily be calculated or measured. Therefore, Legendre-Fenchel transformation can be used to define the complementary potential corresponding to dissipation potential. Dissipation variables for thermo-mechanical system are given in Table 3.2.

The corresponding potential$\varphi^*\left(\sigma, A_k, \vec{g}\right)$ is the dual of the dissipation potential $\varphi\left(\dot{\epsilon}, \dot{V}_k, \vec{q}/T\right)$. For details of the Legendre-Fenchel transformation of the dissipation, potential readers are referred to Lemaitre and Chaboche (1990).

The corresponding potential must be differentiable. The normality rule also applies to the corresponding potential. Therefore, constitutive relation can be given by

$$\dot{\epsilon} = \frac{\partial \varphi^*}{\partial \sigma} \tag{3.108}$$

$$-\dot{V}_k = \frac{\partial \varphi^*}{\partial A_k} \tag{3.109}$$

$$-\frac{\vec{q}}{T} = \frac{\partial \varphi^*}{\partial \vec{g}} \tag{3.110}$$

The potentials φ and φ^* must be nonnegative, convex, and zero at the origin to satisfy Clausius-Duhem inequality. According to Lemaitre and Chaboche (1990), the normality rule is sufficient to ensure the satisfaction of the second principle of thermodynamics, but it is not a necessary condition. However, this assumption cannot be substantiated mathematically because it ignores entropy generation due to other mechanisms. This rule applies to "generalized standard materials" under thermo-mechanical loads only. Lemaitre and Chaboche (1985) define standard material as that for which only the first of the above three rules $\dot{\epsilon}^P = \frac{\partial \varphi^*}{\partial \sigma}$ applies. This first relation yields the plasticity or viscoplasticity. The standard material here would only include metals under thermo-mechanical loading.

Determination of dissipation potential in Newtonian mechanics is an empirical process. The dissipation potential represents the energy dissipated in the system by heat, mechanical work at the lattice level, and all other mechanisms. While potential cannot be directly measured, state variables that define the potential can be measured or indirectly calculated. Therefore, these potentials are defined in terms of state variables $\dot{\epsilon}$, $\dot{V}_k$, $\frac{-\vec{q}}{T}$ or dual variables $\sigma, A_k, \vec{g}$.

The dissipation potentials can be written as a function of rate of state variables or as a function of state variables themselves:

$$\varphi\left(\dot{\epsilon}, \dot{V}_k, \frac{\vec{q}}{T}\right) \quad or \quad \varphi(\epsilon, T, V_k) \tag{3.111a}$$

$$\varphi^*\left(\sigma, A_k, \vec{g}\right) \quad or \quad \varphi^*(\epsilon, T, V_k) \tag{3.111b}$$

Decoupling of Intrinsic and Thermal Dissipation

Dissipation process involves reversal mechanisms. Because thermal loading in an unconstrained system does not lead to any dissipation, thermal dissipation is usually treated separately. Therefore, the dissipation potential can be written as the sum of two terms: one representing intrinsic dissipation (such as mechanical work) and the other thermal dissipation:

$$\varphi^* = \varphi_1^*(\sigma, A_k) + \varphi_2^*\left(\vec{q}\right) \tag{3.112}$$

The second law of thermodynamics must be satisfied by each dissipation term. The following inequalities now must be satisfied:

$$\sigma : \epsilon^P - A_k \dot{V}_k = \sigma : \frac{\partial \varphi_1^*}{\partial \sigma} + A_k \frac{\partial \varphi_1^*}{\partial A_k} \geq 0 \tag{3.113a}$$

$$-\vec{g}\frac{\vec{q}}{T} - \vec{g}\frac{\partial \varphi_2^*}{\partial \vec{g}} \geq 0 \tag{3.113b}$$

Time-Independent Dissipation (Instantaneous Dissipation)

When we were discussing laws of thermodynamics, it was stipulated that energy exchange could not happen instantaneously. However, constitutive modeling of materials assuming that the material response is independent of the rate of loading simplifies the modeling and analysis significantly. Therefore, such a simplification yields the theory of plasticity. As stated by Lemaitre and Chaboche (1990) when the dissipation potential $\varphi\left(\dot{\epsilon}^P, \dot{V}_k\right)$ is a positive, homogeneous function of degree one, its dual function $\varphi^*(\sigma, A_k)$ is non-differentiable.

Convexity of the dissipation potential $\varphi^*(\sigma, A_k)$ is checked with an indicator function of

$$\varphi^* = 0 \quad \text{if} \quad f < 0 \quad \dot{\epsilon}^P = 0 \tag{3.114a}$$

$$\varphi^* = +\infty \quad \text{if} \quad f = 0 \quad \dot{\epsilon}^P \neq 0 \tag{3.114b}$$

The indicator function (also referred to as yield function) f also must be convex and defined by the same dual variables, A_k $f(\sigma, A_k)$. Based on normality rule, it is possible to show that

$$\dot{\epsilon}^P = \frac{\partial F}{\partial \sigma}\dot{\lambda} \quad if \begin{cases} \dot{f} = 0 \\ f = 0 \end{cases} \tag{3.115}$$

where F is a potential function (yield surface) which is equal to f in the case of associative plasticity theories and $\dot{\lambda}$ is a scalar multiplier determined by the consistency condition of $\dot{f} = 0$.

Equations describing normality then are given by

$$\dot{\epsilon}^P = \dot{\lambda}\frac{\partial F}{\partial \sigma} = \dot{\lambda}\frac{\partial f}{\partial \sigma} \tag{3.116}$$

$$-\dot{V}_k = \dot{\lambda}\frac{\partial F}{\partial A_k} = \dot{\lambda}\frac{\partial f}{\partial A_k} \tag{3.117}$$

When the plastic strain increment is not normal to F function (yield surface), then the plastic strain is normal to a new function Q, the plastic potential. However, this is called nonassociative plasticity, because the increment of the plastic strain is not normal to the yield function F.

Nonassociative plasticity is governed by

$$\dot{\epsilon}^P = \dot{\lambda}\frac{\partial Q}{\partial \sigma} \quad if \begin{cases} f = 0 \\ \dot{f} = 0, if < 0 \rightarrow \dot{\epsilon}^P = 0 \end{cases} \tag{3.118}$$

The theory of plasticity fundamentals are discussed in Chap. 4, in the literature survey section.

Dissipation Power and Onsager Reciprocal Relations

Earlier in the chapter, we discussed separating internal entropy production into two parts, namely, mechanical work dissipation and the dissipation due to heat conduction.

Specific entropy, *s*, equation was earlier derived as

$$\frac{ds}{dt} + \frac{1}{\rho} div \left(\frac{q}{T}\right) - \frac{1}{T}\left(\frac{du}{dt} - \sigma_{ij}\, D_{ij} + q_{i,i}\right) \geq 0 \tag{3.119}$$

Remembering from the first law of thermodynamics

$$\rho \frac{du}{dt} = \sigma_{ij} D_{ij} + \rho r - q_{i,i} \tag{3.120}$$

We can write the specific entropy rate as

$$\frac{ds}{dt} = \frac{1}{\rho T}\sigma_{ij}\, D_{ij} + \frac{r}{T} - \frac{1}{\rho T} q_{i,i} \tag{3.121}$$

Earlier we also derived the specific entropy production as

$$\gamma \equiv \frac{ds}{dt} - \frac{r}{T} + \frac{1}{\rho T} q_{i,i} - \frac{q_i}{\rho T^2} \cdot grad\ T \tag{3.122}$$

Hence, we can write

$$\gamma = \frac{1}{\rho T}\sigma_{ij}\, D_{ij} - \frac{q}{\rho T^2} \cdot grad\ T \tag{3.123}$$

Thus, internal entropy generation is separated into two parts; the first part is due to mechanical work dissipation and the second part due to irreversible heat conduction. The strong form of Clausius-Duhem inequality requires

$$\frac{1}{\rho T}\sigma_{ij}\, D_{ij} > 0 \tag{3.124}$$

$$-\frac{q_i}{\rho T^2} \cdot grad\ T > 0 \tag{3.125}$$

From here onward, we will use Malvern (1969) description of Onsager reciprocal relations, as follows:

The terms in the internal entropy production are called **generalized irreversible forces X** and **fluxes J.** Selection of force and flux term is arbitrary. However, it is assumed that the dot product of the generalized irreversible force X and the generalized flux vector J must give the dissipation power, per unit mass. Applying this concept to the formulation above, we can assign

Generalized irreversible forces, $\frac{1}{\rho T}\sigma_{ij}$ and $-\frac{1}{\rho T^2} \cdot \text{grad}\ T$

Generalized fluxes D_{ij} *and* q_i

$$\gamma = \boldsymbol{X}^i \cdot \boldsymbol{J} = \boldsymbol{X}^i_m\, \boldsymbol{J}_m \tag{3.126}$$

Constitutive equations (also called phenomenological equations) give the fluxes as functions of the forces or vice versa. Hence, we can write a phenomenological relation between generalized irreversible fluxes and forces as follows:

$$J_m = L_{mk} X^i_k \tag{3.127a}$$

$$X^i_k = a_{km}\, J_m \tag{3.127b}$$

where it is assumed that coefficients satisfy the **Onsager reciprocal relations** (Onsager 1931)

$$L_{mk} = L_{km} \; and \quad a_{km} = a_{mk} \tag{3.128}$$

"It is pointed out that in certain cases the Onsager reciprocal relations must be replaced by $a_{mk}(\boldsymbol{b}) = a_{km}(-\boldsymbol{b})$**,** or $a_{mk} = -a_{km}$, for example in systems effected by a magnetic field **b**" (De Groot 1952; Malvern 1969). Symmetry of the constitutive matrix may also be affected by other factors such as material anisotropy or length-scale affects. Substitution of the phenomenological constitutive relations into the entropy production yields two quadratic equations:

$$\gamma = \frac{1}{T}\, L_{mk} X^i_m X^i_k \geq 0 \text{ and } \gamma = a_{km} J_k J_m \geq 0 \tag{3.129}$$

These quadratic forms must be positive-definite. A necessary and sufficient condition for the positive definiteness of a quadratic equation having a symmetric coefficient matrix with real elements is that all the eigenvalues $|L_{mk} - \lambda\delta_{mk}| = 0 \;\; or \;|$ $a_{km} - \lambda\delta_{km}| = 0$ be positive. The second quadratic equation is called a **dissipation function,** Q:

$$Q = a_{km} J_k J_m \tag{3.130}$$

Based on the above definition, generalized irreversible force can be given by the following relation:

$$X^i_k = \frac{1}{2} \frac{\partial Q}{\partial J_k} \tag{3.131}$$

Ziegler (1963) has shown that these phenomenological relations (Eqs. 3.130 and 3.127a,b and Onsager relation if they have symmetric dissipation function) follow from a principle of maximum rate of entropy production or maximum dissipation power, which he deduces for "quasi-static processes" by a statistical mechanics approach modified to include irreversible processes (Malvern 1969). Table 3.3 lists some of the thermodynamic forces and fluxes for some dissipation mechanisms

Table 3.3 Examples of different thermodynamic forces and corresponding fluxes in some dissipative processes. (After Imanian and Modarres (2018))

Primary mechanism	Thermodynamic force, X	Thermodynamic flux	Examples
Heat conduction	Temperature gradient	Heat flux	Fatigue, creep, wear
Plastic deformation of solids	Stress	Plastic strain	Fatigue, creep, wear
Chemical reaction	Reaction affinity	Reaction rate	Corrosion, wear
Mass diffusion	Chemical potential	Diffusion flux	Wear, creep
Electrochemical reaction	Electrochemical potential	Corrosion rate density	Corrosion

Here we conclude our discussion of some fundamental concepts of thermodynamics as it relates to the unified mechanics theory. For further understanding of thermodynamics concepts, readers are encouraged to refer to thermodynamics' books listed in the references section of this chapter.

References

Boltzmann, L. (1877). Sitzungberichte der Kaiserlichen Akademie der Wissenschaften. Mathematisch-Naturwissen Classe. Abt. II, LXXVI, pp 373–435 (Wien. Ber. 1877, 76:373–435). Reprinted in Wiss. Abhandlungen, Vol. II, reprint 42, p. 164–223, Barth, Leipzig, 1909.

Callen, H. B. (1985). *Thermodynamics and an introduction to Thermostatistics* (2nd ed.). New York: John Wiley & Sons.

Callister, W. D., Jr., & Rethwisch, D. (2010). *Materials science and engineering and introduction*. John Wiley & Sons.

Carnot, S. (1824). Réflexions sur la puissance motrice du feu et sur les machines propres à développer cette puissance (in French). Paris: Bachelier. (First Edition 1824) and (Reissue of 1878).

Clausius, R. (1850). Ueber die bewegende Kraft der Wärme und die Gesetze, welche sich daraus für die Wärmelehre selbst ableiten lassen. *Annalen der Physik, 79*(4), 368–397. https://doi.org/10.1002/andp.18501550403. 500–524. Bibcode:1850AnP...155..500C. See English Translation: On the Moving Force of Heat, and the Laws regarding the Nature of Heat itself which are deducible therefrom. Phil. Mag. (1851), series 4, 2, 1–21, 102–119.

Clausius, R. (1856). On a modified form of the second fundamental theorem in the mechanical theory of heat. *Philosophical Magazine, 12*(77), 81–98. https://doi.org/10.1080/14786445608642141. Retrieved 25 June 2012.

Coleman, B. D. (1964). Thermodynamics of materials with memory. *Archive for Rational Mechanics and Analysis, 17*, 1–46.

Coleman, B. D., & Mizel, V. J. (1964). Existence of caloric equations of state in thermodynamics. *The Journal of Chemical Physics, 40*, 1116–1125.

Coleman, B. D., & Mizel, V. J. (1967). Existence of entropy as a consequence of asymptotic stability. *Archive for Rational Mechanics and Analysis, 25*, 243.

De Groot, S. R. (1952). *Thermodynamics of irreversible processes*. North Holland Publishing.

DeHoff, R. T. (1993). *Thermodynamics in materials science*. McGraw Hill.

Eddington, A. S. (1958). *The nature of the physical world*. University of Michigan Press.

Halliday, D., & Resnick, R. (1966). *Physics*. John Wiley & Sons, Inc.

Imanian, A., & Modarres, M. (2018). A thermodynamic entropy-based damage. Assessment with applications to prognosis and health management. *Structural Health Monitoring, 17*(2), 240–254.

Lemaitre, J., & Chaboche, J. L. (1985). *Mechanics of solid materials*. Cambridge University Press.

Lemaitre, J., & Chaboche, J. L. (1990). *Mechanics of solid materials*. Cambridge University Press.

Malvern, L. E. (1969). *Introduction to the Mechanics of continuous medium*. Prentice-Hall.

Naderi, M., Amiri, M., & Khonsari, M. M. (2009). On the thermodynamic entropy of fatigue fracture. *Proceedings of the Royal Society A: Mathematical, Physical and Engineering, 466* (2114), 423–438.

Onsager, L. (1931). Reciprocal relations in irreversible processes I. *Physics Review, 37*, 405–426.

Planck, M. (1900a). Über eine Verbesserung der Wienschen Spektralgleichung. *Verhandlungen der Deutschen Physikalischen Gesellschaft, 2*, 202–204. Translated in ter Haar, D. (1967). "On an Improvement of Wien's Equation for the Spectrum" (PDF). The Old Quantum Theory. Pergamon Press. pp. 79–81. LCCN 66029628.

Planck, M. (1900b). Zur Theorie des Gesetzes der Energieverteilung im Normalspectrum. *Verhandlungen der Deutschen Physikalischen Gesellschaft, 2*, 237. Translated in ter Haar, D. (1967). "On the Theory of the Energy Distribution Law of the Normal Spectrum" (PDF). The Old Quantum Theory. Pergamon Press. p. 82. LCCN 66029628.

Planck, M. (1900c). Entropie und Temperatur strahlender Wärme [entropy and temperature of radiant heat]. *Annalen der Physik, 306*(4), 719–737. https://doi.org/10.1002/andp.19003060410.

Planck, M. (1900d). Über irreversible Strahlungsvorgänge [on irreversible radiation processes] (PDF). *Annalen der Physik, 306*(1), 69–122. https://doi.org/10.1002/andp.19003060105.

Ramsey, N. F. (1956). Thermodynamics and statistical Mechanics at negative absolute temperatures. *Physics Review, 103*(1), 20.

Sharp, K., & Matschinsky, F. (2015). "Translation of Ludwig Boltzmann's Paper, On the Relationship between the Second Fundamental Theorem of the Mechanical "Theory of Heat and Probability Calculations Regarding the Conditions for Thermal Equilibrium" Sitzungberichte der Kaiserlichen Akademie der Wissenschaften. Mathematisch-1031 Naturwissen Classe. Abt. II, LXXVI 1877, pp 373–435 (Wien. Ber. 1877, 76:373–435)] 1032 Reprinted in Wiss. Abhandlungen, Vol. II, reprint 42, p. 164–223, Barth, Leipzig, 1909" 1033, Entropy, 2015, 17, 1971–2009.

Thomson, W. (1851). On the dynamical theory of heat; with numerical results deduced from Mr. Joule's equivalent of a thermal unit and M. Regnault's observations on steam. *Mathematics and Physics Papers, 1*, 175–183.

Truesdell, C., & R. Toupin, R. (1960). *The Classical Field Theories. In S. Flügge (Ed.), Principles of classical mechanics and field theory/Prinzipien der Klassischen Mechanik und Feldtheorie* (pp. 226–858). Springer.

Truesdell, C., & Noll, W. (1985). *The non-linear field theories of Mechanics*. Springer.

Tu, K. N., & Gusak, A. M. (2019). A unified model of mean-time-to-failure for electromigration, thermomigration, and stress-migration based on entropy production. *Journal of Applied Physics, 126*, 075109.

Yun, H., & Modarres, M. (2019). Measures of entropy to characterize fatigue Damage in metallic materials. *Entropy, 21*(8), 804.

Ziegler, (1963). Some extremum principles in irreversible thermodynamics with application to continuum mechanics. In I. N. Sneddon and R. Hill (Eds.), *Progress in solid mechanics* (pp. 91–193). North Holland, Amsterdam.

Chapter 4
Unified Mechanics Theory

The term unified is used to describe the unification of universal laws of motion of Newton and laws of thermodynamics at ab-initio level. As we discussed in earlier chapters, Newton's laws do not account for energy loss. They only govern what happens to a system in initial moment a load is applied. However, the laws of thermodynamics control what happens after the initial moment over time. Historically, continuum mechanics is based on the laws of Newton only and phenomenological test data fit empirical models that are supposed to satisfy thermodynamics' laws introduce energy loss... In the next section, we will discuss the earlier work on using thermodynamics in continuum mechanics.

4.1 Literature Review of Use of Thermodynamics in Continuum Mechanics

Efforts to unify Newtonian mechanics and thermodynamics have been attempted since the first formulation of laws of thermodynamics. Because Newton's laws do not account for dissipation of energy, researchers used different methods to incorporate energy dissipation into mechanics formulations to be able to formulate degradation, fracture, and fatigue and life span predictions. Most of these efforts have been based on phenomenological curve fitting methods, where a polynomial potential is used to curve fit the fracture, or degradation test data. There are no unifying theme among these models other than fitting a polynomial to a test data and using it in conjunction with Newton's laws. For the sake of completeness, we will summarize some of them. While we do not claim to include every effort, we will summarize the most well-known work in the field.

Lagrangian mechanics (1781) [long before Thermodynamics laws were written] can be considered the first attempt to include nonconservative forces into Newton's laws. Quoting from Haddad's (2017) seminal work on literature review of

C. Basaran, *Introduction to Unified Mechanics Theory with Applications*,
https://doi.org/10.1007/978-3-030-57772-8_4

thermodynamics, "In an attempt to generalize classical thermodynamics to nonequilibrium thermodynamics, Onsager (1931, 1932) developed reciprocity theorems for irreversible processes based on the concept of a local equilibrium that can be described in terms of state variables that are predicated on linear approximations of thermodynamic equilibrium variables. Onsager's theorem pertains to the thermodynamics of linear systems, wherein a symmetric reciprocal relation applies between forces and fluxes.

In particular, the force exerted by the thermal gradient causes a flow or flux of matter in thermo-diffusion. Conversely, a concentration gradient causes a heat flow, an effect that has been experimentally verified for linear transport processes involving thermos-diffusion, thermoelectric, and thermomagnetic effects. Classical irreversible thermodynamics as originally developed by Onsager characterizes the rate of entropy production of irreversible processes as a sum of the product of fluxes with their associated forces, postulating a linear relationship between the fluxes and forces. The thermodynamic fluxes in the Onsager formulation include the effects of heat conduction, flow of matter (i.e., diffusion), mechanical dissipation (i.e., viscosity), and chemical reactions. This thermodynamic theory, however, is only correct for near thermodynamic equilibrium processes wherein a local and linear instantaneous relation between the fluxes and forces holds.

Building on Onsager's classical irreversible thermodynamic theory, Prigogine (1955, 1961, 1968) developed a thermodynamic theory of dissipative nonequilibrium structures. This theory involves kinetics describing the behavior of systems that are away from equilibrium states. Prigogine's thermodynamics lacks [fundamental] functions of the system state, and hence his concept of entropy for a system away from [thermodynamic] equilibrium does not have a total differential. Furthermore, Prigogine's characterization of dissipative structures is predicated on a linear expansion of the entropy function about a particular [thermodynamic] equilibrium, and hence is limited to the neighborhood of the [thermodynamic] equilibrium. This is a severe restriction on the applicability of this theory. In addition, his entropy cannot be calculated nor determined (Prigogine 1971, 1977; Haddad 2017).

Prigogine's work (1954, 1955, 1957, 1961, 1971, 1977) on dissipative structures and their role in thermodynamic systems far from equilibrium won him the Nobel Prize in Chemistry in 1977. In engineering mechanics, most of our states are near thermodynamic equilibrium point. Therefore, Prigogine's work is important to understand. Quoting from Prigogine's own Wikipedia webpage, "Prigogine proved that dissipation of energy in chemical systems result in the emergence of new structures due to internal self-re-organization. In his 1955 text, Prigogine drew connections between dissipative structures and the Rayleigh-Bénard instability, and the Turing mechanism, Turing (1952), which describes the way in which patterns in nature such as stripes and spots can arise naturally out of a homogeneous uniform state. Rayleigh-Bénard instability is a type of natural convection, occurring in a plane horizontal layer of fluid heated from below, in which the fluid develops a regular pattern of convection cells known as Bénard cells (Getling 1998; Koschmieder 1993). Turing mechanism describes the way in which patterns in nature such as stripes and spots can arise naturally out of a homogeneous uniform state (Turing 1952)."

"Prigogine's dissipative structure theory led to pioneering research in self-organizing systems, as well as philosophical inquiries into the formation of complexity on biological entities and the quest for a creative and irreversible role of time in the natural sciences. Prigogine's formal concept of self-organization was used also as a "complementary bridge" between General Systems Theory and thermodynamics, conciliating the cloudiness of some important systems theory concepts with scientific rigor. Quoting from Prigogine's own Wikipedia page, definitions of self-organization and general systems theory are described as, "self-organization", also called (in the social sciences) spontaneous order, is a process where some form of overall order arises from local interactions between parts of an initially disordered system. The process can be spontaneous when sufficient energy is available, not needing control by any external agent. It is often triggered by random fluctuations, amplified by positive feedback. The resulting organization is wholly decentralized, distributed over all the components of the system. As such, the organization is typically robust and able to survive or self-repair substantial perturbation. Chaos theory discusses self-organization in terms of islands of predictability in a sea of chaotic unpredictability. General systems theory is the interdisciplinary study of systems. A system is a cohesive conglomeration of interrelated and interdependent parts that is either natural or fabricated. Every system is delineated by its spatial and temporal boundaries, surrounded and influenced by its environment, described by its structure and purpose or nature and expressed in its functioning. In terms of its effects, a system can be more than the sum of its parts if it expresses synergy or emergent behavior. Changing one part of the system usually affects other parts and the whole system, with predictable patterns of behavior. For systems that are self-learning and self-adapting, the growth and adaptation depend upon how well the system is adjusted with its environment. Some systems function mainly to support other systems by aiding in the maintenance of the other system to prevent failure. The goal of general systems theory is systematically discovering a system's dynamics, constraints, conditions, and elucidating principles (purpose, measure, methods, tools, etc.) that can be discerned and applied to systems at every level of nesting, and in every field for achieving optimized equifinality (Beven 2006)." Quoting from Wikipedia to define, "equifinality which is the principle that in open systems a given end state can be reached by many potential means. Also meaning that a goal can be reached by many ways. The same final state may be achieved via many different loading paths. However, in closed systems, a direct cause-and-effect relationship exists between the initial condition and the final state of the system. Biological and social systems are open systems, however, operate quite differently". Simply, the idea of equifinality suggests that similar results may be achieved with different initial conditions and in many different ways.

Entropy has also been used beyond positive sciences. The seminal paper on the topic was published by Jaynes (1957) who proved that statistical mechanics, which is based on Boltzmann's equation, could be generalized to information theory independent of experimental verification. Jaynes (1957) using von Neumann–

Shannon definition of entropy as a measure of uncertainty represented by a probability distribution, showed that entropy becomes the primitive concept more fundamental than energy. Jaynes (1957) also proved that thermodynamic entropy is identical with the information-theory entropy of the probability distribution except for the presence of Boltzmann's constant.

Valanis (1971) established irreversibility and existence of entropy as a state function in Newtonian mechanics formulation. Valanis (1971) was able to prove this for reversible and irreversible systems and processes, irrespective of the constitutive equations of the system. He postulated that, in the case of a reversible system, entropy is a function of the deformation gradients (strains) and temperature; in the case of an irreversible system, it is also a function of n internal variables necessary to describe the irreversibility of the system. Valanis based his proof of the existence of entropy as a consequence of the integrability of the differential form of the first law using an extended form of the Caratheodory conjecture, to the effect that in the neighborhood of a thermodynamic state there exist other states, which are not accessible by processes, which are reversible and adiabatic. Based on this work Valanis (1971) proposed the endochronic plasticity theory, which deals with the plastic response of materials by means of memory integrals, expressed in terms of memory kernels. Formulation of this theory is based on thermodynamics concepts and provides a unified point of view to describe the elastic-plastic behavior of material, since it places no requirement for a yield surface and "loading function" to distinguish between loading and unloading. A key ingredient of the theory is that the deformation history is defined with respect to a deformation memory scale called intrinsic time. In the original version of the endochronic theory, proposed by Valanis (1971), the intrinsic time was defined as the path length in the total strain space. The so-called endochronic theory violates the second law of thermodynamics and leads to constitutive relations, which characterize inherently unstable materials (Rivlin 1981). Aiming at the correction of this deficiency a new version of the endochronic theory was developed by Valanis and Komkov (1980) in which the intrinsic time was defined as the path length in the plastic strain space.

Rice (1971) used internal-variable thermodynamic formalism for description of the microstructural rearrangements to characterize metal plasticity. Rice (1971) postulated that the theoretical foundations of constitutive relations at finite strain for metals exhibiting inelasticity is a consequence of specific structural rearrangements on the microscale of the material, such as metals deforming plastically through dislocation motion. Rice (1971) is actually a generalization of his earlier work Kestin and Rice (1970), where it is assumed that each microstructural arrangement proceeds at a rate governed by its associated thermodynamic force. This work became the framework for the time-dependent inelastic behavior in terms of a flow potential, and reduces to statements on normality of plastic strain increment to yield surface in the time-independent case. Rice (1971) approach assumes inelastic deformation under macroscopically homogeneous strain and temperature as a sequence of constrained equilibrium states. Using equilibrium thermodynamic formalism and thermodynamic potentials, Rice (1971) relates changes in the local structural rearrangements to corresponding changes in the macroscopic stress or strain states. However, Rice (1971) assumes that a discrete set of scalar internal variables characterize the state of internal rearrangement. Each internal variable characterizes a specific structural rearrangement. Unfortunately, this classification does not satisfy the second law of thermodynamics. According to second law of thermodynamics, not the individual internal variables like stress or strain, but only the fundamental relation [entropy generation rate] determines the new structural rearrangement. Rice (1971) also assumes that if various equilibrium states are considered each corresponding to the same set of values for the thermodynamic internal variables, then neighboring states are related by the usual laws of thermo-elasticity. This later assumption also violates the second law of thermodynamics, because according second law of thermodynamics even thermo-elastic deformation leads to irreversible entropy generation. Easiest way to explain this is fatigue under elastic loading or molecular dynamics simulations under elastic loading. According to traditional thermo-elasticity assumption if there is no inelastic deformation, thermodynamics internal state variables do not change. As a result, materials could never fatigue under elastic loading. Of course, this violates the second law of thermodynamics, because there is no reversible thermodynamic process in metals.

Rice (1971) postulates that "if neighboring constrained equilibrium states corresponding to different sets of internal variables are considered, we must write" [using Rice's variables without adopting to our notation]

$$V^0S : \delta E - f_\alpha \delta\xi_\alpha + \theta\delta\left(V^0\eta\right) = \delta\left(V^0u\right) \tag{4.1}$$

In the formulation V^0 denotes volume at some reference state at a temperature θ_0, S is Kirchhoff stress (symmetric), δE is increment of macroscopic Lagrange (or material) strain tensor, f_α defines thermodynamic forces ($f_1, f_2, \ldots, f_n$ acting on internal variables, $\delta\xi_\alpha$ is set of (total number is unspecified as n) thermodynamic internal state variables that characterize the state of internal rearrangement, θ is temperature, $\delta(V^0\eta)$ is the increment of entropy, and $\delta(V^0u)$ is the increment of internal energy. This is the most important equation in Rice (1971) theoretical framework. The rest of the theoretical framework is based on this equation. This equation finally leads to defining a yield surface as a potential, which is the most important ingredient for the theory of incremental plasticity. Unfortunately, in this equation given by Rice (1971) there is no relation between entropy and thermodynamic forces ($f_1, f_2, \ldots, f_n$) acting on internal variables in the formulation. Derivative of entropy with respect to these thermodynamic forces are considered zero and entropy does not change as result of these forces. As a result, in practice defining these thermodynamic forces and internal variables becomes a trial-error process (or even an art). This point was also emphasized in Chap. 3, when we discussed thermodynamic potential. Because of this problem yield surface is an empirical function with different coefficients for different materials for different loading paths for different temperatures, different strain rates, and different length scales or geometries of the materials. Of course, it is possible to come up with different yield surfaces for the same material, using different constants. Therefore, it would be more accurate to call these thermodynamic forces as pseudo-thermodynamics forces. Since they are empirical and many are not directly related to laws of thermodynamics. While Rice (1971) interprets $f_\alpha \delta\xi_\alpha$ as the work of increments of internal variables, in practice when we obtain a yield surface from experiments; many of our exponents or "material constants" have no physical meaning to justify as an internal thermodynamic variable. Because, two different scientists can have significantly different number of constants to define a yield surface for the same material. Our argument becomes clearer in the following formulation by Rice (1971)

Free energy ϕ and its Legendre transform ψ (complementary energy) are given by the following relations

$$\phi = \phi(E,\theta,\xi) = u - \theta\eta \quad \text{and} \quad \psi = \psi(S,\theta,\xi) = E : \frac{\partial \phi}{\partial E} - \phi = E : \frac{\partial \phi}{\partial E} - u + \theta\eta \tag{4.2}$$

Based on definitions in Eq. (4.2) and assuming that entropy is constant, the variation of complementary energy can be written as,

$$\delta\psi = E : \delta S + \frac{1}{V^0} f_\alpha \delta\xi_\alpha + \eta\delta\theta \tag{4.3}$$

Of course, in real life when variation is applied in addition to stress δS, internal variables $\delta\xi_\alpha$, and temperature $\delta\theta$, entropy will also change. Nevertheless, here entropy change is assumed zero, $\theta\delta\eta = 0$. Assuming that internal variables $\delta\xi_\alpha = 0$ are also constant, Rice (1971) derives at thermo-elastic constitutive equation based on Newtonian mechanics, where it is assumed that no elastic deformation can generate entropy.

$$E = \frac{\partial \psi(S,\theta,\xi)}{\partial S} \quad \text{and} \quad S = \frac{\partial \phi(E,\theta,\xi)}{\partial E} \tag{4.4}$$

Thermodynamics forces associated with internal variables are defined by

$$f_\alpha = V^0 \frac{\partial \psi(S,\theta,\xi)}{\partial \xi_\alpha} = -V^0 \frac{\partial \phi(E,\theta,\xi)}{\partial \xi_\alpha} \tag{4.5}$$

Using Maxwell's relations, which are set of equations derivable from the symmetry of second derivatives of potentials, Rice (1971) arrives at

$$\frac{\partial E(S,\theta,\xi)}{\partial \xi_\alpha} = \frac{1}{V^0} \frac{\partial f_\alpha(S,\theta,\xi)}{\partial S} \tag{4.6}$$

Equation (4.6) serves as a central equation in the rest of the paper in the development of the theoretical foundations of inelastic constitutive theory. A flow potential for the inelastic strain rate is proposed by postulating that at any given temperature and pattern of internal rearrangement within the material, the rate at which any specific structural rearrangement occurs is fully determined by the thermodynamic forces. The concept, used by Rice (1971), is identical to basis for the unified mechanics theory where pseudo-dynamic forces are replaced with the fundamental equation. It is assumed that any specific microstructural configuration can be fully determined by the thermodynamic force associated with that microstructural rearrangement, as follows:

$$\dot{\xi}_\beta \text{ is a function of } f_\beta, \theta, \xi \text{ (for } \beta = 1, 2, \ldots, n) \tag{4.7}$$

Rice (1971) further postulates that the current temperature and pattern of internal microstructural rearrangement may enter the kinetic equations as parameters, but the influence of the macroscopic stress state on a given microstructural rearrangement appears only through the fact that the associated force is dependent on stress. Author acknowledges that this is not the most general class of kinetic equations; however, it does represent conventional metal plasticity behavior where the associated shear stress governs slip on a crystallographic system or at the discrete dislocation. The force on a given segment of dislocation line governs its motion. Rice (1971) recast the kinematic equations in an integral form as follows,

$$\dot{\xi}_\beta = \frac{\partial}{\partial f_\beta} \int_0^f \dot{\xi}_\alpha(f, \theta, \xi)\, df_\alpha \tag{4.8}$$

where the integral is carried out at fixed values of θ and ξ and defines a point function since each term in the integrand is an exact differential. Rice (1971) further postulates that thermodynamic forces may be viewed as functions of the macroscopic stress, S, θ,and ξ and then defines a function

$$\Omega(S, \theta, \xi) = \frac{1}{V^0} \int_0^f \dot{\xi}_\alpha(f, \theta, \xi)\, df_\alpha \tag{4.9}$$

Stress derivative of this function is given by

$$\frac{\partial \Omega(S, \theta, \xi)}{\partial S} = \frac{1}{V^0} \dot{\xi}_\alpha(f, \theta, \xi) \frac{\partial f_\alpha(S, \theta, \xi)}{\partial S} \tag{4.10}$$

Rice (1971) elegantly shows that the right-hand side of this equation is equal to the inelastic strain rate in the following way.

Let δE be the difference (variation) in strain between the neighboring constrained equilibrium states, differing by δS, $\delta\theta$, $\delta\xi$. It is assumed that the variations in the internal variables correspond to variations in the macroscopic strain. An inelastic or plastic portion $(\delta E)^p$ of the strain difference is defined as that part which would result from the change in internal variables if stress and temperature were held fixed:

$$(\delta E)^p = \frac{\partial E(S,\theta,\xi)}{\partial \xi_\alpha}\delta\xi_\alpha = \frac{1}{V^0}\frac{\partial f_\alpha(S,\theta,\xi)}{\partial S}\delta\xi_\alpha \tag{4.11}$$

Similarly, an elastic (or thermo-elastic) portion $(\delta E)^e$is defined as that which would result from the change in stress and temperature, if the other internal variables were held fixed. As a result, the following relations can be written

$$\delta E = (\delta E)^e + (\delta E)^p \tag{4.12}$$

From complementary energy we can write

$$(\delta E)^e = \frac{\partial^2 \psi}{\partial S \partial S} : \delta S + \frac{\partial^2 \psi}{\partial S \partial \theta}\delta\theta \tag{4.13}$$

Based on the viewpoint of the classical theory of irreversible processes, e.g., De Groot and Mazur (1962) and Rice (1971) assumes that macroscopically homogeneous deformation processes may be suitably approximated as sequence of constrained thermodynamic equilibrium states, each fully characterized by values for E, θ, ξ at the corresponding instant. As a result, Rice (1971) assumes that all the preceding relations are valid during this process. Thus, the following relations can be written

$$\dot{E} = (\dot{E})^e + (\dot{E})^p \tag{4.14}$$

$$(\dot{E})^p = \frac{1}{V^0}\frac{\partial f_\alpha(S,\theta,\xi)}{\partial S}\delta\dot{\xi}_\alpha \tag{4.15}$$

An analogous expression for the elastic portion can be written from Eq. (4.13).

Therefore, Ω is called a flow potential (or more commonly known as a yield surface). According to normality rule, it can be shown that the inelastic portion of the strain-rate vector is normal to the surface of constant flow potential in stress space.

$$(\dot{E})^p = \frac{\partial \Omega(S,\theta,\xi)}{\partial S} \tag{4.16}$$

Rice (1971) defines thermodynamics restriction on the formulation in the following way. Following the earlier stated assumption that a material is taken from one constrained thermodynamic equilibrium state to another by an irreversible process extending from time t^1–t^2, then the first and second laws of thermodynamics in classical form dealing exclusively with the comparison of constrained equilibrium states are given by

$$\int_{t^1}^{t^2}\left\{\int_A T\cdot\dot{z}\,dA+\int_A q dA\right\}dt=V^0\left[u^2-u^1\right] \tag{4.17}$$

$$\int_{t^1}^{t^2}\left\{\int_A (q/\theta)dA\right\}dt\leq V^0\left[\eta^2-\eta^1\right] \tag{4.18}$$

where T is the surface stress vector (traction), q is the heat supplied to the surface per unit area (positive inward), $\dot{z}$ is the deformation rate, u is the internal energy, and η is the entropy. Time integral from t^1 to t^2 follow the irreversible process. It is important to point out that elastic deformation here is considered a reversible process; hence, fatigue under elastic stresses is not possible. Alternatively, during a molecular dynamics simulations under elastic loading atoms return to their original lattice site. Based on earlier work by Kestin and Rice (1970), following the classical approach to irreversible processes for a macroscopically homogeneous system with internal variables, Rice (1971) views all actual processes as a sequence of constrained equilibrium states. The work of surface stresses can be written in terms of internal stresses and strains as $V^0 S:\dot{E}$ for contrained equilibrium states and the temperature is assumed uniform in the material. Then the first and second laws of equilibrium thermodynamics can be generalized for any actual homogenous process as follows

$$S:E+\dot{Q}=\dot{u}$$
$$Q=\frac{1}{V^0}\int_A q\,dA\leq\theta\dot{\eta} \tag{4.19}$$

where Q is the total heat supply rate per unit reference volume. Rice (1971) further defines entropy generation rate due to internal variables and conjugate thermodynamic forces as

$$\sigma=\frac{1}{\theta V^0}f_\alpha\dot{\xi}_\alpha\geq 0 \tag{4.20}$$

Then, Rice (1971) defines the total change in entropy as summation of Q, the total heat supply rate per unit reference volume plus the entropy generation due to internal variable and thermodynamic forces as

$$Q+\theta\sigma=\theta\dot{\eta} \tag{4.21}$$

Q is positive inward, in Rice (1971) formulation, which we are quoting without modification. It is assumed that heat entering from outside does not increase the irreversible entropy generation. On the other hand, heat generated in the system and dissipating outward leads to irreversible entropy generation.

The second law requirement that non-negative work rate of associated forces on internal variables is given by

$$\dot{\xi}_\alpha \frac{\partial \phi(E, \theta, \xi)}{\partial \xi_\alpha} \leq 0 \tag{4.22}$$

Remember in Eq. (4.5) f_α was defined as

$$f_\alpha = -V^0 \frac{\partial \phi(E, \theta, \xi)}{\partial \xi_\alpha} \tag{4.23}$$

Rice (1971) concludes by, "That is, the rates of internal arrangement actually occurring during a process must be such that the free energy would decrease if strain and temperature were held fixed at current values."

In Newtonian mechanics, Eqs. (4.22) and (4.23) are primarily used to impose the requirement that flow potential must be convex and smooth surface, and but are not actually included in Newtonian laws to degrade the energy. Degradation is imposed by empirical formulation with a damage potential obtained from experimental data.

Later, Rice (1977) derived thermodynamic restrictions on the quasi-static growth, or healing of Griffith cracks. However as presented they were nothing beyond global restriction on the detailed molecular kinetics of crack growth. Therefore, there was no attempt to integrate these restrictions into Newton's laws. Other work by Rice (1977) on use of thermodynamics in theory of plasticity and fracture mechanics was based on or a derivative of his work discussed here.

Bazant (1972) used thermodynamics for modeling mechanics of interfaces in concrete structure, where a thermodynamically consistent formulation of interacting continua with surfaces based on surface thermodynamics was proposed. He accounted for creep, shrinkage and delayed thermal dilation and their interaction with adsorbed water layers confined between two solid adsorbent surfaces

Kijalbaev and Chudnovsky (1970) and Chudnovsky (1973, 1984) proposed a probabilistic model of the fracture process unifying the phenomenological study of long-term strength degradation of materials, fracture mechanics and statistical approach to fracture mechanics. Chudnovsky (1984) used irreversible thermodynamics to model the deterministic side of the failure phenomenon and stochastic calculus to account for the failure mechanisms controlled by "chance," particularly the random roughness of fracture surfaces. Kijalbaev and Chudnovsky (1970) and Chudnovsky (1973, 1984) derived the entropy production for an elastic medium with damage. Citing Swalin (1972) on the invariance of the entropy jump during a phase transition with respect to stresses and temperature, Chudnovsky (1973) hypothesized that the entropy jump invariance can be used to predict the local failure. Author proposed pairing of a phenomenological critical damage parameter and entropy jump, ΔS, requirement as a local failure criterion. It is important to point out that Chudnovsky proposed calculating entropy generation in a fracture mechanics problem for all mechanisms, damage, translation, rotation, isotropic expansion, and distortion of the active crack zone.

Klamecki (1980a, b, 1984) proposed an entropy-based model of plastic deformation energy dissipation in sliding. In the entropy production rate author included both microstructural changes and internal heat generation. This is the first paper in the literature we could find where partial derivative of entropy was taken with respect to internal energy, microstructure generated entropy, surface area, and mass. As a result, Klamecki (1980a, 1984) postulated that temperature component of internal energy U, microstructural energy, G, surface energy, and chemical potential are represented in entropy calculations, and defined entropy by

$$S = S(U, G, A, M) \tag{4.24}$$

$$dS = \frac{\partial S}{\partial U} dU + \frac{\partial S}{\partial G} dG + \frac{\partial S}{\partial A} dA + \frac{\partial S}{\partial M} dM \tag{4.25}$$

And

$$T\, dS = dU + \phi dG + \gamma dA + \mu dM \tag{4.26}$$

where the temperature, internal energy, microstructural energy, surface energy, and chemical potential are represented by T, U, ϕ, γ, and μ, respectively. Klamecki (1980a, b, 1984) defined the conservation of energy by

$$dU = \delta Q + \delta W \tag{4.27}$$

where Q represents heat and W represents work. δ is used to imply the equation is for an incremental process [variation]. Work, W, is assumed to be due to plastic deformation only, and then taken as stress multiplied by plastic strain. The heat, Q, has two components, one is assumed to be due to internal heat generation, R, in the system and heat flow into the system due to temperature gradient. As a result, internal energy can be written as

$$dU = K_i (\nabla T)_i + R + M\rho^{-1}\tau_{ij} d\varepsilon_{ij} \tag{4.28}$$

where K_i is the thermal conductivity, ρ is the mass density. Substituting internal energy given by Eq. (4.28) in entropy Eq. (4.26) leads to

$$TdS = K_i(\nabla T)_i + R + M\rho^{-1}\tau_{ij}d\varepsilon_{ij} + \phi dG + \gamma dA + \mu dM \tag{4.29}$$

As Klamecki (1984) points out this definition of entropy is very valuable since it contains, or may be further generalized to contain, all the energy dissipation mechanisms in sliding bodies. Of course, there is nothing in the equation that limits it to siding bodies only. Klamecki (1984) speculated that there are three regimes of system behavior, which are of interest when the intent is to describe changes in the system by the use of entropy generation. At thermodynamic equilibrium, the system entropy is at maximum, and the change in entropy with time is zero. When the system is not in equilibrium because of a continual supply of energy to it, it can be near equilibrium or far from equilibrium state. Klamecki (1984) asserted that the process occurring in the nonequilibrium states could be analyzed by studying the entropy production in the system. The entropy created in the system is given by

$$TdS = R + M\rho^{-1}\tau_{ij}d\varepsilon_{ij} + \phi dG + \gamma dA + \mu dM \tag{4.30}$$

These terms have the general form of a thermodynamic force F, multiplied by a corresponding flux, J. Then entropy production is

$$S = \sum_{i=1}^{n} F_i J_i \tag{4.31}$$

where n is the number of internal entropy generation mechanisms. Entropy production rate is defined by

$$\dot{S} = \frac{dS}{dt} = \sum_{i=1}^{n}\left(F_i\frac{dJ_i}{dt} + J_i\frac{dF_i}{dt}\right) \tag{4.32}$$

Here, Klamecki (1984) postulates that the first term in the summation represents the entropy generation rate at far from the equilibrium state where thermodynamic force does not change by time. The second term represents the entropy generation near equilibrium. As a result, the summation can be summarized as follows:

$$\dot{S} = \dot{S}_J + \dot{S}_F \tag{4.33}$$

In near equilibrium, entropy production rate $\dot{S}_F$ will continuously get smaller, according to second law of thermodynamics. Thermodynamic forces change in such a way that a stationary state of minimum rate of entropy generation results at equilibrium, as stipulated by second law of thermodynamics.

$$\dot{S}_F = 0 \tag{4.34}$$

Klamecki (1984) attributes this principle, the second law, to Glansdorff and Prigogone (1971), for reasons beyond our understanding.

As stated by Klamecki (1984) far from thermodynamic equilibrium, no widely accepted entropy production evolution criterion has been formulated so far. Laws of thermodynamics are valid around equilibrium points. Since in mechanics all our problems are around equilibrium points and solved as a sequence of constrained equilibrium points, not having laws governing far from equilibrium is not a concern.

Recently, Evans et al. (1993) proposed the fluctuation theorem (FT). Laboratory experiment that verified the validity of the FT was carried out in 2002. Where a plastic bead was pulled through a solution by a laser. Fluctuations in the velocity were recorded that were opposite to what the second law of thermodynamics would dictate for macroscopic systems Wang et al. (2002), Chalmers (2016), and Searles and Evans (2004).

Fluctuation theorem does not state or prove that the second law of thermodynamics is wrong or invalid. The second law of thermodynamics is valid for macroscopic systems at equilibrium or near equilibrium. Rivas and Martin-Delgado (2017) have also encountered a partial violation of the second law of thermodynamics in a quantum system known as Hofstadter lattice. Ostoja-Starzewski (2016) and Ostoja-Starzewski and Raghavan (2016) proved violations of the second law are relevant as the length and/or time scales become very small. The second law then needs to be replaced by the fluctuation theorem, and mathematically the irreversible entropy is a sub martingale. As indicated above these are far from thermodynamic equilibrium cases. This partial violation has no place within the framework of classical thermodynamics because it is a spontaneous event that does not effect the laws of thermodynamics at the macroscale at equilibrium or near equilibrium.

Klamecki (1984) summarized these concepts in a very simple figure. Author postulates that entropy production is a function of l thermodynamic variables and so can be represented by a surface in $(l + 1)$ dimensional space. For ease of plotting only two variables are actually used in the Fig. 4.1. The equilibrium state is the state of

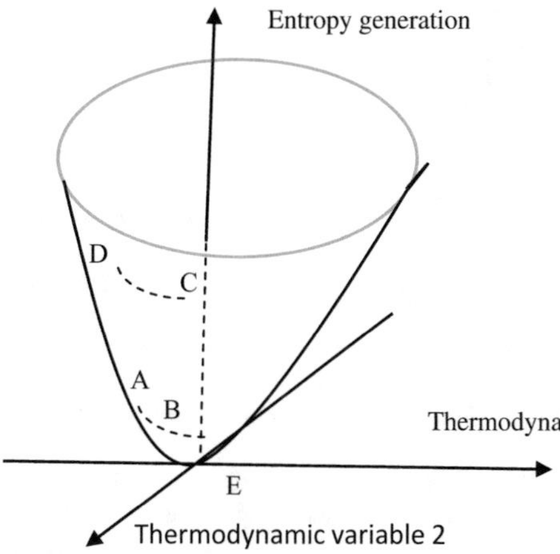

Fig. 4.1 Graphical representation of entropy generation associated with system states at equilibrium (state E), near equilibrium (state B), and far from equilibrium (state D) as a function of thermodynamic state variables 1 and 2. After Klamecki (1984)

maximum entropy and zero entropy generation and is represented by point E in the figure. If the system is maintained in a nonequilibrium state by supplying energy to it, two processes are usually considered. Near equilibrium, the system will evolve from its initial state A to a state B of minimum rate of entropy generation. If the system is far from equilibrium, it may move from its initial state C to a state D, but at present, no description of this state or the process in terms of entropy production is available. The only accepted governing principle is energy conservation.

Whaley (1983) proposed a model for fatigue crack nucleation using the irreversible thermodynamics to quantify the damage caused by plastic strain. Whaley (1983) model is based on the hypothesis that entropy gain which results from dynamic irreversible plastic strain is a material constant. This was later proven experimentally by Naderi et al. (2010) and, Imani and Modarres (2015). Whales (1983) postulated that structural damage by fatigue could be quantified by a single material parameter, the critical entropy threshold of fracture. Whaley (1983) postulated that the critical entropy threshold of fracture is therefore related to the irreversible part of the fracture energy. The critical entropy threshold of static fracture is a random variable and the variability can be quantified by a confidence interval. According Whaley (1983), the confidence interval for the entropy threshold just comes from the variance of the plastic strain. It is noteworthy that Whaley (1983) assumed that only plastic strain generates irreversible entropy, of course that is not accurate.

Naderi et al. (2010) were able to prove experimentally that total cumulative entropy generation is constant at the time of failure and is independent of sample geometry, load, and loading frequency. They named this critical entropy value Fatigue Fracture Entropy (FFE). Figure 4.2 shows experimental fatigue fracture

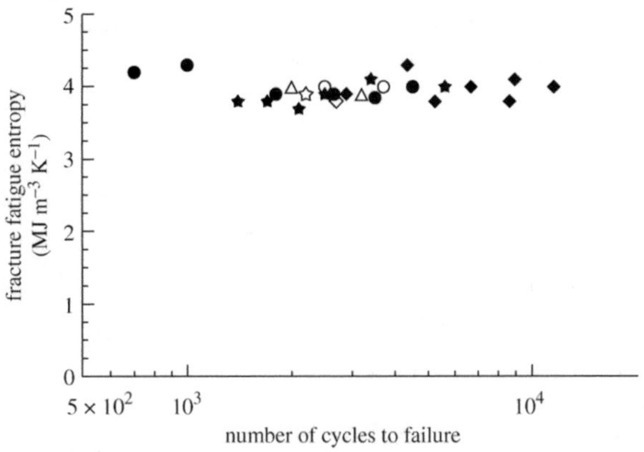

Fig. 4.2 Fatigue fracture entropy versus the number of cycles to failure for different bending fatigue tests of Al 6061-T6 with different specimen thickness, frequencies, and displacement amplitudes. Fatigue fracture entropy remains at roughly 4 MJ m^{-3} K^{-1}, regardless of thickness, load, and frequency. Displacement amplitude varied from 25 mm to 50 mm. Filled circle, thickness = 6.35 mm, f = 10 Hz; filled diamond, thickness = 3.00 mm, f = 10 Hz; filled star, thickness = 4.82 mm, f = 10 Hz; unfilled circle, thickness = 6.35 mm, f = 6.5 Hz; unfilled triangle, thickness = 4.82 mm, f =12.5 Hz; unfilled star, thickness = 6.35 mm, f = 6.5 Hz; unfilled diamond, thickness = 6.35 mm, f = 12.5 Hz. After Naderi et al. (2010)

entropy invariance with respect to geometry and loading frequency. According to the total accumulated entropy of metals, undergoing repeated cyclic load as it reaches the point of fracture is a constant value, independent of load amplitude, geometry, size of specimen, frequency, and stress state; Naderi et al. (2010) experimentally validated invariance of experimental fatigue fracture entropy with respect to loading path and displacement loading amplitude for Al 6061-T6.

Fracture fatigue entropy remains at about 4 MJ m^{-3} K^{-1} for both tension-compression and bending fatigue. Displacement amplitude is varied from 25 to 50 mm. Filled square, tension-compression; filled circle, bending; filled star, torsion. After Naderi et al. (2010).

Amiri and Khonsari (2012) and Jang and Khonsari (2018) also did extensive experiments to show that fatigue fracture entropy remains constant independent of loading condition, loading frequency, and sample geometry. Their samples were AISI 1018 carbon steel and Al 7075-T6. Liakat and Khonsari (2015) measured fatigue fracture entropy of un-notched and V-notched specimens. They observed that fatigue fracture entropy remains constant.

Imanian and Modarres (2015, 2018) and Yun and Modarres (2019) experimentally proved the concept of using entropy as a degradation metric, and authors discussed the entropic characterization of the corrosion-fatigue degradation mechanism. They proposed an entropy-based damage prognostics and health management technique for integrity assessment and remaining useful life prediction of aluminum 7075-T651 specimens. Their experimental validations proved that using entropy as a thermodynamic state function for damage characterization is an effective way of handling the endurance threshold uncertainties for life prediction purposes. Authors also derived the formulation of entropy generation during corrosion-fatigue.

Imanian and Modarres (2015, 2018) and Yun and Modarres (2019) also proved experimentally that multiple entropic-damage tests show the evolution of corrosion-fatigue volumetric entropy for different loading conditions. They measured the cumulative final value of fracture corrosion-fatigue entropy. The final entropy value is between 0.7 MJ m^{-3} K^{-1} and 1.5 MJ m^{-3} K^{-1}. Authors also observed that there is a narrow band distribution of entropy to failure data points [fracture entropies] irrespective of loading condition. They concluded that the entropy has the ability to quantify the uncertainties associated with microstate variables. Furthermore, they stated that it reveals the independence of entropy to the loading condition (i.e., failure path). Fracture entropy's slim distribution can be interpreted due to uncertainties, such as instrumental measurement errors, the legitimacy of the assumptions considered in entropy evaluation, weak control of the experimental, operational and environmental conditions, and human error.

Sosnovskiy and Sherbkov (2015, 2016) published a generalized theory of evolution based on the concept of tribo-fatigue entropy. The essence of the proposed approach is that tribo-fatigue entropy is determined by the process of degradation of any system due to thermodynamic mechanical effects causing the change in the state. Sosnovskiy and Sherbkov (2016) provided a mathematical framework for law of entropy increase in a general form. They also provided extensive experimental validation for mechanothermodynamics theory. They stated, "It is shown that the mechanothermodynamics—a generalized physical discipline—is possible by constructing a bridge between Newtonian mechanics and thermodynamics. The entropy is the bridge between Thermodynamics and Mechanics." The first and the second principles of mechanothermodynamics were presented. They formulated foundation of the general theory of degradation evolution of mechanothermodynamic systems for

- Energy theory of limiting states
- Energy theory of damage
- Foundations of the theory of electrochemical damage"

They provided mathematical fundamentals of the theory of interaction between damage caused by loads of different nature (mechanical, thermodynamic, etc.). Authors proposed a single function for critical damage (limiting) states of metals and polymer materials operating in different conditions. The analysis of 136 laboratory experiment results showed that this single logarithmic function is fundamental: it is valid for low, average, and high-strength pure metals, alloys, and polymers over a wide range of temperatures of medium (from helium temperature to 0.8 TS, where TS is the material melting temperature) and mechanical stresses (up to the strength limit for single static loading) while the fatigue life was of the order of 106–108 cycles.

Mechanothermodynamics uses the same idea as unified mechanics theory in order to unify Newtonian mechanics and thermodynamics, that is, using entropy as a bridge to connect Newtonian mechanics and thermodynamics, which was first published in Basaran and Yan (1998). Sosnovskiy and Sherbakov (2015, 2016) used a logarithmic function for evolution of degradation; it is not clear why they do not use Boltzmann equation, since their logarithmic function can be obtained directly from Boltzmann equation. The term tribo-fatigue-entropy in their work refers to entropy generation in tribology [which is their focus] and fatigue process. They surmise that if an analogy between light and strain energy is justified, then strain energy absorption law may be similar to Bouger's light absorption law. This law, which is also exponential, becomes basis for their degradation function formulation.

Haddad (2017) published probably the most comprehensive review of the history of thermodynamics from its classical to its postmodern forms. Haddad et al. (2005) and Haddad (2019) also provided general systems theory framework for thermodynamics which attempts to harmonize thermodynamics with classical Newtonian mechanics. The main idea used by Haddad (2019) to unify mechanics and thermodynamics is attributed to Basaran and Yan (1998). Haddad stated that, "The dynamical system notions of entropy proposed Haddad (2005, 2019), Basaran and Yan (1998), Basaran and Nie (2004), Sosnovskiy and Sherbakov (2016) involving an analytical description of an objective property of matter can potentially offer a conceptual advantage over the subjective quantum expressions for entropy proposed in the literature (e.g., Daróczy entropy, Hartley entropy, Rényi entropy, von Neumann entropy, infinite-norm entropy) involving a measure of information. An even more important benefit of the dynamical systems representation of thermodynamics is the potential for developing a unified classical and quantum theory that encompasses both mechanics and thermodynamics without the need for statistical (subjective or informational) probabilities."

Cuadras et al. (2015) proposed a method to characterize electrical resistor damage based on entropy measurements. They postulated that irreversible entropy and the rate at which it is generated are more convenient parameters than resistance for describing damage because they are essentially positive in virtue of the second law of thermodynamics, whereas resistance may increase or decrease depending on the degradation mechanism. They tested commercial resistors in order to characterize the damage induced by power surges. Resistors were biased with constant and pulsed voltage signals, leading to power dissipation in the range of 4–8 W, which is well above the 0.25 W nominal power to initiate failure. Entropy was inferred from the added power and temperature evolution. They studied the relationship among resistance, entropy, and damage. They stated that power surge dissipates into heat (Joule effect) and damages the resistor. They observed that there is a correlation between entropy generation rate and resistor failure. Cuadras et al. (2015) concluded that damage could be conveniently assessed from irreversible entropy generation.

Cuadras (2016) proposed a method to monitor the aging and damage of capacitors based on their irreversible entropy generation rate (Fig. 4.3). Authors overstressed several electrolytic capacitors in the range of 33 mFe–100 mF and monitored their entropy generation rate. They observed a strong relationship between capacitor degradation and entropy generation rate. Therefore, they proposed a threshold entropy generation rate as an indicator of capacitor time-to-failure. This magnitude is related to both capacitor parameters and to a damage indicator such as entropy. They validated the model as a function of capacitance, geometry, and rated voltage. Moreover, they identified different failure modes, such as heating, electrolyte dry-up, and gasification from the dependence of entropy generation rate with temperature.

Cuadras et al. (2017) proposed a method to assess the degradation and aging of light emitting diodes (LEDs) based on irreversible entropy generation rate. Authors

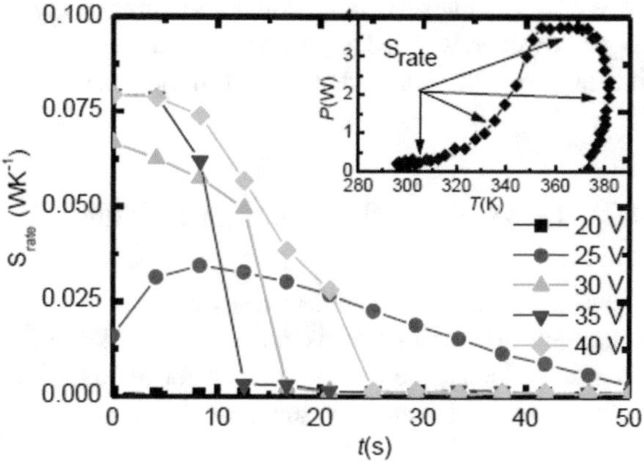

Fig. 4.3 Capacitor entropy generation rate versus time (seconds) for different voltage levels, After Cuadras (2016)

degraded several LEDs and monitored their entropy generation rate in accelerated tests. They compared the thermos-electrical results with the optical light emission evolution during degradation. They found a good relationship between aging and entropy generation rate because they both are related to device parameters and optical performance. They proposed a threshold of entropy generation rate as a reliable damage indicator of LED end-of-life that can avoid the need to perform optical measurements to assess optical aging. The method is far more physics based and beyond the typical statistical empirical models based on curve fitting to a test data.

Cuadras et al. (2017) tested different LED colors and electrical stresses to validate the electrical LED model and we analyzed the degradation mechanisms of the devices to validate the model.

In the last few years, there has been significant interest in using entropy generation rate as a metric in order to predict degradation and fatigue life. Some of them are Basaran and Chandaroy (2002); Basaran and Tang (2002); Basaran et al. (2003, 2005, 2008, 2008b); Tang and Basaran (2003); Gomez and Basaran (2005, 2006); Lin and Basaran (2005); Gomez et al. (2006); Li et al. (2008); Li and Basaran (2009); Gunel and Basaran (2010, 2011a, b); Basaran and Lin (2007a, b, 2008); Basaran and Nie (2007); Bin et al. (2020); Pauli (1973); Planck (1906); Sherbakov and Sosnovskiy (2010); Sosnovskiy (1987, 1999, 2004, 2005, 2007, 2009); Sosnovskiy and Sherbakov (2012, 2017, 2019); Temfack and Basaran (2015); Wang and Yao (2019); Yao and Basaran (2012, 2013a, b, c); Yun and Modarres (2019); Wang and Yao (2017), Guo et al. (2018), Zhang et al. (2018), Wang et al. (2019), and Osara and Bryant (2019a, b). Young and Subbarayan (2019a, b) and Suhir (2019) used Boltzmann-Arrhenius-Zhurkov equation to predict evolution of time to failure. While this approach essentially is based on Basaran and Yan (1998) concept, Suhir advocates using the Boltzmann equation independent of entropy and Newtonian mechanics. Unfortunately, this approach reduces to using Boltzmann distribution just as an evolution function for an empirical model based on test data. Hsiao and Liang (2018) developed a sensor which monitors entropy generation in real time and that can give real-time information on system aging and prediction for further estimating the failure of electrical or any mechanical system.

We do not claim to have done a comprehensive survey of the literature on the topic; however, we tried to give a historical perspective of the efforts. Unfortunately, it is not possible to include them all in this chapter.

4.2 Laws of Unified Mechanics Theory

Newton's first law is about at rest state where externally applied forces are zero hence entropy generation is not possible. While this also includes a motion at constant velocity, the law is intended for hypothetical case in vacuum. Because no other forces can act on the system, according to this law, laws of thermodynamics do not change anything. Therefore, we will start numerating the laws of the Unified Mechanics with the second law.

4.2.1 Second Law of Unified Mechanics Theory

Degradation of initial momentum of a body takes place according to the second law of thermodynamics. The degradation rate is directly proportional to entropy generation rate along the path followed. Entropy generation can be mapped on a nondimensional coordinate called Thermodynamic State Index (TSI), axis, Fig. 4.4, between zero and one. The second law of unified mechanics theory is nothing but just a marriage of Newton's second law and second law of thermodynamics in a single equation. Thermodynamic State Index (Φ) is normalized nondimensional form of the second law of thermodynamics; as a result, combination of the second laws can be given by,

$$F(1-\Phi)dt = d(mv) \tag{4.35}$$

Assuming a constant mass system

$$a = \frac{F}{m}(1-\Phi) \tag{4.36}$$

Combining the second law of Newton and thermodynamics requires the modification of Newtonian space-time coordinate system. A new axis must be added to be able to define the thermodynamic state of the point (Fig. 4.4). As results, the motion of any particle can be defined only in a five-dimensional space that has five linearly independent axes. None of these axes can represent the information of other axes.

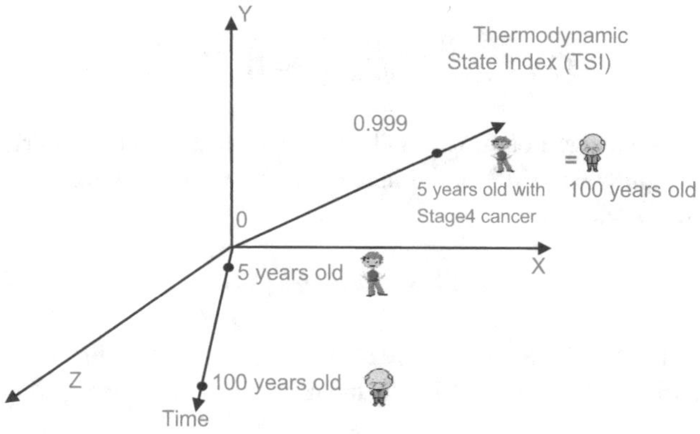

Fig. 4.4 Coordinate system of the unified mechanics theory

Thermodynamic state index is the normalized form of the second law of thermodynamics. Derivation of the $(1 - \Phi)$ term will be provided in the following section. The additional axes Thermodynamics State Index (TSI) is necessary to locate the thermodynamic state of the particle. Coordinates of a point can be defined by Newton's laws of motion in the space-time coordinate system. However, thermodynamic state index axis coordinate cannot be defined by space-time coordinate system. We find it necessary to give the following example. Let us assume there is a 5-year-old boy in Istanbul and 100-year-old man in New York. Using the space-time Cartesian coordinate system, we can define their location by x, y, z coordinates and age on the time axis. However, this does not give any information about their thermodynamic state. Let us assume 5-year-old boy has stage 4 cancer and is expected to die in a few days and 100-year-old is also expected to die in few days. This information cannot be represented in x, y, z-time-space coordinate system. However, on TSI axis both will be at $\Phi = 0.999$ coordinate (Fig. 4.4).

Another example is, going back to Newton's second law, if a soccer ball is given an initial acceleration with a kick force of F, it will move but eventually will come to a stop. Assume the ball was stationary initially in the terrain shown in Fig. 4.6. Depending on the path it follows, it will come to a stop at one of the valleys. Again the initial acceleration of the ball is governed by the second law of Newton and slowing process is governed by the laws of thermodynamics, which is represented by $(1 - \Phi)$ term. Laws of thermodynamics govern the motion of the ball after the initial kicking second.

4.2.2 Third Law of Unified Mechanics Theory

All forces between two objects exist in equal magnitude and opposite direction. However, resulting deformation in two objects will change over time because of irreversible entropy generation. The resulting equation can be given by,

$$F_{12} = \frac{dU_{21}}{du_{21}} = \frac{d}{du_{21}}\left[\frac{1}{2}k_{21}[1 - \Phi]u_{21}^2\right] \tag{4.37}$$

If we assume change in stiffness and TSI are smaller than other derivative terms by order of magnitude as differential in displacement goes to zero, we can write the following simple relation

$$F_{12} = F_{21} = k_{21} \cdot u_{21}(1 - \Phi) \tag{4.38}$$

where U_{21} is the strain energy of the reactionary member, Φ is the Thermodynamic State Index, k is the stiffness and U_{21} is the displacement in the reacting member.

4.3 Evolution of Thermodynamic State Index (Φ)

Materials, under externally applied loading, change their thermodynamic state. This process will follow the laws of thermodynamics. Simply put, first law of thermodynamics will govern the conservation of energy. The second law of thermodynamics will govern the entropy generation rate according to the fundamental equation of the system. Evolution of entropy and disorder in the system is given by Boltzmann equation. When the entropy is maximum and entropy generation rate is zero for a closed and isolated system, the change in TSI will come to a stop.

While author of this book and many others have proven validity of Boltzmann's equation in solids in the last 20+ years extensively, Boltzmann's (1877) original paper and its interpretation on disorder was misinterpreted until recently as being only applicable to gasses. Recently there has been a great deal of interest in translating Boltzmann's original papers into English. One recent such paper is by Sharp and Matschinsky (2015). We find it essential to include their translation in this section. It is important to point out that, Sharp and Matschinsky (2015) also state that "what Boltzmann actually wrote on these subjects is rarely quoted directly, his methods are not fully appreciated and key concepts have been misinterpreted." The following section about Boltzmann's work is a direct quotation from their paper.

"The translation provided here is for Boltzmann (1877). Previous work of Maxwell and Boltzmann's derivations was based on mechanical laws of motion and particle interaction of gasses. However, this work by Boltzmann is much more general. His formulation require only that particles can exchange kinetic energy, but they do not specify the mechanism. As Boltzmann predicted in the final sentences of this paper, his approach was applicable not just gasses, but to liquids and solids of any composition. Indeed the Boltzmann distribution has also passed almost unchanged into the quantum world."

Boltzmann developed the theoretical basis for statistical mechanics with great clarity. Boltzmann used three levels of hierarchy to describe the processes. At the highest level, there is macro-state, where the thermodynamic state variables such as temperature and pressure can be observed directly. The second level in the hierarchy is where energy or velocity components of each molecule can be specified. He calls this Komplexion, which Sharp and Matschinky (2015) translate as complexion. Finally there is the third level at which the number of molecules with each its energy level, velocity, and their position is specified. Here Boltzmann does not make any assumption about the type of the molecules (Boltzmann'sw_0, w_1, etc.). Boltzmann calls a particular set of w-values a Zustandeverteilung, translated as a "state distribution" or "distribution of states." It is important to point out, as discussed by Sharp and Matschinsky (2015) microstate is divided into two separate hierarchies in Boltzmann's formulation, both complexion and state distribution. "Boltzmann then uses permutation mathematics to determine distribution of states, which he denotes by p." "Boltzmann then shows how to find the distribution of states $(w_0^{\max}, w_1^{\max}, \cdots)$ with the largest number of complexious $p^{\max}$ subject to constraints on the temperature and number of molecules. Boltzmann's postulate is that $(w_0^{\max}, w_1^{\max}, \ldots)$ is the most probable state distribution and that this corresponds to thermal equilibrium." Boltzmann introduces discrete energy levels in Sect. 4.1 of his paper as a convenience and considers them unphysical. In Sect. 4.2 he shows that there is no material difference compared to using continuous energy levels.

The statistical mechanics formulation of entropy is presented in Section 5 of Boltzmann (1877). "Boltzmann shows that the statistical mechanics quantity he denotes by Ω (multiplied by 2/3) is equal to the thermodynamic quantity entropy (*S*) as defined by Clausius, with an additive constant. Boltzmann called Ω the permutabilities, translated here literally as "permutability measure." Boltzmann defines permutability measure Ω as follows: A state distribution is specified by the number of molecules having velocity components within some small interval *u* and *u* + *du*, *v* and *v* + *dv*, *w* and *w* + *dw* for every *u*, *v*, and *w* velocity values in the − ∞ , + ∞ range, and having position coordinates in a differential volume *dqdp* ranging over the total volume *V*. For each state distribution, there are a number of possible complexions. [Combination of molecules with certain level of energy or velocity]. One particular state distribution has the most complexions and so therefore is the most probable. Ω is given by the logarithm of the number of complexions for that state with the most complexions."

"Boltzmann (1877) also clearly demonstrates that these are two distinct contributions to entropy generation, one arising from the distribution of heat (kinetic energy) and then the other due to the distribution in space of atoms or molecules." In the initial effort to understand the nature of entropy, Carnot, Clausius, Maxwell Kelvin, and others focused almost entirely on the contribution from heat. "Boltzmann unified the entropy due to thermal aspects with special distribution entropy into one statistical mechanics formulation." Boltzmann also discovered the third fundamental contribution to entropy, namely radiation by deriving the Stefan-Boltzmann Law (1884).

"Careful reading of Boltzmann (1877) is enlightening with regard to a number of apparent paradoxes. Subsequently, encountered in the development of statistical mechanics." It is unfortunate that terms in Boltzmann equations are misinterpreted in many physics textbooks. "First, regarding permutability measure, Ω, is not the logarithm of a probability." That would be obtained by dividing the number of complexions *P* for a given state distribution by the total number of complexions. Boltzmann gives this a different symbol, *w*, from the first letter of the German word for probability, [wahrscheinlichkeit] but he does not use it. Confusingly, Planck later chose to write Boltzmann's equation for entropy as Planck (1901).

$$S = k \ln\ w + \text{constant} \tag{4.39}$$

w is the probability that the system will exist in the state it is in relative to all the possible states it could be in Halliday and Resnick (1966). On the other hand, in most thermodynamics textbooks, Boltzmann's hypothesis is given by

$$S = k \ln\ \Omega \tag{4.40}$$

where k is the Boltzmann's constant and Ω is the number of microstates corresponding to a given state with the macroscopic constraints (Callen 1985; DeHoff 1993).

Sharp and Matschinky (2015) point out that with Boltzmann's (permutabilitat) measure of permutability method for counting possible microstates, there is no need for a posteriori division by $N!$ "to correct" the derivation using the "somewhat mystical arguments of Gibbs (1902) and Planck." Van Kampen (1984), Ehrenfest and Tikal (1921), Van kampen (1984), Jaynes (1992), and Swendsen (2006) have pointed out that correct counting of microstates a la Boltzmann precludes the need for the spurious indistinguishability term $N!$. "This fact has been ignored in most text books."

Translation by Sharp and Matchinsky (2015) also clarifies one very important point about Boltzmann's derivation regarding nonequilibrium states. $\ln\Omega$ is not the logarithm of a volume in momentum-coordinate-phase space [Boltzmann here refers to a five dimensional space, the fifth axis being momentum], occupied by the system. Boltzmann notes, using Liouville's theorem, that dv remains constant in time. Therefore, it cannot describe the entropy increase upon approach to the equilibrium that Boltzmann was so concerned with. He thus avoids at the outset the considerable difficulty Gibbs had been accounting for changes in entropy with time. Boltzmann gave us for the first time a definition of entropy applicable to every state at equilibrium or not. "Then the entropy of the initial and final states is not defined, but one can still calculate the quantity which we have called the permutability measure." By extension, every complexion can then be assigned an entropy, using the permutability measure of the state distribution to which that complexion belongs (Lebowitz 1993), opening the door to the statistical mechanics of nonequilibrium states and irreversible processes.

Boltzmann's work has historically been misinterpreted, assumed applicable to gasses only and ignored in continuum mechanics field. Therefore, we feel compelled to include his original paper in this book. The following section is the English translation of Boltzmann (1877) by Sharp and Matschinky (2015).

4.3.1 On the Relationship Between the Second Fundamental Theorem of the Mechanical Theory of Heat and Probability Calculations Regarding the Conditions for Thermal Equilibrium, by Ludwig Boltzmann (1877)

The relationship between the second fundamental theorem and calculations of probability became clear for the first time when I demonstrated that the theorem's analytical proof is only possible based on probability calculations. (I refer to my publication "Analytical proof of the second fundamental theorem of the mechanical theory of heat derived from the laws of equilibrium for kinetic energy" Wien. Ber. 63, p 8 reprinted ass Wiss. Abhand. Vol I, reprint 20, pp 295 and my "Remarks about several problems in the mechanical theory of heat" 3rd paragraph, Wiss. Abhand, Vol II reprint 39.) [Boltzmann refers to thermodynamics by the term "The mechanical theory of heat"]. [Wiss. Abhand refers to Boltzmann's collected works]. This relationship is also confirmed by demonstrating that an exact proof of the fundamental theorems of the equilibrium of heat is most easily obtained if one demonstrates that a certain quantity—which I wish to define again as E—has to decrease as a result of the exchange of the kinetic energy among the gaseous molecules and therefore reaching its minimum value for the state of the equilibrium of heat. (Compare my "Additional studies about the equilibrium of heat among gaseous molecules" Wiss. Abhand, vol I, reprint 22, p 316). The relationship between the second fundamental theorem and the laws of the equilibrium of heat is made even more compelling in light of the developments in the second paragraph of my "Remarks about several problems of the mechanical theory of heat." There, I mentioned for the first time the possibility of a unique way of calculating the equilibrium of heat using the following formulation. "It is clear that every single uniform state distribution which establishes itself after a certain time given a defined initial state is equally as probable as every single non-uniform state distribution, comparable to the situation in game of Lotto [a board game] where every single quintet is as improbable as the quintet 12345. The higher probability that the state distribution becomes uniform with time arises only because there are far more uniform than non-uniform state distributions". Furthermore: "It is even possible to calculate the probabilities from the relationships of the number of different state distributions. This approach would perhaps lead to an interesting method for the calculation of the equilibrium of heat." It is thereby indicated that it is possible to calculate the state of the equilibrium of heat by finding the probability of the different possible states of the system. The initial state in most cases is bound to be highly improbable, and from it the system will always rapidly approach a more probable state until it finally reaches the most probable state, i.e., that of the heat equilibrium. If we apply this to the second basic theorem, we will be able to identify that quantity which is usually called entropy with the probability of the particular state. Let us assume a system of bodies which are in a state of isolation with no interaction with other bodies, e.g., one body with higher and one body with lower temperature and one so-called intermediate body which accomplishes the heat transfer between the two bodies; or choosing another example by assuming a vessel with absolutely even and rigid walls, one half of which is filled with air of low temperature and pressure whereas the other half is filled with air of high temperature and pressure. The hypothetical system of particles is assumed to have a certain state at time zero. Through the interaction between the particles, the state is changed. According to the second fundamental theorem, this change has to take place in such a way that the total entropy of the particles increases. This means according to our present interpretation that nothing changes except that the probability of the overall state for all particles will get larger and larger. The system of particles always changes from an improbable state to a probable state. It will become clear later what this means. After the publication of my last treatise regarding this topic, the same idea was taken up and developed further by Mr. Oskar Emil Meyer totally independent of me (Die

Kinetische Theorie der Gase, Breslau 1877, Seite 262). He attempts to interpret, in the described manner, the equations of my continued studies concerning the equilibrium of heat particles. However, the line of reasoning of Mr. Meyer remained entirely unclear to me and I will return to my concerns with his approach on page 172 (Wiss. Abhand Vol. II).

We have to take a totally different approach because it is our main purpose not to limit our discussion to thermal equilibrium, but to explore the relationship of this probabilistic formulation to the second theorem of the mechanical theory of heat. We want first to solve the problem which I referred to above and already defined in my "Remarks on some problems of the mechanical theory of heat" [Wiss. Abhand, reprint 39], namely to calculate the probability of state distributions from the number of different distributions. We want first to treat as simple a case as possible, namely a gas of rigid elastic spherical molecules trapped in a container with absolutely elastic walls (which interact with central forces only within a certain small distance, but not otherwise, the latter assumption, which includes the former as a special case, does not change the calculations in the least). Even in this case, the application of probability theory is not easy. The number of molecules is not infinite, in a mathematical sense, yet the number of velocities each molecule is capable of is effectively infinite. Given this last condition, the calculations are very difficult to facilitate understanding. I will, as in earlier work, consider a limiting case.

4.3.2 *Kinetic Energy Has Discrete Values*

We assume initially each molecule is only capable of assuming a finite number of velocities, such as

$$0, \frac{1}{q}, \frac{2}{q}, \frac{3}{q}, \cdots \frac{P}{q} \tag{4.41}$$

where P and q are arbitrary finite numbers. Upon colliding, two molecules may exchange velocities.

The fact that Boltzmann allows exchange of energy between molecules automatically makes his formulation general, i.e., irrespective of gas, liquid, or solid state. But after the collision, both molecules still have one of the above velocities, namely,

$$0, \text{or } \frac{1}{q}, \text{or } \frac{2}{q}, \text{etc.till } \frac{P}{q} \tag{4.42}$$

This assumption does not correspond to any realistic mechanical model, but it is easier to handle mathematically and the actual problem to be solved is re-established by letting P and q go to infinity.

Even if at first sight, this seems a very abstract way of treating the problem, it rapidly leads to the desired objective, and when you consider that in nature all infinities are [nothing] but limiting cases one assumes each molecule can behave in this fashion only in the limiting case where each molecule can assume more and more values of the velocity.

To continue, however, we will consider the kinetic energy, rather than the velocity of the molecules. Each molecule can have only a finite number of values for its kinetic energy. As a further simplification, we assume that the kinetic energies of each molecule form an arithmetic progression, such as the following:

$$0, \epsilon, 2\epsilon, 3\epsilon, \ldots, p\epsilon \tag{4.43}$$

We call p the largest possible value of the kinetic energy,$p\epsilon$. Before impact, each of two colliding molecules shall have a kinetic energy of

$$0, \text{or } \epsilon, \quad \text{or } \ 2\epsilon, \quad \text{etc. } \ p\epsilon \tag{4.44}$$

Which means that after the collision, each molecule still has one of the above values of kinetic energy. The number of molecules in the vessel is n. If we know how many of these n molecules have a kinetic energy of zero, how many have a kinetic energy of ϵ, and so on, then we know the kinetic energy distribution. If at the beginning there is some state distribution among the gas molecules, this will in general be changed by the collisions.

The laws governing this change have already been the subject of my previous investigations. But right way, I note that this is not my intention here, instead I want to establish the probability of a state distribution, regardless of how it is created or, more specifically, I want to find all possible combinations of the $P + 1$ kinetic energy values allowed to each of the n molecules and then establish how many of these combinations correspond to each state distribution. [The term "state distribution" is better translated/interpreted as the distribution of a state into English. However, I am keeping the original translation, because the term refers to what Boltzmann refers to "complexion", which is the distribution of kinetic energies of molecules]. The latter number $(p + 1)$ then determines the likelihood of the relevant state distribution, as I have already stated in my published "Remarks about several problems in the mechanical theory of heat" (Wiss. Abhand. Vol II, reprint 39, p 121).

As a preliminary, we will use a simpler schematic approach to the problem, instead of the exact case. Suppose we have n molecules. Each of them is capable of having kinetic energy

$$0, \epsilon, 2\epsilon, 3\epsilon, \ldots, p\epsilon \tag{4.45}$$

Moreover, suppose these energies are distributed in all possible ways among the n molecules, such that the total energy is a constant, e.g., $\lambda\epsilon = L$. Any such distribution, in which the first molecule may have a kinetic energy of, e.g., 2ϵ, the second may have 6ϵ, and so on, upto the last molecule, we call a complexion and so that each individual complexion can be easily enumerated. We write them in sequence (for convenience we divide through by ϵ), specifying the kinetic energy of each molecule. We seek the number p of complexions where w_0 [number of] molecules have kinetic energy 0, w_1 molecules have kinetic energy ϵ, w_2 have kinetic energy 2ϵ, upto the w_p which have kinetic energy $p\varepsilon$. We said, earlier, that given how many molecules have kinetic energy 0, how many have kinetic energy ϵ, etc., this distribution among the molecules specifies the number of P of complexions for that distribution; in other words, it determines the likelihood of that state distribution. Dividing the number P by the number of all possible complexions, we get the probability of the state distribution. Boltzmann actually never performs this division. It is done later by Max Planck (1901).

Since a distribution of states does not determine kinetic energies exactly, the goal is to describe the state distribution by writing as many zeros as molecules with zero kinetic energy (w_0), w_1 ones for those with kinetic energy ϵ, etc. All these zeros, ones, etc. are the elements defining the state distribution. It is now immediately clear that the number P for each state distribution is exactly the same as the number of permutations of which the elements of the state distribution are capable and that is why the number P is the desired measure of the permutability of the corresponding distribution of states (Table 4.1). Once we have specified every possible complexion, we have also all possible state distributions, the latter differing from the former only by immaterial permutations of molecular labels. All those complexions which contain the same number of zeros, the same numbers of ones, etc., differing from each other merely by different arrangements of elements, will result in the same state distribution; the number of complexions forming the same state distribution and which we have denoted by P must be equal to the number of permutations which the elements of the state distribution are capable of. In order to give a simple numerical example, take $n = 7, \lambda = 7\ p = 7$ so $L = 7\epsilon, P = 7\epsilon$. With seven molecules, there are eight possible values for the kinetic energy 0, ϵ, 2ϵ, 3ϵ, 4ϵ, 5ϵ, 6ϵ, 7ϵ to distribute in any possible way such that the total kinetic energy is 7ϵ. There are then 15 possible state distributions. We enumerate each of them in the above manner, producing the numbers listed in the second column of the following table of state distributions. The numbers in the first column label the different state distributions.

In the last column, Table 4.1 under the heading P is the number of possible permutations of members for each state. The first state distribution, for example, has six molecules with zero kinetic energy and the seventh has kinetic energy 7ϵ. So

Table 4.1 Possible permutations for different states

		P			P			P
1.	0000007	7	6.	0000124	210	11.	0001222	140
2.	0000016	42	7.	0000133	105	12.	0011113	105
3.	0000025	42	8.	0000223	105	13.	0011122	210
4.	0000034	42	9.	0001114	140	14.	0111112	42
5.	0000115	105	10.	0001123	420	15.	1111111[a]	1

[a]The state distributions are so arranged that, read as a number, the rows are arranged in increasing order

$w_0 = 6, w_1 = 1$ $w_2 = w_3 = w_4 = w_5 = w_6 = 0$ [w_1 is also zero. However Boltzmann forgets to include it]. It is immaterial which molecule has kinetic energy 7ϵ.

There are seven possible complexions, which represent this state distribution. Denoting the sum of all possible complexions, 1716, by J, then the probability of the first state distribution is $7/J$; similarly, the probability of the second state distribution is $42/J$; the most probable state distribution is the tenth, as its elements permit the greatest number of permutations. Hereon, we call the number of permutations the relative likelihood of the state distribution; this can be defined in a different way, which we next illustrate with a specific numerical example, since generalization is straightforward. Suppose we have an urn containing an infinite number of paper slips. On each slip is one of the numbers 0, 1, 2, 3, 4, 5, 6, 7; each number is on the same amount of slips and has the same probability of being picked. We now draw the first septet of slips, and note the numbers on them. This septet provides a sample state distribution with a kinetic energy of ϵ times the number written on the first slip for molecule *l* and *s* forth. We return the slips to the urn and draw a second septet, which gives us a second state distribution, etc. After we draw a very large number of septets, we reject all those for which the total does not equal to seven. This still leaves a large number of septets. Since each number [digit] has the same probability of occurrence and the same elements in a different order from different complexions, each possible complexion will occur equally often. By ordering the numbers within each septet by size, we can classify each into one of the fifteen cases tabulated above. It is not clear why Boltzmann does not work with probability of state distribution. Since probability would not change with the sample size, urn with infinite slips example would be unnecessary. Therefore, the number of septets which fall into the class 0000007 relative to the 0000016 class will be 7:42. Similarly for all the other septets. The most likely state distribution is the one which produces the most septets, namely the 10th.

Boltzmann's Footnote: If we divide the number of septets corresponding to a particular state by the total number of septets, we obtain the probability of distribution. Instead of discarding all septets whose total is not 7, we could after drawing a slip remove from the urn all those other slips for which a total of 7 is now impossible, e.g., on drawing a slip with 6 on it, all other slips except those with 0 or l would be removed. If the first six slips all had zero on them, only slips with 7 on them would be left in the urn. One more thing should be noted at this point. We construct all possible complexions. If we denote by $\overline{W_0}$ the arithmetic mean of all values of w_0 which belong to different complexions and form analogous expressions $\overline{W_1}$, $\overline{W_2}\cdots$, in the limit these quantities would also form the same state distribution.

Translators Sharp and Matschinsky's note: Boltzmann's comments on the results of Mr. Oskar Meyer beginning "Ich will hier einige worte *ü*ber die von. Hrn. Oskar Meyer" on p 172 (Wiss. A6) and ending with "Bearbeitung des allgemeinen problems zur*ü*ckkehren" on p 175 (Wiss. A6) are of historical interest only and are omitted.

The first task is to determine the permutation number, previously designated by*P*, for any state distribution. Denoting by *J* the sum of the permutations *P* for all possible state distributions, the quotient *P*/*J* is the state distribution's probability, henceforth denoted by w. Boltzmann uses W for the initial of German word, Wahrscheinlichkeit. We would first like to calculate the permutations P for the state distribution characterized by w_0 molecules with kinetic energy 0, w_1 molecules with the kinetic energy ϵ, etc. It must be understood that

$$w_0 + w_1 + w_2 + \cdots + w_p = n \tag{4.46}$$

$$w_1 + 2w_2 + 3w_3 + \cdots + pw_p = \lambda \tag{4.47}$$

Because the total number of molecules is n, and the total kinetic energy is $\lambda\epsilon = L$. Describing the state distribution as before, a complexion has w_0 molecules with zero energy, w_1 [is the number of molecules] with one unit [of energy], and so on. The permutations, P, arise since of the n elements W_0 are mutually identical. Similarly with w_1, w_2, etc. elements. The total number of permutations is well known.

$$P = \frac{n!}{w_0!w_1!} \tag{4.48}$$

The most likely state distribution will be for those w, w_1, ... values for which P is a maximum or since the numerator is a constant, for which the denominator is a minimum. The values w, w_1 must simultaneously satisfy the two constraints (4.46) and (4.47). Since the denominator of P is a product, it is easiest to determine the minimum of its logarithm that is the minimum of

$$M = \ln\left[w_0!\right] + \ln\left[w_1\right]! + \cdots \tag{4.49a}$$

Here ln is the natural logarithm.[1]

It is natural in our problem that only integer values of w_0, w_1, ... are meaningful. However to apply differential calculus, we will allow non-integer values and so find the minimum of the expression

$$M_1 = \ln\Gamma(w_0 + 1) + \ln\Gamma(w_1 + 1) + \cdots \tag{4.49b}$$

which is identical to (4.49a) for integer values of w_0, w_1, ... We then get the non-integer values which for constraints (4.46) and (4.47) maximize M_1.[2] The solution to the problem will in any case be obtained if for w_0, w_1, etc. we select the closest set of integer values. If here and there a deviation of a few integers is required, the nearest complexion is easily found.

The minimum of M_1 is found by adding to both sides of the equation for M_1 Eq. (4.46) multiplied by the constant h and Eq. (4.47) multiplied by the constant

[1] Translators footnote: The ambiguous symbol "l" [used by Boltzmann] for [natural] logarithm in the original text has been replaced throughout by "ln".

[2] Translators footnote: The original text reads as "maximized but should mean minimized". [Because Boltzmann's objective is to maximize P].

k and setting the partial derivatives with respect to each of the variables w_0, w_1, w_2, … to zero. We thus obtain the following equations

$$\frac{dln\Gamma(w_0+1)}{dw_0}+h=0, \tag{4.50a}$$

$$\frac{dln\Gamma(w_1+1)}{dw_1}+h+k=0, \tag{4.50b}$$

$$\frac{dln\Gamma(w_2+1)}{dw_2}+h+2k=0 \tag{4.50c}$$

$$\vdots \quad \vdots \quad \vdots\, z \; \frac{dln\Gamma(w_p+1)}{dw_p}+h+pk=0, \tag{4.50d}$$

which leads to

$$\frac{dln\Gamma(w_1+1)}{dw_1}-\frac{dln\Gamma(w_0+1)}{dw_0} \tag{4.50e}$$

$$=\frac{dln\Gamma(w_2+1)}{dw_2}-\frac{dln\Gamma(w_1+1)}{dw_1} \tag{4.50f}$$

$$=\frac{dln\Gamma(w_3+1)}{dw_3}-\frac{dln\Gamma(w_2+1)}{dw_2} \tag{4.50g}$$

Exact solution of the problem through evaluation of the gamma function integral is very difficult; fortunately the general solution for arbitrary finite values of p and n does not interest us here, but only the solution for the limiting case of larger and larger number of molecules. Then the numbers w_0, w_1, w_2, etc. become larger and larger, so we introduce the function

$$\phi(x)=\ln\Gamma(x+1)-x(\ln x-1)-\frac{1}{2}\ln 2\pi. \qquad (4.51)^3$$

Then we can write the first equation of (4.50a), (4.50b), (4.50c), (4.50d), (4.50e), (4.50f), (4.50g) as follows

[3]Boltzmann approximates $lnx!$ by $xlnx-x+\frac{1}{2}\ln(2\pi)$ rather than $\left(x+\frac{1}{2}\right)\ln\, -x+\frac{1}{2}\ln(2\pi)$ as is now usual. For x $\gg$ 30 the relative difference is small.

$$\ln w_1 + \frac{d\phi(w_1)}{dw_1} - \ln w_0 + \frac{d\phi(w_0)}{dw_0} = \ln w_2 + \frac{d\phi(w_2)}{dw_2} - \ln w_1 - \frac{d\phi(w_1)}{dw_1} \quad (4.52)$$

Similarly, for the other equations of (4.50a), (4.50b), (4.50c), (4.50d), (4.50e), (4.50f), (4.50g). It is also well known that

$$\phi(x) = -\frac{1}{2}\ln x + \frac{1}{12x} + \ldots \quad (4.53a)$$

This series is not valid for $vx = 0$, but here $x!$ and $\sqrt{2\pi}(x/e)^x$ should have the same value, and $\phi(x) = 0$. Therefore, the problem of finding the minimum of $w_0\,!\,w_1\,!\,w_2\,!\ldots$ is replaced by the easier problem of finding the minimum of

$$\sqrt{2\pi}\left(\frac{w_0}{e}\right)^{w_0}\sqrt{2\pi}\left(\frac{w_1}{e}\right)^{w_1}\sqrt{2\pi}\left(\frac{w_2}{e}\right)^{w_2}$$

Providing w is not zero, even at moderately large values of p and n both problems have matching solutions. From Eqs. (4.53a) and (4.53b) it follows

$$\frac{d\phi(w_0)}{dw_0} = -\frac{1}{2w_0} - \frac{1}{12w_0^2} - \ldots \quad (4.53b)$$

which for larger and larger values of w_0 or $\ln w_0$ vanishes, the same also applies to the other $w's$, so Eq. (4.52) can be written as follows

$$\ln w_1 - \ln w_0 = \ln w_2 - \ln w_1 \quad (4.54a)$$

Or

$$\frac{w_1}{w_0} = \frac{w_2}{w_1} \quad (4.54b)$$

Likewise the equations for the remaining $w's$ are

$$\frac{w_2}{w_1} = \frac{w_3}{w_2} = \frac{w_4}{w_3} = \ldots \quad (4.54c)$$

One sees immediately that by neglecting the expression (4.53b) the minimum of the denominator of

$$\frac{\sqrt{2\pi\left(\frac{n}{2}\right)^n}}{\sqrt{2\pi\left(\frac{w_0}{e}\right)^{w_0}}\sqrt{2\pi\left(\frac{w_1}{e}\right)^{w_1}}\ldots}$$

is found instead of the minimum of the denominator of (4.48). Therefore, for problems involving $w!$, use of a well-known approximation (see Schlömilch's Comp. S. 438) amounts to substitution of $\sqrt{2\pi}(w/e)^w$ for $w!$.

If we denote the common value of the quotient (4.54c) by x, we obtain

$$w_1 = w_0 x, w_2 = w_0 x^2, \quad w_3 = w_0 x^3, etc. \tag{4.55}$$

The two Eqs. (4.46) and (4.47) become

$$w_0\left(1 + x + x^2 + \ldots + x^p\right) = n \tag{4.56}$$

$$w_0\left(x + 2x^2 + 3x^3 + \ldots + px^p\right) = \lambda \tag{4.57}$$

[One sees immediately that these equations differ negligibly from Eq. (4.42) and the preceding ones from my earlier work "Study of the thermal equilibrium of gas molecules." Boltzmann refers to Eq. (4.42) in his earlier study]

We can use the last equation to write

$$w_0 \cdot \frac{x^{p+1} - 1}{x - 1} = n \tag{4.58a}$$

$$w_0 x \cdot \frac{d}{dx}\left[\frac{x^{p+1} - 1}{x - 1}\right] = \lambda \tag{4.58b}$$

Carrying out the differentiation in the last equation

$$w_0 x \frac{px^{p+1} - (p+1)X^P + 1}{(x-1)^2} = \lambda \tag{4.59}$$

Dividing this equation by Eqs. (4.58a) and (4.58b) gives

$$\frac{px^{p+2} - (p+1)x^{p+1} + x}{(x^{p+1} - 1)(x - 1)} = \frac{\lambda}{n} \tag{4.60a}$$

Or

$$(pn - \lambda)x^{p+2} - (pn + n - \lambda)x^{p+1} + (n + \lambda)x - \lambda = 0 \tag{4.60b}$$

One can see immediately from Descartes' theorem[4] that this equation cannot have more than three real positive roots, of which two are=1. Again it is easy to see that both roots are not solutions of Eqs. (4.56) and (4.57) and also do not solve the problem, but that they showed up in the final equation merely as a result of

[4]Cardano's formula.

multiplying by the factor$(x - 1)^2$.[5] To be convinced of this, one need only derive the final equation directly by dividing the Eqs. (4.56) and (4.57). Following this division and having removed the variable x from the denominator and collecting powers of x throughout, we get the equation

$$(np - 1)x^p + (np - n - \lambda)x^{p-1} + (np - 2n - \lambda)x^{p-2} + \ldots (n - \lambda)x - \lambda \\ = 0, \tag{4.61}$$

Which is an equation of pth degree, and whose roots supply the solution to the problem. Thus, Eqs. (4.60a) and (4.60b) cannot have more positive roots than the solution requires. Negative or complex roots have no meaning for the solution to the problem. We note again that the largest allowed kinetic energy $P = p\epsilon$ is very large compared to the mean kinetic energy of a molecule

$$\frac{L}{n} = \frac{\lambda\epsilon}{n} \tag{4.62}$$

From which it follows that p is very large compared to λ/n. The polynomial Eq. (4.61), which shares the same real roots with Eqs. (4.60a), (4.60b), is negative for $x = 0$, $x = 1$; however it has the value

$$n(p + 1)\left(\frac{p}{2} - \frac{\lambda}{n}\right), \tag{4.63}$$

which is positive and very large, since p is very large compared to n. The only positive root occurs for x between zero and one, and we obtain it from the more convenient Eq. (4.60b). Since x is a proper fraction, then the pth [This appears to be a typographic error. The $(p + 2)^{\text{th}}$ power makes mathematical sense] and $(p + 1)^{\text{th}}$ powers are smaller and can be neglected, in which case we obtain

$$x = \frac{\lambda}{n + \lambda} \tag{4.64}$$

This is the value to which x tends for large p, and one can see the important fact that for reasonably large values of p the value of x depends almost exclusively on the ratio λ/n, and varies little with either λ or n providing their ratio is constant. Once one has found x, it follows from Eq. (4.58a) that

$$w_0 = \frac{1 - x}{1 - x^{p+1}} n \tag{4.67}$$

[5]This appears to be a typographic error. The $(p + 2)^{\text{th}}$ power makes mathematical sense.

Moreover, Eq. (4.55) gives the values of the remaining $w's$. It is seen from the quotients

$$\frac{w_0}{n}, \frac{w_1}{n}, \frac{w_2}{n}, \text{etc.}$$

That the probabilities of the various kinetic energy values for larger p are again dependent almost exclusively on the mean energy of the molecule. For infinitely large p we obtain the following limiting values

$$w_0 = \frac{n^2}{n+\lambda}, \quad w_1 = \frac{n^2\lambda}{(n+\lambda)^2}, \quad w_2 = \frac{n^2\lambda^2}{(n+\lambda)^3}, \text{etc.} \tag{4.68}$$

To establish whether we have a maximum or have a minimum, we need to examine the second variation of Eq. (4.49b). We note that w_0, w_1, w_2, etc. are very large, so we can use the approximation formula for $\ln\Gamma(w + 1)$

$$w \cdot (\ln w - 1) - \frac{1}{2}\ln w - \frac{1}{2}\ln(2\pi) + \frac{1}{12w} + \text{etc.}$$

Moreover, neglecting terms, which have second or higher powers of w in the denominator, obtain

$$\delta^2 M = \frac{(\delta w_0)^2}{w_0} + \frac{(\delta w_1)^2}{w_1} + \ldots \tag{4.69}$$

Therefore, we do in fact have a minimum. I also want to remark on the size of the term previously designated J. One easily finds that J is given by the following binomial coefficient

$$J = \binom{\lambda + n - 1}{\lambda}; \tag{4.70}$$

When you neglect terms that diminish with increasing λ or n

$$J = \frac{1}{\sqrt{2\pi}} \frac{(\lambda + n - 1)^{\lambda + n - \frac{1}{2}}}{(n-1)^{n-\frac{1}{2}} \lambda^{\lambda + \frac{1}{2}}} \tag{4.71}$$

Now $\lambda\epsilon/n$ is equal to the average kinetic energy μ of a molecule; therefore

$$\frac{\lambda}{n} = \frac{\mu}{\epsilon}, \tag{4.72}$$

Therefore, for large numbers one has

$$(\lambda + n - 1)^{\lambda+n-\frac{1}{2}} == \lambda^{\lambda+n-\frac{1}{2}}\left[\left(1+\frac{n-1}{\lambda}\right)^{\lambda}\right]^{1+\frac{2n-1}{2\lambda}} = \lambda^{\lambda+n-\frac{1}{2}}\cdot e^{n-1} \tag{4.73}$$

[Note: double equal is relational operator used to compare two variable values whether they are equal or not.] Therefore

$$J = \frac{1}{\sqrt{2\pi}}\frac{\lambda^{n-1}e^{n-1}}{(n-1)^{n-\frac{1}{2}}}, \tag{4.74}$$

Therefore, neglecting diminishing terms

$$\ln J = n\ln\frac{\lambda}{n} + n - \ln\lambda + \frac{1}{2}\ln n - 1 - \frac{1}{2}\ln(2\pi) \tag{4.75}$$

It goes without saying that these formulas are not derived here solely for finite p and n values, because these are unlikely to be of any practical importance, but rather to obtain formulas which provide the correct limiting values when p and n become infinite.

Nevertheless, it may help to demonstrate with specific examples of only moderately large values of p and n that these formulas are quite accurate, and though approximate, are of some value even here.

We first consider the earlier example, where $n = \lambda = 7$, *i.e.*, the number of molecules is seven, and the total kinetic energy is 7ϵ, and so the mean kinetic energy is ϵ. Suppose first that $p = 7$, so each molecule can only have $0, \epsilon, 2\epsilon, 3\epsilon, \ldots 7\epsilon$ of kinetic energy. Then Eq. (4.60b) becomes

$$6x^9 - 7x^8 + 2x - 1 = 0, \tag{4.76}$$

From which it follows

$$x = \frac{1}{2} + \frac{7}{2}x^8 - 3x^9. \tag{4.77}$$

Since x is close to $\frac{1}{2}$, we can set $x = \frac{1}{2}$ in the last two very small terms on the right-hand side and obtain

$$x = \frac{1}{2} + \frac{1}{2^9}(7-3) = \frac{1}{2} + \frac{1}{2^7} = 0.5078125 \tag{4.78}$$

You could easily substitute this value for x back into the right side of Eq. (4.77) and obtain a better approximation for$x$; since we already have an approximate value for x, a more rapid approach is to apply the ordinary Newton iteration method to Eq. (4.76) which results in

$$x = 0.5088742\ldots$$

From this, one finds in accordance with Eq. (4.55)

$$\begin{aligned} w_0 &= 3.4535 & w_4 &= 0.2316 \\ w_1 &= 1.7574 & w_5 &= 0.1178 \\ w_2 &= 0.8943 & w_6 &= 0.0599 \\ w_3 &= 0.4551 & w_7 &= 0.0304 \end{aligned}$$

These numbers satisfy the condition that

$$\sqrt{2\pi}\left(\frac{w_0}{e}\right)^{w_0} \cdot \sqrt{2\pi}\left(\frac{w_1}{e}\right)^{w_1} \ldots etc.$$

is minimized, while the minimized variables w obey the two constraints

$$w_0 + w_1 + w_2 + w_3 + w_4 + w_5 + w_6 + w_7 = 7, \tag{4.79a}$$

$$w_1 + 2w_2 + 3w_3 + 4w_4 + 5w_5 + 6w_6 + 7w_7 = 7, \tag{4.79b}$$

which minimum, incidentally because of the first of Eqs. (4.79a) and (4.79b), coincides with the minimum of $(w_0)^{w_0}(w_1)^{w_1}\ldots$ This provides only an approximate solution to our problem, which asks for so many (w_0) zeros, so many (w_1) ones, etc. with as many permutations as the resulting complexion permits, while the $w's$ simultaneously satisfy the constraints Eqs. (4.79a, 4.79b). Since p and n here are very small, one hardly expects any great accuracy, yet you already get the solution to the permutation problem by taking the nearest integer for each w, with the exception of w_3, for which you have to assign the value of 1 instead of 0.4551. In this manner, it is apparent

$$w_0 = 3, w_1 = 2, w_2 = w_3 = 1, w_4 = w_5 = w_6 = w_7 = 0$$

In addition, in fact we saw in the previous table that the complexion of 0001123 has the most permutations. We now consider the same special case with $n = \lambda = 7$, but set $p = \infty$; that is, the molecules may have kinetic energies of 0, 1, 2, 3…∞. We know then that the values of the variables w will vary little from those of the former case. In fact, we obtain

$$x = \frac{1}{2}; w_0 = \frac{7}{2} = 3.5, w_1 = \frac{w_0}{2} = 1.75, w_2 = \frac{w_1}{2} = 0.875 \; etc.$$

We consider a little more complicated example. Take $n = 13$, $\lambda = 19$, but we only treat the simpler case where $p = \infty$. Then we have

$$
\begin{array}{ll}
x = 19/32 & w_4 = 0.65605 \\
w_0 = 5.28125 & w_5 = 0.38950 \\
w_1 = 3.13574 & w_6 = 0.23133 \\
w_2 = 1.86815 & : \\
w_3 = 1.10493 & :
\end{array}
$$

Substituting here for the $w's$ the nearest integers, we obtain

$$w_0 = 5, w_1 = 3, w_2 = 2, w_3 = 1, w_4 = 1, w_5 = w_6 \ldots 0,$$

Already from the fact that $w_0 + w_1 + \ldots$ should=13, it is seen that again one of the $w's$ must be increased by one unit. From those $w's$ that are set=0, w_5 differs least from the next highest integer. We want therefore $w_5 = 1$, and obtain the complexion

0000011122345

whose digit sum is in fact=19. The number of permutations this complexion is capable of is

$$\frac{13!}{5!3!2!} = \frac{13!}{4!3!2!} \cdot \frac{1}{5}$$

A complexion whose sum of digits is also=19, and which one might suppose is capable of very many permutations, would be the following:

0000111222334.

The number of permutations is

$$\frac{13!}{4!3!3!2!} = \frac{13!}{4!3!2!} \cdot \frac{1}{6}$$

This is less than the number of permutations of the first complexion we found from the approximate formula. Likewise, we expect that the number of permutations of the two complexions

0000111122335

and

0000111122344

is smaller still. This is, for both complexions

$$\frac{13!}{4!4!2!2!} = \frac{13!}{4!3!2!} \cdot \frac{1}{8}$$

Other possible complexions are capable of still less permutations and it would be quite superfluous to follow these up here. It is seen from the examples given here that the above formula, even for very small values of p and n, gives values of w within one or two units of the true values. In the mechanical theory of heat, we are always dealing with extremely large numbers of molecules, so such small differences disappear, and our approximate formula provides an exact solution to the problem. We see also that the most likely state distribution is consistent with that known from gases in thermal equilibrium. According to Eq. (4.68) the probability of having a kinetic energy $s\epsilon$ is given by

$$w_s = \frac{n^2}{n+\lambda} \cdot \left(\frac{\lambda}{n+\lambda}\right)^s \tag{4.80}$$

Since $\lambda\epsilon/n$ is equal to the average kinetic [energy] of a molecule μ, which is finite, so n is very small compared to λ. So the following approximations

$$\frac{n^2}{n+\lambda} = \frac{n^2}{\lambda} = \frac{n\epsilon}{\mu}, \frac{\lambda}{n+\lambda} = 1 - \frac{n}{\lambda} = e^{-\frac{n}{\lambda}} = e^{-\frac{\epsilon}{\mu}} \tag{4.81}$$

hold, from which it follows that

$$w_s = \frac{n\epsilon}{\mu} e^{-\frac{\epsilon s}{\mu}}, \tag{4.82}$$

To achieve a mechanical theory of heat, these formulas must be developed further, particularly through the introduction of differentials and some additional considerations.

4.3.2.1 Kinetic Energies Exchange in a Continuous Manner

In order to introduce differentials into our formula we wish to illustrate the problem in the same manner as indicated on p171 (Wiss. Abhand. vol II) because this seems to be the best way to clarify the matter. Here each molecule was only able to have one of 0, ϵ, 2ϵ, ...$p\epsilon$ values for kinetic energy. We generated all possible complexions, *i.e.*, all the ways of distributing $1 + p$ values of the kinetic energy among the molecules, yet subject to the constraints of the problem, using a hypothetical urn containing infinitely many paper slips. Equal numbers of paper slips have kinetic energy values 0, ϵ, etc. written on them. To generate the first complexion, we draw a slip of paper for each molecule, and note the value of the kinetic energy assigned in this way to each molecule. Very many complexions are generated in the same way, they are assigned to this or that state distribution, and then we determine the most probable one. That state distribution which has the most complexions, we consider as the most likely, or corresponding to thermal equilibrium. Proceeding to the continuous kinetic energy case the most natural approach is as follows:

Taking ϵ to be some very small value, we assume that in the urn are very many slips of paper labeled with kinetic energy values between zero and ϵ. In the urn are also equal numbers of paper slips labeled with kinetic energy values between ϵ and 2ϵ, 2ϵ and 3ϵ up to infinity. Since ϵ is very small, we can regard all molecules with kinetic energy between x and $x + \epsilon$ as having the same kinetic energy. The rest of the calculation proceeds as in Sect. 4.1 above. We assume some complexion has been drawn; w_0 molecules have kinetic energy between zero and ϵ, w_1 molecules have values between ϵ and 2ϵ, w_2 have values between 2ϵ and 3ϵ, etc.

Here, because the variables w_0, w_1, w_2, etc. will be infinitely small, of the order of magnitude of ϵ, we prefer to write them as

$$w_0 = \epsilon f(0); \qquad w_1 = \epsilon f(\epsilon); \qquad w_2 = \epsilon f(2\epsilon) \; etc. \tag{4.83}$$

The probability of the state distribution in question is given, exactly as in Sect. 4.1, by the number of permutations that the elements of the state distribution are capable of, e.g., by the number

$$\frac{n!}{w_0!w_1!w_2!\ldots}$$

Again, the most likely state distribution, which corresponds to thermal equilibrium, is defined by the maximum of this expression, that is, when the denominator is minimized. We use again the reasonable approximation of Sect. 4.1, replacing $w!$ by the expression

$$\sqrt{2\pi}\left(\frac{w}{e}\right)^{w}$$

We can omit the term $\sqrt{2\pi}$ since it is a constant factor in the minimization; the key again is to replace minimization of the denominator with minimization of its logarithm; then we obtain the condition for thermal equilibrium, that

$$M = w_0 \ln w_0 + w_1 \ln w_1 + w_2 \ln w_2 + \ldots - n \tag{4.84}$$

is a minimum; while again satisfying the two constraints

$$n = w_0 + w_1 + w_2 + \ldots \tag{4.85}$$

$$L = \epsilon w_1 + 2\epsilon w_2 + 3\epsilon w_3 + \ldots \tag{4.86}$$

which are identical with Eqs. (4.46) and (4.47) of Sect. 4.1. Using Eq. (4.83) here, we replace the variables w by the function f and obtain thereby

$$M = \epsilon[f(0)\ln f(0) + f(\epsilon)\ln f(\epsilon) + f(2\epsilon)\ln f(2\epsilon) + \ldots] + \epsilon\ln\epsilon[f(0) + f(\epsilon) + f(2\epsilon) + \ldots] - n \tag{4.87}$$

Moreover, Eqs. (4.85) and (4.86) become

$$n = \epsilon[f(0) + f(\epsilon) + f(2\epsilon) + \ldots], \tag{4.88}$$

$$L = \epsilon[\epsilon f(\epsilon) + 2\epsilon f(2\epsilon) + 3\epsilon f(3\epsilon)\ldots]. \tag{4.89}$$

Using Eq. (4.88) the expression for M can also be written as

$$M = \epsilon[f(0)\ln f(0) + f(\epsilon)\ln f(\epsilon) + f(2\epsilon)\ln f(2\epsilon) + \ldots] - n + n\ln\epsilon \tag{4.90}$$

Since n and ϵ are constant (because ϵ has the same value for all possible complexions, and is constant between different state distribution), one can minimize

$$M' = \epsilon[f(0)\ln f(0) + f(\epsilon)\ln f(\epsilon) + f(2\epsilon) ln f(2\epsilon) + \ldots] \tag{4.91}$$

instead. As ϵ is made still smaller, the allowed values of kinetic energy approach a continuum. For vanishingly small ϵ, various sums in Eqs. (4.88), (4.89), (4.91) can be written in the form of integrals, leading to the following equations

$$M' = \int_0^\infty f(x)\ \ln f(x)dx \tag{4.92}$$

$$n = \int_0^\infty f(x)dx, \tag{4.93}$$

$$L = \int_0^\infty xf(x)dx, \tag{4.94}$$

The functional form of $f(x)$ is sought which minimizes expression (4.92) subject to the constraints (4.93) and (4.94), so one proceeds as follows: To the right side of Eq. (4.92) one adds Eq. (4.93) multiplied by a constant k, and Eq. (4.94) multiplied by a constant h. The resulting integral is

$$\int_0^\infty [f(x)\ln f(x) + kf(x) + hxf(x)]dx$$

where x is the independent variable, and f is the function to be varied. This results in

$$\int_0^\infty [lnf(x) + k + hx]\delta f(x)dx$$

Setting the quantity, which has been multiplied by $\delta f(x)$ in square brackets=0, and solving for the function $f(x)$, we obtain

$$f(x) = Ce^{-hx} \tag{4.95}$$

Here the constant e^{-k-1} is denoted by C for brevity. The second variation of M'

$$\delta^2 M' = \int_0^{\infty} \frac{[\delta f(x)]^2}{f(x)} \cdot dx, \tag{4.96}$$

is necessarily positive, since $f(x)$ is positive for all values of x lying between zero and ∞. By the calculus of variations M' is a minimum. From Eq. (4.95), the probability that the kinetic energy of a molecule lies between x and $x + dx$ at thermal equilibrium is

$$f(x)dx = Ce^{-hx}dx \tag{4.97}$$

The probability that the velocity of a molecule lies between ω and $\omega + d\omega$ would be

$$Ce^{-\frac{hm\omega^2}{2}} \cdot m\omega d\omega \tag{4.98}$$

where m is the mass of a molecule. Equation (4.98) gives the correct state distribution for elastic disks moving in two dimensions, for elastic cylinders with parallel axis moving in space, but not for elastic spheres, which move in space. For the latter the exponential function must be multiplied by $\omega^2 2d\omega$ not $\omega d\omega$. To get the right state distribution for the latter case we must set up the initial distribution of paper slips in our urn in a different way. To this point we assumed that the number of paper slips labeled with kinetic energy values between 0 and ϵ is the same as those between ϵ and 2ϵ. As also for slips with kinetic energies between 2ϵ and 3ϵ, 3ϵ and 4ϵ, etc.

Now, however, let us assume that the three velocity components along the three coordinate axes, rather than the kinetic energies, are written on the paper slips in the urn. The idea is the same: There are the same number of slips with u between 0 and ϵ, v between 0 and ξ, and w between 0 and η. The number of slips with u between ϵ and 2ϵ, v between zero and ξ, and w between zero and η is the same. Similarly, the number for which u is between ϵ and 2ϵ, v is between ξ and 2ξ, w is between zero andη. Generally, the number of slips for which u, v, w are between the limits u and $u + \epsilon$, v and $v + \xi$, w and $w + \eta$ are the same. Here u, v, w have any magnitude, while ϵ, ξ, η are infinitesimal constants. With this one modification of the problem, we end up with the actual state distribution established in gas molecules.

(LB footnote: We can of course, instead of using finite quantities ϵ, ξ, η and then taking the limit as they go to zero, write du, dv, dw from the outset, then the distribution of paper slips in the urn must be such that the number for which u, v, w are between u and $u + du$, v and $v + dv$, w and $w + dw$ are proportional to the product *dudvdw* and independent of *u*, *v*, and *w*. The earlier distribution of slips in the urn is characterized by the fact that although ϵ could be replaced by *dx*, kinetic energies between zero and *dx*, *dx* and 2*dx*, 2*dx* and 3*dx*, etc. occurred on the same number of slips.)

If we now define

$$w_{abc} = \epsilon\zeta\eta f(a\epsilon, b\zeta, c\eta) \tag{4.99}$$

As the number of molecules of any complexion for which the velocity components lie between the limits $a\epsilon$ and$(a + 1)\epsilon$, $b\zeta$ and $(b + 1)\zeta$, $c\eta$ and $(c + 1)\eta$, the number of permutations, or complexions of these elements for any state distribution, becomes

$$\mathcal{P} = \frac{n!}{\prod_{a=-p}^{a=+p}\prod_{b=-q}^{b=+q}\prod_{c=-r}^{c=+r} w_{abc}!} \tag{4.100}$$

where we first assume u adopts only values between $-p\epsilon$ and $+p\epsilon$, v between $-q\zeta$ and $+q\zeta$, w between $-r\eta$ and$+r\eta$. Where again, the most likely state distribution occurs when this expression, or if you will, its logarithm, is maximum. We again substitute

$$n!\text{by } \sqrt{2\pi}\left(\frac{n}{e}\right)^n \text{ and } w!\text{by } \sqrt{2\pi}\left(\frac{w}{e}\right)^w$$

where you can again immediately omit the factors of $\sqrt{2\pi}$ as they simply contribute additive constants $-\frac{1}{2}\ln 2\pi$ to $\ln \mathcal{P}$; omitting also the constant n ln n term, the requirement for the most probable state distribution is that the sum

$$-\sum_{a=-p}^{a=+p}\sum_{b=-q}^{b=+q}\sum_{c=-r}^{c=+r} w_{abc} \ln w_{abc}$$

is a maximum, which only differs from $\ln \mathcal{P}$ by an additive constant. The constraints that the number of molecules=n and that the total kinetic energy=L take the following form

$$n = \sum_{a=-p}^{a=p}\sum_{b=-q}^{b=q}\sum_{c=-r}^{c=r} w_{abc} \tag{4.101}$$

$$L = \frac{m}{2}\cdot\sum_{a=-p}^{a=p}\sum_{b=-q}^{b=q}\sum_{c=-r}^{c=r}\left(a^2\epsilon^2 + b^2\zeta^2 + c^2\eta^2\right) w_{abc} \tag{4.102}$$

Substituting for w_{abc} using Eq. (4.99), one immediately sees that the triple sums can in the limit be expressed as definite integrals; omitting an additive constant, the quantity to be maximized becomes

$$\Omega = -\int_{-\infty}^{+\infty}\int_{-\infty}^{+\infty}\int_{-\infty}^{+\infty} f(u,v,w)\ln f(u,v,w)\,dudvdw, \tag{4.103}$$

The two constraint equations become

$$n = \int_{-\infty}^{+\infty}\int_{-\infty}^{+\infty}\int_{-\infty}^{+\infty} f(u,v,w)dudvdw, \tag{4.104}$$

$$L = \frac{m}{2}\int_{-\infty}^{+\infty}\int_{-\infty}^{+\infty}\int_{-\infty}^{+\infty} \left(u^2+v^2+w^2\right)f(u,v,w)dudvdw, \tag{4.105}$$

The variable Ω, which differs from the logarithm of the number of permutations only by an additive constant, is of special importance for this work and we call it the permutability measure [bold emphasize by L.B.]. I note, incidentally, that suppression of the additive constants has the advantage that the total permutability measure of two bodies is equal to the sum of the permutability measures of each body. Thus, it is the maximum of the quantity (4.103) subject to the constraints (4.104) and (4.105) that is sought. No further explanation of this problem is needed here; it is a special case of the problem I have already discussed in my treatise "On the thermal equilibrium of gases on which external forces act"[6] in the section which immediately precedes the appendix. There I provided evidence that this state distribution corresponds to the condition of thermal equilibrium. Boltzmann's use of term thermal equilibrium is important. In mechanics, this corresponds to equilibrium of energy. Thus, one is justified in saying that the most likely state distribution corresponds with the condition of thermal equilibrium. For if an urn is filled with slips of paper labeled in the manner described earlier, the most likely sampling will correspond to the state distribution for thermal equilibrium. We should not take this for granted, however, without first defining what is meant by the most likely state distribution. For example, if the urn were filled with slips labeled in the original manner, then the statement would be incorrect.

The reasoning needed to arrive at the correct state distribution will not escape those experienced in working with such problems. The same considerations apply to the following circumstance: If we group all the molecules whose coordinates at a particular time lie between the limits

[6]Wien. Ber. (1875) 72:427-457 (Wiss. Abhand. Vol. II, reprint 32).

$$\xi \text{ and } \xi + d\xi, \eta \text{ and } \eta + d\eta, \zeta \text{ and } \zeta + d\zeta, \tag{4.106}$$

In addition, whose velocity components lie between the limits

$$u \text{ and } u + du, v \text{ and } v + dv, w \text{ and } w + dw, \tag{4.107}$$

Moreover, let these molecules collide with other molecules under specific conditions; after a certain time their coordinates will lie between the limits

$$\Xi \text{ and } \Xi + d\Xi, H \text{ and } H + dH, Z \text{ and } Z + dZ, \tag{4.108}$$

Moreover, their velocity components will lie between the limits

$$U \text{ and } U + dU, V \text{ and } V + dV, W \text{ and } W + dW, \tag{4.109}$$

Then at any time

$$d\xi \cdot d\eta \cdot d\zeta \cdot du \cdot dv \cdot dw = d\Xi \cdot dH \cdot dZ \cdot dU \cdot dV \cdot dW. \tag{4.110}$$

This is a general result. If at time zero the coordinates and velocity components of arbitrary molecules (material points) lie between the limits (4.106) and (4.107) and unspecified forces act between these molecules, [Because Boltzmann does not make any assumption about the type of forces acting between the molecules therefore shear forces between molecules are not excluded.] So that at time t the coordinates and velocity components lie between the limits (4.108) and (4.109), then Eq. (4.110) is still satisfied.

If, instead of the velocity components, one uses the kinetic energy x and the velocity direction defined by the two angles α and β, to describe the action of the forces, these variables would initially lie between the limits

$$\xi \text{ and } \xi + d\xi, \eta \text{ and } \eta + d\eta, \zeta \text{ and } \zeta + d\zeta,$$

$$x \text{ and } x + dx, \alpha \text{ and } \alpha + d\alpha, \beta \text{ and } \beta + d\beta,$$

Then after the action of the forces lie between the limits

$$\Xi \text{ and } \Xi + d\Xi, H \text{ and } H + dH, Z \text{ and } Z + dZ,$$

$$X \text{ and } X + dX, A \text{ and } A + dA, B \text{ and } B + dB,$$

And so

$$\begin{aligned} d\xi \cdot d\eta \cdot d\zeta \cdot \sqrt{x} \cdot dx \cdot \varphi(\alpha, \beta) d\alpha \cdot d\beta = d\Xi \cdot dH \cdot dZ \cdot \sqrt{X} \cdot dX \\ \cdot \varphi(A, B) dA \cdot dB \end{aligned} \tag{4.111}$$

So the product of the differentials $du \cdot dv \cdot dw$ becomes $dU \cdot dV \cdot dW$. Therefore, the list of slips in the urn must be labeled uniformly with velocity components lying between u and $u + du$, v and $v + dv$, w and $w + dw$, whatever values u, v, w have. Given a certain value of the coordinates, the velocities must be described by the corresponding "moments." On the other hand, $\sqrt{x}dx$ goes over to $\sqrt{X}dX$. With the introduction of kinetic energy, slips must be labeled so that you have the same number with kinetic energy between x and $x + \sqrt{x}dx$ where dx is constant but x is completely arbitrary. This last sentence is in agreement with my "Remarks on some problems of the mechanical theory of heat" (Wiss. Abhand. Vol II, reprint 39 p121), where I demonstrated that this is the only valid way to find the most likely state distribution corresponding to the actual thermal equilibrium; here we have demonstrated *a posteriori* that this leads to the correct state distribution for thermal equilibrium, that which is the most likely in our sense.

Of course, it is easy to analyze those cases where other conditions exist besides the principle of conservation of kinetic energy. Suppose, for example, a very large number of molecules for whom (1) the total kinetic energy is constant; (2) the net velocity of the center of gravity in the directions of the x-axis; (3) y-axis; and (4) z-axis are given. The question arises, what is the most probable distribution of the velocity components among the molecules, using the term in the previous sense. We then have exactly the same problem, except with four constraints instead of one. The solution gives us the most probable state distribution

$$f(u, v, w) = Ce^{-h\left[(u-\alpha)^2+(v-\beta)^2+(w-\gamma)^2\right]}, \tag{4.112}$$

where C, h, α; β; γ are constants. This is in fact the state distribution for a gas at thermal equilibrium at a certain temperature, not at rest, but moving with a constant net velocity. You can treat similar problems such as the rotation of a gas in the same manner, by adding in the appropriate constraint equations, which I have discussed in my essay "On the definition and integration of the equations of molecular motion in gases." (Wiss. Abhand. Vol II, reprint 36)

Some comment regarding the derivation of Eq. (4.103) from Eq. (4.100) is required here. The formula for $x!$ is

$$\sqrt{2\pi x}\left(\frac{x}{e}\right)^x e^{\frac{1}{12x}+\cdots}.$$

The substitution of this into Eq. (4.100) gives

$$\mathcal{P} = \frac{\sqrt{2\pi}n^{n+\frac{1}{2}} \cdot e^{\frac{1}{12n}+\cdots}}{(2\pi)^{p+q+r+\frac{3}{2}}\Pi_{a=-p}^{a=+p}\Pi_{b=-q}^{b=+q}\Pi_{c=-r}^{c=+r}(w_{abc})^{w_{abc}+\frac{1}{2}} \cdot e^{\frac{1}{12w_{abc}}+\cdot}} \tag{4.113}$$

From which it follows

$$\ln \mathcal{P} = \left(n + \frac{1}{2}\right) \ln n + \frac{1}{12n} + \cdots - (p + q + r + 1) \ln 2\pi - \sum_{a=-p}^{a=+p} \sum_{b=-q}^{b=+q}$$
$$\times \sum_{c=-r}^{c=+r} \left[\left(w_{abc} + \frac{1}{2}\right) \ln w_{abc} + \frac{1}{12w_{abc}} + \cdots \right] \tag{4.114}$$

First note that in determining the magnitude of $\mathcal{P}$, in the limit of a very large number of molecules, n (and thus also of w_{abc}), other small quantities such as ϵ; ζ; η can be treated as infinitesimals. So all terms which have n or w_{abc} in the denominator can be neglected, and the $\frac{1}{2}$ in the term $w_{abc} + \frac{1}{2}$. The terms containing w_{abc} scale with the total mass of the gas, while the related $\frac{1}{2}$ terms refer only to a single molecule. So the latter quantities can be neglected as the number of molecules increases. We then get

$$\ln \mathcal{P} = n \ln n - (p + q + r + 1) \ln 2\pi - \sum_{a=-p}^{a=+p} \sum_{b=-q}^{b=+q} \sum_{c=-r}^{c=+r} w_{abc} \ln w_{abc}. \tag{4.115}$$

Substituting $\epsilon\zeta\eta f(a\epsilon, b\zeta, c\eta)$, for w_{abc} we obtain

$$\ln \mathcal{P} = n \ln n - (p + q + r + 1) \ln 2\pi - n \ln (\epsilon\zeta\eta) - \sum_{a=-p}^{a=+p} \sum_{b=-q}^{b=+q}$$
$$\times \sum_{c=-r}^{c=+r} \epsilon\zeta\eta f(a\epsilon, b\zeta, c\eta) \ln \mathrm{f}(a\epsilon, b\zeta, c\eta). \tag{4.116}$$

One sees that aside from the triple sum, the terms on the right-hand side are constant, and so can be omitted. We also let ϵ, ζ, η decrease while p, q, r increases infinitely, so the triple sum goes over into a triple integral over limits $-\infty$ to $+\infty$ and from $ln\, \mathcal{P}$ we arrive immediately at the expression given by Eq. (4.103) for the permutability measure Ω. The critical condition is that the number of molecules is very large; this means that w_{abc} is large compared to $\frac{1}{2}$; also that the velocity components between the limits $a\epsilon$ and $(a + 1)\epsilon$, $b\zeta$ and $(b + 1)\zeta$, $c\eta$ and $(c + 1)\eta$ are identical to those between the limits u and $u + du$, v and $v + dv$, w and $w + dw$. This may appear strange at first sight, since the number of gas molecules is finite albeit large, whereas *du*,*dv*, *dw* are mathematical differentials. However, on closer deliberation this assumption is self-evident. For all applications of differential calculus to the theory of gases are based on the same assumption, namely: diffusion, ***internal friction*** [bold emphasize by this author], heat conduction, etc. In each infinitesimal volume element *dxdydz* there are still infinitely many gas molecules whose velocity components lie between the limits u and $u + du$, v and $v + dv$, w and$w + dw$. The above assumption is nothing more than that very many molecules have velocity components lying within these limits for every u, v, w.

4.3.2.2 Consideration of Polyatomic Polyatomic Gas Molecules and External Forces

I will now generalize the formulas obtained so far, by first extending them to so-called polyatomic gas molecules and then including external forces and thereby finally beginning to extend the discussion to any **solid** and **liquid** [bold emphasize by C.B.]. In order not to consider too many examples, I will in each case deal with the most important case, where, aside from the equation for kinetic energy, there is no other constraint.

The first generalization can be applied to our formulas without difficulty. So far, we assumed each molecule was an elastic sphere or a material point, so that its position in space was entirely defined by three variables (e.g., three orthogonal coordinates). We know that this is not the case with real gas molecules. We shall therefore assume that three coordinates are insufficient to completely specify the position of all parts of a molecule in space; rather r variables will be necessary

$$p_1, p_2, p_3 \cdots p_r,$$

the so-called generalized coordinates. Three of them, p_1, p_2, p_3, are the orthogonal coordinates of the center of mass of the molecule, the others can be either the coordinates of the individual atoms relative to the center of mass, the angular direction, or whatever specifies the location of every part of the molecule. We will also remove the restriction that only one type of gas molecule is present. We assume instead, there exists a second type whose every molecule has the generalized coordinates

$$p'_1, p'_2, p'_3, \ldots\ldots\ldots\ldots p'_r$$

For the third type the generalized coordinates are

$$p''_1, p''_2, p''_3 \cdots p''_{r''},$$

If there are $v + 1$ types of molecules, the generalized coordinates of the final type are

$$p_1^{(v)}, p_2^{(v)}, p_3^{(v)} \cdots p_{r^{(v)}}^{(v)}.$$

The first three coordinates are always the orthogonal coordinates of the center of mass. Of course, the necessary assumption is that many molecules of each type are present. Let l be the total kinetic energy of the first type of gas; χ is its potential energy[7] (so that $\chi + l$ is constant if internal forces only are acting). Furthermore

[7]The quantity χ called "Kraftfunktion" or "Ergal" by Boltzmann is translated as potential energy.

$$q_1, q_2, q_3 \ldots q_r$$

are the momentum coordinates corresponding to p_1, $p_2 \ldots p_r$ [Note that here Boltzmann defines momentum as a new additional axis]. We can think of l in terms of the coordinates $p_1, p_2 \ldots p_r$ and their derivatives with respect to time

$$\dot{p}_1, \dot{p}_2, \dot{p}_3 \ldots \dot{p}_r,$$

Moreover, denote the quantities $c_1(dl/d\dot{p}_1); c_2(dl/d\dot{p}_2)$ by q_1, $q_2 \ldots$, where c_1, $c_2 \ldots$ are arbitrary constants. I would like to note here that in my essay "Remarks on some problems in the mechanical theory of heat" Sect. 4.3 (Wiss. Abhand. Vol II, reprint 39) the indexed variables designated p_i referred to coordinate derivatives, while here they have been designated by q; this mistake would probably not have caused any misunderstanding.

We denote with the appropriate accents the analogous quantities for other types of molecules. According to the calculations of Maxwell, Watson, and myself, in a state of thermal equilibrium, the number of molecules for which the magnitudes of p_4; $p_5 \ldots p_r$, q_1, q_2, $q_3 \ldots q_r$ lie between the limits

$$p_4 \text{ and } p_4 + dp_4; p_5 \text{ and } p_5 + dp_5, \text{etc.} q_r \text{ and } q_r + dq_r \tag{4.117}$$

is given by

$$Ce^{-h(\chi+1)} dp_4 dp_5 \ldots dq_r \tag{4.118}$$

where C and h are constants, independent of p and$q$. Analogous expressions hold of course for the other molecular species with the same value of h, but differing values of C. Exactly the same equation as (4.118) is also obtained using the methods of Sects. 4.1 and 4.2. Consider all those molecules of the first type, for which the variables p_4, $p_5 \ldots q_r$ at some time 0 lie between the limits (4.117), after a lapse of sometime t, the values of the same variables lie between the limits

$$P_4 \text{ and } P_4 + dP_4; P_5 \text{ and } P_5 + dP_5 \textit{ etc.} \quad Q_r \text{ and } Q_r + dQ_r \tag{4.119}$$

The general principle already invoked gives the following equation

$$dp_4 \cdot dp_5 \ldots dq_r = dP_4 \cdot dP_5 \ldots dQ_r \tag{4.120}$$

There is of course also

$$dp_1 \cdot dp_2 \ldots dp_3 = dP_1 \cdot dP_2 \ldots dP_3 \tag{4.120a}$$

So that, in fact, for the variables

$$p_{4,}\ \ p_5, \ldots q_r$$

the product of their differentials does not change during a constant time interval. Therefore, we must now imagine $v + 1$ urns. In the first are slips of paper, upon which are written all possible values of the variables $p_4, p_5 \ldots q_r$; and the number of slips which have values within the limits of Eq. (4.117) is such that, when divided by the product $dp_4, \cdot dp_5 \ldots q_r$ it is a constant.

Similarly, for the labeling of the slips with the variables $p'_4, p'_5 \ldots q'_r$ in the second urn, except that for the latter the constant can have a different value. The same applies to the other urns. We draw from the first urn for each molecule of the first type, from the second urn for molecules of the second type, etc. We now suppose that the values of the variables for each molecule are determined by the relevant drawings. It is of course entirely chance that determines the state distributions for the gas molecules, and we must first discard those state distributions which do not have the prescribed value for total kinetic energy. It will then be most likely that the state distribution described by Eq. (4.118) will be drawn, *i.e.*, that one corresponding to thermal equilibrium. The proof of this is straightforward. Therefore, the results found in the first two sections can be readily generalized to this case.

We want to generalize the problem further, assuming that the gas is composed of molecules specified exactly as before. But now so-called external forces are acting, e.g., those like gravity, which originate outside the gas. For details on the nature of these external forces, and how to treat them, see my treatise "On the thermal equilibrium of gases on which external forces act".[8] The essence of the solution to the problem remains the same. Only now, the state distribution will no longer be the same at all points of the vessel containing the gas; therefore $dp_1 \cdot dp_2 \cdot dp_3 = dP_1 \cdot dP_2 \cdot dP_3$ will no longer hold. We will now understand the generalized coordinates $p_1, p_2 \ldots p_r$ more generally to determine the absolute position of the molecule in space and the relative position of its constituents.

The notion that p_1, p_2, p_3 are just the orthogonal coordinates of the center of gravity is dropped. The same is true for the molecules of all the other types of gas. There is one further point to notice. Previously the only necessary condition was that throughout the vessel very many molecules of each type were present; now it is required that even in a small element of space, over which the external forces do not vary significantly in either size or direction, very many molecules are present (a condition, incidentally, which must hold for any theoretical treatment of problems where external forces on gases come into play). This is because our method of sampling presupposes that the states of many molecules can be considered equivalent, in the sense that the state distribution is not changed when the states of these molecules are exchanged. The probability of a state distribution is then determined by the number of complexions of which this state distribution is capable of.

This is why, for the case just considered, with $v + 1$ molecular species present, $v + 1$ urns must be constructed.

[8]Wien. Ber. (1875) 72:427–457.

We assume first that a complexion has been drawn where

$$w_{000\ldots} = f(0, 0, 0, \ldots)\alpha\beta\gamma\ldots \tag{4.121}$$

molecules whose variables $p_1, p_2, \ldots q_r$ lie between limits 0 and α, 0 and β, 0 and γ, etc. Furthermore, exactly

$$w_{10000\ldots} = f(\alpha, 0, 0, \ldots)\alpha\beta\gamma\ldots \tag{4.122}$$

molecules have the same variables within the limits α and 2α, 0 and β, 0 and γ, etc. and generally

$$w_{abc} = f(a\alpha, b\beta, c\gamma\cdots k\kappa)\alpha\beta\gamma\cdots\kappa \tag{4.123}$$

molecules with variables $p_1, p_2, \ldots q_r$ between limits

$$a\alpha \text{ and } (a+1)\alpha, b\beta \text{ and } (b+1)\beta \ldots k\kappa \text{ and } (k+1)\kappa, \tag{4.123}$$

These limits are so close that we can equate all the values in between, then it is as if the variable p_1 could only take the values 0, α, 2α, 3α, etc., variable p_2 could take the values 0, β, 2β, 3β, etc. Let n be the total number of molecules of the first type. We again distinguish the variables for the other gases by the corresponding accents, so that

$$\mathcal{P} = \frac{n!n'!n''!\ldots n^{(v)}!}{\Pi w_{abc\ldots k}!\Pi w'_{a'b'\ldots k'}\Pi w''_{a''b''\ldots k''}\Pi w^{(v)}_{a^{(v)}b^{(v)}\ldots k^{(v)}}!} \tag{4.124}$$

is the possible number of permutations of the elements of this complexion, which we call the permutability. The products are to be read so that the indices $a, b\ldots, a'. b'\ldots$ etc. run over all possible values, *i.e.*, $-\infty$ to $+\infty$ for orthogonal coordinates, zero to 2π for angular coordinates, and so on. Consider first the case where p_1 really can take only the values 0, α, 2α, 3α, …, and similarly with the other variables; then expression (4.124) is just the number of complexions this state distribution could have; this number is, according to the assumptions made above, a measure of the probability of the state distribution. The variables w and n are all very large; we can again therefore replace $w!$ with $\sqrt{2\pi}(w/e)^w$. We also denote the sum

$$n \ln n + n' \ln n' + \ldots n^{(v)} \ln n^{(v)}$$

by N, so we can also replace $n!$ by $\sqrt{2\pi}(n/e)^n$ and then immediately take the logarithm

$$\ln \mathcal{P} = N - C \ln 2\pi$$
$$- \left[\sum w_{ab\ldots} \cdot \ln w_{ab\ldots} + \sum w'_{a'b'\ldots} \cdot \ln w'_{a'b'\ldots} + \ldots \right]. \tag{4.125}$$

The sums are to be understood in the same sense as the products above. $2C$ is the number of factorials in the denominator of Eq. (4.124) minus $v + 1$. Let us now substitute the expression (4.123) for the variables w into Eq. (4.125) and then take the limit of infinitesimal $\alpha, \beta, \gamma \ldots$

Omitting unnecessary constants, the magnitude we obtain for the permutability measure, denoted by Ω, is

$$\Omega = -\left[\int \int \ldots f(p_1, p_2 \ldots q_r) \ln f(p_1, p_2 \ldots q_r) dp_1 dp_2 \ldots dq_r + \right.$$
$$\left. \int \int \ldots f(p'_1, p'_2 \ldots q'_{r'}) \ln f(p'_1, p'_2 \ldots q'_{r'}) \mathrm{d}p'_1 \mathrm{d}p'_2 \ldots \mathrm{d}q'_{r'} + \ldots \right]. \tag{4.126}$$

The integration is to extend over all possible values of the variables. I have in my paper "On the thermal equilibrium of gases on which external forces act" demonstrated that the expression in the square brackets is at a minimum for a gas in a state of thermal equilibrium, including, of course, the kinetic energy constraint equation.

4.3.2.3 On the Conditions for the Maximum of the Power-Exponent Free Product Determining the State Distribution Function

Before I go into the treatment of the second law, I want to concisely treat a problem whose importance I believe I have shown in Sect. 4.1, in the discussion of the work of Mr. Oskar Emil Meyer's on this subject, namely the problem of finding the maximum of the product of the probabilities of all possible states. However, I want to deal with this problem only for mono-atomic gases, and with no other constraint than the equation for the kinetic energy. We first consider the simplest case where only a discrete number of kinetic energy values, $0, \epsilon, 2\epsilon \ldots p\epsilon$ are possible, and to start we use kinetic energies, not velocity components, as variables. We again denote by $w_0, w_1, w_2 \ldots w_v$ the number of molecules with kinetic energy $0, \epsilon, 2\epsilon \ldots p\epsilon$.

If we treat the subject in the usual way, the following relationship holds: The quantity

$$B = w_0 \cdot w_1 \cdot w_2 \ldots w_p \tag{4.127}$$

or, if you prefer, the quantity

$$\ln B = \ln w_0 + \ln w_1 + \ln w_2 + \ldots \ln w_p \tag{4.128}$$

must be a maximum, with the constraints

$$n = w_0 + w_1 + w_2 + w_2 \ldots + w_p \tag{4.129}$$

and

$$L = \left(w_1 + 2w_2 + 3w_3 + \cdots + pw_p\right)\epsilon. \tag{4.130}$$

If to Eq. (4.128) we add Eq. (4.129) multiplied by h, and add Eq. (4.130) multiplied by k, then set the partial derivatives of the sum with respect to w_0, w_1, w_2... equal to zero, we obtain the equations

$$\frac{1}{w_0} + h = 0, \quad \frac{1}{w_1} + h + k = 0, \quad \frac{1}{w_2} + h + 2k = 0 \; etc. \tag{4.131}$$

from which, by elimination of the constants h and k

$$\frac{1}{w_1} - \frac{1}{w_0} = \frac{1}{w_2} - \frac{1}{w_1} = \frac{1}{w_3} - \frac{1}{w_2} = \ldots \tag{4.132}$$

or

$$\frac{1}{w_0} = a, \quad \frac{1}{w_1} = a + b, \quad \frac{1}{w_2} = a + 2b, \quad \ldots \frac{1}{w_p} = a + pb. \tag{4.133}$$

Substituting these values into Eqs. (4.129) and (4.130) the two constants a and b can be determined:

$$n = \frac{1}{a} + \frac{1}{a+b} + \frac{1}{a+2b} + \cdots + \frac{1}{a+pb}, \tag{4.134}$$

$$L = \frac{\epsilon}{a+b} + \frac{2\epsilon}{a+2b} + \frac{3\epsilon}{a+3b} + \cdots + \frac{p\epsilon}{a+pb}. \tag{4.135}$$

The direct determination of the two unknowns a and b from these equations would be extremely lengthy. The method of *Regula Falsi* would provide a more rapid solution for each special case; I have not troubled myself with such calculations, but will give here only a general discussion of how the expected solutions can be easily obtained, keeping in mind that these methods can only provide an approximation solution to the problem, since only positive integers are allowed, but fractional values are not. The first point to note is that the problem ceases to have any meaning as soon as the product $p \cdot (p + 1)/2$ is greater than L/ϵ. Because then it necessarily follows that one of the $w's$, and so the product B, is zero.

Then there is no question of a maximum value for B. For the problem to make any sense, an excessive value for the kinetic energy cannot be possible. If

$$p \cdot \frac{p+1}{2} = \frac{L}{\epsilon} \tag{4.136}$$

Then all the $w's$ from w_0 onwards must be equal to one for B to be nonzero. A greater variation in values can occur only if smaller values of p are chosen. Then, when n is large the above equations provide usable approximations. First, a will be significantly smaller than b, so w_0 is very large, and w_1 will be much smaller; w_2 will be close to $w_1/2$, w_3 will be close to $2w_2/3$, etc. In general, the decrease in the variable w with increasing index will be fairly insignificant when the maximum of $w_0 \cdot w_1 \cdot w_2 \ldots$ is sought, rather than the maximum of $w_0^{w_0} w_1^{w_1} w_2^{w_2} \ldots$. Given much smaller p values, the value of a is not much less than b, so w_0 is also not that much larger than the other $w's$; then w_2 is greater than $w_1/2$, w_3 is greater than $(2/3)w_2$, etc. The decrease of w with increasing index is even less. Decreasing p still further, a will dominate, and there will be hardly any decrease in w with increasing index. Finally b becomes negative, and the size of w will even increase with increasing index. The following cases provide examples; for each the integer values of the $w's$ which maximize B are given.

$$\begin{aligned}
&n = 30, L = 30\epsilon, p = 5, w_0 = 17, w_1 = 5, w_2 = 3, w_3 = 2, w_4 = 2, w_5 = 1.\\
&n = 31, L = 26\epsilon, p = 4, w_0 = 18, w_1 = 6, w_2 = 3, w_3 = 2, w_4 = 2.\\
&n = 40, L = 40\epsilon, p = 5, w_0 = 23, w_1 = 7, w_2 = 3, w_3 = 3, w_4 = 2, w_5 = 2.\\
&n = 40, L = 40\epsilon, p = 6, w_0 = 24, w_1 = 6, w_2 = 3, w_3 = 3, w_4 = 2, w_5 = 1, w_6 = 1.\\
&n = 18, L = 45\epsilon, p = 5, w_0 = 3, w_1 = 3, w_2 = 3, w_3 = 3, w_4 = 3, w_5 = 3.\\
&n = 23, L = 86\epsilon, p = 5, w_0 = 1, w_1 = 2, w_2 = 2, w_3 = 3, w_4 = 4, w_5 = 11.
\end{aligned} \tag{4.137}$$

Let us now turn to the case where the value of the kinetic energy is continuous; first, consider the kinetic energy x as the independent variable, so the problem, in our view is the following: The expression

$$Q = \int_0^P \ln f(x) dx \tag{4.138}$$

becomes a maximum, while at the same time

$$n = \int_0^P f(x) dx \quad \text{and} \quad L = \int_0^P x f(x) dx \tag{4.139}$$

are constant. P is also constant. I have purposefully set the upper integration limit to P, not ∞. It is then still straightforward to allow P to increase more and more. Proceeding accordingly, we obtain:

$$\delta \int_{0}^{P} [\ln x + hf(x) + kxf(x)]dx = \int_{0}^{P} \left[\frac{1}{f} + h + kx\right] dx\delta f = 0 \tag{4.140}$$

From which it follows

$$f = -\frac{1}{h + kx} = \frac{1}{a + bx}, \tag{4.141}$$

If we set $h = \ - a$; $k = \ - b$. To determine these two constants we use the equations

$$n = \int_{0}^{P} f(x)dx = \frac{1}{b} \ln \frac{a + bP}{a}, \tag{4.142}$$

$$L = \int_{0}^{P} xf(x)dx = \frac{P}{b} - \frac{a}{b_2} \ln \frac{a + bP}{a}, \tag{4.143}$$

which, writing a/b as α, leads to:

$$L + \alpha n = \frac{P}{b}, \quad bn = \ln \left(1 + \frac{P}{\alpha}\right) \tag{4.144}$$

and also

$$(L + \alpha n) \ln \left(1 + \frac{P}{\alpha}\right) = Pn. \tag{4.145}$$

From this transcendental equation α must be determined, from which it is easy to obtain a and b. Since Pn is the kinetic energy which the gas would have if every molecule in it had the maximum possible kinetic energy P, we see immediately that Pn is infinitely greater than L. L/n is the average kinetic energy of a molecule. It is then easy to verify that P/α cannot be finite because then $Pn/\alpha n$ would be finite, and in the expression $L + \alpha nL$ could be neglected. But then in Eq. (4.145) only P/α terms would remain, and only vanishingly small values of this term could satisfy the equation, which is inconsistent with the original assumption. Nor can P/α be vanishingly small because then L would again be vanishingly small compared to αn. Furthermore

$$\ln\left(1+\frac{P}{\alpha}\right) \tag{4.146}$$

could be expanded in powers of P/α, and Eq. (4.145) would yield a finite value for P/α. There remains only the possibility that P/α is very large. Since

$$\frac{\alpha n}{Pn}\ln\frac{Pn}{\alpha n} \tag{4.146}$$

vanishes, Eq. (4.145) gives

$$\alpha = \frac{a}{b} = pe^{-\frac{nP}{L}}, \tag{4.147}$$

from which follows:

$$b = \frac{P}{L}, \qquad a = \frac{p^2}{L}e^{-\frac{nP}{L}} \tag{4.148}$$

By the approach used in this section, using the mean kinetic energy of a molecule, these equations show that in the limit of increasing p, W, L the probability of dispersion in kinetic energy remains indeterminate. We now want to consider a second, more realistic problem. We take the three velocity components u, v, w parallel to the three coordinate axes as the independent variables, and find the maximum of the expression

$$Q = \int_{-\infty}^{+\infty}\int_{-\infty}^{+\infty}\int_{-\infty}^{+\infty} \ln f(u,v,w)dudvdw, \tag{4.149}$$

while simultaneously the two expressions

$$n = \int_{-\infty}^{+\infty}\int_{-\infty}^{+\infty}\int_{-\infty}^{+\infty} f(u,v,w)dudvdw, \tag{4.150}$$

$$L = \int_{-\infty}^{+\infty}\int_{-\infty}^{+\infty}\int_{-\infty}^{+\infty} (u^2+v^2+w^2)f(u,v,w)dudvdw \tag{4.151}$$

remain constant, integrating over u, v, w. If the velocity magnitude is

$$\omega = \sqrt{u^2+v^2+w^2} \tag{4.152}$$

In addition, its direction is given by the two angles θ and φ (length and breadth[9]); we have as is well known

$$dudvdw = \omega d\omega \sin\theta d\theta d\varphi, \tag{4.153}$$

Hence

$$Q = 4\pi \int_0^P \ln f(\varrho\Sigma\sigma)\varrho\Sigma\sigma^2 d\varrho\Sigma\sigma, 5 \tag{4.154}$$

$$n = 4\pi \int_0^P \varrho\Sigma\sigma^2 f(\varrho\Sigma\sigma)d\varrho\Sigma\sigma, \tag{4.155}$$

$$L = 4\pi \int_0^P \varrho\Sigma\sigma^2 f(\varrho\Sigma\sigma)d\varrho\Sigma\sigma \tag{4.156}$$

If there are no external forces, then clearly $f(u, v, w)$ is independent of direction of the velocity. Instead of integrating to infinity, we intentionally integrate to a finite value of P. Evaluating $f(\varrho\Sigma\sigma)$ just as we did for $f(x)$ earlier, we obtain

$$f(\varrho\Sigma\sigma) = -\frac{1}{h + k\varrho\Sigma\sigma^2} = \frac{1}{a^2 + b^2\varrho\Sigma\sigma^2}, \tag{4.157}$$

where we set $-h = a^2$ and $-k = b^2$. The two constant a and b are to be determined from Eqs. (4.155) and (4.156) which become, given the value of $f(\varrho\Sigma\sigma)$:

$$n = 4\pi \int_0^P \frac{\varrho\Sigma\sigma^2 d\varrho\Sigma\sigma}{a^2 + b^2\varrho\Sigma\sigma^2} = 4\pi\left(\frac{P}{b^2} - \frac{a}{b^3}arctg\frac{bP}{a}\right) \tag{4.158}$$

$$L = 4\pi \int_0^P \frac{\varrho\Sigma\sigma^4 d\varrho\Sigma\sigma}{a^2 + b^2\varrho\Sigma\sigma^2} = \frac{4\pi P^3}{3b^2} - \frac{a^2}{b^3}n \tag{4.159}$$

From the last equation we get

[9]Translator note: Altitude and azimuth?

$$\frac{4\pi}{b^2} = \frac{3}{P^3}\left(L + \frac{a^2}{b^2}n\right). \tag{4.160}$$

Substituting this equation into the first equation of (4.159) gives

$$n = 3\left(\frac{L}{P^2} + \frac{a^2}{b^2P^2}n\right)\left(1 - \frac{a}{Pb}arctg\frac{bP}{a}\right). \tag{4.161a}$$

If however b_2 is negative, we put $-b_2$ instead of b_2 and obtain:

$$f(\varrho\Sigma\sigma) = \frac{1}{a^2 - b^2\varrho\Sigma\sigma^2},$$

$$n = 4\pi\left(-\frac{P}{b^2} + \frac{a}{2b^3}n\left(\frac{a+bP}{a-Pb}\right)\right)$$

$$L = \frac{4\pi P^3}{3b^2} + \frac{a^2}{b^2}n.$$

$$n = 3\left(\frac{L}{P^2} - \frac{a^2n}{b^2P^2}\right)\left(1 - \frac{a}{2bP}\ln\left(\frac{a+bP}{a-Pb}\right)\right) \tag{4.161b}$$

From Eqs. (4.161a) and (4.161b), one first has to calculate the ratio a/b; and by the same means by which Eq. (4.145) was analyzed, we first determine whether bP/a is infinitely small, finite, or infinitely large, the only difference being in Eq. (4.161a) every infinitesimal variation of L/nP^2 occurs. However, I will not discuss the point further, except to note that as n and P grow larger, one also cannot get a result, which depends only on the average kinetic energy. In addition, I will not discuss in detail those cases where there are other constraint equations besides the equation for kinetic energy, as this would lead me too far afield.

In order to provide a demonstration of how general the concept of the most probable state distribution of gas molecules is, here I supply another definition for it. Suppose again that each molecule can only have a discrete number of values for the kinetic energy, $0, \epsilon, 2\epsilon, 3\epsilon \ldots \infty$. The total kinetic energy is $L = \lambda\epsilon$. We want to determine the kinetic energy of each molecule in the following manner: We have in an urn just as many identical balls (n) as molecules present. Every ball corresponds to a certain molecule. We now make λ draws from this urn, returning the ball to the urn each time. The kinetic energy of the first molecule is now equal to the product of ϵ and the number of times the ball corresponding to this molecule is drawn. The kinetic energies of all other molecules are determined analogously. We have produced a distribution of the kinetic energy L among the molecules (a complexion). We again make λ draws from the urn and produce a second complexion, then a third, etc. many times (J), and produce J complexions. We can define the most probable state distribution in two ways: First, we find how often in all J complexions a molecule has kinetic energy 0, how often the kinetic energy is ϵ, 2ϵ, etc., and say that the ratios of these numbers should provide the probabilities that a molecule has kinetic energy0, ϵ, 2ϵ, etc. at thermal equilibrium. Second, for each complexion we form the corresponding state distribution. If some state distribution is composed of $\mathcal{P}$ complexions, we then denote the quotient $\mathcal{P}/J$ as the probability of the state distribution. At first glance, this definition of a state distribution seems very plausible. However, we shall presently see that this should not be used, because under these conditions, the distribution whose probability is the greatest would not correspond to thermal equilibrium. [Boltzmann does not justify this statement. However, later Planck does show that actually highest probability is equilibrium.] It is easy to cast the hypothesis that concerns us into formulas. First of all, we want to discuss the first method of probability determination. We consider the first molecule, and assume that λ draws were made; the probability that the first molecule was picked in the first draw is $1/n$; however the probability that another ball was drawn is $(n-1)/n$. Thus, the probability that on the 1st, 2d, 3rd ... kth draws the molecule corresponding to the first ball has been picked, and then a different ball for each of the following, is given by

$$\left(\frac{1}{n}\right)^k \left(\frac{n-1}{n}\right)^{\lambda-k} = \left(\frac{n-1}{n}\right)^{\lambda} \cdot \left(\frac{1}{n-1}\right)^k \tag{4.162}$$

Likewise is the probability that the ball corresponding to the first molecule is picked on the 1st, 2d, 3rd ... $(k-1)^{\text{th}}$, and then $(k+1)$th draws etc. The probability that the ball corresponding to the first molecule is picked for any arbitrary k draws and not for the others is

$$w_k = \frac{\lambda!}{(\lambda-k)!k!}\left(\frac{n-1}{n}\right)^{\lambda}\left(\frac{1}{n-1}\right)^k. \tag{4.163}$$

This probability that a molecule has then kinetic energy $k\epsilon$ is exactly the same for all the other molecules. Using again the approximation formula for the factorial, we obtain

$$w_k = \sqrt{\frac{1}{2\pi}} \cdot \left(\lambda\frac{n-1}{n}\right)^{\lambda} \cdot \frac{1}{\sqrt{\left(1-\frac{k}{\lambda}\right)\cdot k}}\left[\frac{\lambda-k}{(n-1)k}\right]^k \cdot (\lambda-k)^{-\lambda}, \tag{4.164}$$

which shows that the probability of the larger kinetic energies is so disproportionately important that the entire expression does not approach a clearly identifiable limit with increasing k; λ, $1/\epsilon$, and n. We will now proceed to the second possible definition of the most probable state distribution. We need to consider all J complexions that we have formed by J drawings of λ balls from our urn. One of the various possible complexions consists of λ drawings of the ball corresponding to the first ball. We want to express this complexion symbolically by $m_1^{\lambda} \cdot m_2^0 \cdot m_3^0 \ldots m_n^0$. A second complexion, with $\lambda - 1$ draws of the ball corresponding to the first molecule, and one draw of the ball corresponding to the second molecule we want to express as

$$m_1^{\lambda-1} \cdot m_2^1 \cdot m_3^0 \ldots m_n^0$$

We see that the different possible complexions are expressed exactly by various components; the sum of these appears as the power series

$$(m_1 + m_2 + m_3 + \cdots + m_n)^\lambda \qquad \text{(A)}$$

that is developed according to the polynomial theorem. The probability of each such complexion is thus exactly proportional to the coefficient of the corresponding power series term, when you first form the product

$$\left(m_1' + m_2' + \cdots + m_n'\right)\left(m_1'' + m_2'' + \cdots + m_n''\right)\cdots\left(m_1^{(\lambda)} + m_2^{(\lambda)} + \cdots + m_n^{(\lambda)}\right)$$

Finally omit from this product the upper indexes, which then generates a term exactly proportional to the polynomial coefficient. Then by the symbol $m_1' \cdot m_3'' \cdot m_7''' \ldots$ we understand that the first pick corresponded to the first molecule, the second pick corresponded to the third molecule, on the third pick the ball corresponding to the seventh molecule was picked out, etc. All possible products of the variables m_1', m_1'', m_2', etc. represent equi-probable complexions. We want to know how often among all the terms of the power series (A) (whose total number is n^λ) there occur terms whose coefficients contain any one state distribution. For example, consider the state distribution where one molecule has all the kinetic energy, all others have zero kinetic energy. This state distribution appears to correspond to the following members of the power series (A)

$$m_1^\lambda \cdot m_2^0 \cdot m_3^0, \ldots, m_1^0 \cdot m_2^\lambda \cdot m_3^0, \ldots, m_1^0 \cdot m_2^0 \cdot m_3^\lambda \ldots etc.$$

with "undivided" λ. Similarly, for the state distribution in which w_0 molecules have kinetic energy zero, w_1 molecules have kinetic energy ϵ, w_2 molecules have kinetic energy 2ϵ, etc. there are

$$\frac{\lambda!}{w_0! w_1! w_2! \ldots w_\lambda!}$$

members of the power series (A). Each of these elements has the same polynomial coefficient, and that is identical to

$$\frac{\lambda!}{(0!)^{w_0!} \cdot (1!)^{w_1!} \cdot (2!)^{w_2!} \ldots (\lambda!)^{w_\lambda!}}$$

In summary, therefore, according to the now accepted definition, the probability of this state distribution is

$$\frac{(\lambda!)^2}{n^\lambda}\cdot\frac{1}{w_0!w_1!w_2!\ldots w_\lambda!}\cdot\frac{1}{(0!)^{w_0}\cdot(1!)^{w_1}\cdot(2!)^{w_2}\ldots(\lambda!)^{w_\lambda!}}$$

However, the maximization of this quantity also does not lead to the state distribution corresponding to thermal equilibrium.

4.3.2.4 Relationship of the Entropy to that Quantity Which I Have Called the Probability Distribution

When considering this relationship, let us initially deal with the simplest and clearest case, by first investigating a monoatomic gas on which no external forces act. In this case, formula (4.103) of Section 2 applies. To give it full generality, however, this formula must also include the x, y, z coordinates of the position of the molecule [Note by C. B.: Here it is clear that LB considers the new axis linearly independent of x, y, z coordinates]. The maximum of such a generalized expression (4.103) then yields not only the distribution of the velocity components of the gas molecules, which was sufficient for the case considered there, but also the distribution of the whole mass of gas in an enclosing vessel, where there it was taken for granted that the gas mass fills the vessel uniformly.

The generalization of (4.103) for the permutability measure can be easily obtained from Eq. (4.126) by substituting x, y, z, u, v, w for $p_1, p_2 \ldots q_r$ and simply omitting the terms with accented variables. It reads as follows:

$$\Omega = -\int\int\int\int\int\int f(x,y,z,u,v,w)\ln f(x,y,z,u,v,w)dxdydzdudvdw, \tag{4.165}$$

where $f(x, y, z, u, v, w)dxdydzdudvdw$ is the number of gas molecules present for which the six variables x, y, z, u, v, w lie between the limits x and $x + dx$, y and $y + dy$, z and $z + dz$... etc. w and $w + dw$ and the integration limits for the velocity are between $-\infty$ and $+\infty$, and for the coordinates over the dimensions of the vessel in which the gas exists. If the gas was not previously in thermal equilibrium, this quantity must grow. We want to compute the value this quantity has when the gas has reached the state of thermal equilibrium. Let V be the total volume of the gas, T be the average kinetic energy of a gas molecule, and N be the total number of molecules of the gas; finally m is the mass of a gas molecule. There is then for the state of thermal equilibrium

$$f(x, y, z, u, v, w) = \frac{N}{V\left(\frac{4\pi T}{3m}\right)^{3/2}} \cdot e^{-\frac{3m}{4T}\left(u^2+v^2+w^2\right)}. \tag{4.166}$$

Substituting this value into Eq. (4.165) gives

$$Q = \frac{3N}{2} + N\ln\left[V\left(\frac{4\pi T}{3m}\right)^{\frac{3}{2}}\right] - N\ln N \tag{4.167}$$

If by dQ we denote the differential heat supplied to the gas, where

$$dQ = NdT + pdV \tag{4.168}$$

and

$$pV = \frac{2N}{3} \cdot T. \tag{4.169}$$

p is the pressure per unit area, the entropy of the gas is then:

$$\int \frac{dQ}{T} = \frac{2}{3}N \cdot \ln\left(V \cdot T^{\frac{3}{2}}\right) + C. \tag{4.170}$$

Since N is regarded as a constant, with a suitable choice of constant

$$\int \frac{dQ}{T} = \frac{2}{3}\Omega \tag{4.171}$$

It follows that for each so-called reversible change of state, wherein in the infinitesimal limit the gas stays in equilibrium throughout the change of state, the increase in the Permutability measure Ω multiplied by $\frac{2}{3}$ is equal to $\int dQ/T$ taken over the change in state, *i.e.*, it is equal to the entropy increment. Whereas in fact when a very small amount of heat dQ is supplied to a gas, its temperature and volume increase by dT and dV. Then it follows from Eqs. (4.168) and (4.169)

$$dQ = NdT + \frac{2N}{3V} \cdot TdV \tag{4.172}$$

while from Eq. (4.167) it is found that:

$$d\Omega = +N\frac{dV}{V} + \frac{3N}{2} \cdot \frac{dT}{T}. \tag{4.173}$$

It is known that if in a system of bodies many reversible changes are taking place, then the total sum of the entropy of all these bodies remains constant. On the other hand, if among these processes are ones that are not reversible, then the total entropy of all the bodies must necessarily increase, as is well known from the fact that $\int dQ/T$ integrated over a nonreversible cyclic process is negative. According to Eq. (4.171) the sum of the permutability measures of the bodies $\Sigma\Omega$ and the total permutability measure must have the same increase. Therefore at thermal equilibrium the magnitude of the permutability measure times a constant is identical to the entropy, to within an additive constant; but it also retains meaning during a nonreversible process, continually increasing.

We can establish two principles: The first refers to a collection of bodies in which various state changes have occurred, at least some of which are irreversible, e.g., where some of the bodies were not always in thermal equilibrium. If the system was in a state of thermal equilibrium before and after all these changes, then the sum of the entropies of all the bodies can be calculated before and after those state changes without further ado, and it is always equal to 2/3 times the permutability measure of all the bodies. The first principle is that the total entropy after the state changes is always greater than before it. The same is of course true of the permutability measure. The second principle refers to a gas that undergoes a change of state without requiring that it begin and end in thermal equilibrium. Then the entropy of the initial and final states is not defined, but one can still calculate the quantity, which we have called the permutability measure, and that is to say that its value is necessarily larger after the state change than before. We shall presently see that the latter proposition can be applied without difficulty to a system of several gases, and it can be extended as well to polyatomic gas molecules, and when external forces are acting. For a system of several gases, the sum of the individual gas permutability measures must be defined to be the permutability measure of the system; if one introduces on the other hand the number of permutations itself, then the number of permutations of the system would be the product of the number of permutations of the constituents. If we assume the latter principle applies to anybody, then the two propositions just discussed are special cases of a single general theorem, which reads as follows:

We consider any system of bodies that undergoes some state changes, without requiring the initial and final states to be in thermal equilibrium. Then the total permutability measure for the bodies continually increases during the state changes, and can remain constant only so long as all the bodies during the change of state remain infinitely close to thermal equilibrium (reversible state changes.)

[C.B. note: it is important to point out that Boltzmann drops the term "gas," in his theorem]

To give an example, we consider a vessel divided into two halves by a very thin partition. The remaining walls of the vessel should also be very thin, so that the heat they absorb can be neglected, and surrounded by a substantial mass of other gas. One-half of the vessel should be completely filled with gas, while the other is initially completely empty. Suddenly pulling away the divider, which requires no significant work, causes that gas to spread at once throughout the vessel. Calculating the permutability measure for the gas, we find that this increases during this process, without changes in any other body. Now the gas is compressed in a reversible manner to its old volume by a piston. To achieve this manipulation, we can if we want assume that the piston is created by a surrounding dense gas enclosed in infinitely thin walls. This gas is unchanged except that it moves down in space.

Since the permutability measure does not depend on the absolute position in space, the permutability measure of the gas driving the piston does not change. That of the gas inside the vessel decreases to the initial value, since this gas has gone through a cyclic process. However, since this was not reversible, $\int dQ/T$ integrated over this cycle is not equal to the difference between the initial and final values of the entropy, but it is smaller due to the uncompensated transformation in the expansion. In contrast, heat is transferred to the surrounding gas. Therefore, for this process the permutability measure of the surrounding gas is increased by as much as that of the enclosed gas in the vessel during the first process. Since the latter mass of gas went through a cyclic process, its entropy decreased by as much during the second process as it increased during the first, but not by $\int dQ/T$; and because the second process was reversible, the entropy of the surrounding gas increased as much as the enclosed gas's decreased. The result is, as it has to be, that the sum of the permutability measures of all bodies of gas has increased. For a gas which moves at a constant speed in the direction of the x-axis[10]

$$f(x, y, z, u, v, w) = V \frac{N}{\sqrt{\left(\frac{4\pi T}{3m}\right)^3}} \cdot e^{-\frac{3m}{4T}\left((u-\alpha)^2 + v^2 + w^2\right)}. \tag{4.174}$$

If we substitute this expression into Eq. (4.165) we get exactly Eq. (4.167) again. Thus, the translational movement of a mass of gas does not increase its permutability measure. In addition, the same is true for the kinetic energy arising from any other net mass movement (molar movement), because it arises from the progression of the individual volume elements and their deformations and rotations which are of a higher order—infinitely small—and therefore entirely negligible. Here we obviously ignore the changes of permutability measure due to internal friction or temperature changes connected with those molecular motions. The temperature T of the moving gas is understood to mean half of the average value of $m[(u - \alpha)^2 + v^2 + w^2]$. So if **frictional** [bold emphasis by this author] and temperature changes are not present (e.g., if a gas, together with its enclosing vessel falls freely), a net mass movement has no effect on the permutability measure, until its kinetic energy is converted into heat, which is why molar motion is known as heat of infinite temperature.

Let us now move on to a mono-atomic gas on which gravity acts. The permutability measure is represented by the same Eq. (4.126), but instead of the generalized coordinates, we again introduce$x$, y, z, u, v, w. Equation (4.126) thus gives us a value for Ω which is exactly the same as Eq. (4.165). In the case of thermal equilibrium one has

[10]Translator Note: Here V should appear in the denominator, of the equation.

$$f(x, y, z, u, v, w) = Ce^{-\frac{3}{2T}\left(gz+\frac{mw^2}{2}\right)}. \tag{4.175}$$

where $\omega^2 = u^2 + v^2 + w^2$. The constant C is determined by the density of the gas. One has, e.g., a prismatic shaped vessel of height h, with a flat, horizontal bottom surface with area $=q$. Further, let N be the total number of gas molecules in the vessel, and z denote the height of a gas molecule from the bottom of the vessel, then

$$C = \frac{N}{\left(\frac{4\pi T}{3m}\right)^{\frac{3}{2}} \cdot q\int_0^h e^{-\frac{3gz}{2T}}dz} = \frac{N}{\left(\frac{4\pi T}{3m}\right)^{\frac{3}{2}} \cdot q \cdot \frac{2T}{3g}\left(1 - e^{-\frac{3gh}{2T}}\right)} \tag{4.176}$$

From which it follows:

$$\Omega = \frac{3N}{2} + N\ln\left(\frac{4\pi T}{3m}\right)^{3/2} + N\ln q + N\ln\frac{2T}{3g}\left(1 - e^{-\frac{3gh}{2T}}\right)$$
$$+ N\left(1 - \frac{3ghe^{-\frac{3gh}{2T}}}{2T\left(1 - e^{\frac{3gh}{2T}}\right)}\right) - N\ln N. \tag{4.177}$$

It is immediately apparent from the last formula that when a mass of gas falls a bit lower, without otherwise undergoing a change, that does not change its Ω value a bit. (Of course, gravity acts as a constant downward force, and its increase with the approach to the earth's center is neglected, which is always the case in heat theory problems.)

Let us go now to the most general case of an arbitrary gas on which any external force acts, so we again apply Eq. (4.126). However, in order that the formulas not be too lengthy, let only one type of gas be present in the vessel. The permutability measure of a gas mixture can then be found without difficulty, since it is simply equal to the sum of the permutability measures each component would have if present in the vessel alone. For thermal equilibrium, then

$$f = \frac{Ne^{-h(\chi+L)}}{\int\int e^{-h(\chi+L)}}dodw, \tag{4.178}$$

where χ is the potential energy, L is the kinetic energy of a molecule, N is the number of molecules in the vessel, where

$$do = dp_1dp_2\ldots dp_r, \qquad dw = dq_1dq_2\ldots dq_r \tag{4.179}$$

It therefore follows:

$$\Omega = -\int\int f \ln f dodw$$
$$= -N \ln N + N \ln \int\int e^{-h(\chi+L)} dodw + hN_{\overline{\chi}} + \frac{rN}{2}. \quad (4.180)$$

In the penultimate term, $\overline{\chi}$ is the average potential energy of a molecule, whose magnitude is

$$\frac{1}{N}\int\int \chi f dodw = \frac{\int\int \chi e^{-h(\chi+L)} dodw}{\int\int e^{-h(\chi+L)} dodw} \quad (4.181)$$

The last term can be found by taking into account that

$$L = \frac{\int\int L e^{-h(\chi+L)} dodw}{\int\int e^{-h(\chi+L)} dodw} = \frac{r}{2h} \quad (4.182)$$

(In respect of the above, see the already cited book of Watson, pp 36 and 37). The second term on the right of Eq. (4.180) can be further transformed, if one introduces instead of $q_1, q_2 \ldots q_r$ the variables $s_1, s_2 \ldots s_r$, which have the property that the term L is reduced to $s_1^2 + s_2^2 + \cdots s_r^2$. We then denote by Δ the Jacobian

$$\sum \pm \frac{dq_1}{ds_1}\frac{dq_2}{ds_2}\cdots\frac{dq_r}{ds_r}.$$

So then

$$\int\int e^{-h(\chi+L)} dodw = \left(\frac{\pi}{h}\right)^{\frac{r}{2}} \int \Delta e^{-h\chi} do, \quad \overline{\chi} = \frac{\int \Delta \chi e^{-h\chi do}}{\int \Delta e^{-h\chi do}} \quad (4.183)$$

And so also

$$\Omega = N \ln \int \Delta e^{-h\chi} do - \frac{Nr}{2} \ln h + hN\overline{\chi} + \frac{rN}{2}(1 + \ln \pi) - N \ln N. \quad (4.184)$$

To this expression, one can compare Eq. (18) of my paper "Analytical proof of the second law of thermodynamics from the approach to equilibrium in kinetic energy" (Wiss. Abhand. Vol I, reprint 20), or Eq. (95) of my "Further Studies" (Wiss. Abhand. Vol I, reprint 22) by replacing $p_1, p_2 \ldots$ with $x_1, y_2 \ldots q_1, q_2 \ldots$ with $u_1, v_2 \ldots; s_1, s_2 \ldots$ with

$$\sqrt{\frac{m}{2}} u_1, \sqrt{\frac{m}{2}} v_1 \ldots,$$

r with $3r$, whereby

$$\Delta = \left(\frac{2}{m}\right)^{\frac{3r}{2}} \tag{4.185}$$

It can be seen that Eq. (4.184) is identical to 3*N*/2 times Eq. (18) of the aforementioned paper, except for an additive constant, wherein multiplying by *N* indicates that Eq. (18) applies to one molecule only. Equation (95) of "Further Studies" is in an opposite manner denoted as Ω and thus also as the entropy. Taken with a negative sign, it is however greater than Ω here by*N* ln *N*. The former is because in the "Further Studies" I was looking for a value, which must decrease, as a result of this, introducing the magnitude of f^* instead of using*f*; this was however less clear. From this agreement, it follows that our statement about the **relationship of entropy to the permutability measure applies to the general case exactly as it does to a monatomic gas** [bold emphasis by C.B.].

Up to this point, these propositions may be demonstrated exactly using the theory of gases. If one tries, however, to generalize to liquid drops and solid bodies, one must dispense with an exact treatment from the outset, since far too little is known about the nature of the latter states of matter, and the mathematical theory is barely developed. Nevertheless, I have already mentioned reasons in previous papers, in virtue of which it is likely that for these two aggregate states, the thermal equilibrium is achieved when Eq. (4.126) becomes a maximum, and that when thermal equilibrium exists, the entropy is given by the same expression. **It can therefore be described as likely that the validity of the principle which I have developed is not just limited to gases, but that the same constitutes a general natural law applicable to solid bodies and liquid droplets, although the exact mathematical treatment of these cases still seems to encounter extraordinary difficulties** [bold emphasis by C.B.].

4.4 Critic of Boltzmann's Mathematical Derivation

Boltzmann assumes that particles are linearly independent rather than as an ensemble. He also assumes that there is no interaction. However, when his formulation is used in conjunction with conservation of energy, conservation of mass, and Newton's laws, it automatically makes a solid basis for statistical mechanics. Because even if the particles he uses in his experiments were not gas atoms and were interacting, the number of microstates could not change for the ensemble. Essentially molecules could be replaced by ensemble of molecules in his formulation and his statistical basis would not change because he does not make any assumptions to preclude forces between molecules. However, internal friction and interaction between ensemble atoms makes his formulation an upper bound for solids. Because of internal interactions and frictions there may be less number of microstates possible in solids but the number of microstates cannot be more than Boltzmann's theory. Finally, Boltzmann's entropy equation is given by

$$S = k \ln \Omega \tag{4.186}$$

where S is thermodynamic entropy, k is Boltzmann's constants, and Ω is number of microstates corresponding to a macrostate. In 1901, Max Planck modifies this equation to be

$$S = k \ln w + \text{constant} \tag{4.187}$$

Here w is probability of a microstate.

While we discussed the difference between Boltzmann's equation and Maxwell's version of Boltzmann equation earlier, unfortunately, with the exception of the physics textbook by Halliday and Resnick (1966), w term defined by Max Planck (1901) has been misreported in the literature as number of microstates. Therefore, we find it necessary to include the entire paper by Max Planck (1901) in this chapter.

Most important contribution of Max Planck's (1901) is that it establishes the relationship between entropy and disorder. Even though his paper is about electromagnetic theory of radiation, the concept is applicable to solids, liquids, and gasses.

The fact that Boltzmann does not account for ensemble makes his denominator bigger and possibility of complexions higher; essentially, he calculates an upper bound for number of complexions. Of course, if we accounted for interaction between the molecules, as constraints, number of complexions may be smaller. Therefore, his permutation give the upper bound solution for the number of complexions. When we calculate the complexions in solids, of course, we take into account the ensemble interaction by all governing laws and conditions imposed on the system. Therefore, Boltzmann's solution is perfectly valid for systems with ensemble interactions.

Moreover, Boltzmann assumes that molecules do not have any potential energy. However, he maintains conservation of energy. Therefore, if we assume that his kinetic energy also includes potential energy of the molecules, like deformation, then his mathematics would still perfectly be true.

4.5 On the Law of Distribution of Energy in the Normal Spectrum, By Max Planck, Annalen Der Physik, Vol. 4, P. 553 Ff (1901)

The recent spectral measurements made by O. Lummer and E. Pringsheim,[11] and even more notable those by H. Rubens and F. Kurlbaum,[12] which together confirmed an earlier result obtained by H. Beckmann,[13] show that the law of energy distribution

[11] O. Lummer and E. Pringsheim, *Transactions of the German Physical Society* 2 (1900), p. 163.

[12] H. Rubens and F. Kurlbaum, *Proceedings of the Imperial Academy of Science*, Berlin, October 25, 1900, p. 929.

[13] H. Beckmann, *Inaugural dissertation*, T¨ubingen 1898. See also H. Rubens, *Weid. Ann.* 69 (1899) p. 582.

in the normal spectrum, first derived by W. Wien from molecular-kinetic considerations and later by me from the theory of electromagnetic radiation, is not valid generally.

In any case, the theory requires a correction, and I shall attempt in the following to accomplish this based on the theory of electromagnetic radiation, which I developed. For this purpose, it will be necessary first to find in the set of conditions leading to Wien's energy distribution law that term which can be changed; thereafter it will be a matter of removing this term from the set and making an appropriate substitution for it.

In my last article[14] I showed that the physical foundations of the electromagnetic radiation theory, including the hypothesis of "natural radiation," withstand the most severe criticism; and since to my knowledge there are no errors in the calculations, the principle persists that the law of energy distribution in the normal spectrum is completely determined when one succeeds in calculating the entropy S of an irradiated, monochromatic, vibrating resonator as a function of its vibrational energy U. Since one then obtains, from the relationship $dS/dU = 1/\theta$, the dependence of the energy U on the temperature θ, and since the energy is related to the density of radiation at the corresponding frequency by a simple relation,[15] one also obtains the dependence of this density of radiation on the temperature. The normal energy distribution is then the one in which the radiation densities of all different frequencies have the same temperature.

Consequently, the entire problem is reduced to determining S as a function of U, and it is to this task that the most essential part of the following analysis is devoted. In my first treatment of this subject, I had expressed S, by definition, as a simple function of U without further foundation, and I was satisfied to show that this form of entropy meets all the requirements imposed on it by thermodynamics. At that time I believed that this was the only possible expression and that consequently Wein's law, which follows from it, necessarily had general validity. In a later, closer analysis,[16] however, it appeared to me that there must be other expressions which yield the same result, and that in any case one needs another condition in order to be able to calculate S uniquely. I believed I had found such a condition in the principle, which at the time seemed to me perfectly plausible, that in an infinitely small irreversible change in a system, near thermal equilibrium, of N identical resonators in the same stationary radiation field, the increase in the total entropy $S_N = NS$ with which it is associated depends only on its total energy $U_N = NU$ and the changes in this quantity, but not on the energy U of individual resonators. This theorem leads again to Wien's energy distribution law. Nevertheless, since the latter is not confirmed by experience, one is forced to conclude that even this principle cannot be generally valid and thus must be eliminated from the theory.[17]

---

[14] M. Planck, *Ann. d. Phys.* 1 (1900), p. 719.

[15] Compare with equation (8).

[16] M. Planck, *loc. cit.*, pp. 730 ff.

[17] Moreover one should compare the critiques previously made of this theorem by W. Wien (*Report of the Paris Congress* 2, 1900, p. 40) and by O. Lummer (*loc. cit.*, 1900, p. 92).

Thus, another condition must now be introduced which will allow the calculation of S, and to accomplish this it is necessary to look more deeply into the meaning of the concept of entropy. Consideration of the untenability of the hypothesis made formerly will help to orient our thoughts in the direction indicated by the above discussion. In the following a method will be described which yields a new, simpler expression for entropy and thus provides also a new radiation equation which does not seem to conflict with any facts so far determined.

4.5.1 *Calculations of the Entropy of a Resonator as a Function of Its Energy*

§1.
Entropy depends on disorder and this disorder, according to the electromagnetic theory of radiation for the monochromatic vibrations of a resonator when situated in a permanent stationary radiation field, depends on the irregularity with which it constantly changes its amplitude and phase, provided one considers time intervals large compared to the time of one vibration but small compared to the duration of a measurement. If amplitude and phase both remained absolutely constant, which means completely homogeneous vibrations, no entropy could exist and the vibrational energy would have to be completely free to be converted into work. The constant energy U of a single stationary vibrating resonator accordingly is to be taken as time average, or what is the same thing, as a simultaneous average of the energies of a large number N of identical resonators, situated in the same stationary radiation field, and which are sufficiently separated so as not to influence each other directly. It is in this sense that we shall refer to the average energy U of a single resonator. Then to the total energy

$$U_N = NU \tag{4.188}$$

Of such a system of N resonators, there corresponds a certain total entropy

$$S_N = NS \tag{4.189}$$

Of the same system, where S represents the average entropy of a single resonator and the entropy S_N depends on the disorder with which the total energy U_N is distributed among the individual resonators.

§2. We now set the entropy S_N of the system proportional to the logarithm of its probability w, within an arbitrary additive constant, so that the N resonators together have the energy E_N:

$$S_N = k \log\, w + \text{constant} \tag{4.190}$$

In my opinion, this actually serves as a definition of the probability w, since in the basic assumptions of electromagnetic theory there is no definite evidence for such a probability. The suitability of this expression is evident from the outset, in view of its simplicity and close connection with a theorem from kinetic gas theory.[18]

§3. It is now a matter of finding the probability w so that the N resonators together possess the vibrational energy U_N. Moreover, it is necessary to interpret U_N not as a continuous, infinitely divisible quantity, but as a discrete quantity composed of an integer number of finite equal parts. Let us call each such part the energy element ϵ; consequently, we must set

$$U_N = P\epsilon \tag{4.191}$$

where P represents a large integer generally, while the value of ϵ is yet uncertain.

(The above paragraph in the original German)

"Es kommt nun darauf an, die Wahrscheinlichkeit W daf¨ur zu finden, dass die N Resonatoren insgesamt die Schwingungsenergie U_N besitzen. Hierzu ist es notwendig, U_N nicht als eine stetige, unbeschr¨ankt teilbare, sondern als eine discrete, aus einer ganzen Zahl von endlichen gleichen Teilen zusammengesetzte Gr¨osse aufzufassen. Nennen wireinen solchen Teil ein Energieelement ϵ, so ist mithin zu setzen

$$U_N = P\epsilon \tag{4.192}$$

wobei P eine ganze, im allgemeinen grosse Zahl bedeutet..."

Now it is evident that any distribution of the P energy elements among the N resonators can result only in a finite, integer definite number. Every such form of distribution we call, after an expression used by L. Boltzmann for a similar idea, a "complex" [Boltzmann actually refers to this as "complexion"]. If one denotes the resonators by the numbers 1, 2, 3, ...N, and writes these side by side, and if one sets under each resonator the number of energy elements assigned to it by some arbitrary distribution, then one obtains for every complex a pattern of the following form:

$$\begin{array}{cccccccccc} 1 & 2 & 3 & 4 & 5 & 6 & 7 & 8 & 9 & 10 \\ 7 & 38 & 11 & 0 & 9 & 2 & 20 & 4 & 4 & 5 \end{array} \tag{4.193}$$

Here we assume $N = 10$, $P = 100$. The number R of all possible complexes is obviously equal to the number of arrangements that one can obtain in this fashion for the lower row, for a given N and P. For the sake of clarity we should note that two complexes must be considered different if the corresponding number patterns contain the same numbers but in a different order.

[18]L. Boltzmann, *Proceedings of the Imperial Academy of Science*, Vienna, (II) 76 (1877), p. 428.

From combination theory, one obtains the number of all possible complexes as:

$$R = \frac{N(N+1)(N+2)\dots(N+P-1)}{1\cdot 2\cdot 3\cdots P} = \frac{(N+P-1)!}{(N-1)!P!} \tag{4.194}$$

Now according to Stirling's theorem, we have in the first approximation:

$$N! = N^N \tag{4.195}$$

Consequently, the corresponding approximation is:

$$R = \frac{(N+P)^{N+P}}{N^N \cdot P^P} \tag{4.196}$$

§4. The hypothesis which we want to establish as the basis for further calculation proceeds as follows: in order for the N resonators to possess collectively the vibrational energy U_N, the probability w must be proportional to the number R of all possible complexes formed by distribution of the energy U_N among the N resonators; or in other words, any given complex is just as probable as any other. Whether this actually occurs in nature one can, in the last analysis, prove only by experience. But should experience finally decide in its favor it will be possible to draw further conclusions from the validity of this hypothesis about the particular nature of resonator vibrations, namely in the interpretation put forth by J. v. Kries[19] regarding the character of the "original amplitudes, comparable in magnitude but independent of each other." As the matter now stands, further development along these lines would appear to be premature.

§5. According to the hypothesis introduced in connection with Eq. (4.190), the entropy of the system of resonators under consideration is, after suitable determination of the additive constant:

$$S_N = k\log R = k\{(N+P)\log(N+P) - N\log N - P\log P\} \tag{4.197}$$

Moreover, by considering (4.191) and (4.188)

$$S_N = kN\left\{\left(1+\frac{U}{\epsilon}\right)\log\left(1+\frac{U}{\epsilon}\right) - \frac{U}{\epsilon}\log\frac{U}{\epsilon}\right\} \tag{4.198}$$

Thus, according to Eq. (4.189) the entropy S of a resonator as a function of its energy U is given by:

$$S = k\left\{\left(1+\frac{U}{\epsilon}\right)\log\left(1+\frac{U}{\epsilon}\right) - \frac{U}{\epsilon}\log\frac{U}{\epsilon}\right\} \tag{4.199}$$

---

[19]Joh. v. Kries, *The Principles of Probability Calculation* (Freiburg, 1886), p. 36.

4.5.2 Introduction of Wien's Displacement Law

§6. Next to Kirchoff's theorem of the proportionality of emissive and absorptive power, the so-called displacement law, discovered by and named after W. Wien,[20] which includes as a special case the Stefan-Boltzmann law of dependence of total radiation on temperature, provides the most valuable contribution to the firmly established foundation of the theory of heat radiation. In the form given by M. Thiesen[21] it reads as follows:

$$E \cdot d\lambda = \theta^5 \psi(\lambda\theta) \cdot d\lambda \tag{4.200}$$

where λ is the wavelength, $E \cdot d\lambda$ represents the volume density of the "black-body" radiation[22] within the spectral region λ to $\lambda + d\lambda$, θ represents temperature, and $\psi(x)$ represents a certain function of the argument x only.

§7. We now want to examine what Wien's displacement law states about the dependence of the entropy S of our resonator on its energy, and its characteristic period, particularly in the general case where the resonator is situated in an arbitrary diathermic medium. For this purpose, we next generalize Thiesen's form of the law for the radiation in an arbitrary diathermic medium with the velocity of light c. Since we do not have to consider the total radiation, but only the monochromatic radiation, it becomes necessary in order to compare different diathermic media to introduce the frequency n instead of the wavelength λ.

Thus, let us denote by $u \cdot d\nu$ the volume density of the radiation energy belonging to the spectral region ν to $\nu + d\nu$; then we write: $u \cdot d\nu$ instead of $E \cdot d\lambda$; c/ν instead of λ, and $c \cdot d\nu/\nu^2$ instead of $d\lambda$. From which we obtain

$$u = \theta^5 \frac{c}{\nu^2} \cdot \psi\left(\frac{c\theta}{\nu}\right) \tag{4.201}$$

Now according to the well-known Kirchhoff-Clausius law, the energy emitted per unit time at the frequency ν and temperature θ from a black surface in a diathermic medium is inversely proportional to the square of the velocity of propagation c^2; hence the energy density u is inversely proportional to c^3 and we have:

$$u = \frac{\theta^5}{\nu^2 c^3} \cdot f\left(\frac{\theta}{\nu}\right) \tag{4.202}$$

[20]W. Wien, Proceedings of the Imperial Academy of Science, Berlin, February 9, 1893, p. 55.

[21]M. Thiesen, Transactions of the German Physical Society 2 (1900), p. 66.

[22]Perhaps one should speak more appropriately of a "white" radiation, to generalize what one already understands by total white light.

where the constants associated with the function f are independent of c. In place of this, if f represents a new function of a single argument, we can write:

$$u = \frac{\nu^3}{c^3} \cdot f\left(\frac{\theta}{\nu}\right) \tag{4.203}$$

In addition, from this we see, among other things, that as is well known, the radiant energy $u \cdot \lambda^3$ at a given temperature and frequency is the same for all diathermic media.

§8. In order to go from the energy density u to the energy U of a stationary resonator situated in the radiation field and vibrating with the same frequency, ν, we use the relation expressed in Eq. (4.34) of my paper on irreversible radiation processes[23]:

$$K = \frac{\nu^2}{c^2} U \tag{4.204}$$

(K is the intensity of a monochromatic linearly, polarized ray), which together with the well-known equation:

$$u = \frac{8\pi K}{c} \tag{4.205}$$

yields the relation:

$$u = \frac{8\pi\nu^2}{c^3} U \tag{4.206}$$

From this and from Eq. (4.203) follows:

$$U = \nu \cdot f\left(\frac{\theta}{\nu}\right) \tag{4.207}$$

where now c does not appear at all. In place of this we may also write:

$$\theta = \nu \cdot f\left(\frac{U}{\nu}\right) \tag{4.208}$$

§9. Finally, we introduce the entropy S of the resonator by setting

$$\frac{1}{\theta} = \frac{dS}{dU} \tag{4.209}$$

[23] M. Planck, *Ann. D. Phys.* 1 (1901), p. 99.

We then obtain:

$$\frac{dS}{dU} = \frac{1}{\nu} \cdot f\left(\frac{U}{\nu}\right) \tag{4.210}$$

And integrated:

$$S = f\left(\frac{U}{\nu}\right) \tag{4.211}$$

That is, the entropy of a resonator vibrating in an arbitrary diathermic medium depends only on the variable U/ν, containing besides this only universal constants. This is the simplest form of Wien's displacement law known to me.

§10. If we apply Wien's displacement law in the latter form to Eq. (4.199) for the entropy S, we then find that the energy element ϵ must be proportional to the frequency ν, thus:

$$\varepsilon = h\nu \tag{4.212}$$

And consequently:

$$S = k\left\{\left(1 + \frac{U}{h\nu}\right) \log\left(1 + \frac{U}{h\nu}\right) - \frac{U}{h\nu} \log \frac{U}{h\nu}\right\} \tag{4.213}$$

Here h [Planck constant] and k [Boltzmann constant] are universal constants. By substitution into Eq. (4.208) one obtains:

$$\frac{1}{\theta} = \frac{k}{h\nu} \log\left(1 + \frac{h\nu}{U}\right) \tag{4.214}$$

$$U = \frac{h\nu}{e^{\frac{h\nu}{k\theta}} - 1} \tag{4.215}$$

Moreover, from Eq. (4.206) there then follows the energy distribution law sought for:

$$u = \frac{8\pi h\nu^3}{c^3} \cdot \frac{1}{e^{\frac{h\nu}{k\theta}} - 1} \tag{4.216}$$

On the other hand, by introducing the substitutions given in (4.203), in terms of wavelength λ instead of the frequency:

$$E = \frac{8\pi ch}{\lambda^5} \frac{1}{\left[e^{ch/k\lambda\theta} - 1\right]} \tag{4.217}$$

I plan to derive elsewhere the expressions for the intensity and entropy of radiation progressing in a diathermic medium, as well as the theorem for the increase of total entropy in nonstationary radiation processes.

4.5.3 *Numerical Values*

§11. The values of both universal constants h [Planck constant] and k [Boltzmann constant] may be calculated rather precisely with the aid of available measurements. F. Kurlbaum,[24] designating the total energy radiating into air from 1 cm^2 of a black body at temperature t°C in 1 s by S_t, found that:

$$S_{100} - S_0 = 0.0731 \cdot \frac{watt}{cm^2} = 7.31 \cdot 10^5 \cdot \frac{erg}{cm^2 \cdot sec} \tag{4.218}$$

From this one can obtain the energy density of the total radiation energy in air at the absolute temperature 1:

$$\frac{4 \cdot 7.31 \cdot 10^5}{3 \cdot 10^{10} \cdot \left(373^4 - 273^4\right)} = 7.061 \cdot 10^{-15} \cdot \frac{erg}{cm^3 \cdot deg^4} \tag{4.219}$$

On the other hand, according to Eq. (4.216) the energy density of the total radiant energy for $\theta = 1$ is:

$$u^* = \int_0^\infty u\, d\nu = \frac{8\pi h}{c^3} \int_0^\infty \frac{\nu^3 d\nu}{e^{h\nu/k} - 1} \tag{4.220a}$$

$$u^* = \frac{8\pi h}{c^3} \int_0^\infty \nu^3 \left(e^{-h\nu/k} + e^{-2h\nu/k} + e^{-3h\nu/k} + \ldots\right) d\nu \tag{4.220b}$$

And by term wise integration:

$$u^* = \frac{8\pi h}{c^3} \cdot 6 \left(\frac{k}{h}\right)^4 \left(1 + \frac{1}{24} + \frac{1}{34} + \frac{1}{44} + \cdots\right) = \frac{48\pi k^4}{c^3 h^3} \cdot 1.0823 \tag{4.220c}$$

If we set this equal to $7.061 \cdot 10^{-15}$, then, since $c = 3 \cdot 10^{10}$ cm/s, we obtain:

$$\frac{k^4}{h^3} = 1.1682 \cdot 10^{15} \tag{4.221}$$

[24] F. Kurlbaum, *Wied. Ann.* 65 (1898), p. 759.

§12. O. Lummer and E. Pringswim[25] determined the product $\lambda_m\theta$, where λ_m is the wavelength of maximum energy in air at temperature θ, to be 2940 μm·degrees. Thus, in absolute measure:

$$\lambda_m = 0.294 \text{ cm} \cdot \text{deg} \tag{4.222}$$

On the other hand, it follows from Eq. (4.217), when one sets the derivative of E with respect to θ equal to zero, thereby finding $\lambda = \lambda_m$

$$\left(1 - \frac{ch}{5k\lambda_m\theta}\right) \cdot e^{ch/k\lambda_m\theta} = 1 \tag{4.223}$$

And from this transcendental equation:

$$\lambda_m\theta = \frac{ch}{4.9651 \cdot k} \tag{4.224}$$

Consequently:

$$\frac{h}{k} = \frac{4.9561 \cdot 0.294}{3 \cdot 10^{10}} = 4.866 \cdot 10^{-11} \tag{4.225}$$

From this and from Eq. (4.221) the values for the universal constants become:

$$h = 6.55 \cdot 10^{-27} \text{erg} \cdot \text{s}$$

$$k = 1.346 \cdot 10^{-16} \frac{erg}{deg}$$

These are the same number that I indicated in my earlier communication. [There are no references at the end of Maxwell's paper]

.

4.6 Thermodynamic State Index (TSI) in Unified Mechanics Theory

We postulate that at any given temperature and pattern of internal rearrangement within the material, the rate at which any specific microstructural rearrangement occurs is fully determined by the thermodynamic forces, this is the same concept, postulated by Rice (1971). Thermodynamic State Index (TSI) axis defines location of any system on thermodynamic axis. The values of TSI are between zero and one.

[25]O. Lummer and Pringsheim, *Transactions of the German Physical Society* 2 (1900), p. 176.

The TSI axis coordinate cannot be defined by space-time coordinates; therefore it is a linearly independent axis. Similar to what Boltzmann defined as an additional axis. The coordinate value of a point on TSI axis defined by laws of thermodynamics and according to the fundamental equation of the system, we defined earlier. Fundamental equation must account for all irreversible entropy generation mechanisms according to our failure definition. For example, if color change in a polymer is our failure definition, then fundamental equation will only have mechanisms that contribute to color change not mechanical failure.

We have already presented second law of thermodynamics in Boltzmann-Planck formulation as

$$S = k \ln w \tag{4.226}$$

where S in entropy, k is Boltzmann's constant, and w is the disorder parameter, which is the probability that the system will exist in the state it is relative to all the possible states it could be in,

In Fig. 4.5, w and s are disorder parameter and entropy, respectively, and subscript o is for initial, and f is for final state.

The second law of thermodynamics states that there is a natural tendency of any isolated system, living or nonliving, to degenerate into a more disordered state. When irreversible entropy generation rate becomes zero, the system reaches an equilibrium point. This is usually a valley in the energy landscape. It is important to point out that not all entropy generation mechanisms contribute to void generation in a solid. Those that do must be identified, if the interest is to define void generating entropy mechanisms...

In statistical mechanics, thermodynamic entropy is considered an intrinsic property of a system. Boltzmann-Planck relates equation entropy to the number of microstates that are consistent with the macroscopic boundary and initial conditions that characterize the system.

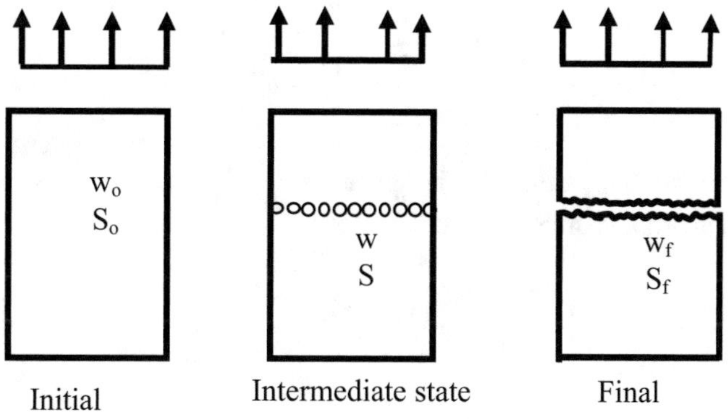

Fig. 4.5 A sample under uniaxial tension, w is disorder parameter, S is entropy

It is important to point out that the final form of Boltzmann equation was derived by Planck 1901. Therefore, we will use Planck's definition for variable w (probability or disorder). However, even if we use Ω, the number of microstates, nothing would change in the formulation of TSI, because we normalize the equation, as a result it is a nondimensional linearly independent axis.

In Eq. (4.226) entropy is for one atom. In order to convert it to unit specific mass (gram/mole) we need to multiply it by Avogadro's number $N_A = 6.022352\ x\ 10^{23}$ *atoms/mole* and divide by specific mass (gram/mole) m_s

$$s = \frac{N_A}{m_s} k \ln w \tag{4.227}$$

Initially, let that probability of a material being in a completely "ordered ground state" [a reference state] be w_o. Under any external loading the system will move from the initial configuration to a new microstate defined by s *and* w

$$s_0 \rightarrow s$$

$$w_0 = e^{\frac{m_s s_0}{N_A k}} \rightarrow \mathrm{w} = e^{\frac{m_s s}{N_A k}} \tag{4.228}$$

According to the second law, in the final stage the system will reach maximum entropy and maximum disorder (zero entropy generation rate) state.

$$s \rightarrow s_{\max}$$

$$w = e^{\frac{m_s s}{N_A k}} \rightarrow \mathrm{w}_{max} = e^{\frac{m_s\ s_{max}}{N_A k}} \tag{4.229}$$

During this travel over the energy terrain, we can define the thermodynamics state of the system at any point as a dimensionless variable that defines the distance from the origin [or any reference state]

$$0 \leq \Phi = \frac{W - W_0}{W} \leq 1 \tag{4.230}$$

$$\Phi = \frac{e^{\frac{m_s s}{N_A k}} - e^{\frac{m_s s_0}{N_A k}}}{e^{\frac{m_s s}{N_A k}}} = \left[1 - e^{-\frac{m_s}{N_A}\left(\frac{s - s_0}{k}\right)}\right] \tag{4.231}$$

Boltzmann constant k can also be given by $k = \frac{\mathrm{R}}{N_A}$where R is the gas constant. Finally, the thermodynamics state index (TSI).

$$\Phi = \left[1 - e^{-\frac{m_s\ \Delta s}{R}}\right] \tag{4.232}$$

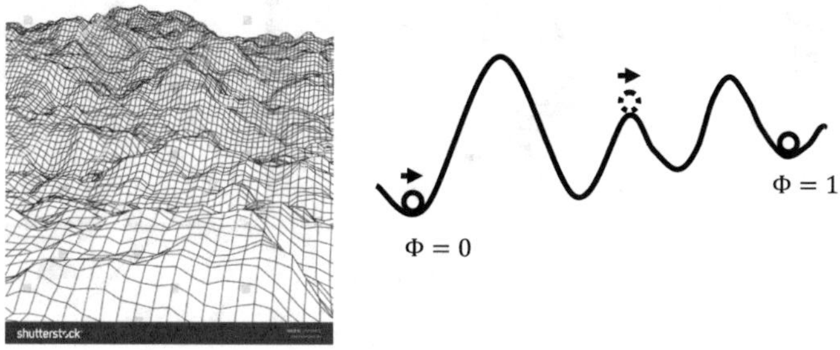

Fig. 4.6 Definition of Thermodynamic State Index

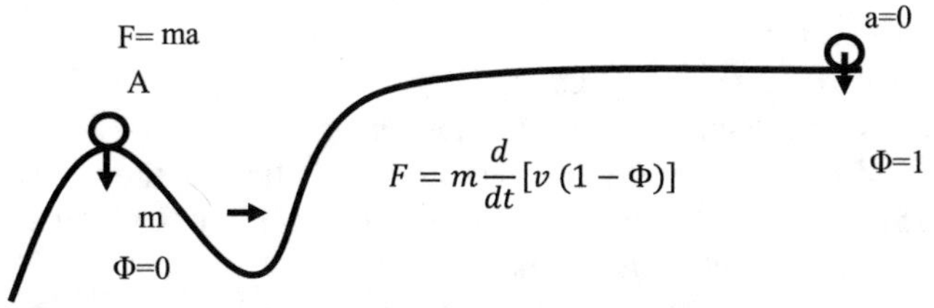

Fig. 4.7 After the initial kick, slowing down of the ball is governed by the laws of thermodynamics

Thermodynamic State Index imposes laws of thermodynamics on Newton's laws (Fig. 4.6). We will explain this with the following simple examples.

For example, if we have a ball with an initial acceleration at A (Fig. 4.7).

The ball will have acceleration of zero at some point B. The energy of the ball during between these 2 points will decrease according to laws of thermodynamics, which is represented by Φ.

Another example would be a simple spring (Fig. 4.8).

If we have a linear-elastic, one-dimensional spring over time stiffness of the spring will decrease and finally the spring will break in two pieces due to fatigue.

The degradation of the energy storage capacity of the spring will follow second law of thermodynamics

$$U = \frac{1}{2}k\delta^2(1 - \Phi) \tag{4.233}$$

Taking first derivative of strain energy with respect to deformation and ignoring second order terms, we obtain,

$$\delta = \frac{F}{K(1 - \Phi)} \tag{4.234}$$

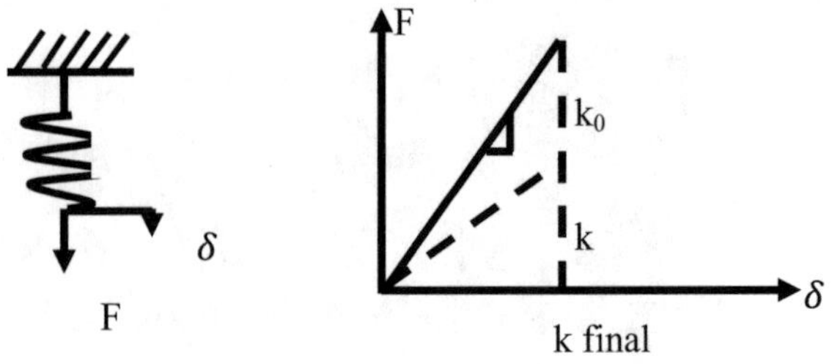

Fig. 4.8 One dimensional spring subjected to axial cycling loading

Of course, the degradation of elastic energy storage capacity is due to degradation of stiffness. However, we should point out that here we assumed that $\frac{\partial \Phi}{\partial \delta}$ is smaller than $\frac{\partial\left[\frac{1}{2}k\delta^2\right]}{\partial \delta}$ by an order of magnitude which might be true for high cycle fatigue but not true for low cycle fatigue.

It is important to point out that entropy is essentially energy unavailable for work. Therefore, TSI index degrades total available energy of a closed-isolated system. Multiplication of stiffness k with $(1 - \phi)$ is due to simplicity of the example chosen. Essentially TSI coordinate maps entropy generation rate [fundamental equation] of any system to a linearly independent axis.

The question may be asked what happens to a system moving from one stable equilibrium valley to another, gets an external energy boost during travel due to an external factor. Because entropy is an additive property, and we have to maintain conservation of energy, we can include the new addition in the system. Actually, in computational mechanics we solve differential equations in incremental format. Therefore using Thermodynamic State Index works very well for incremental solution procedures (Fig. 4.8).

Ramification of TSI coordinate is the fact that entropy generation rate becomes a nodal unknown in addition to other nodal unknowns. For example, displacement in a mechanical analysis. However, in order to reduce the amount of computation it is easier to calculate the TSI at Gauss integration points based on the results of the previous step. The amount of error is negligible.

4.6.1 Experimental Verification Example

Now that the Thermodynamic State Index (TSI) is defined, we will use two cases to calculate TSI using the experimental data. First a fully reversed uniaxial cyclic loading on a steel sample is considered and secondly a monotonic traction again on a steel sample. The experiments were performed in the same conditions and on the identical specimens (Fig. 4.9) for the two cases.

In order to evaluate Thermodynamic State Index (TSI) in both cases, only the effects of plastic deformation are taken into account, for the sake of simplicity. We assume that heat generated during cycling loading is insignificant and the

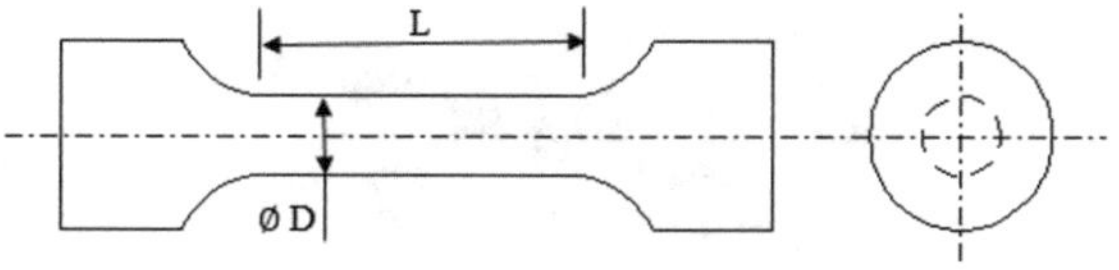

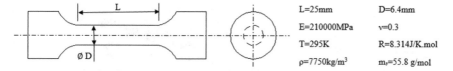

Fig. 4.9 Specimen dimensions and properties

temperature of the specimen remains constant. [Of course, this is not true.] Furthermore, because the load is uniaxial, the plasticity is reduced to one dimension. The internal entropy generation is then reduced to

$$\Delta s = \int_{t_0}^{t} \frac{\boldsymbol{\sigma} : \dot{\boldsymbol{\varepsilon}}^p}{\rho T} dt \tag{4.235}$$

For numerical computation, Eq. (4.235) is simplified to:

$$\Delta s_i = \sum \frac{(\sigma * \Delta \varepsilon^p)}{\rho T}. \tag{4.236}$$

4.6.1.1 Tension-Compression Cyclic Loading

The experiment consists of a displacement-controlled test conducted in a material characterization unit, which applies uniaxial tension and compression in repeated and fully reversed cycles to a sample. The experimental data obtained from the digital output consist of forces (F) and displacements (u). Here the fatigue failure occurs after 80 cycles. Since we assume small strain in the model, plastic strain is calculated as follows:

$$\varepsilon^p = \varepsilon - \varepsilon^e, \quad \varepsilon^p = \varepsilon - \frac{\sigma}{E}. \tag{4.237}$$

Figure 4.10 shows the entire engineering stress and strain diagram obtained from the experiment, which is a fully reversed uniaxial cyclic loading. Figure 4.11 presents the TSI evolution as a function of number cycles. As expected, TSI is initially zero and finally reaches the value of one.

Readers can find more examples in the papers cited in the references section

4.6.1.2 Monotonic Loading Test

For this case, the sample is identical to the one used for the cyclic loading. The only difference is that this time it is continuously loaded until failure occurs. The loading is a uniaxial tension; the experimental data is obtained in the similar fashion and

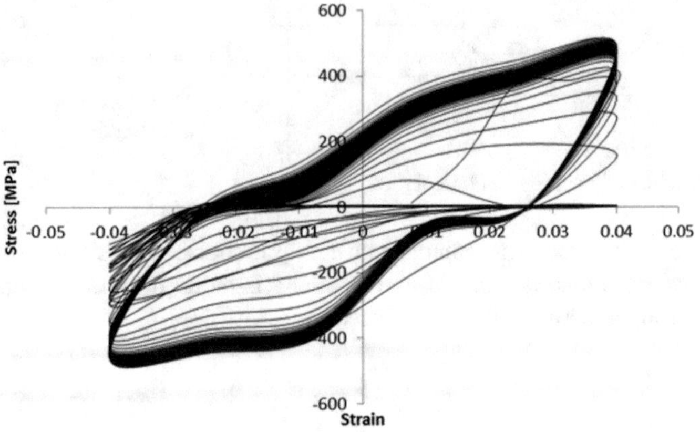

Fig. 4.10 Stress–strain diagram of cyclic loading

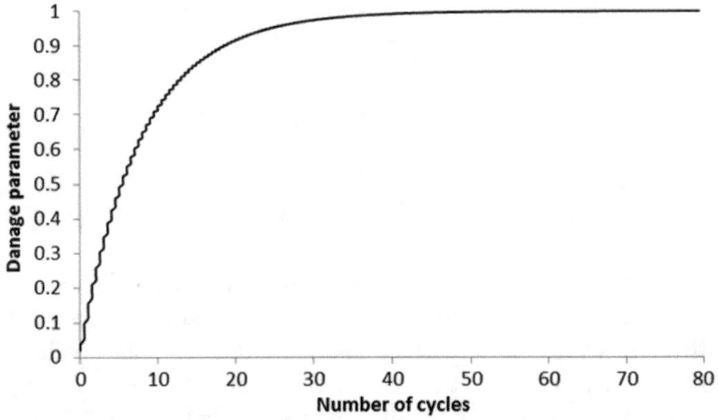

Fig. 4.11 Thermodynamic State Index evolution for cyclic loading

treated in the same way as before. Figure 4.12 shows the stress-strain diagram obtained from monotonic tension tests. Figure 4.13 presents the evolution of the TSI for the specimen. TSI starts from zero and reaches one at the end. We can see that the TSI remains very small for a while before it starts to increase. This domain corresponds to the elastic range of the material. In fact, because we assumed that only plastic deformation could cause degradation in the material, there is no entropy generated in the elastic range according our assumption.

As of this writing, there is a recent special issue of journal of Entropy dedicated to this topic. The journal has excellent examples where readers can download the papers free of charge

https://www.mdpi.com/journal/entropy/special_issues/fatigue

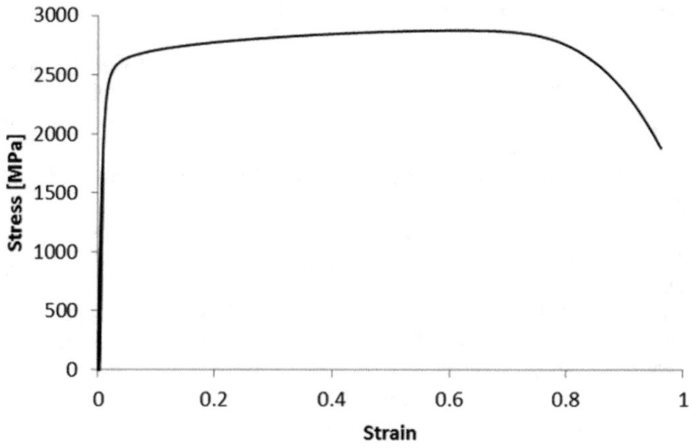

Fig. 4.12 Stress–strain diagram of cyclic loading

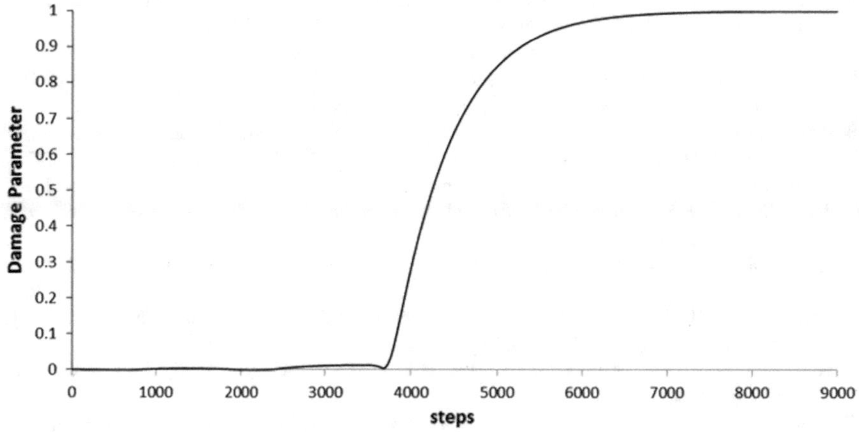

Fig. 4.13 Thermodynamic State Index evolution for monotonic loading

References

Amiri, M., & Khonsari, M. M. (2012). On the role of entropy generation in processes involving fatigue. *Entropy, 14*(1), 24–31.

Basaran, C., & Chandaroy, R. (2002). Thermomechanical analysis of solder joints under thermal and vibrational loading. *Transactions of ASME Journal of Electronic Packaging, 124*(1), 60–67.

Basaran, C., Li, S., & Abdulhamid, M. (2008). Thermomigration induced degradation in solder alloys. *Journal of Applied Physics, 103*, 123520.

Basaran, C., & Lin, M. (2007a). Damage mechanics of electromigration in microelectronics copper interconnects. *International Journal of Materials and Structural Integrity, 1*(1/2/3), 16–39.

Basaran, C., & Lin, M. (2007b). Electromigration induced strain field simulations for nanoelectronics lead-free solder joints. *International Journal of Solids and Structures, 44*, 4909–4924.

Basaran, C., & Lin, M. (2008). Damage mechanics of electromigration induced failure. *Mechanics of Materials, 40*, 66–79.

Basaran, C., Lin, M., & Ye, H. (2003). A thermodynamic model for electrical current induced damage. *International Journal of Solids and Structures, 40*(26), 7315–7327.

Basaran, C., & Nie, S. (2004). An irreversible thermodynamic theory for damage mechanics of solids. *International Journal of Damage Mechanics, 13*(3), 205–224.

Basaran, C., & Nie, S. (2007). A thermodynamics based damage mechanics model for particulate composites. *International Journal of Solids and Structures, 44*, 1099–1114.

Basaran, C., Nie, S., & Hutchins, C. (2008b). Time dependent behavior of a particle filled composite PMMA/ATH at elevated temperatures. *Journal of Composite Materials, 42*(19), 2003–2025.

Basaran, C., & Tang, H. (2002). Implementation of a thermodynamic framework for damage mechanics of solder interconnects in microelectronic packaging. *International Journal of Damage Mechanics, 11*(1), 87–108.

Basaran, C., & Yan, C. Y. (1998). A thermodynamic framework for damage mechanics of solder Joints. *Transactions of ASME Journal of Electronic Packaging, 120*, 379–384.

Basaran, C., Zhao, Y., Tang, H., & Gomez, J. (2005). A damage mechanics based unified constitutive model for solder alloys. *Transactions of ASME Journal of Electronic Packaging, 127*(3), 208–214.

Bazant, Z. P. (1972). Thermodynamics of interacting continua with surfaces and creep analysis of concrete structures. *Nuclear Engineering Structures and Design, 20*, 477–505.

Beven, K. (2006). A manifesto for the equifinality thesis. *Journal of Hydrology, 320*(1), 18–36.

Bin, N., Jamal, M., Kumar, A., Rao, C. L., & Basaran, C. (2020). Low cycle fatigue life prediction using unified mechanics theory in Ti–6Al–4V alloys. *Entropy, 22*(1), 24.

Boltzmann, L. (1877). Sitzungberichte der Kaiserlichen Akademie der Wissenschaften. Mathematisch-Naturwissen Classe. Abt. II, LXXVI, pp 373–435 (Wien. Ber. 1877, 76:373–435). Reprinted in Wiss. Abhandlungen, Vol. II, reprint 42, p. 164–223, Barth, Leipzig, 1909.

Callen, H. B. (1985). *Thermodynamics and an introduction to Thermostatistics* (2nd ed.). New York: John Wiley & Sons.

Chalmers, M. (2016). Second law of thermodynamics "broken". *New Scientist*. Retrieved 2016-02-09.

Chudnovsky, A. (1973). Fracture of solids. In the book Scientific Papers on Elasticity and Plasticity, N9, pp. 3–43, Leningrad 1973, (Russian).

Chudnovsky, A. (1984). Statistics and thermodynamics of fracture. *International Journal of Engineering Science, 22*(8–10), 989–997.

Cuadras, A., Crisóstomo, J., Ovejas, V. J., & Quilez, M. (2015). Irreversible entropy model for damage diagnosis in resistors. *Journal of Applied Physics, 118*, 165103.

Cuadras, A., Romero, R., & Ovejas, V. J. (2016). Entropy characterisation of overstressed capacitors for lifetime prediction. *Journal of Power Sources, 336*, 272–278.

Cuadras, A., Yao, J., & Quilez, M. (2017). Determination of LEDs degradation with entropy generation rate. *Journal of Applied Physics, 122*, 145702.

de Groot, S. R., & Mazur, P. (1962). *Non-equilibrium thermodynamics*. Amsterdam/New York: North-Holland/Wiley.

DeHoff, R. T. (1993). *Thermodynamics in materials science*. McGraw Hill.

Ehrenfest, P., & Trkal, V. (1921). Deduction of the dissociation-equilibrium from the theory of quanta and a calculation of the chemical constant based on this. *Ann. Phys., 65*, 609–628.

Euler, L. (1736). Mechanica sive motus scientia analytice exposita.

Evans, D. J., Cohen, E. G. D., & Morriss, G. P. (1993). Probability of second law violations in shearing steady states. *Physical Review Letters, 71*(15), 2401–2404.

Getling, A. V. (1998). *Rayleigh–Bénard convection: Structures and dynamics*. Singapore: World Scientific. ISBN 978-981-02-2657-2.

Glansdorff, P., & Prigogine, I. (1971). *Thermodynamics theory of structure, stability and fluctuations*. London: Wiley-Interscience.

Gomez, J., & Basaran, C. (2005). A thermodynamics based damage mechanics constitutive model for low cycle fatigue analysis of microelectronics solder joints incorporating size effect. *International Journal of Solids and Structures, 42*(13), 3744–3772.

Gomez, J., & Basaran, C. (2006). Damage mechanics constitutive model for Pb/Sn solder joints incorporating nonlinear kinematic hardening and rate dependent effects using a return mapping integration algorithm. *Mechanics of Materials, 38*, 585–598.

Gomez, J., Lin, M., & Basaran, C. (2006). Damage mechanics modeling of concurrent thermal and vibration loading on electronics packaging. *Multidiscipline Modeling in Materials and Structures, 2*(3), 309–326.

Gunel, E. M., & Basaran, C. (2010). Stress whitening quantification in thermoformed of mineral filled acrylics. *ASME Journal of Engineering Materials and Technology, 132*, 031002–031011.

Gunel, E. M., & Basaran, C. (2011a). Damage characterization in non-isothermal stretching of acrylics: Part I theory. *Mechanics of Materials, 43*(12), 979–991.

Gunel, E. M., & Basaran, C. (2011b). Damage characterization in non-isothermal stretching of acrylics: Part II experimental validation. *Mechanics of Materials, 43*(12), 992–1012.

Guo, Q., Zaõri, F., & Guo, X. (2018). An intrinsic dissipation model for high-cycle fatigue life prediction. *International Journal of Mechanical Sciences, 140*, 163–171.

Haddad, W. M. (2017). Thermodynamics: The unique universal science. *Entropy, 19*, 621.

Haddad, W. M. (2019). *A dynamical systems theory of thermodynamics*. Princeton, NJ: Princeton University Press.

Haddad, W. M., Chellaboina, V., & Nersesov, S. G. (2005). *Thermodynamics: A dynamical systems approach*. Princeton, NJ: Princeton University Press.

Halliday, D., & Resnick, R. (1966). *Physics*. John Wiley & Sons, Inc.

Hsiao, C.-C., & Liang, B.-H. (2018). The generated entropy monitored by pyroelectric sensors. *Sensors, 18*, 3320.

Imanian, A., & Modarres, M. (2015). A thermodynamic entropy approach to reliability assessment with applications to corrosion fatigue. *Entropy, 17*(10), 6995–7020.

Imanian, A., & Modarres, M. (2018). A thermodynamic entropy-based damage. Assessment with applications to prognosis and health management. *Structural Health Monitoring, 17*(2), 240–254.

Jang, J. Y., & Khonsari, M. M. (2018). On the evaluation of fracture fatigue entropy. *Theoretical and Applied Fracture Mechanics, 96*, 351–361.

Jaynes, E. (1992). The Gibbs paradox. In C. Smith, G. Erickson, & P. Neudorfer (Eds.), *Maximum entropy and Bayesian methods* (pp. 1–22). Dordrecht, The Netherlands: Kluwer Academic Publishers.

Jaynes, E. T. (1957). Information theory and statistical mechanics. *Physics Review, 106*, 620–615.

Kestin, J., & Rice, J. R. (1970). Paradoxes in the application of thermodynamics to strained solids. In E. B. Stuart et al. (Eds.), *A critical review of thermodynamics* (p. 275). Baltimore: Mono Book Corp..

Kijalbaev, D., & Chudnovsky, A. (1970). On fracture of deformable solids. *Journal of Applied Mechanics and Technical Physics, N3*, 105.

Klamecki, B. E. (1980a). A thermodynamic model of friction. *Wear, 63*, 113–120.

Klamecki, B. E. (1980b). Wear—An entropy production model. *Wear, 58*, 325–330.

Klamecki, B. E. (1984). An entropy-based model of plastic deformation energy dissipation in sliding. *Wear, 96*, 319–329.

Koschmieder, E. L. (1993). *Bénard cells and taylor vortices*. Cambridge: Cambridge University Press. ISBN 0521-40204-2.

Lebowitz, J. (1993). Boltzmann's entropy and time's arrow. *Physics Today, 46*, 32–38.

Li, S., Abdulhamid, M., & Basaran, C. (2008). Simulating damage mechanics of electromigration and thermomigration. *Simulation: Transactions of the Society for Modeling and Simulation International, 84*(8/9), 391–401.

Li, S., & Basaran, C. (2009). A computational damage mechanics model for thermomigration. *Mechanics of Materials, 41*(3), 271–278.

Liakat, M., & Khonsari, M. M. (2015). Entropic characterization of metal fatigue with stress concentration. *International Journal of Fatigue, 70*, 223–234.

Lin, M., & Basaran, C. (2005). Electromigration induced stress analysis using fully coupled mechanical-diffusion equations with nonlinear material properties. *Computational Materials Science, 34*(1), 82–98.

Naderi, M., Amiri, M., & Khonsari, M. M. (2010). On the thermodynamic entropy of fatigue fracture. *Proceedings of the Royal Society A, 466*, 423–438.

Onsager, L. (1931). Reciprocal relations in irreversible processes I. *Physics Review, 37*, 405–426.

Onsager, L. (1932). Reciprocal relations in irreversible processes II. *Physics Review, 38*, 2265–2279.

Osara, J. A., & Bryant, M. D. (2019a). A thermodynamic model for lithium-ion battery degradation: Application of the degradation-entropy generation theorem. *Inventions, 4*, 23.

Osara, J. A., & Bryant, M. D. (2019b). Thermodynamics of fatigue: Degradation-entropy generation methodology for system and process characterization and failure analysis. *Entropy, 21*(7), 685.

Ostoja-Starzewski, M., & Raghavan, B. V. (2016). Continuum mechanics versus violations of the second law of thermodynamics. *Journal of Thermal Stresses, 39*(6), 734–749.

Ostoja-Starzewski, M. (2016). Second law violations, continuum mechanics, and permeability. *Continuum Mechanics and Thermodynamics, 28*(1–2), 489–501.

Pauli, W. (1973). *Statistical mechanics*. Cambridge, MA: MIT Press.

Planck, M. (1901). On the Law of Distribution of Energy in the Normal Spectrum, Annalen der Physik, vol. 4., p 553

Planck, M. (1906). Section 134: Entropie und Warscheinlichkeit. In *Vorlesungen uber die theorie der wurmestrahlung*. Leipzig, Germany: J.A. Barth.

Prigogine, I. (1955). *Introduction to thermodynamics of irreversible processes*. Springfield, IL: Charles C. Thomas Publisher.

Prigogine, I. (1957). *The molecular theory of solutions*. Amsterdam: North Holland Publishing Company.

Prigogine, I. (1961). *Introduction to thermodynamics of irreversible processes* (2nd ed.). New York: Interscience. OCLC 219682909.

Prigogine, I. (1968). *Introduction to thermodynamics of irreversible processes*. New York, NY: Wiley-Interscience.

Prigogine, I., & Defay, R. (1954). *Chemical thermodynamics*. London: Longmans Green and Co..

Prigogine, I., & Herman, R. (1971). *Kinetic theory of vehicular traffic*. New York: American Elsevier. ISBN 0-444-00082-8.

Prigogine, I., & Nicolis, G. (1977). *Self-organization in non-equilibrium systems*. New York: Wiley. ISBN 0-471-02401-5.

Reprinted in Wiss. Abhandlungen, Vol. II, reprint 42, p. 164–223.

Rice, J. R. (1971). Inelastic constitutive relations for solids: An internal-vraibale theory and its application to metal plasticity. *Journal of the Mechanics and Physics of Solids, 19*, 433–455.

Rice, J. R. (1977). Thermodynamics of the quasi-static growth of griffith cracks. *Journal of the Mechanics and Physics of Solids, 26*, 61–78.

Rivas, A., & Martin-Delgado, M. A. (2017). Topological heat transport and symmetry-protected boson currents. *Scientific Reports, 7*(1), 6350. https://doi.org/10.1038/s41598-017-06722-x.

Rivlin, R. S. (1981). Some comments on the endochronic theory of plasticity. *International Journal of Solids and Structures, 17*(2), 231–248.

Searles, D. J., & Evans, D. J. (2004). Fluctuations relations for nonequilibrium systems. *Australian Journal of Chemistry, 57*(12), 1119–1123. https://doi.org/10.1071/ch04115.

Sharp, K. and Matschinsky, F. (2015) Translation of Ludwig Boltzmann's paper "on the relationship between the second fundamental theorem of the mechanical theory of heat and probability calculations regarding the conditions for thermal equilibrium". Entropy, 17, 1971–2009.

Sherbakov, S. S., & Sosnovskiy, L. A. (2010). *Mechanics of tribo-fatigue systems* (p. 407). Minsk, Belarus: BSU.

Sitzungberichte der Kaiserlichen Akademie der Wissenschaften. Mathematisch-Naturwissen Classe. Abt. II, LXXVI 1877, pp 373–435.

Sosnovskiy, L. A. (1987). *Statistical mechanics of fatigue damage* (p. 288). Minsk, Belarus: Nauka i Tekhnika. (In Russian).

Sosnovskiy, L. A. (1999). *Tribo-fatigue: The dialectics of life* (2nd ed., p. 116). Gomel, Belarus: BelSUT Press. (In Russian).

Sosnovskiy, L. A. (2004). *L-risk (mechanothermodynamics of irreversible damages)* (p. 317). Gomel, Belarus: BelSUT Press. (In Russian).

Sosnovskiy, L. A. (2005). *Tribo-fatigue: Wear-fatigue damage and its prediction (foundations of engineering mechanics)* (p. 424). Berlin/Heidelberg, Germany: Springer.

Sosnovskiy, L. A. (2007). *Mechanics of wear-fatigue damage* (p. 434). Gomel, Belarus: BelSUT Press.

Sosnovskiy, L. A. (2009). Life field and golden proportions. *Nauka i Innovatsii, 79*, 26–33. (In Russian).

Sosnovskiy, L. A., & Sherbakov, S. S. (2012). Mechanothermodynamical system and its behavior. *Continuum Mechanics and Thermodynamics, 24*, 239–256.

Sosnovskiy, L. A., & Sherbakov, S. S. (2015). *Mechanothermodynamics*. Springer.

Sosnovskiy, L. A., & Sherbakov, S. S. (2016). Mechanothermodynamic entropy and analysis of damage state of complex systems. *Entropy, 18*(7), 268.

Sosnovskiy, L. A., & Sherbakov, S. S. (2017). A model of mechanothermodynamic entropy in tribology. *Entropy, 19*, 115.

Sosnovskiy, L. A., & Sherbakov, S. S. (2019). On the development of mechanothermodynamics as a new branch of physics. *Entropy, 21*(12), 1188.

Suhir, E. (2019). Failure Oriented Accelerated Testing (FOAT) Boltzmann Arrnhenius Zhurkov Equation (BAZ) and theor application in aerospace microelectronics and photonics reliability engineering. *International Journal of Aeronautical Science and Aerospace Research, 6*(3), 185–191.

Swalin, R. A. (1972). *Thermodynamic of solids*. New York, NY: John Wiley & Sons.

Swendsen, R. H. (2006). Statistical mechanics of colloids and Boltzmann's definition of the entropy. *American Journal of Physics, 74*, 187–190.

Tang, H., & Basaran, C. (2003). A damage mechanics based fatigue life prediction model. *Transactions of ASME, Journal of Electronic Packaging, 125*(1), 120–125.

Temfack, T., & Basaran, C. (2015). Experimental verification of a thermodynamic fatigue life prediction model. *Materials Science and Technology, 31*(13), 2015.

Turing, A. M. (1952). The chemical basis of morphogenesis (PDF). *Philosophical Transactions of the Royal Society of London B, 237*(641), 37–72. https://doi.org/10.1098/rstb.1952.0012. Bibcode: 1952RSPTB.237...37T. JSTOR 92463.

Valanis, K. C. (1971). *International Journal of Non-Linear Mechanics, 6*(3), 337–360.

Valanis, K. C., & Komkov, V. (1980). Irreversible thermodynamics from the point of view of internal variable theory/A Lagrangian formulation, – Archiwum Mechaniki Stosowanej, rcin.org.pl.

Van Kampen, N. G. (1984). The Gibbs paradox. In W. E. Parry (Ed.), *Essays in theoretical physics in honour of Dirk ter Haar* (pp. 303–312). Oxford, UK: Pergamon.

Wassim M Haddad, (2019). A dynamical systems theory of thermodynamics, Princeton University press.

Wassim M Haddad, Qing Hui, Sergey G Nersesov, Vijaysekhar Chellaboina, (2005). Thermodynamic modeling, energy equipartition, and nonconservation of entropy for discrete-time dynamical systems, J. of Advances in Difference Equations, Volume 2005, Issue 3, Pp. 248040, Springer International Publishing.

Wang, J., & Yao, Y. (2017). An entropy based low-cycle fatigue life prediction model for solder materials. *Entropy, 19*(10), 503.

Wang, J., & Yao, Y. (2019). An entropy-based failure prediction model for the creep and fatigue of metallic materials. *Entropy, 21*(11), 1104.

Wang, S., Mittag, S., & Evans. (2002). Experimental demonstration of violations of the second law of thermodynamics. *Physical Review Letters*.

Wang, T., Samal, S. K., Lim, S. K., & Shi, Y. (2019). Entropy production based full-chip fatigue analysis: From theory to mobile applications. *IEEE Transactions on Computer-Aided Design of Integrated Circuits and Systems, 38*(1).

Whales, P. W. (1983). Entropy production during fatigue as a criterion for failure: The critical entropy threshold: A mathematical model for fatigue. U.S. Office of Naval Research Technical Report No. 1, Govt. Accession No. A134767.

Whaley, P. W. (1983). A thermodynamic approach to metal fatigue. *In Proceedings of the ASME International Conference of Advances in Life Prediction Methods*, Albany, NY (pp. 18-21).

Yao, W., & Basaran, C. (2012). Electromigration analysis of solder joints under ac load: A mean time to failure model. *Journal of Applied Physics, 111*(6), 063703.

Yao, W., & Basaran, C. (2013a). Electromigration damage mechanics of lead-free solder joints under pulsed DC loading: A computational model. *Computational Materials Science, 71*, 76–88.

Yao, W., & Basaran, C. (2013b). Electrical pulse induced impedance and material degradation in IC chip packaging. *Electronic Materials Letters, 9*(5), 565–568.

Yao, W., & Basaran, C. (2013c). Computational damage mechanics of electromigration and thermomigration. *Journal of Applied Physics, 114*, 103708.

Young, C., & Subbarayan, G. (2019b). Maximum entropy models for fatigue damage in metals with application to low-cycle fatigue of aluminum 2024-T351. *Entropy, 21*(10), 967.

Young, C., & Subbarayan, G. (2019a). Maximum entropy models for fatigue damage aluminum 2024-T351. *Entropy, 21*, xx.

Yun, H., & Modarres, M. (2019). Measures of entropy to characterize fatigue damage in metallic materials. *Entropy, 21*(8), 804.

Zhang, M.-H., Shen, X.-H., He, L., & Zhang, K.-S. (2018). Application of differential entropy in characterizing the deformation inhomogeneity and life prediction of low-cycle fatigue of metals. *Materials, 11*, 1917.

Chapter 5
Unified Mechanics of Thermo-mechanical Analysis

5.1 Introduction

This chapter describes the implementation of unified mechanics theory (UMT) for thermo-mechanical analysis. Implementation will use viscoplasticity, because thermo-viscoplastic model is necessary for metals/alloys operating at $0.3T_{\mathrm{m}}$ or higher where T_{m} is the melting temperature in Kelvin.

5.2 Unified Mechanics Theory-Based Constitutive Model

5.2.1 Flow Theory

5.2.1.1 Newtonian Mechanics: Elastic Constitutive Relationship (Hooke's Law)

For a classical von Mises rate-independent plasticity, the elastic constitutive relationship is given by Hooke's law in the rate form as follows:

$$\dot{\boldsymbol{\sigma}} = \boldsymbol{C} : \dot{\boldsymbol{\varepsilon}}^{\mathbf{e}} \tag{5.1a}$$

$$\dot{\boldsymbol{\varepsilon}}^{\mathbf{e}} = \left(\dot{\boldsymbol{\varepsilon}} - \dot{\boldsymbol{\varepsilon}}^{\mathbf{p}} - \dot{\boldsymbol{\varepsilon}}^{\boldsymbol{\theta}}\right) \tag{5.1b}$$

where $\dot{\boldsymbol{\varepsilon}}, \dot{\boldsymbol{\varepsilon}}^{\mathbf{p}}$, and $\dot{\boldsymbol{\varepsilon}}^{\boldsymbol{\theta}}$ are total strain rate, inelastic strain rate, and thermal strain rate vectors, respectively, and $\boldsymbol{C}$ is the elastic constitutive tensor for the virgin material. In Eqs. (5.1a) and (5.1b) ":" represents the inner product between the fourth-order constitutive tensor $\boldsymbol{C}$ and the elastic strain rate vector, $\dot{\boldsymbol{\varepsilon}}^{\mathbf{e}}$.

C. Basaran, *Introduction to Unified Mechanics Theory with Applications*,
https://doi.org/10.1007/978-3-030-57772-8_5

5.2.1.2 Yield Surface

There is no uniformity in terminology in the plasticity literature; "yield function," "yield surface," "plastic potential function," and "plastic flow function" are all used interchangeably. For most metals an elastoplastic domain is defined according to the following yield function of the von Mises type:

$$F(\boldsymbol{\sigma}, \alpha) = \|\boldsymbol{S} - \boldsymbol{X}\| - \sqrt{\frac{2}{3}}K(\alpha) \equiv \|\boldsymbol{S} - \boldsymbol{X}\| - R(\alpha) \tag{5.2}$$

where $F(\boldsymbol{\sigma}, \alpha)$ is a yield surface separating the elastic from the inelastic domain, $\boldsymbol{\sigma}$ is the second-order stress tensor, α is a hardening parameter which specifies the evolution of the radius of the yield surface, $\boldsymbol{X}$ is the deviatoric component of the back-stress tensor describing the position of the center of the yield surface in stress space, $\boldsymbol{S}$ is the deviatoric component of the stress tensor given by $S = \sigma - \frac{1}{3}p\boldsymbol{I}$ where p is the hydrostatic pressure and $\boldsymbol{I}$ is the second-order identity tensor, $R(\alpha) \equiv \sqrt{\frac{2}{3}}K(\alpha)$ is the radius of the yield surface in stress space, and $\| \ \|$ represents the norm operator. A schematic representation of the yield surface is described in Fig. 5.1.

5.2.1.3 Flow Rule

The evolution of the plastic strain is represented by a general associative flow rule of the form

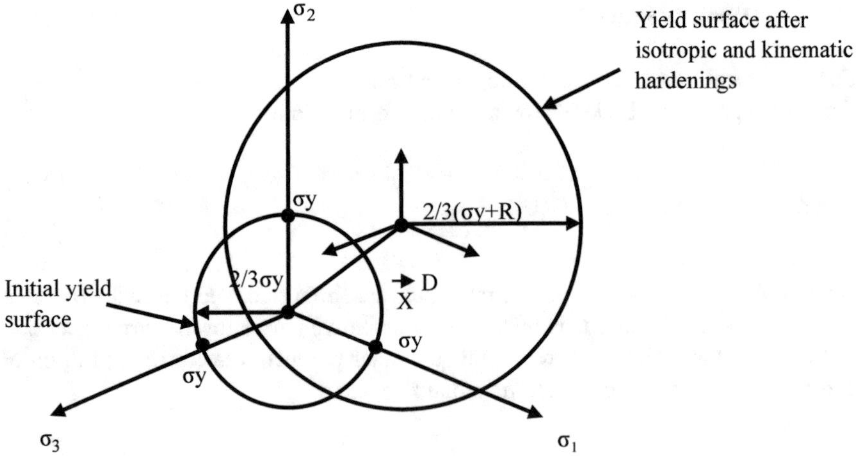

Fig. 5.1 Yield surface in the principal stress space

$$\dot{\boldsymbol{\varepsilon}}^{\mathrm{p}} = \gamma \frac{\partial F}{\partial \sigma} \equiv \gamma \underline{\widehat{n}} \tag{5.3}$$

with $\underline{\widehat{n}} = \frac{\partial F}{\partial \sigma}$ being a vector normal to the yield surface in stress space and specifying the direction of plastic flow [this is also referred to as the normality rule], $\dot{\varepsilon}^{\mathrm{p}}$ has already been defined as the plastic strain rate, and γ is a nonnegative consistency parameter, which is derived later in the chapter.

5.2.1.4 Isotropic Hardening

Isotropic hardening describes the increasing radius of the yield surface in Eq. (5.2). Chaboche (1989) proposed the following evolution function for isotropic hardening in metals:

$$K(\alpha) = \sqrt{\frac{2}{3}} Y_0 + R_\infty \left(1 - \mathrm{e}^{-c\alpha}\right) \tag{5.4a}$$

where α is a plastic hardening parameter or plastic strain trajectory evolving according to Eq. (5.4b), Y_0 is the initial yield stress in uniaxial tension, R_∞ is an isotropic hardening saturation value, and c is an isotropic hardening rate material parameter:

$$\dot{\alpha} = \sqrt{\frac{2}{3} \dot{\boldsymbol{\varepsilon}}^{\mathrm{p}} \dot{\boldsymbol{\varepsilon}}^{\mathrm{p}}} \tag{5.4b}$$

Using Eqs. (5.3) and (5.4b), we can write the standard definition of equivalent plastic strain as follows:

$$\alpha = \int_{t_0}^{t_1} \sqrt{\frac{2}{3} \dot{\boldsymbol{\varepsilon}}^{\mathrm{p}} \dot{\boldsymbol{\varepsilon}}^{\mathrm{p}}}\, dt \tag{5.4c}$$

5.2.1.5 The Nonlinear Kinematic Hardening (NLK) Rule

For implementation, the NLK rule describing the movement of the center of the yield surface in stress space empirical equation proposed Chaboche (1989) is used. The model is based on the work of Armstrong and Frederick (1966). Nonlinearities are introduced as a recall term to the Prager (1955) linear hardening rule given in Eq. (5.5) and where c_1 and c_2 are material parameters:

$$\dot{\boldsymbol{X}} = c_1 \dot{\boldsymbol{\varepsilon}}^{\mathrm{p}} - c_2 \boldsymbol{X} \dot{\alpha} \tag{5.5}$$

In Eq. (5.5) the first term represents the linear kinematic hardening rule as defined by Prager (1955). The second term is a recall term, often called a dynamic recovery term, which introduces the nonlinearity between the back-stress X and the actual plastic strain. When $c_2 = 0$, Eq. (5.5) reduces to the Prager (1955) linear kinematic hardening rule. The NLK equation describes the rapid changes due to the plastic flow during cyclic loadings and plays an important role even under stabilized conditions (after saturation of cyclic hardening). According to Chaboche (1989), Eq. (5.5) takes into account the transient hardening effects in each stress-strain loop, and after unloading, dislocation remobilization is implicitly described due to the back-stress effect and the larger plastic modulus at the beginning of the reverse plastic flow.

5.2.1.6 Consistency Parameter, γ

In Eqs. (5.3) and (5.4b), γ is a nonnegative consistency parameter representing the irreversible character of plastic flow. Consistency parameter must satisfy the following requirements:

1. For a rate-independent plasticity material model, γ obeys the so-called loading/unloading and consistency condition:

$$\gamma \geq 0 \quad \text{and} \quad F(\sigma, \alpha) \leq 0 \tag{5.6}$$

$$\gamma \dot{F}(\sigma, \alpha) = 0 \tag{5.7}$$

2. For a rate-dependent plasticity material model, conditions specified by Eqs. (5.6) and (5.7) are replaced by a constitutive equation of the form:

$$\gamma = \frac{\langle \varnothing(F) \rangle}{\eta} \tag{5.8}$$

where η represents a viscosity material parameter, $\langle\rangle$ is Macauley bracket, and $\varnothing(F)$ is a material-specific function defining the character of the viscoplastic flow. When $\eta \rightarrow 0$, the constitutive model approaches the rate-independent case (Simo and Hughes 1997). In the case of a rate-independent material, F satisfies conditions specified by Eqs. (5.6) and (5.7), and additionally stress states such $F(\boldsymbol{\sigma}, \alpha) > 0$ are ruled out. On the other hand, in the case of a rate-dependent material, the magnitude of the viscoplastic flow is proportional to the distance of the stress state to the surface defined by $F(\boldsymbol{\sigma}, \alpha) = 0$.

5.2.1.7 Viscoplastic Creep Rate Law

The relation between γ and η expressed in Eq. (5.8) is a general constitutive equation, and different forms of the constitutive relationship describing the material-specific evolution of the viscoplastic strain can be implemented. The creep law proposed by Kashyap and Murty (1981) and extended to the multiaxial case by Basaran et al. (2005) is used:

$$\dot{\varepsilon}^{\mathrm{vp}} = \frac{AD_0Eb}{k\theta}\left(\frac{\langle F\rangle}{E}\right)^n\left(\frac{b}{d}\right)^p \mathrm{e}^{-Q/\widehat{R\theta}}\left(\frac{\partial F}{\partial \sigma}\right) \tag{5.9}$$

where $\boldsymbol{A}$ is a dimensionless material parameter, which is temperature and rate dependent, $D_i = D_0 e^{-Q/\widehat{R\theta}}$ is a diffusion coefficient with D_0 representing a frequency factor, $\boldsymbol{Q}$ is the creep activation energy, R is the universal gas constant, θ is the absolute temperature in Kelvin, $E(\theta)$ is a temperature-dependent Young's modulus, b is the characteristic length of crystal dislocation (magnitude of Burger's vector), k is Boltzman's constant, d is the average grain size, p is a grain size exponent, and n is a stress exponent for viscoplastic deformation rate, where $1/n$ defines the strain sensitivity. In Eq. (5.9) we can identify

$$\langle \varnothing(F)\rangle = \langle F\rangle^n \quad \text{and} \quad \eta = \frac{k\theta}{AD_0E^{n-1}b}\left(\frac{d}{b}\right)^p e^{Q/\widehat{R\theta}} \tag{5.10}$$

For a more in-depth study of different viscoplastic models and comparison of most models published in the literature, readers are referred to Lee and Basaran (2011).

5.2.2 *Effective Stress Concept and Strain Equivalence Principle*

The derivation that follows is due to Lemaitre (1996) and explains the effective stress concept. In order to introduce the effective stress concept, it is useful to consider a representative volume element (RVE) of material loaded by a force $\vec{F}$. At a point M oriented by a plane defined by its normal direction $\vec{n}$ and its abscissa x along the direction $\vec{n}$ (Fig. 5.2), the nominal uniaxial stress is $\sigma = \vec{F}/\delta S$, where $\vec{F} = \vec{n}F$; δS is the area of the intersection of the plane with the RVE.

The effective area of the intersections of all micro-cracks or micro-cavities that lie in δS is represented by δS_{Dx}. No micro-forces are acting on the surface of micro-cracks and micro-cavities. It is convenient to introduce an effective stress concept related to the surface that effectively resists the load, namely, $(\delta S - \delta S_{Dx})$ (Rabotnov 1969):

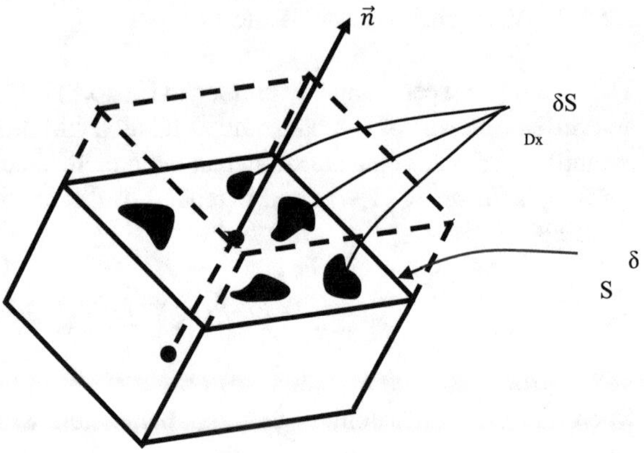

Fig. 5.2 Representative volume element. (After Lemaitre and Chaboche 1990)

$$\widetilde{\sigma} = \frac{\vec{F}}{\delta S - \delta S_{Dx}} \tag{5.11}$$

The effective stresses $\widetilde{\sigma}$ are higher than the nominal stresses because local stresses are redistributed to the undamaged material.

Thermodynamics enables us to define the state at any point by a set of continuous state variables. This postulate means that the constitutive equations written for the surface of $(\delta S - \delta S_{Dx})$ are not modified by the damage in the material. The true stress is the effective stress and not the nominal stress. As a result we can postulate that the following strain equivalence principle (Lemaitre and Chaboche 1990): "Any strain constitutive equation for a damaged material may be derived in the same way as for a virgin material except that the usual stress is replaced by the effective stress."

5.2.2.1 Unified Mechanics Theory (UMT) Implementation

Based on UMT strain energy density rate can be given by

$$\dot{U} = \frac{1}{2}\dot{\boldsymbol{\sigma}} : \dot{\boldsymbol{\varepsilon}^{\mathrm{e}}} = \frac{1}{2}(1 - \Phi)\boldsymbol{C} : [\dot{\boldsymbol{\varepsilon}}^{\mathrm{e}}]^2 \tag{5.12}$$

where Φ is the thermodynamic state index, $\boldsymbol{C}$ is the tangential constitutive tensor, and $\dot{\boldsymbol{\varepsilon}}^{\mathrm{e}}$ is the elastic strain rate vector. Taking the derivative with respect to strain rate and ignoring the derivative of strain energy rate with respect to thermodynamic state index, Φ, yield

$$\dot{\boldsymbol{\sigma}} = (1 - \Phi)\ \boldsymbol{C} : \dot{\boldsymbol{\varepsilon}}^{\mathrm{e}} \tag{5.13}$$

Of course ignoring the derivative of Φ with respect to strain rate is not accurate. However, for any incremental procedure, the ignored portion is smaller by order of magnitude because of the square over strain rate. Moreover, our objective here is to introduce the simplest implementation of UMT. Elastic strain rate can be calculated from the total strain rate vector by subtracting the inelastic and thermal components:

$$\dot{\boldsymbol{\sigma}} = (1 - \Phi)\boldsymbol{C} : \left(\dot{\boldsymbol{\varepsilon}} - \dot{\boldsymbol{\varepsilon}}^{\text{vp}} - \dot{\boldsymbol{\varepsilon}}^{\theta}\right) \tag{5.14}$$

$$F = \left\|\boldsymbol{S} - \boldsymbol{X}^{\Phi}\right\| - (1 - \Phi)\sqrt{\frac{2}{3}}K(\alpha) \equiv \left\|\boldsymbol{S} - \boldsymbol{X}^{\Phi}\right\| - (1 - \Phi)R(\alpha) \tag{5.15}$$

where Φ is the TSI and the evolution of the back-stress due to degradation is given by

$$\dot{\boldsymbol{X}}^{\Phi} = (1 - \Phi)(c_1\dot{\boldsymbol{\varepsilon}}^{\text{vp}} - c_2\boldsymbol{X}\dot{\alpha}) \tag{5.16}$$

It is important to point out that in unified mechanics theory, we only care about degradation in thermodynamic state index axis space. Kinematic hardening function given in Eq. (5.15) can be defined in many other forms.

5.3 Return Mapping Algorithm

Material nonlinear finite element method implementation requires more sophisticated solution methods compared to linear elastic analysis. A return mapping algorithm is a particular form of an algorithm based on a combination between an explicit method and an implicit method. The explicit method makes an initial approximation to the solution. The approximated solution is then used in the implicit method to improve the prediction. The following trial (elastic predictor), the state can be written as

$$\boldsymbol{S}_{n+1}^{\text{tr}} = \boldsymbol{S}_n + (1 - \Phi)2\mu\Delta\boldsymbol{e}_{n+1} \tag{5.17}$$

where μ is the shear modulus and $\Delta\boldsymbol{e}_{n+1}$ is the deviatoric strain increment vector. The increment of the back-stress can then be computed using Eq. (5.16):

$$d\boldsymbol{X}_{n+1}^{\Phi} = (1 - \Phi)\left\{c_1 d\boldsymbol{\varepsilon}_{n+1}^{\text{vp}} - c_2\Delta\gamma\left[\beta\boldsymbol{X}_n^{\Phi} + (1 - \beta)\boldsymbol{X}_{n+1}^{\Phi}\right]\right\} \tag{5.18}$$

where $c_2' = \sqrt{\frac{2}{3}}c_2$ and a generalized midpoint rule for the recall term with the extreme values of $\beta = 0$ and $\beta = 1$ correspond to the backward and forward Euler methods, respectively. Utilizing Eqs. (5.3) and (5.14), the incremental form of the viscoplastic strain can be written as

$$d\boldsymbol{\varepsilon}_{n+1}^{\text{vp}} = \Delta\gamma\frac{\boldsymbol{S}_{n+1} - \boldsymbol{X}_{n+1}^{\Phi}}{\left\|\boldsymbol{S}_{n+1} - \boldsymbol{X}_{n+1}^{\Phi}\right\|} \tag{5.19}$$

Substitution of Eq. (5.19) into Eq. (5.18) yields for any value of the integration parameter β between 0 and 1:

$$d\boldsymbol{X}_{n+1}^{\Phi} = a_{n+1}\Delta\gamma\left(\frac{\boldsymbol{S}_{n+1} - \boldsymbol{X}_{n+1}^{\Phi}}{\left\|\boldsymbol{S}_{n+1} - \boldsymbol{X}_{n+1}^{\Phi}\right\|} - \frac{c_2'}{c_1}\boldsymbol{X}_n^{\Phi}\right) \tag{5.20}$$

with $a_{n+1} = \frac{c_1(1-\Phi)}{1+c_2'(1-\Phi)(1-\beta)\Delta\gamma}$.

Using the flow rule expressed in Eq. (5.3) allows us to express Eq. (5.17) as

$$\boldsymbol{S}_{n+1} = \boldsymbol{S}_{n+1}^{\mathrm{tr}} - \Delta\gamma(1-\Phi)2\mu\frac{\boldsymbol{S}_{n+1} - \boldsymbol{X}_{n+1}^{\Phi}}{\left\|\boldsymbol{S}_{n+1} - \boldsymbol{X}_{n+1}^{\Phi}\right\|} \tag{5.21}$$

And introducing the relative stress tensor in $\boldsymbol{\xi}_{n+1}^{D} = \boldsymbol{S}_{n+1} - \boldsymbol{X}_{n+1}^{\Phi}$ yields

$$\boldsymbol{\xi}_{n+1}^{\Phi} = \boldsymbol{S}_{n+1} - \boldsymbol{X}_{n+1}^{\Phi} \equiv \boldsymbol{S}_{n+1}^{\mathrm{tr}} - \Delta\gamma(1-\Phi)2\mu\frac{\boldsymbol{S}_{n+1} - \boldsymbol{X}_{n+1}^{\Phi}}{\boldsymbol{S}_{n+1} - \boldsymbol{X}_{n+1}^{\Phi}} - \boldsymbol{X}_n^{\Phi} - d\boldsymbol{X}_{n+1}^{\Phi} \tag{5.22}$$

Using Eq. (5.20) into Eq. (5.22) gives

$$\boldsymbol{S}_{n+1} - \boldsymbol{X}_{n+1}^{\Phi} + \Delta\gamma[(1-\Phi)2\mu + a_{n+1}]\frac{\boldsymbol{S}_{n+1} - \boldsymbol{X}_{n+1}^{\Phi}}{\left\|\boldsymbol{S}_{n+1} - \boldsymbol{X}_{n+1}^{\Phi}\right\|} = \boldsymbol{B}_n \tag{5.23}$$

which results after letting $\boldsymbol{B}_n = \boldsymbol{S}_{n+1}^{\mathrm{tr}} - \boldsymbol{X}_n^{\Phi} + b_{n+1}\Delta\gamma\boldsymbol{X}_n$ and $b_{n+1} = \frac{c_2'}{c_1}a_{n+1}$.

The normal to the yield surface can be expressed in terms of the initial values of the stress; the state variables and the strain increment at each step are as follows:

$$\underline{\dot{n}}_{n+1} \equiv \frac{\boldsymbol{S}_{n+1} - \boldsymbol{X}_{n+1}^{\Phi}}{\left\|\boldsymbol{S}_{n+1} - \boldsymbol{X}_{n+1}^{\Phi}\right\|} = \frac{\boldsymbol{B}_n}{\|\boldsymbol{B}_n\|} \tag{5.24}$$

Taking the trace product of Eq. (5.22) with itself yields

$$\begin{aligned} &\left\|\boldsymbol{S}_{n+1} - \boldsymbol{X}_{n+1}^{\Phi}\right\| + \Delta\gamma[(1-\Phi)2\mu + a_{n+1}] \\ &\quad = \Big\{\left\|\boldsymbol{S}_n - \boldsymbol{X}_n^{\Phi}\right\|^2 + \left\|(1-\Phi)2\mu\Delta\boldsymbol{e}_{n+1} + b_{n+1}\Delta\gamma\boldsymbol{X}_n^{\Phi}\right\|^2 \\ &\qquad +2\left(\boldsymbol{S}_n - \boldsymbol{X}_n^{\Phi}\right) : \left[(1-\Phi)2\mu\Delta\boldsymbol{e}_{n+1} + b_{n+1}\Delta\gamma\boldsymbol{X}_n^{\Phi}\right]\Big\}.^{1/2} \end{aligned} \tag{5.25}$$

Using Eq. (5.4b) for the rate-independent case or Eq. (5.8) for the rate-dependent case results in the following nonlinear scalar equation for the consistency parameter γ which can be solved by a local Newton method:

$$
\begin{aligned}
g(\Delta\gamma) = \Big\{ & \left\| \boldsymbol{S}_n - \boldsymbol{X}_n^{\Phi} \right\|^2 + \left\| (1-\Phi)2\mu\Delta\boldsymbol{e}_{n+1} + b_{n+1}\Delta\gamma\boldsymbol{X}_n^{\Phi} \right\|^2 \\
& +2\left(\boldsymbol{S}_n - \boldsymbol{X}_n^{\Phi}\right) : \left[(1-\Phi)2\mu\Delta\boldsymbol{e}_{n+1} + b_{n+1}\Delta\gamma\boldsymbol{X}_n^{\Phi}\right] \Big\}.^{1/2} \\
& -(1-\Phi)\sqrt{\frac{2}{3}}K\left(\alpha_n + \sqrt{\frac{2}{3}}\Delta\gamma\right) \\
& -\Delta\gamma[(1-\Phi)2\mu + a_{n+1}] - \Theta\left(\frac{\Delta\gamma\eta}{\Delta t}\right)
\end{aligned} \tag{5.26}
$$

Once Eq. (5.26) is solved for $\Delta\gamma$,the following updating scheme can be used

$$\alpha_{n+1} = \alpha_n + \sqrt{2/3}\Delta\gamma \tag{5.27}$$

$$\boldsymbol{\varepsilon}_{n+1}^{\mathrm{vp}} = \boldsymbol{\varepsilon}_n^{\mathrm{vp}} + \Delta\gamma\frac{\boldsymbol{B}_n}{\|\boldsymbol{B}_n\|} \tag{5.28}$$

$$\boldsymbol{X}_{n+1}^{\Phi} = \boldsymbol{X}_n^{\Phi} + a_{n+1}\Delta\gamma\left(\frac{\boldsymbol{S}_{n+1} - \boldsymbol{X}_{n+1}^{\Phi}}{\left\|\boldsymbol{S}_{n+1} - \boldsymbol{X}_{n+1}^{\Phi}\right\|} - \frac{c_2'}{c_1}\boldsymbol{X}_n^{\Phi}\right) \tag{5.29}$$

$$\boldsymbol{\xi}_{n+1}^{\Phi} = (1-\Phi)K(\alpha_{n+1})\frac{\boldsymbol{B}_n}{\|\boldsymbol{B}_n\|} \tag{5.30}$$

$$\boldsymbol{S}_{n+1} = \boldsymbol{\xi}_{n+1}^{\Phi} + \boldsymbol{X}_{n+1}^{\Phi} \tag{5.31}$$

$$
\begin{aligned}
\boldsymbol{\sigma}_{n+1} = {} & \kappa(1-\Phi)\mathrm{tr}(\boldsymbol{\varepsilon}_{n+1})\,\mathrm{I} + 2\mu(1-\Phi) \\
& \times\left(\boldsymbol{e}_{n+1} - \boldsymbol{\varepsilon}_n^{\mathrm{vp}} - \gamma_{n+1}\frac{\boldsymbol{B}_n}{\|\boldsymbol{B}_n\|} - e_{n+1}^{\theta}\right)
\end{aligned} \tag{5.32}
$$

5.3.1 *Linearization (Consistent Jacobian)*

Differentiating Eq. (5.32) with respect to the total strain but not entropy at the end of the step yields

$$d\boldsymbol{\sigma}_{n+1} = (1-\Phi)\left[\boldsymbol{C} - 2\mu\underline{\widehat{n}}_{n+1} \otimes \frac{\partial \Delta\gamma}{\partial \boldsymbol{\varepsilon}_{n+1}} - 2\mu\Delta\gamma \otimes \frac{\partial \underline{\widehat{\boldsymbol{n}}}_{n+1}}{\partial \boldsymbol{\varepsilon}_{n+1}}\right] : d\boldsymbol{\varepsilon}_{n+1} \tag{5.33}$$

Ignoring differentiation with respect to entropy generation term in the thermodynamic state index rate is not correct. If the strain increments are small, then the ignored part of the differentiation is smaller by an order of magnitude. However, our objective here is to present an earlier and simpler version of the UMT formulation.

The operator $\otimes$ represents the ordered (dyadic) outer product and $\underline{\widehat{\boldsymbol{n}}}_{n+1}$ is the normal vector to the yield surface as defined in Eq. (5.3) evaluated at the end of the increment, and $\frac{\partial \Delta\gamma}{\partial \boldsymbol{\varepsilon}_{n+1}}$ can be calculated from Eq. (5.25):

$$\frac{\partial \Delta\gamma}{\partial \boldsymbol{\varepsilon}_{n+1}} = \frac{\underline{\widehat{\boldsymbol{n}}}_{n+1}}{\overline{K}_3} \tag{5.34}$$

Using

$$\overline{K}_3 = \overline{K}_1 + \overline{K}_2$$

$$\overline{K_1} = 1 + \frac{K'}{3\mu} + \frac{a_{n+1}}{2\mu(1-\Phi)}$$

$$\overline{K_2} = \frac{a'_{n+1}\Delta\gamma}{2\mu(1-\Phi)} + \frac{\underline{\widehat{\boldsymbol{n}}}_{n+1} b_{n+1}}{2\mu(1-\Phi)}\left[b_{n+1}(1-\beta)\Delta\gamma - 1\right] : \boldsymbol{X}_n^{\Phi} + \frac{1}{2\mu(1-\Phi)}\frac{\partial \Theta}{\partial \Delta\gamma}$$

In addition, $\frac{\partial \underline{\widehat{\boldsymbol{n}}}_{n+1}}{\partial \boldsymbol{\varepsilon}_{n+1}}$ can be obtained from Eq. (5.24) as

$$\frac{\partial \underline{\widehat{n}}_{n+1}}{\partial \boldsymbol{\varepsilon}_{n+1}} = \frac{\partial \underline{\widehat{\boldsymbol{n}}}_{n+1}}{\partial \vec{B}_n}\frac{\partial B_n}{\partial \boldsymbol{\varepsilon}_{n+1}} \equiv \frac{1}{\|\boldsymbol{B}_{n+1}\|}\left(\hat{I} - \underline{\widehat{\boldsymbol{n}}}_{n+1} \otimes \underline{\widehat{\boldsymbol{n}}}_{n+1}\right) : \frac{\partial \boldsymbol{B}_n}{\partial \boldsymbol{\varepsilon}_{n+1}} \tag{5.35a}$$

and

$$\frac{\partial \boldsymbol{B}_n}{\partial \boldsymbol{\varepsilon}_{n+1}} = 2\mu(1-\Phi)\left(\widehat{\prod} - \frac{1}{3}\boldsymbol{I} \otimes \boldsymbol{I}\right) + \left(b'_{n+1}\Delta\gamma + b_{n+1}\right)\boldsymbol{X}_n \otimes \frac{\partial \Delta\gamma}{\partial \boldsymbol{\varepsilon}_{n+1}}$$

where $\widehat{\prod}$ is a fourth-order unit tensor mapped into the unit matrix and **I** a second-order identity tensor mapped into unit column vector and K' denotes the derivative with respect to the argument in $K(\alpha)$.

Letting $\bar{K}_4 = b'_{n+1}\Delta\gamma + b_{n+1}$ and substituting $\frac{\partial \boldsymbol{B}_n}{\partial \boldsymbol{\varepsilon}_{n+1}}$ in Eq. (5.35a) yields

$$\frac{\partial \underline{\widehat{\boldsymbol{n}}}_{n+1}}{\partial \boldsymbol{\varepsilon}_{n+1}} = \frac{\partial \underline{\widehat{\boldsymbol{n}}}_{n+1}}{\partial \mathbf{B}_n} \frac{\partial \boldsymbol{B}_n}{\partial \boldsymbol{\varepsilon}_{n+1}}$$

$$\equiv \frac{2\mu(1-\Phi)}{\|\boldsymbol{B}_n\|}\left(\widehat{\prod} - \underline{\widehat{\boldsymbol{n}}}_{n+1} \otimes \underline{\widehat{\boldsymbol{n}}}_{n+1} - \frac{1}{3}\mathbf{I}\otimes\mathbf{I}\right) + \frac{1}{\|\boldsymbol{B}_n\|}$$

$$\times \left(\mathbf{I} - \underline{\widehat{\boldsymbol{n}}}_{n+1} \otimes \underline{\widehat{\boldsymbol{n}}}_{n+1}\right)$$

$$: \left(\frac{\overline{K}_4}{\overline{K}_3} \underline{\widehat{\boldsymbol{n}}}_{n+1} \otimes \boldsymbol{X}_n^{\Phi}\right) \tag{5.35b}$$

Using Eqs. (5.34) and (5.35b) in Eq. (5.33) results in the following algorithmic version of the material Jacobian coupling the effects of degradation and rate dependency:

$$\boldsymbol{C}_{n+1}^{\mathrm{EVPD}} = (1-\Phi)\kappa\mathbf{I}\otimes\mathbf{I} + 2\mu(1-\Phi)\delta_{n+1}\left(\widehat{\prod} - \frac{1}{3}\mathbf{I}\otimes\mathbf{I}\right)$$

$$- 2\mu(1-\Phi)\overline{\theta}_{n+1}\underline{\widehat{\boldsymbol{n}}}_{n+1} \otimes \underline{\widehat{\boldsymbol{n}}}_{n+1}$$

$$- \frac{2\mu(1-\Phi)}{\|B_n\|}\Delta\gamma\left(\widehat{\prod} - \underline{\widehat{\boldsymbol{n}}}_{n+1} \otimes \underline{\widehat{\boldsymbol{n}}}_{n+1}\right)$$

$$: \left(\frac{\overline{K}_4}{\overline{K}_3} \underline{\widehat{\boldsymbol{n}}}_{n+1} \otimes \boldsymbol{X}_n^{\Phi}\right) \tag{5.35c}$$

where

$$\delta_{n+1} = 1 - \frac{\Delta\gamma 2\mu(1-\Phi)}{\|\boldsymbol{B}_n\|} \quad \text{and} \quad \overline{\theta}_{n+1} = \frac{1}{\overline{K}_3} - \frac{\Delta\gamma 2\mu(1-\Phi)}{\|\boldsymbol{B}_n\|}$$

The pseudocode corresponding to the above algorithm is detailed in Table 5.1. Moreover, Table 5.2 shows the local Newton-Raphson algorithm used to solve for the consistency parameter, which preserves the quadratic rate of convergence of the overall Newton scheme.

5.4 Thermodynamic Fundamental Equation in Thermo-mechanical Problems

We have defined thermodynamic fundamental relations [equation] in the earlier chapter on thermodynamics. However we feel it is necessary to restate some basics again to make it easier to understand the derivation for the readers. We find it helpful to quote these basics directly from DeHoff (1993).

In irreversible thermodynamics, the so-called balance equation for the entropy plays a central role. This equation expresses the fact that the entropy of a volume element changes with time for two reasons. First, it changes because entropy flows

Table 5.1 Return mapping algorithm classical theory rate-dependent model

Update strain $\boldsymbol{\varepsilon}_{n+1} = \boldsymbol{\varepsilon}_n + \nabla \Delta \boldsymbol{\mu}$

Compute trial state $\boldsymbol{S}^{tr}_{n+1} = \boldsymbol{S}_n + (1 - \boldsymbol{\Phi}_n) 2\mu \boldsymbol{\Delta} \boldsymbol{e}_{n+1}$

Compute trial yield function $\boldsymbol{F}^{tr}_{n+1} = \left\| \boldsymbol{S}^{tr}_{n+1} - \boldsymbol{X}^{D}_{n} \right\| - (1 - \boldsymbol{\Phi}) \sqrt{\tfrac{2}{3}} \boldsymbol{K}(\boldsymbol{\alpha}_n)$

IF $\boldsymbol{F}^{tr}_{n+1} > \mathbf{0}$**THEN**

Call Newton local and solve $\mathbf{g}(\boldsymbol{\Delta\gamma}) = \mathbf{0}$ for $\boldsymbol{\Delta\gamma}$

$$\mathrm{g}(\boldsymbol{\Delta\gamma}) = \Big\{ \left\| \boldsymbol{S}_n - \boldsymbol{X}^{\boldsymbol{\Phi}}_n \right\|^2 + \left\| (1 - \boldsymbol{\Phi}) 2\mu \boldsymbol{\Delta} \boldsymbol{e}_{n+1} + \boldsymbol{b}_{n+1} \boldsymbol{\Delta\gamma} \boldsymbol{X}^{\boldsymbol{\Phi}}_n \right\|^2 +$$
$$2\left(\boldsymbol{S}_n - \boldsymbol{X}^{\boldsymbol{\Phi}}_n\right) : \left[(1 - \boldsymbol{\Phi}) 2\mu \boldsymbol{\Delta} \boldsymbol{e}_{n+1} + \boldsymbol{b}_{n+1} \boldsymbol{\Delta\gamma} \boldsymbol{X}^{\boldsymbol{\Phi}}_n \right] \Big\}^{1/2}$$
$$-(1 - \boldsymbol{\Phi}) \sqrt{\frac{2}{3}} \boldsymbol{K} \left(\boldsymbol{\alpha}_n + \sqrt{\frac{2}{3}} \boldsymbol{\Delta\gamma} \right) - \boldsymbol{\Delta\gamma} [(1 - \boldsymbol{\Phi}) 2\mu + \boldsymbol{a}_{n+1}] - \boldsymbol{\Theta} \left(\frac{\boldsymbol{\Delta\gamma\eta}}{\boldsymbol{\Delta t}} \right)$$

where $\boldsymbol{a}_{n+1} = \frac{c_1 (1 - \boldsymbol{\Phi})}{1 + c_2' (1 - \boldsymbol{\Phi})(1 - \beta) \boldsymbol{\Delta\gamma}}$, $\boldsymbol{b}_{n+1} = \frac{c_2}{c_1} \boldsymbol{a}_{n+1}$

Update

$$\boldsymbol{n}_{n+1} \equiv \frac{\boldsymbol{S}_{n+1} - \boldsymbol{X}^{\boldsymbol{\Phi}}_{n+1}}{\left\| \boldsymbol{S}_{n+1} - \boldsymbol{X}^{\boldsymbol{\Phi}}_{n+1} \right\|} = \frac{\boldsymbol{B}_n}{\|\boldsymbol{B}_n\|} \text{ where } \boldsymbol{B}_n = \boldsymbol{S}^{tr}_{n+1} - \boldsymbol{X}^{\boldsymbol{\Phi}}_n + \boldsymbol{b}_{n+1} \boldsymbol{\Delta\gamma} \boldsymbol{X}^{\boldsymbol{\Phi}}_n$$
$$\boldsymbol{\alpha}_{n+1} = \boldsymbol{\alpha}_n + \sqrt{\tfrac{2}{3}} \boldsymbol{\Delta\gamma}$$
$$\boldsymbol{\varepsilon}^{vp}_{n+1} = \boldsymbol{\varepsilon}^{vp}_n + \boldsymbol{\Delta\gamma} \frac{\boldsymbol{B}_n}{\|\boldsymbol{B}_n\|}$$
$$\boldsymbol{X}^{\boldsymbol{\Phi}}_{n+1} = \boldsymbol{X}^{\boldsymbol{\Phi}}_n + \boldsymbol{a}_{n+1} \boldsymbol{\Delta\gamma} \left(\frac{\boldsymbol{S}_{n+1} - \boldsymbol{X}^{\boldsymbol{\Phi}}_{n+1}}{\left\| \boldsymbol{S}_{n+1} - \boldsymbol{X}^{\boldsymbol{\Phi}}_{n+1} \right\|} - \frac{c_2'}{c_1} \boldsymbol{X}^{\boldsymbol{\Phi}}_n \right)$$
$$\boldsymbol{\xi}^{\boldsymbol{\Phi}}_{n+1} (1 - \boldsymbol{\Phi}) \boldsymbol{K}(\boldsymbol{\alpha}_{n+1}) \frac{\boldsymbol{B}_n}{\|\boldsymbol{B}_n\|}$$
$$\boldsymbol{S}_{n+1} = \boldsymbol{\xi}^{\boldsymbol{\Phi}}_{n+1} + \boldsymbol{X}^{\boldsymbol{\Phi}}_{n+1}$$
$$\boldsymbol{\sigma}_{n+1} = \boldsymbol{\kappa}(1 - \boldsymbol{\Phi}) \boldsymbol{tr}(\boldsymbol{\varepsilon}_{n+1}) \widehat{\boldsymbol{I}} + 2\boldsymbol{\mu}(1 - \boldsymbol{\Phi}) \left(\boldsymbol{e}_{n+1} - \boldsymbol{\varepsilon}^{vp}_n - \boldsymbol{\gamma}_{n+1} \frac{\boldsymbol{B}_n}{\|\boldsymbol{B}_n\|} - \boldsymbol{e}^{\theta}_{n+1} \right)$$

Compute consistent Jacobian

$$\boldsymbol{C}^{EVPD}_{n+1} = (1 - \boldsymbol{\Phi}) \boldsymbol{\kappa} \widehat{\boldsymbol{I}} \otimes \widehat{\boldsymbol{I}} + 2\boldsymbol{\mu}(1 - \boldsymbol{\Phi}) \boldsymbol{\delta}_{n+1} \left(\prod - \tfrac{1}{3} \widehat{\boldsymbol{I}} \otimes \widehat{\boldsymbol{I}} \right) - 2\boldsymbol{\mu}(1 - \boldsymbol{\Phi}) \overline{\boldsymbol{\theta}}_{n+1}$$
$$\widehat{\boldsymbol{n}}_{n+1} \otimes \widehat{\boldsymbol{n}}_{n+1} - \frac{2\mu(1 - \boldsymbol{\Phi})}{\|\boldsymbol{B}_n\|} \boldsymbol{\Delta\gamma} \left(\prod - \widehat{\boldsymbol{n}}_{n+1} \otimes \widehat{\boldsymbol{n}}_{n+1} \right) : \left(\frac{\overline{\boldsymbol{K}}_4}{\overline{\boldsymbol{K}}_3} \widehat{\underline{\boldsymbol{n}}}_{n+1} \otimes \boldsymbol{X}^{\boldsymbol{\Phi}}_n \right)$$

Where $\boldsymbol{\delta}_{n+1} = \mathbf{1} - \frac{\Delta\gamma 2\mu(1 - \boldsymbol{\Phi})}{\|\boldsymbol{B}_n\|}$ and $\overline{\boldsymbol{\theta}}_{n+1} = \frac{1}{\overline{K}_3} - \frac{\Delta\gamma 2\mu(1 - \boldsymbol{\Phi})}{\|\boldsymbol{B}_n\|}$with

$$\overline{\boldsymbol{K}}_3 = \overline{\boldsymbol{K}}_1 + \overline{\boldsymbol{K}}_2$$
$$\overline{K_1} = 1 + \frac{K'}{3\mu} + \frac{a_{n+1}}{2\mu(1 - \Phi)}$$
$$\overline{K_2} = \frac{a'_{n+1} \Delta\gamma}{2\mu(1 - \Phi)} + \frac{\widehat{n}_{n+1} b_{n+1}}{2\mu(1 - \Phi)} [b_{n+1}(1 - \beta)\Delta\gamma - 1] : \boldsymbol{X}^{\boldsymbol{\Phi}}_n + \frac{1}{2\mu(1 - \Phi)} \frac{\partial \Theta}{\partial \Delta\gamma}$$
$$\overline{\boldsymbol{K}}_4 = \boldsymbol{b}'_{n+1} \boldsymbol{\Delta\gamma} + \boldsymbol{b}_{n+1}$$

ELSE

Elastic step $(\_)_{n+1} = (\_)^{tr}_{n+1}$ (Exit)

END IF

EXIT

into the volume element; second, it changes because there is an entropy source due to irreversible phenomena inside the volume element. The entropy source is always a nonnegative quantity, since entropy can only be created, never destroyed, unlike energy.

> Entropy is the measure of how much energy is unavailable for work. Imagine an isolated and closed system with some hot objects and some cold objects. Work can be done as heat is transferred from the hot to the cooler objects; however, once this transfer has occurred, it is impossible to extract additional work from them alone. Energy is always conserved, but

Table 5.2 Local Newton iteration for the consistency parameter classical theory

Let $\Delta\gamma^{(0)} \leftarrow \mathbf{0}$

$\alpha_{n+1}{}^{(0)} \leftarrow \alpha_n$

Start Iterations

DO_UNTIL $|g(\Delta\gamma)| < tol$

$k \leftarrow k + 1$

Compute $\Delta\gamma^{(k+1)}$

$$g(\Delta\gamma) = \Big\{ \| S_n - X_n^{\Phi} \|^2 + \| (1-\Phi) 2\mu \Delta e_{n+1} + b_{n+1} \Delta\gamma X_n^{\Phi} \|^2 + 2\left(S_n - X_n^{\Phi}\right) : \left[(1-\Phi)2\mu\Delta e_{n+1} + b_{n+1}\Delta\gamma X_n^{\Phi}\right] \Big\}^{1/2} - (1-\Phi)\sqrt{\frac{2}{3}}K\left(\alpha_n + \sqrt{\frac{2}{3}}\Delta\gamma\right) - \Delta\gamma[(1-\Phi)2\mu + a_{n+1}] - \Theta\left(\frac{\Delta\gamma\eta}{\Delta t}\right)$$

$$\mathbf{dg}\left(\Delta\gamma^{(k)}\right) = \frac{\partial g\left(\Delta\gamma^{(k)}\right)}{\partial \Delta\gamma^{(k)}}$$

$$\Delta\gamma^{(k+1)} \leftarrow \Delta\gamma^{(k)} - \frac{g\left(\Delta\gamma^{(k)}\right)}{Dg\left(\Delta\gamma^{(k)}\right)}$$

END DO_UNITL

when all objects have the same temperature, the energy is no longer available for conversion into work.

The entropy [energy unavailable for work] of the universe increases or remains constant in all natural processes. It is possible to find a system for which entropy decreases but only due to a net increase in a related system. For example, the originally hot objects and cooler objects reaching thermal equilibrium in an isolated system may be separated, and some of them put in a refrigerator. The objects would again have different temperatures after a period of time, but now the system of the refrigerator would have to be included in the analysis of the complete system. No net decrease in entropy of all the related system occurs. This is yet another way of stating the second law of thermodynamics (DeHoff 1993).

The concept of entropy has far-reaching implications that tie the order of our universe to probability and statistics. Imagine a new deck of cards in order by suits, with each unit in numerical order. As the deck is shuffled, no one would expect the original order to return. There is a probability that the randomized order of the shuffled deck would return to the original format, but it is exceedingly small. An ice cube melts, and the molecules in the liquid form have less order than in the frozen form. An infinitesimally small probability exists that all of the slower moving molecules will aggregate in one space so that the ice cube will reform from the pool of water. The entropy, or disorder, of the universe increases as hot bodies cool and cold bodies warm. Eventually, the entire universe will be at the same temperature so the energy will be no longer usable (DeHoff 1993).

In order to relate the entropy generation directly to various irreversible processes that occur in a system, one needs the macroscopic conservation laws of mass, momentum, and energy in local, i.e., differential form. These conservation laws contain a number of quantities such as the diffusion flows, the heat flow, and the stress tensor, which are related to the transport of mass, exchange of energy, and exchange of energy momentum. Then, the entropy generation can be calculated by using the thermodynamic Gibbs relation, which connects the rate of the change in entropy in the medium to the rate of the change in energy and work. Entropy generation rate has a relatively simple formula: it is a sum of all entropy generating micro-mechanism terms, each being a product of a **flux** characterizing an irreversible process, and a quantity called **thermodynamic force**, which is related to the gradient [nonuniformity] in the system (Mazur and De Groot 1962). We discussed this topic earlier in Chap. 2 in detail in Onsager relations, as well. The complete entropy generation rate equation can thus serve as a basis for the systematic description of the irreversible processes occurring in a system. We refer to this equation as the thermodynamic fundamental equation.

"As yet the set of conservation laws, together with the entropy balance equation and the equations of state are to a certain extent empty, since this set of equations contain the irreversible fluxes as unknown parameters and can therefore not be solved with the given initial and boundary conditions for the state of the system" (Mazur and De Groot 1962). At this point we must therefore supplement the equations by an additional set of phenomenological relationships, which relate the irreversible fluxes and the thermodynamic forces appearing in the entropy source strength. Irreversible thermodynamics, in its present form, is mainly restricted to the study of the linear relationship between the fluxes and the thermodynamic forces as well as possible cross-effects between various phenomena. This is not a very serious restriction however, since even rather extreme physical situations are still described by linear laws."

> Together with the phenomenological equations, the original set of conservation laws may be said to be complete in the sense that one now has a consistent set of partial differential equations for the state parameters of a material system, which may be solved with the proper initial and boundary conditions (Mazur and De Groot 1962).

5.4.1 Conservation Laws

Thermodynamics is based on two fundamental laws: the first law of thermodynamics or the law of conservation of energy, and the second law of thermodynamics or the entropy law. A systematic macroscopic scheme for the description of irreversible processes must also be built upon these two laws. However, it is necessary to formulate these laws in a suitable way. Since we wish to develop a theory applicable to systems of which the properties are continuous functions of space coordinates and time, we shall give a local formulation of the law of conservation of energy. As the local momentum and mass densities may change in time, we will also need local formulations of the laws of conservation of momentum and conservation of mass. In solid mechanics, the thermodynamic system is usually chosen as a collection of continuous matter, i.e., the system is a closed system not interchanging matter with its surroundings; the bounding surface of the system in general moves with the flow of matter.

5.4.1.1 Conservation of Mass

Consider an arbitrary volume V fixed in space, bounded by surface $\mathbf{\Omega}$. The rate of change of the mass within the volume V is (Malvern 1969)

$$\frac{d}{dt}\int^{V}\rho dV=\int^{V}\frac{\partial\rho}{\partial t}dV \tag{5.36}$$

where ρ is the density (mass per unit volume). If no mass is created or destroyed inside V, this quantity must be equal to the rate of the material flow into the volume V through its surface $\mathbf{\Omega}$ (Malvern 1969):

$$\int^{V}\frac{\partial\rho}{\partial t}dV=\int^{\Omega}\rho\boldsymbol{v}\cdot d\mathbf{\Omega} \tag{5.37}$$

where $\boldsymbol{v}$ is the velocity and $d\mathbf{\Omega}$ is a vector with magnitude $|d\mathbf{\Omega}|$ normal to the surface and counted positive from the inside to the outside. The quantities ρ and $\boldsymbol{v}$ are all functions of time and of space coordinates. Applying Gauss's theorem to the surface integral in Eq. (5.37), we obtain

$$\frac{\partial\rho}{\partial t}=-\text{div}\,\rho\boldsymbol{\nu} \tag{5.38}$$

Equation (5.38) is valid for an arbitrary volume V, which expresses the fact that the total mass is conserved, i.e., that the total mass in any volume element of the system can only change if matter flows into (or out of) the volume element. This equation has the form of a so-called balanced equation: the local change of the density is equal to the negative divergence of the flow of mass. The continuity equation in the vector form of Eq. (5.38) is independent of any choice of coordinates.

The conservation of mass equation can also be written in an alternative form by introducing the substantial time derivative (Mazur and De Groot 1962):

$$\frac{d}{dt}=\frac{\partial}{\partial t}+\boldsymbol{\nu}\cdot\text{grad} \tag{5.39}$$

With the help of Eq. (5.39), Eq. (5.38) becomes

$$\frac{d\rho}{dt}=-\rho\,\text{div}\,\boldsymbol{v} \tag{5.40}$$

With the specific volume $\boldsymbol{v}=\rho^{-1}$, Eq. (5.40) may also be written as

$$\rho \frac{dv}{dt} = \text{div}\, \boldsymbol{v} \tag{5.41}$$

Finally the following time derivative relation is valid for an arbitrary local property a that may be a scalar or a component of a vector or tensor:

$$\rho \frac{da}{dt} = \frac{\partial a\rho}{\partial t} + \text{div}\, a\rho\boldsymbol{v} \tag{5.42}$$

which is a consequence of Eqs. (5.38) and (5.39). We can verify Eq. (5.42) directly. According to Eq. (5.39), the left-hand side of Eq. (5.42) is

$$\rho \frac{da}{dt} = \rho \frac{\partial a}{\partial t} + \rho\boldsymbol{v} \cdot \text{grad}\, a \tag{5.43}$$

According to Eq. (5.38), the right side of Eq. (5.42) is

$$\begin{aligned} \frac{\partial a\rho}{\partial t} + \text{div}\, a\rho\boldsymbol{v} &= a\frac{\partial \rho}{\partial t} + \rho\frac{\partial a}{\partial t} + a\,\text{div}\rho\boldsymbol{v} + \rho\boldsymbol{v} \cdot \text{grad}\, a \\ &= a(-\text{div}\,\rho\boldsymbol{v}) + \rho\frac{\partial a}{\partial t} + a\,\text{div}\,\rho\boldsymbol{v} + \rho\boldsymbol{v} \cdot \text{grad}\, a \\ &= \rho\frac{\partial a}{\partial t} + \rho\boldsymbol{v} \cdot \text{grad}\, a \end{aligned} \tag{5.44}$$

So Eq. (5.42) is true.

5.4.1.2 Momentum Principle in Newtonian Mechanics

The momentum principle for a collection of particles states that the time rate of the change in the total momentum for a given set of particles equals to the vector sum of all the external forces acting on the particles of the set, provided Newton's third law of action and reaction governs the initial forces (Malvern 1969). Consider a given mass of the medium, instantaneously occupying a volume V bounded by surface Ω and acted upon by external surface **t** and body force **b**. Then the momentum principle can be expressed as (Malvern 1969)

$$\int^{\Omega} \boldsymbol{\sigma}^{(n)} d\Omega + \int^{V} \rho \mathbf{b} dV = \frac{d}{dt} \int^{V} \rho \boldsymbol{v} dV \tag{5.45}$$

or in rectangular coordinates

$$\int^{\Omega} \sigma_i d\Omega + \int^{V} \rho b_i dV = \frac{d}{dt}\int^{V} \rho v_i dV \tag{5.46}$$

Substituting $\sigma_i = \sigma_{ji} n_j$ and transforming the surface integral by using the divergence theorem, we obtain (Malvern 1969):

$$\int^{V} \left(\frac{\partial \sigma_{ji}}{\partial x_j} + \rho b_i - \rho \frac{dv_i}{dt}\right) dV \tag{5.47}$$

for an arbitrary volume V. Whence at each point we have (Malvern 1969)

$$\rho \frac{dv_i}{dt} = \frac{\partial \sigma_{ji}}{\partial x_j} + \rho b_i \tag{5.48}$$

where n_j is the component of the normal unit vector $\boldsymbol{n}$, $v_i (i = 1, 2, 3)$ is a Cartesian component of $\boldsymbol{v}$, and $x_j (j = 1, 2, 3)$ is the Cartesian coordinates. The quantities $\sigma_{ji}(i, j = 1, 2, 3)$ and $b_i (i = 1, 2, 3)$ are the Cartesian components of the stress tensor $\boldsymbol{\sigma}$ and body force $\mathbf{b}$, respectively. For a nonpolar case, the stress tensor $\boldsymbol{\sigma}$ is symmetric, namely

$$\sigma_{ij} = \sigma_{ji} \quad (i, j = 1, 2, 3) \tag{5.49}$$

In tensor notation Eq. (5.48) is written as (Mazur and De Groot 1962)

$$\rho \frac{d\boldsymbol{v}}{dt} = \operatorname{div} \boldsymbol{\sigma} + \rho \mathbf{b} \tag{5.50}$$

From a microscopic point of view, the stress tensor $\boldsymbol{\sigma}$ results from the short-range interactions between the particles of the system, whereas $\mathbf{b}$ contains the external forces as well as a possible contribution from long-range interactions in the system.

Using relation (5.46), the equation of motion (5.50) can also be written as

$$\frac{\partial \rho \boldsymbol{v}}{\partial t} = -\operatorname{div}(\rho \boldsymbol{v}\boldsymbol{v} - \boldsymbol{\sigma}) + \rho \mathbf{b} \tag{5.51}$$

where $\boldsymbol{v}\boldsymbol{v} = \boldsymbol{v} \otimes \boldsymbol{v}$ is an ordered (dyadic) product. This equation also has the form of a balance equation for the momentum density $\rho \boldsymbol{v}$. In fact one can interpret the quantity $(\rho \boldsymbol{v}\boldsymbol{v} - \boldsymbol{\sigma})$ as a momentum flow with a convective part $\rho \boldsymbol{v}\boldsymbol{v}$ and the quantity $\rho \mathbf{b}$ as a source of momentum, but no entropic part.

It is also possible to derive from Eq. (5.48) a balance equation for the kinetic energy of the center of gravity by multiplying both members with the component v_i of $\boldsymbol{v}$ and summing over i:

$$\rho \frac{d\frac{1}{2}v^2}{dt} = \sum_{i,j=1}^{3} \frac{\partial}{\partial x_j}(\sigma_{ji}v_i) - \sum_{i,j=1}^{3} \sigma_{ji} \frac{\partial}{\partial x_j} v_i + \rho b_i v_i \quad (i = 1, 2, 3) \tag{5.52}$$

or in tensor notation

$$\rho \frac{d\frac{1}{2}\boldsymbol{v}^2}{dt} = \mathrm{div}(\boldsymbol{\sigma} \cdot \boldsymbol{v}) - \boldsymbol{\sigma} : \mathbf{L} + \rho \mathbf{b} \cdot \boldsymbol{v} \tag{5.53}$$

where $\mathbf{L} =$ grad $\boldsymbol{v}$ is the spatial gradient of the velocity. $\mathbf{L}$ can be written as the sum of a symmetric tensor $\mathbf{D}$ called the rate of deformation tensor or the stretching tensor and a skew-symmetric tensor $\mathbf{W}$ called the spin tensor or the vorticity tensor as follows (Malvern 1969):

$$\mathbf{L} = \mathbf{D} + \mathbf{W} \tag{5.54}$$

where $\mathbf{D} = \frac{1}{2}(\mathbf{L} + \mathbf{L}^T)$, $\mathbf{W} = \frac{1}{2}(\mathbf{L} - \mathbf{L}^T)$.

Since $\mathbf{W}$ is skew-symmetric, while $\boldsymbol{\sigma}$ is symmetric, it follows that

$$\boldsymbol{\sigma} : \mathrm{Grad}\, \boldsymbol{v} = \sigma_{ij} L_{ij} = \sigma_{ij} D_{ij} = \boldsymbol{\sigma} : \mathbf{D} \tag{5.55}$$

We can also establish the relationship between the strain rate $d\boldsymbol{\varepsilon}/dt$ and the rate of the deformation tensor $\mathbf{D}$ (Malvern 1969):

$$\frac{d\boldsymbol{\varepsilon}}{dt} = \mathbf{F}^T \cdot \mathbf{D} \cdot \mathbf{F} \tag{5.56}$$

where $\mathbf{F}$ is the deformation gradient tensor referring to the undeformed configuration. When the displacement gradient components are small compared to unity, Eq. (5.56) is reduced to (Malvern 1969)

$$\frac{d\boldsymbol{\varepsilon}}{dt} \approx \mathbf{D} \tag{5.57}$$

With the help of Eqs. (5.42) and (5.53)

$$\frac{\partial \frac{1}{2}\rho \boldsymbol{v}^2}{\partial t} = -\mathrm{div}\left(\frac{1}{2}\rho \boldsymbol{v}^2 \cdot \boldsymbol{v} - \boldsymbol{\sigma} \cdot \boldsymbol{v}\right) - \boldsymbol{\sigma} : \mathbf{D} + \rho \mathbf{b} \cdot \boldsymbol{v} \tag{5.58}$$

For the conservative body forces which can be derived from a potential Ψ independent of time (Mazur and De Groot 1962)

$$\mathbf{b} = -\text{grad }\Psi, \qquad \frac{\partial \Psi}{\partial t} = 0 \tag{5.59}$$

We can now establish an equation for the rate of change of the potential energy density $\rho\Psi$. It follows from Eqs. (5.38) and (5.59) that

$$\begin{aligned}\frac{\partial \rho\Psi}{\partial t} &= \Psi\frac{\partial \rho}{\partial t} + \rho\frac{\partial \Psi}{\partial t} = \Psi(-\text{div}\,\rho\boldsymbol{v}) \\ &= -\,\text{div}\,\rho\Psi\boldsymbol{v} + \rho\boldsymbol{v}\cdot\text{grad}\Psi = -\text{div}\,\rho\Psi\boldsymbol{v} - \rho\mathbf{b}\cdot\boldsymbol{v}\end{aligned} \tag{5.60}$$

Adding Eqs. (5.59) and (5.60) for the rate of change of the kinetic energy $\frac{1}{2}\rho\boldsymbol{v}^2$ and the potential energy $\rho\Psi$

$$\frac{\partial \rho\left(\frac{1}{2}\boldsymbol{v}^2 + \Psi\right)}{\partial t} = -\text{div}\left\{\rho\left(\frac{1}{2}\boldsymbol{v}^2 + \Psi\right)\boldsymbol{v} - \boldsymbol{\sigma}\cdot\boldsymbol{v}\right\} - \boldsymbol{\sigma}\cdot\mathbf{D} \tag{5.61}$$

This equation shows that the sum of kinetic and potential energy is not conserved, since an entropy source term appears at the right-hand side.

5.4.1.3 Conservation of Energy

The first law of thermodynamics relates the work done on the system and the heat transfer into the system to the change in total energy of the system. Suppose that the only energy transferred to system is by mechanical work done on the system by surface tractions and body forces, by heat exchange through the boundary, and the heat generated within the system by external agencies (e.g., inductive heating). According to the principle of conservation of energy, the total energy content within an arbitrary volume V in the system can only change if energy flows into (or out of) the volume considered through its boundary Ω, which can be expressed as (Malvern 1969)

$$\frac{d}{dt}\int^{V}\rho e dV = \int^{V}\frac{\partial \rho e}{\partial t}dV = -\int^{S}\mathbf{J}_e\cdot d\boldsymbol{\Omega} + \int^{V}\rho r dV \tag{5.62}$$

where e is the energy per unit mass, $\mathbf{J}_e$ is the energy flux per unit surface and unit time, and r is the distributed internal heat source of strength per unit mass. We shall refer to e as the total specific energy, because it includes all forms of energy in the system. Similarly we shall call $\mathbf{J}_e$ the total energy flux. With the help of Gauss's theorem, we can obtain the differential [local form] of the law of conservation of energy:

$$\frac{\partial \rho e}{\partial t} = -\text{div}\,\mathbf{J}_e + \rho r \tag{5.63}$$

In order to relate this equation to the previously obtained Eq. (5.61) for the kinetic and potential energy, we must specify what are the various contributions to the energy e and the total energy flux $\mathbf{J}_e$. The total specific energy e includes the specific kinetic energy $\frac{1}{2}v^2$, the specific potential energy Ψ, and the specific internal energy u (Mazur and De Groot 1962):

$$e = \frac{1}{2}v^2 + \Psi + u \tag{5.64}$$

From a macroscopic point of view, this relation can be considered as the definition of internal energy, u. From a microscopic point of view, u represents the energy of thermal agitation [atomic vibrations] as well as the energy due to the short-range atomic interactions.

Similarly, the total energy flux includes a convective term $\rho e \boldsymbol{v}$, an energy flux $\boldsymbol{\sigma} \cdot \boldsymbol{v}$ due to the mechanical work performed on the system, and finally a heat flux $\mathbf{J}_q$ (Mazur and De Groot 1962):

$$\mathbf{J}_e = \rho e \boldsymbol{v} - \boldsymbol{\sigma} \cdot \boldsymbol{v} + \mathbf{J}_q \tag{5.65}$$

This equation may be also considered as defining the heat flux $\mathbf{J}_q$. Then the heat flowing rate per unit mass is

$$\rho \frac{dq}{dt} = -\text{div}\,\mathbf{J}_q \tag{5.66}$$

where q is the heat flowing into the system per unit mass. If we subtract Eq. (5.61) from Eq. (5.63), we obtain, using also Eqs. (5.64) and (5.65), the balance equation for the internal energy u:

$$\frac{\partial \rho u}{\partial t} = -\text{div}\{\rho u \boldsymbol{v} + \mathbf{J}_q\} + \boldsymbol{\sigma} : \mathbf{D} + \rho r \tag{5.67}$$

It is apparent from Eq. (5.67) that the internal energy u is not conserved. In fact a source term appears which is equal but of opposite sign to the source term of the balance equation (5.61) for kinetic and potential energy.

With the help of Eq. (5.42), Eq. (5.67) may be written in an alternative form:

$$\rho \frac{du}{dt} = -\text{div}\,\mathbf{J}_q + \boldsymbol{\sigma} : \mathbf{D} + \rho r \tag{5.68}$$

The total stress tensor $\boldsymbol{\sigma}$ can be split into a scalar hydrostatic pressure part p and a deviatoric stress tensor $\mathbf{S}$:

$$\boldsymbol{\sigma} = -p\mathbf{I} + \mathbf{S} \tag{5.69}$$

where $\mathbf{I}$ is the unit matrix with element $\delta_{ij}(\delta_{ij} = 1, \text{if } i = j; \delta_{ij} = 0, \text{if } i \neq j), p = -\frac{1}{3}\sigma_{kk}$. With the help of Eq. (5.69), Eq. (5.68) becomes

$$\rho \frac{du}{dt} = -\text{div } \mathbf{J}_q - p \text{ div } \boldsymbol{v} + \mathbf{S} : \mathbf{D} + \rho r \tag{5.70}$$

where the following equality is utilized

$$\mathbf{I} : \mathbf{D} = \mathbf{I} : \text{Grad } \boldsymbol{v} = \sum_{i,j=1}^{3} \delta_{ij} \frac{\partial}{\partial x_j} v_i = \sum_{i=1}^{3} \frac{\partial}{\partial x_i} v_i = \text{div } \boldsymbol{v} \tag{5.71}$$

Utilizing Eq. (5.41), the first law of thermodynamics can finally be written in the following form:

$$\frac{du}{dt} = \frac{1}{\rho} \boldsymbol{\sigma} : \mathbf{D} + r - \frac{1}{\rho} \text{ div } \mathbf{J}_q \tag{5.72}$$

5.4.1.4 Entropy Law and Entropy Balance

Historically, thermodynamics in the traditional sense was concerned with the study of reversible transformations. For an irreversible process in which the thermodynamic state of a solid changes from some initial state to a current state, it can be assumed that such a process can occur along an imaginary reversible isothermal path. The processes defined in this way will be thermodynamically admissible if, at any instant of evolution, the Clausius-Duhem inequality is satisfied. According to the principles of thermodynamics, two more new variables, temperature T and entropy S, are introduced for any macroscopic system. The entropy [energy unavailable for work] of the universe, taken as a system plus whatever surroundings are involved in producing the change within the system, can only increase. Changes in the real world are always irreversible processes, which result in the production of entropy and thus a permanent change in the universe (DeHoff 1993). Another ad-hoc definition of entropy is, it is not possible to make a 100% efficient engine [or any mechanism].

The variation of the entropy dS may be written as the sum of two and only two terms for a closed system (Mazur and De Groot 1962):

$$dS = dS_{\text{e}} + dS_{\text{i}} \tag{5.73}$$

where dS_{e} is the entropy derived from the transfer of heat from external sources across the boundary of the system, and dS_{i} is the entropy produced inside the system. The second law of thermodynamics states that dS_{i} must be zero for any reversible process (or equilibrium) and positive for irreversible transformation of the system, namely (Mazur and De Groot 1962)

$$dS_{\mathrm{i}} \geq 0 \tag{5.74}$$

The entropy supplied, dS_{e}, on the other hand, may be positive, zero, or negative, depending on the interaction of the system with its surroundings.

For an irreversible process in which the thermodynamic state of a solid changes from some initial state to a current state, it is assumed that such a process can occur along an imaginary reversible isothermal path which consists of a two-step sequence (Krajcinovic 1996). This is the so-called local equilibrium assumption, which postulates that the thermodynamic state of a material medium at a given point and instant is completely defined by the knowledge of the values of a certain number of variables at that instant. The method of local state implies that the laws which are valid for the macroscopic system remain valid for infinitesimally small parts of it, which is in agreement with the point of view currently adopted in the macroscopic description of a continuous system. This method also implies, on a microscopic model, that the local macroscopic measurements performed on the system are really measurements of the properties of small parts of the system, which still contain a large number of the constituting particles. It is assumed that the representative volume element is the smallest unit of continuum. This hypothesis of "local equilibrium" can, from a macroscopic point of view, only be justified by virtue of the validity of the conclusions derived from it. Ultrarapid phenomena for which the time scales of the evolutions are at the same order as the atomic relaxation time for a return to thermodynamic equilibrium are excluded from this theory's field of application (Lemaitre and Chaboche 1990). All physical processes can be described with precision utilizing the proper number of thermodynamic state variables. The processes defined in this way will be thermodynamically admissible if, at any instant of evolution, the Clausius-Duhem inequality is satisfied.

In irreversible thermodynamics, one of the important objectives is to relate dS_{i}, the internal entropy production, to the various irreversible phenomena which may occur inside the system. Before calculating the entropy production in terms of quantities which characterize the irreversible phenomena, we can rewrite Eqs. (5.73) and (5.74) in a form which is more suitable for the description of the systems in which the densities of the extensive properties (such as mass and energy considered in conservation laws) are continuous functions of spatial coordinates (Mazur and De Groot 1962):

$$S = \int^{V} \rho s dV \tag{5.75}$$

$$\frac{dS_e}{dt} = -\int^{\Omega} \mathbf{J}_{S,\text{tot}} \cdot d\boldsymbol{\Omega} \tag{5.76}$$

$$\frac{dS_i}{dt} = \int^{V} \gamma dV \tag{5.77}$$

where s is the entropy per unit mass, $\mathbf{J}_{S,\text{ tot}}$ is the total entropy flux which is a vector that coincides with the direction of entropy flow and has a magnitude equal to the entropy crossing unit area perpendicular to the direction of flow per unit time, and γ is the entropy generation rate per unit volume and unit time.

Utilizing Eqs. (5.75)–(5.77), Eq. (5.73) may be written, using Gauss's theorem, in the form (Mazur and De Groot 1962):

$$\int^{V} \left(\frac{\partial \rho s}{\partial t} + \text{div } \mathbf{J}_{S,\text{tot}} - \gamma\right) dV = 0 \tag{5.78}$$

where the divergence of $\mathbf{J}_{S,\text{ tot}}$ simply represents the net entropy leaving unit volume per unit time. From this relation, it follows, since Eq. (5.40) must be held for an arbitrary volume V, that

$$\frac{\partial \rho s}{\partial t} = -\text{div } \mathbf{J}_{S,\text{tot}} + \gamma \tag{5.79}$$

$$\gamma \geq 0 \tag{5.80}$$

These two formulations are the local forms of Eqs. (5.73) and (5.74), i.e., the local mathematical expressions for the second law of thermodynamics. Equation (5.79) is formally a balance equation for the entropy density ρs with a source γ which satisfies the important inequality (5.80). With the help of Eq. (5.42), Eq. (5.79) can be rewritten in a slightly different form as follows:

$$\rho \frac{ds}{dt} = -\text{div } \mathbf{J}_S + \gamma \tag{5.81}$$

where the entropy flux $\mathbf{J}_S$ is the difference between the total entropy flux $\mathbf{J}_{S,\text{ tot}}$ and a convective term $\rho s \boldsymbol{v}$

$$\mathbf{J}_S = \mathbf{J}_{S,\text{tot}} - \rho s \boldsymbol{v} \tag{5.82}$$

For applications in continuum mechanics, we must relate the changes in the properties of the system to the entropy generation rate. This requires, us to obtain explicit expressions for the entropy flux $\mathbf{J}_S$ and the entropy generation rate γ that appears in Eq. (5.81). This explicit equation is called the fundamental equation.

We postulate the existence of a thermodynamic potential from which the state laws can be derived. Let us assume that a function, with a scalar value, concave with respect to temperature and convex with respect to other variables, allows us to satisfy a priori the conditions of thermodynamic stability imposed by Clausius-Duhem inequality. The specific Helmholtz free energy, φ, is defined as the difference between the specific internal energy density u and the product between the absolute temperature T and specific entropy s:

$$\varphi = u - Ts \tag{5.83a}$$

Differentiating this and with the help of the law of conservation of energy, we have the following relations:

$$d\varphi = du - Tds - sdT \tag{5.83b}$$

$$= (\delta q + \delta w) - Tds - sdT \tag{5.83c}$$

$$= \delta q + \left(\delta w^{\mathrm{d}} + \delta w^{\mathrm{e}}\right) - Tds - sdT \tag{5.83d}$$

$$= \left(\delta q + \delta w^{\mathrm{d}} - Tds\right) + \left(\delta w^{\mathrm{e}} - sdT\right) \tag{5.83e}$$

where q is the total heat flowing into the system per unit mass, including the conduction through the surface and the distributed internal heat source; w is the total work done on the system per unit mass by external loads and body forces; w^{d} is the lost energy associated with the total work, which is generally dissipated in the form of heat; and w^{e} is the elastic energy [available for work] associated with the total work. For the quantitative treatment of entropy for irreversible processes, let's introduce the definition of entropy for irreversible processes:

$$ds = \frac{\delta q + \delta w^{\mathrm{d}}}{T} \tag{5.84}$$

With the help of Eq. (5.84), we can write the following relation:

$$Tds = du - dw^{\mathrm{e}} \tag{5.85}$$

This is the Gibbs relation which combines the first and second laws of thermodynamics. From the definition of the entropy, we also have

$$dw^{\mathrm{e}} = d\varphi + sdT \quad \text{or} \quad d\varphi = dw^{\mathrm{e}} - sdT \tag{5.86}$$

The Helmholtz free energy, φ, is the isothermal recoverable elastic energy available for work. It should be pointed out that the specific elastic energy w^{e}, namely, the work stored in the system per unit mass during a process, is not a function of the process path. It depends only on the end state of the process for a given temperature. The elastic energy is frequently also referred to as the available energy of the process. The elastic energy is the maximum amount of work that could be produced by a device between any given two states. If the device is work absorbing, the elastic energy work of the process is the minimum amount of work that must be supplied (Li 1989).

In order to find the explicit form of the entropy balance equation (5.81), we insert the expressions (5.72) for $\frac{du}{dt}$ into Eq. (5.85) with the time derivatives given by Eq. (5.39):

$$\rho \frac{ds}{dt} = -\frac{\operatorname{div} \mathbf{J}_q}{T} + \frac{1}{T} \boldsymbol{\sigma} : \mathbf{D} - \frac{\rho}{T} \frac{dw^{\mathrm{e}}}{dt} + \frac{\rho r}{T} \tag{5.87a}$$

Noting that

$$\frac{\operatorname{div} \mathbf{J}_q}{T} = \operatorname{div} \frac{\mathbf{J}_q}{T} + \frac{1}{T^2} \mathbf{J}_q \cdot \operatorname{grad} T \tag{5.87b}$$

It is easy to cast Eq. (5.87) into the form of a balance equation (5.81):

$$\rho \frac{ds}{dt} = -\operatorname{div} \frac{\mathbf{J}_q}{T} - \frac{1}{T^2} \mathbf{J}_q \cdot \operatorname{grad} T + \frac{1}{T} \boldsymbol{\sigma} : \mathbf{D} - \frac{\rho}{T} \frac{dw^{\mathrm{e}}}{dt} + \frac{\rho r}{T} \tag{5.88}$$

From comparison with Eq. (5.81), it follows that the expressions for the entropy flux and the entropy production rate are given by

$$\mathbf{J}_S = \frac{\mathbf{J}_q}{T} \tag{5.89}$$

$$\gamma = \frac{1}{T} \boldsymbol{\sigma} : \mathbf{D} - \frac{\rho}{T} \frac{dw^{\mathrm{e}}}{dt} - \frac{1}{T^2} \mathbf{J}_q \cdot \operatorname{grad} T + \frac{\rho r}{T} \tag{5.90}$$

Equation (5.90) represents the entropy generation rate by the internal dissipations. The sum of the first two terms is called the intrinsic dissipation or mechanical dissipation. It consists of plastic dissipation plus the dissipation associated with the evolution of other internal variables. The last two terms are the thermal dissipation due to the conduction of heat and the internal heat source. The structure of the expression for γ is that of a bilinear form: it consists of a sum of products of two factors. One of these factors in each term is a **flux** quantity (heat flow $\mathbf{J}_q$, $\boldsymbol{\sigma}$ stress tensor) already introduced in the conservation laws. The other factor in each term is related to a gradient of an intensive state variable (gradients of temperature and velocity). These quantities which multiply the **fluxes** in the expression for the entropy production are called **thermodynamic forces**. As we discussed earlier, actually the assignment of **flux** and **thermodynamic force** is rather arbitrary. However, their multiplication must yield the entropy generation rate.

The entropy generation rate γ must be zero if the thermodynamic equilibrium conditions are satisfied within the system. Another requirement which Eq. (5.90) must satisfy is that it is invariant under the transformation of different reference frames, since the notions of reversible and irreversible behavior must be invariant under such a transformation. It can be seen that Eq. (5.90) satisfies these requirements. Finally it may be noted that Eq. (5.88a) also satisfies the Clausius-Duhem inequality:

$$\boldsymbol{\sigma} : \mathbf{D} - \rho\left(\frac{d\varphi}{dt} + s\frac{dT}{dt}\right) - \mathbf{J}_q \cdot \frac{\text{grad}\, T}{T} \geq 0 \tag{5.91}$$

Between two particles which are at different temperatures, heat is transferred only by conduction, a process which takes place at the molecular and atomic levels. The law of heat conduction for isotropic bodies may be stated as follows:

$$\mathbf{J}_q = -k \, \text{grad}\, T \tag{5.92}$$

where k, with units of Btu/ft h °F, is the thermal conductivity and where $\mathbf{J}_q$ is the heat flux.

This law of heat conduction was stated first by Fourier who based it on experimental observation. Fourier's law expresses a linear relation between the heat flux vector $\mathbf{J}_q$ and its dual variable grad T. Since solid, opaque bodies are of primary interest here, heat is transferred from point to point within this body solely by conduction. The field equation of the boundary value problem will, therefore, always be some form of the Fourier heat conduction equation. Of course, heat may be transferred to the surface of the body by other modes of heat transfer which correspond to various thermal boundary conditions. Then the expression for the internal entropy generation rate for thermo-mechanical problems can be simplified as

$$\gamma = \frac{1}{T}\boldsymbol{\sigma} : \mathbf{D} - \frac{\rho}{T}\frac{dw_{\mathrm{e}}}{dt} + \frac{k}{T^2}|\text{grad}\, T|^2 + \frac{\rho\, r}{T} \tag{5.93a}$$

Total entropy generation is of course time integration of Eq. (5.93a):

$$s = \frac{1}{\rho}\int_{t_1}^{t_2} \gamma\, dt \tag{5.93b}$$

The specific entropy production rate for small strain thermo-mechanical problems in metals can be simplified as

$$\frac{ds_i}{dt} = \frac{\gamma}{\rho} = \frac{\boldsymbol{\sigma} : \dot{\varepsilon}^{\mathrm{p}}}{T\rho} + \frac{k}{T^{2\rho}} |\mathrm{grad}\ T|^2 + \frac{r}{T} \tag{5.93c}$$

However, this simplification is based on an assumption that entropy generation due to elastic deformation and all other mechanisms are negligible. Of course this would be not true for high strain rate loading, elastic fatigue, and most composite materials where significant entropy is generated by internal relative elastic deformations of constituents.

5.4.1.5 Fully Coupled Thermo-mechanical Equations

The formalism of continuum mechanics and thermodynamics requires the existence of a certain number of state variables. For thermo-mechanical problems in pure metals at low strain rates of loading, there are two observable variables: the temperature T and the total strain $\boldsymbol{\varepsilon}$. For dissipative phenomena the current state also depends on the past history (load trajectory) that is represented, in the local state, by the values of internal variables at each instant. Plasticity and viscoplasticity require the introduction of the plastic (or viscoplastic) strain $\boldsymbol{\varepsilon}^{\mathrm{p}}$ as a state variable. Other phenomena, such as softening, hardening, degradation, and fracture, require the introduction of other internal variables of less obvious nature. These variables represent the internal state of matter (density of dislocations, crystal structure of lattice, material phase, polycrystalline grain size, configuration of micro-cracks and cavities, etc.) Lemaitre and Chaboche (1990) state that "there is no objective way to choose the internal state variables best suited to the study a phenomenon." However, this is only true if Newtonian mechanics is used in conjunction with a phenomenological damage potential curve fit to a test data. In unified mechanics theory, internal state variables are defined by the entropy generating mechanisms, which all must be taken into account.

For general study, here state variables will be denoted by $V_k(k = 1, 2, \ldots)$ representing either a scalar or a tensorial variable.

For small strain formulation, total strain can be written as a summation of elastic and plastic components:

$$\boldsymbol{\varepsilon} = \boldsymbol{\varepsilon}^{\mathrm{e}} + \boldsymbol{\varepsilon}^{\mathrm{p}} \tag{5.94}$$

The relations between the energy, stress tensor, and strain tensor can be obtained using the formalism of thermodynamics. Here we choose the specific Helmholtz free energy, φ, which depends on observable variables and internal state variables:

$$\varphi = (\boldsymbol{\varepsilon}, T, \boldsymbol{\varepsilon}^{\mathrm{e}}, \boldsymbol{\varepsilon}^{\mathrm{p}}, V_k) \tag{5.95}$$

For small strain formulation, the strain appears only in the form of their additive decomposition, so that

$$\varphi((\boldsymbol{\varepsilon} - \boldsymbol{\varepsilon}^{\mathrm{p}}), T, V_k) = \varphi(\boldsymbol{\varepsilon}^{\mathrm{e}}, T, V_k) \tag{5.96}$$

which shows that (Lemaitre and Chaboche 1990)

$$\frac{\partial \varphi}{\partial \boldsymbol{\varepsilon}^{\mathrm{e}}} = \frac{\partial \varphi}{\partial \boldsymbol{\varepsilon}} = -\frac{\partial \varphi}{\partial \boldsymbol{\varepsilon}^{\mathrm{p}}} \tag{5.97}$$

And the following expressions define the relation between Helmholtz free energy and thermodynamic state variables:

$$\boldsymbol{\sigma} = \rho \frac{\partial \varphi}{\partial \boldsymbol{\varepsilon}^{\mathrm{e}}} \tag{5.98}$$

$$s = -\frac{\partial \varphi}{\partial T} \tag{5.99}$$

$$A_k = \rho \frac{\partial \varphi}{\partial V_k} \tag{5.100}$$

where A_k is a thermodynamic force associated with the internal variables V_k. The vector formed by the variables is the gradient of the function φ in the space of the variables T, $\boldsymbol{\varepsilon}^{\mathrm{e}}$, and V_k. This vector is normal to the surface $\varphi = \text{constant}$.

Let us return to the equation of the conservation of energy for small strains (from Eqs. (5.57) and (5.68)):

$$\rho \dot{u} = -\mathrm{div}\, \mathbf{J}_q + \boldsymbol{\sigma} : \dot{\boldsymbol{\varepsilon}} + \rho r \tag{5.101}$$

and replace $\rho \dot{u}$ by the expression derived from Eqs. (5.83a), (5.83b), (5.83c), (5.83d) and (5.83e):

$$\rho \dot{u} = \rho \dot{\varphi} + \rho \dot{s} T + \rho s \dot{T} \tag{5.102}$$

And utilizing $\dot{\varphi}$ and $\dot{s}$ by their expression as a function of the state variables with the help of Eqs. (5.98)–(5.100)

$$\dot{\varphi} = \frac{\partial \varphi}{\partial \boldsymbol{\varepsilon}^{\mathrm{e}}} : \dot{\boldsymbol{\varepsilon}}^{\mathrm{e}} + \frac{\partial \varphi}{\partial T} \dot{T} + \frac{\partial \varphi}{\partial V_k} \dot{V}_k = \frac{1}{\rho} \boldsymbol{\sigma} : \dot{\boldsymbol{\varepsilon}}^{\mathrm{e}} - s \dot{T} + A_k \dot{V}_k \tag{5.103}$$

$$\dot{s} = -\frac{\partial^2 \varphi}{\partial \boldsymbol{\varepsilon}^{\mathrm{e}} \partial T} : \dot{\boldsymbol{\varepsilon}}^{\mathrm{e}} - \frac{\partial^2 \varphi}{\partial T^2} \dot{T} - \frac{\partial^2 \varphi}{\partial V_k \partial T} \dot{V}_k = -\frac{1}{\rho} \frac{\partial \boldsymbol{\sigma}}{\partial T} : \dot{\boldsymbol{\varepsilon}}^{\mathrm{e}} + \frac{\partial s}{\partial T} \dot{T} - \frac{1}{\rho} \frac{\partial A_k}{\partial T} \dot{V}_k \tag{5.104}$$

We obtain

$$-\text{div}\,\mathbf{J}_q = \rho T \frac{\partial s}{\partial T}\dot{T} - \boldsymbol{\sigma}$$
$$: (\dot{\boldsymbol{\varepsilon}} - \dot{\boldsymbol{\varepsilon}}^{\mathrm{e}}) + A_k V_k - \rho r - T\left(\frac{\partial \boldsymbol{\sigma}}{\partial T} : \dot{\boldsymbol{\varepsilon}}^{\mathrm{e}} + \frac{\partial A_k}{\partial T}\dot{V}_k\right) \tag{5.105}$$

By introducing the specific heat defined by

$$C = T\frac{\partial s}{\partial T} \tag{5.106}$$

and taking into account Fourier's Law for isotropic materials

$$\text{div}\,\mathbf{J}_q = -k\,\text{div}\,(\text{grad}\,T) = -k\nabla^2 T \tag{5.107}$$

where ∇^2 denotes the Laplacian operator. Using $\dot{\boldsymbol{\varepsilon}}^{\mathrm{p}} = \dot{\boldsymbol{\varepsilon}} - \dot{\boldsymbol{\varepsilon}}^{\mathrm{e}}$, we obtain

$$k\nabla^2 T = \rho C\dot{T} - \boldsymbol{\sigma} : \dot{\boldsymbol{\varepsilon}}^{\mathrm{p}} + A_k V_k - \rho r - T\left(\frac{\partial \boldsymbol{\sigma}}{\partial T} : \dot{\boldsymbol{\varepsilon}}^{\mathrm{e}} + \frac{\partial A_k}{\partial T}\dot{V}_k\right) \tag{5.108}$$

This is the fully coupled thermo-mechanical equation, which can simulate the evolution of temperature influenced by the mechanical work with properly imposed boundary conditions. $A_k\dot{V}_k$ represents the non-recoverable energy in the materials corresponding to other dissipation phenomena. If we simplify the problem to thermo-mechanical loading at small strain rates on pure metals, we may be able to ignore other dissipation terms, and then we can write

$$A_k\dot{V}_k \approx 0 \tag{5.109}$$

which results in the fully coupled elastoplastic thermo-mechanical equation

$$k\nabla^2 T = \rho C\dot{T} - \boldsymbol{\sigma} : \dot{\boldsymbol{\varepsilon}}^{\mathrm{p}} - \rho r - T\frac{\partial \boldsymbol{\sigma}}{\partial T} : \dot{\boldsymbol{\varepsilon}}^{\mathrm{e}} \tag{5.110}$$

Equation (5.110) also allows us to calculate heat flux $\mathbf{J}_q$ generated due to elastic and/or inelastic work in a solid body.

For the isotropic linear thermoelastic materials, the stress-strain relationship in Newtonian mechanics is given by

$$\sigma_{ij} = \lambda\delta_{ij}\varepsilon_{kk} + 2\mu\varepsilon_{ij} - (3\lambda + 2\mu)\delta_{ij}\alpha(T - T_0) \tag{5.111}$$

where T_0 is the reference temperature, α is the isotropic thermal expansion coefficient, and λ and μ are the Lame's coefficients

$$\lambda = \frac{\nu E}{(1+\nu)(1-2\nu)}; \quad \mu = \frac{E}{2(1+\nu)} \tag{5.112}$$

If the internal heat generation is neglected, Eq. (5.108) defines the response of isotropic linear thermoelastic materials:

$$k\nabla^2 T = \rho C\dot{T} + (3\lambda + 2\mu)\alpha T\dot{\varepsilon}_{kk} \tag{5.113}$$

The last term represents the interconvertibility of the thermal and mechanical energy.

Thermodynamic state index earlier was given by

$$\phi = \left[1 - \mathrm{e}^{-\frac{m_s(\Delta s)}{R}}\right] \tag{5.114}$$

With the help of Eqs. (5.87) and (5.108), the total specific entropy generation rate for small strains is given by

$$\frac{ds}{dt} = \frac{c}{T}\frac{\partial T}{\partial t} - \frac{1}{\rho}\left(\frac{\partial \boldsymbol{\sigma}}{\partial T} : \dot{\boldsymbol{\varepsilon}}^{\mathrm{e}} + \frac{\partial A_k}{\partial T} : V_k\right) \tag{5.115}$$

where it is assumed $\boldsymbol{\sigma} : \dot{\boldsymbol{\varepsilon}} - \rho\dot{w}^{\mathrm{e}} = \boldsymbol{\sigma} : \dot{\boldsymbol{\varepsilon}}^{\mathrm{p}} - A_k\dot{V}_k$ holds true, which represents the total mechanical dissipation rate, and $A_k\dot{V}_k$ is assumed to be zero. Of course this assumption is not true for elastic fatigue loading or for any problem where plastic work dissipation is not the dominant entropy generation mechanism. In the presence of plastic dissipation, elastic dissipation is small for quasi-static loading.

With the help of Eqs. (5.93a), (5.93b) and (5.93c), the specific entropy production rate for small strains becomes

$$\frac{ds_{\mathrm{i}}}{dt} = \frac{\gamma}{\rho} = \frac{\boldsymbol{\sigma} : \dot{\boldsymbol{\varepsilon}}^{\mathrm{p}}}{T\rho} + \frac{r}{T} - \frac{k}{T^2\rho}|\mathrm{grad}\ T|^2 \tag{5.116}$$

The fundamental equations governing the temperature, stresses, deformation, and the entropy production rate in a continuum have been derived in the previous chapters. However, these quantities are all interrelated and have to be determined again simultaneously.

In most practical problems, the effect of the stresses and deformations upon the temperature distribution is quite small and may be neglected. This procedure allows the determination of the temperature distribution in the solid resulting from prescribed thermal conditions to become the first step of a thermal stress analysis; the second step is then the determination of the stresses, deformations, and TSI field due to this temperature distribution. Of course, performing thermo-mechanical analysis in one batch possible, but it is far more computationally intensive.

Entropy change caused by the heat transfer between the system and its surroundings has no influence on the degradation of the material, if the temperature field in the body is uniform as a result of this exchange. If this heat exchange leads to a temperature field that is nonuniform, then heat coming from outside can lead to irreversible entropy generation in the material as thermomigration and other scattering mechanisms. Only the internal entropy generation, namely, the entropy created in the system, should be used as a basis for the systematic description of the irreversible processes, which can be given by

$$\Delta s = \Delta s_{\mathrm{i}} = \int_{t_0}^{t} \frac{\boldsymbol{\sigma} : \dot{\boldsymbol{\varepsilon}}^{\mathrm{p}}}{T\rho} dt - \int_{t_0}^{t} \left(\frac{k}{T^2 \rho} |\mathrm{grad}\, T|^2 \right) dt + \int_{t_0}^{t} \frac{r}{T} dt \tag{5.117}$$

Equation (5.117) shows that the entropy generation is not only a function of the loading or straining process but also of the temperature. However, a uniform increase in temperature in a stress-free field does not cause any TSI. While this fundamental equation accounts for strain rate, it does not account for all mechanisms under thermo-mechanical loading. If the strain rate is very high, phase change, elastic dissipation, melting, and other mechanisms can be significant entropy generation sources that this equation does not account for.

5.5 Numerical Validation of the Thermo-mechanical Constitutive Model

Two problems are selected to validate the constitutive model described in this chapter. First the model results are compared with testing performed on specimens of thin layer solder joints of Pb37/Sn63 under monotonic and fatigue shear testing at different strain rates and temperatures.

5.5.1 Thin Layer Solder Joint-Monotonic and Fatigue Shear Simulations

Table 5.3 lists the strain rate and temperature ranges for testing and simulations. The monotonic and mechanical shear testing was performed on thin layer of solder joint (Pb37/Sn63) in pure shear loading. Figure 5.3 shows the specimen attached to copper plates. Material parameters used for the numerical simulation are shown in Table 5.4. Figure 5.4 shows monotonic shear testing response under strain rate of 1.67×10^{-3}/s at different temperatures. Figure 5.5 shows monotonic shear testing response under strain rate of 1.67×10^{-3}/s at 22 °C. Figures 5.6, 5.7, 5.8, 5.9 and 5.10 show cyclic stress-strain response. Figure 5.11 shows evaluation of the damage parameter, [TSI] as a function of number of cycles. The computational simulations are in good agreement with the experimental results. The differences are due to the imperfect microstructure (such as voids) of the solder while the numerical model assumes a perfect continuum. Moreover, entropy generation due to internal heat generation was ignored for the sake of simplicity.

More examples of unified mechanics theory for thermo-mechanical problems are provided in detail in the references listed at the end of the chapter.

5.6 Thermo-mechanical Analysis of Cosserat Continuum: Length-Scale Effects

5.6.1 *Introduction*

Classical continuum mechanics formulation does not include a term for the size effect. Strain gradient plasticity is needed when traditional continuum mechanics formulation is unable to represent the material stress-strain behavior due to size effects. Traditionally, continuum mechanics formulations are independent of size. However, in some mechanics problems, this is not true, and material response is size

Table 5.3 Loading scheme used in the analysis

Case I	**Monotonic shear loading**		
1	Strain rate	1.67e−3/s	Temperature
		a	−40 °C
		b	22 °C
		c	60 °C
		d	100 °C
2	Strain rate	1.67e−1/s	22 °C
3	Strain rate	1.67e−2/s	22 °C
4	Strain rate	1.67e−3/s	22 °C
5	Strain rate	1.67e−4/s	22 °C
Case II	**Cyclic shear loading**		
1	Temp = 22 °C Strain rate	 1.67e−3/s	 ISR
		a	0.005
		b	0.012
		c	0.02
Case III	**Fatigue shear**		
1	Strain rate	1.67e−4/s	ISR
		a	0.004
		b	0.012
		c	0.022

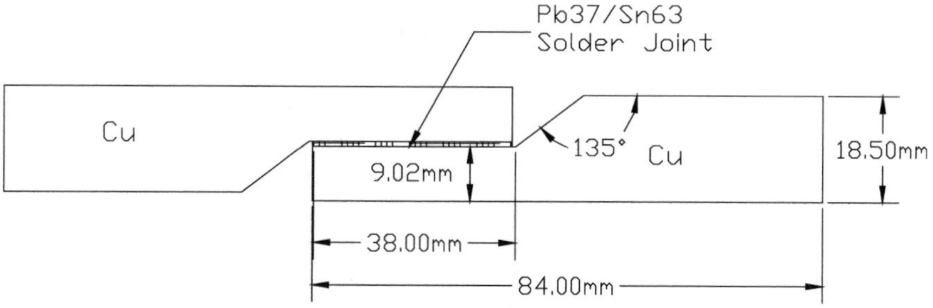

Fig. 5.3 Thin layer solder joint attached to copper plates

Table 5.4 Material parameters

Elastic (θ temperature in K)	
Young's modulus (GPa)	52.10–0.1059θ
Shear modulus (GPa)	19.44–0.0395θ
Isotropic hardening	
R_{00} (MPa)	37.47–0.0748θ
c (Dimensionless)	383.3
σ_y (MPa)	60.069–0.140θ
Kinematic hardening	
c_1 (MPa)	2040
c_2 (Dimensionless)	180
Creep strain rate	
A (Dimensionless)	7.60E+09
D_0 (mm/s^2)	48.8
b (mm)	3.18E−07
d (mm)	1.06E−02
n	1.67
p	3.34
Q (mJ/mol)	4.47E+07
Φcritical	0.1

dependent. A class of plasticity theories where size effect is considered are usually called strain gradient plasticity theories.

In strain gradient plasticity theories, it is assumed that classic plastic behavior is due to slip of statistically stored dislocations and length-scale effects are due to slip of geometrically necessary dislocations. As a result, the dislocation term in Taylor's flow stress equation is modified to add the geometrically necessary dislocation density:

$$\tau = \alpha G b \sqrt{\rho_{\mathrm{SSD}} + \rho_{\mathrm{GND}}} \tag{5.118}$$

where α is geometrical factor that depends on the type and arrangement of the interacting dislocations, G is the shear modulus, b is Burger's vector, ρ_{SSD} is the

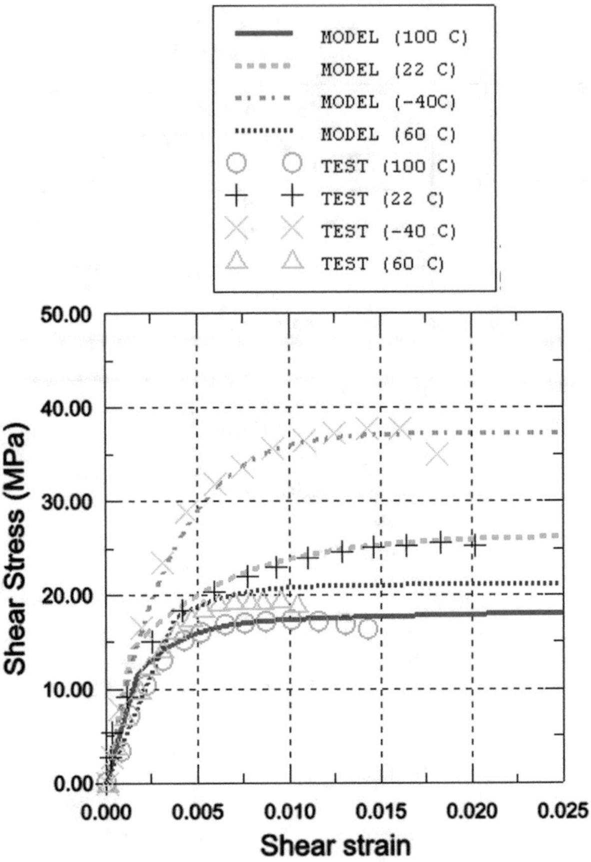

Fig. 5.4 Monotonic shear testing under strain rate 1.67×10^{-3}/s at different temperatures

density of statistically stored dislocations, and ρ_{GND} is the density of geometrically necessary dislocations.

There are different methods to account for size effect in continuum mechanics. One of the methods is the use of Cosserat continuum. In this section we will discuss (Cosserat continuum) couple stress-based strain gradient theory without restrictions to any particular constitutive model. Next we describe a particular constitutive model incorporating coupling between degradation and rate dependency and considering size effects by means of a couple stress theory material length scale. We provide two distinct theories within the same framework, namely, a general couple stress theory where translational and rotational degrees of freedom are treated as independent kinematic variables and a more restrictive reduced couple stress theory where translational and rotational degrees of freedom are constrained to satisfy a given kinematic relation. The kinematic constraint allows the implementation of the reduced theory into a displacement-based finite element formulation. The use of

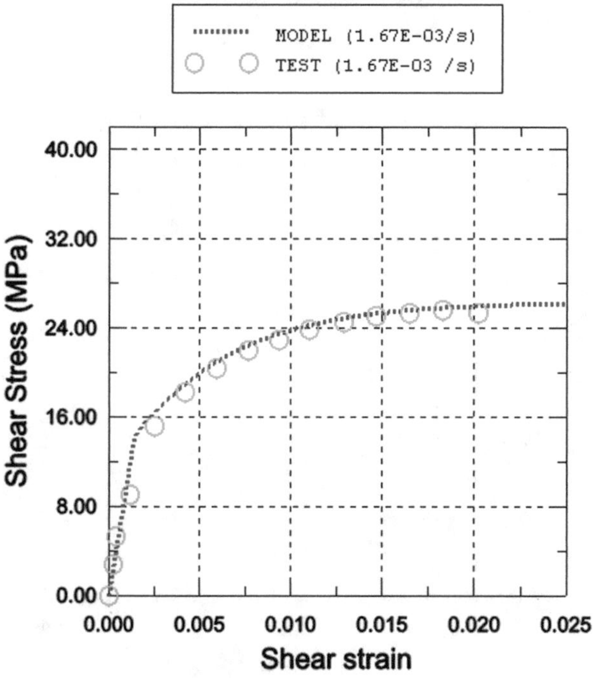

Fig. 5.5 Monotonic shear testing under strain rate 1.67×10^{-3}/s at 22 °C

this approach relaxes the strong C^1 shape function continuity requirements and allows the use of C^0 elements thus keeping the convergence features of the finite element method. Here the kinematic constraint is enforced via a penalty function/reduced integration scheme that allows treatment of the reduced theory based on ideas motivated by the general theory. The current description progressively arrives at this formulation after exploring different possibilities within the context of the variational calculus. Throughout all the discussion, the objective should be clear that the final goal is to implement the reduced couple stress theory framework in the form of a user element subroutine within the finite element code ABAQUS which imposes additional restrictions.

5.6.2 Cosserat Couple Stress Theory

The motivation behind Cosserat's couple stress theory, also referred to as Cosserat continuum, Fig. 5.12, Gomez (2006) stems from the following arguments. In classical continuum mechanics, it is assumed that a material point occupies no space (has zero volume) and then the finite size of material microstructure is ignored. This is equivalent to the definition of representative volume element. In order to

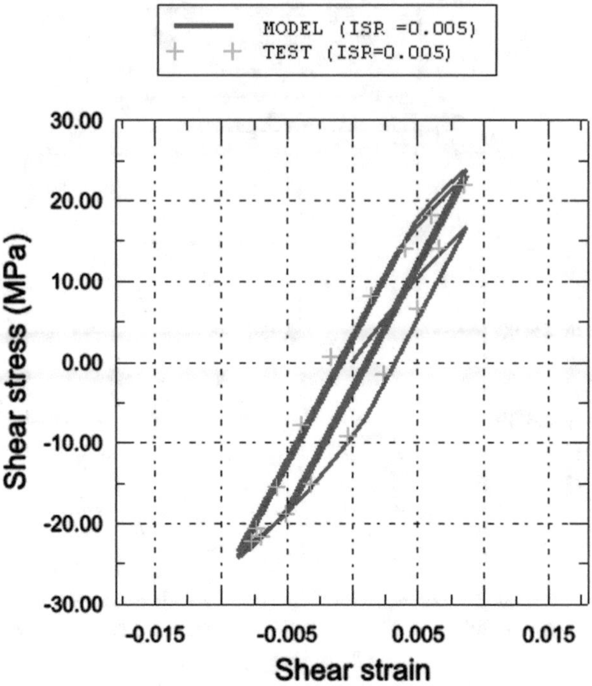

Fig. 5.6 Cyclic shear simulation vs. test data at 22 °C, strain rate 1.67×10^{-3}/s, and ISR = 0.005

include the effects of the finite size of material's microstructure while retaining a mathematical continuum, classical continuum mechanics have been extended or generalized. In Cosserat continuum, each material particle is enriched with higher-order kinematic degrees of freedom. In particular, the specification of position requires also the definition of a rotation. In the more general Cosserat theory, this new kinematic quantity, the micro rotation, is independent of the classical continuum mechanics theory rotation vector $\boldsymbol{\theta}_k = \frac{1}{2} \boldsymbol{e}_{ijk} \boldsymbol{u}_{j,k}$ where $\boldsymbol{e}_{ijk}$ is the alternating pseudo-tensor and a coma represents a derivative with respect to rectangular Cartesian coordinates. This type of theory has been termed as "indeterminate couple stress theory" by Eringen (1968) and "generalized couple stress theory" by Fleck and Hutchinson (1997). The following discussion elaborates further upon this theory. There are different interpretations of a Cosserat continuum. In the Cosserat and Cosserat (1909), couple stress theory, a differential material element, is subjected to not only normal and shear stresses but also couple stress components as shown in Fig. 5.12. For linear elastic behavior, the usual stress components are functions of the strains, and the couple stresses are functions of the strain gradients. Two distinct theories are identified in the work of the Cosserat and Cosserat (1909). First, there is a reduced couple stress theory with the kinematic degrees of freedom being the displacement $\boldsymbol{u}_i$ and an associated material rotation $\boldsymbol{\theta}_i$ tied to the displacements by

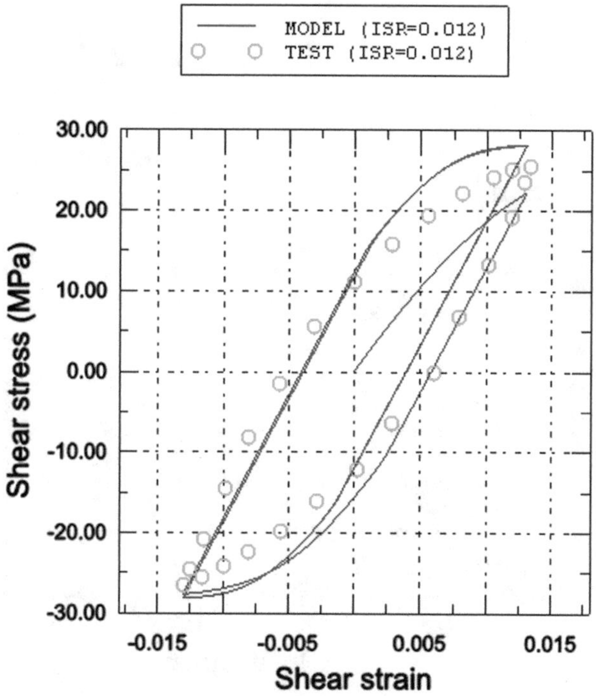

Fig. 5.7 Cyclic shear simulation vs. test data at 22 °C, strain rate 1.67 × 10^{-3}/s, and ISR = 0.012

the kinematic constraint in Eq. (5.119) and with the definitions of small strain and curvatures expressed in Eqs. (5.120) and (5.121):

$$\theta_i = \frac{1}{2} e_{ijk} u_{k,j} \tag{5.119}$$

$$\varepsilon_{ij} = \frac{1}{2}\left(u_{i,j} + u_{j,i}\right) \tag{5.120}$$

$$x_{ij} = \theta_{i,j} \tag{5.121}$$

Second, there is a general couple stress theory in terms of a micro rotation ω_i, which is regarded as an independent kinematic degree of freedom. The rotation and micro rotation are related by a relative rotation tensor $\alpha_{ij} = e_{ijk}\omega_k - e_{ijk}\theta_k$. For the particular choice of $\omega_k = \theta_k$, the general couple stress theory reduces to the reduced couple stress theory. Equilibrium of the differential element shown in Fig. 5.12, after neglecting body forces and body couples, yields, [in Newtonian Mechanics]

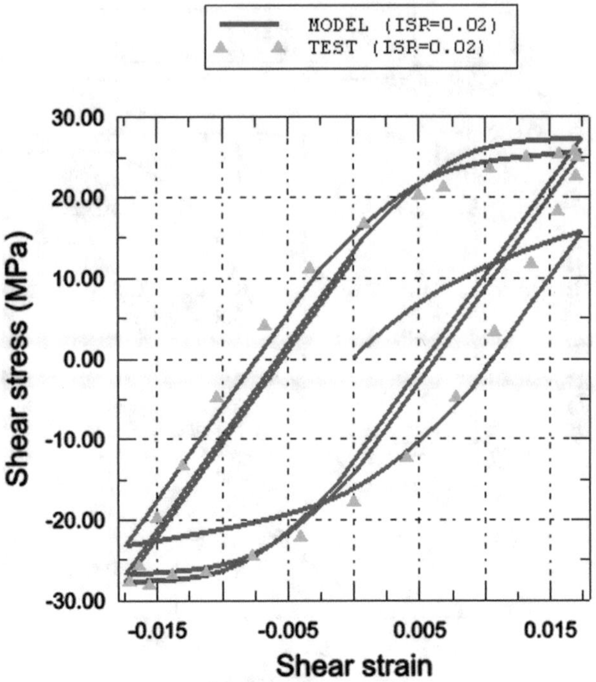

Fig. 5.8 Cyclic shear simulation vs. test data at 22 °C, strain rate 1.67 × 10^{-3}/s, and ISR = 0.02

$$\boldsymbol{\sigma_{ji,j} + \tau_{ji,j} = 0} \tag{5.122}$$

$$\boldsymbol{\tau_{jk} + \frac{1}{2} e_{jik} m_{pi,p} = 0} \tag{5.123}$$

where σ_{ij} and τ_{ij} are the symmetric and antisymmetric components of the Cauchy stress tensor, respectively, and $\boldsymbol{m_{ij}}$ is the couple stress tensor. Surface stress tractions and surface couple stress tractions are given by

$$\sigma_i^{(n)} = \left(\sigma_{ij} + \tau_{ij}\right) n_j \tag{5.124}$$

$$q_i = m_{ij} n_j \tag{5.125}$$

where $\sigma_i^{(n)}$ is the traction vector, q_i is the couple tractions vector, and n_i is a surface outward normal vector. For a linear elastic solid, the strain energy density is given by Eq. (5.126) where l is a length scale associated to elastic processes and may be regarded to be of the order of interatomic distances (Koiter 1964):

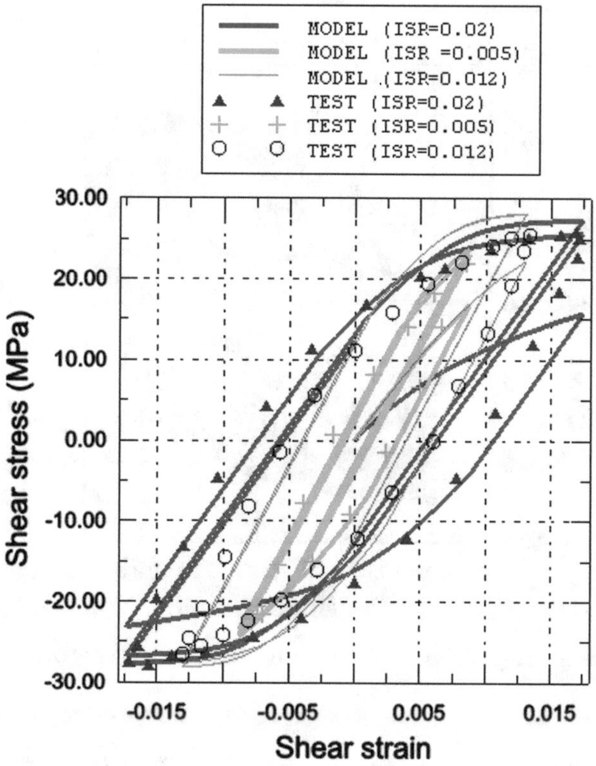

Fig. 5.9 Cyclic shear simulation vs. test data at 22 °C, strain rate 1.67 × 10^{-3}/s, and different inelastic strain ranges

$$W = \frac{E}{2(1+\nu)}\left(\frac{\nu}{1-2\nu}\varepsilon_{ii}\varepsilon_{jj} + \varepsilon_{ij}\varepsilon_{ij} + l^2 x_{ij}x_{ij}\right) \tag{5.126}$$

A constitutive relationship can be directly derived out of Eq. (5.126) as follows:

$$\sigma_{ij} = \frac{\partial W}{\partial \varepsilon_{ij}} \equiv C_{ijkl}\varepsilon_{kl} \tag{5.127}$$

$$m_{ij} = \frac{\partial W}{\partial x_{ij}} \equiv D_{ijkl}x_{kl} \tag{5.128}$$

where C_{ijkl} and D_{ijkl} are elastic constitutive tensors relating stress and couple stress components to elastic strains and elastic curvatures, respectively. The principle of virtual work for the reduced Cosserat theory can be given by, [in Newtonian mechanics]

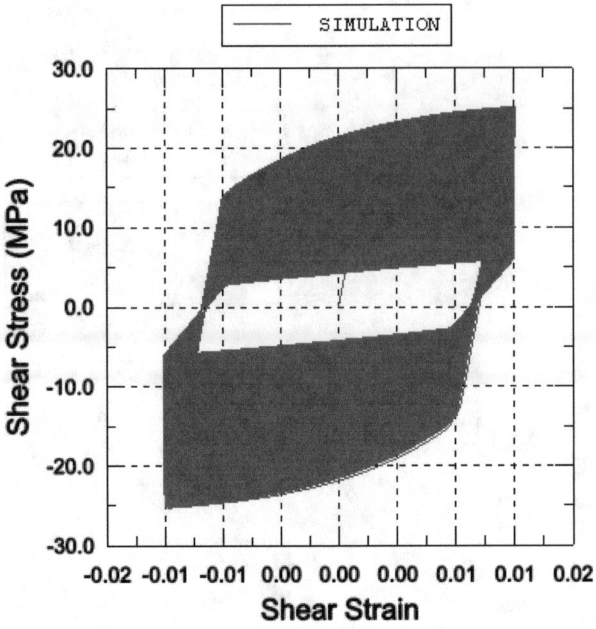

Fig. 5.10 Isothermal fatigue at strain rate 1.67 × 10^{-4}/s at 22 °C with inelastic strain range 0.022

$$\iiint_V \sigma_{ij}\delta\varepsilon_{ij}dV + \iiint_V m_{ij}\delta x_{ij}dV = \iint_S \sigma_i^{(n)}\delta u_i dS + \iint_S q_i\delta\theta_i dS \tag{5.129}$$

The concept of "reduced Cosserat" will be discussed later in the chapter.

5.6.3 *Toupin-Mindlin Higher-Order Stress Theory*

The Cosserat couple stress theory introduces curvatures as gradients of rotation. Toupin (1962), Mindlin and Tiersten (1962) and Mindlin (1964, 1965) extended the original Cosserat formulation by considering also gradients of the normal strain components, therefore, including all the components of the second gradient of displacement. The additional gradients of strain as variables give rise to new work conjugates in the form of force couples per unit area or higher-order stresses. The physical interpretation of these force couples and higher-order stresses is explained in Fig. 5.12. In this type of solid, Newtonian mechanics equilibrium (neglecting body forces) is given by

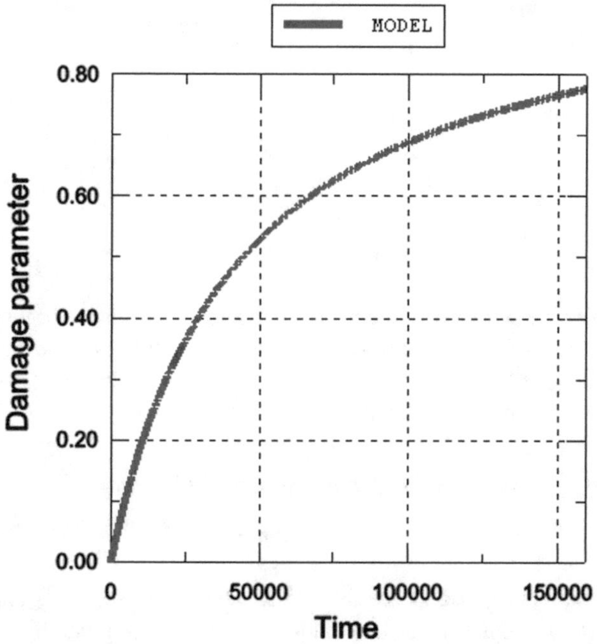

Fig. 5.11 Evolution of the damage parameter per Eq. (5.114) for $D_{cr} = 1.0$ under fatigue loading with ISR = 0.022

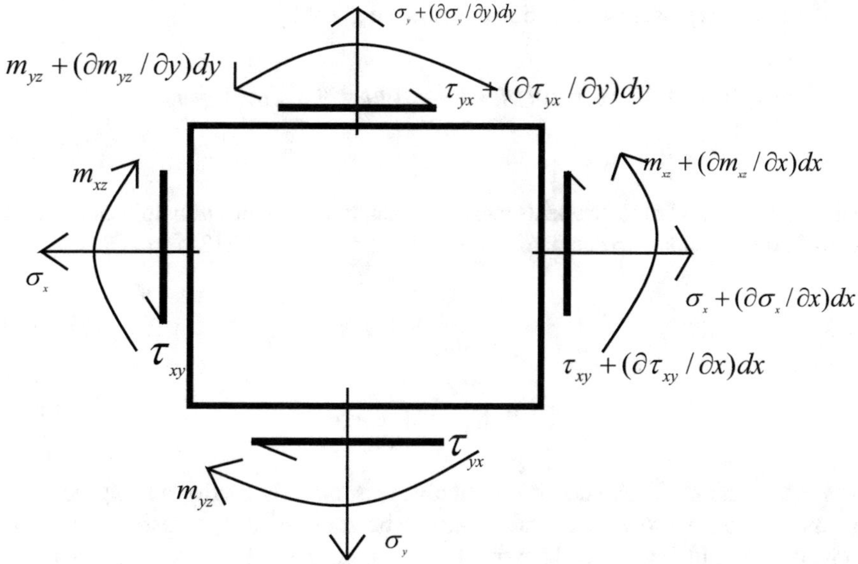

Fig. 5.12 Cosserat continuum stress point [After Gomez (2006)]

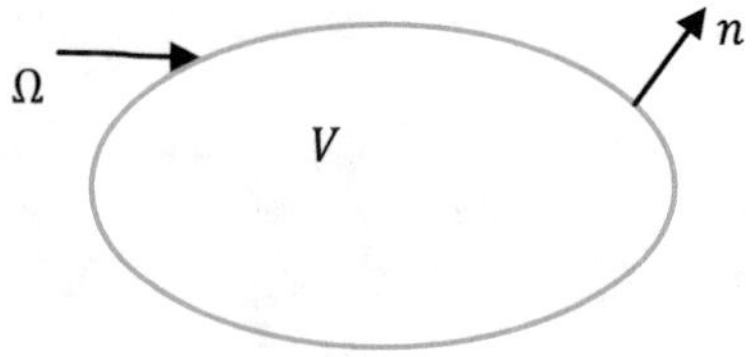

Fig. 5.13 Schematic of a general solid body

$$\sigma_{ik,i} - \tau_{ijk,ij} = 0 \tag{5.130}$$

where σ_{ij} is now the symmetric Cauchy stress tensor and τ_{ijk} is the higher-order stress tensor Fig. 5.13. Surface tractions and surface couple stresses [distributed moment] are given in Eqs. (5.124) and (5.125) respectively. The higher-order stresses correspond to force couples per unit area:

$$\sigma_i^{(n)} = \left(\sigma_{ik} - \tau_{ijk.j}\right)n_i + n_i n_j \tau_{ijk}\left(D_p n_p\right) - D_j\left(n_i \tau_{ijk}\right) \tag{5.131a}$$

$$r_k = n_i n_j \tau_{ijk} \tag{5.131b}$$

where the operators D_j and D are defined as $D_j = (\delta_{jk} - n_j n_k)\partial_k$ and $D = n_k \partial_k$. Generalized strain-displacement kinematic relations can be given by

$$\varepsilon_{ij} = \frac{1}{2}\left(u_{i.j} + u_{j,i}\right) \tag{5.132a}$$

$$\eta_{ijk} = u_{k,ij} \tag{5.132b}$$

A strain energy density function can now be given by

$$W = \frac{1}{2}\lambda\varepsilon_{ii}\varepsilon_{jj} + \mu\varepsilon_{ij}\varepsilon_{ij} + a_1\eta_{ijj}\eta_{ikk} + a_2\eta_{iik}\eta_{kjj} + a_3\eta_{iik}\eta_{jjk} + a_4\eta_{ijk}\eta_{ijk} + a_5\eta_{ijk}\eta_{kji} \tag{5.133}$$

where λ and μ are Lamè constants and a_i are material constants with units of force. A constitutive relationship can be directly derived out of Eq. (5.133) such that

$$\sigma_{ij} = \frac{\partial W}{\partial \varepsilon_{ij}} \equiv C_{ijkl}\varepsilon_{kl} \tag{5.134a}$$

$$\tau_{ijk} = \frac{\partial W}{\partial \eta_{ijk}} \equiv D_{ijklmn}\eta_{lmn} \tag{5.134b}$$

where C_{ijkl} and D_{ijklmn} are elastic constitutive tensors relating stress and higher-order stress components to elastic strains and elastic strain gradients, respectively. The principle of virtual work can be written by equating an external work increment and an internal increment of strain energy and is given by [in Newtonian mechanics]

$$\iiint_V \sigma_{ij}\delta\varepsilon_{ij}dV + \iiint_V \tau_{ijk}\delta\eta_{ijk}dV = \iint_S \sigma_i^{(n)}\delta u_i dS + \iint_S r_i(D\delta u_i)dS \tag{5.135}$$

5.6.4 Equilibrium Equations and Problem Formulation

The description that follows for the two different classes of Cosserat continuum is for a general solid occupying a volume Ω and bounded by a surface $\partial\Omega$ defined by an outward normal vector n as schematically shown in Fig. 5.13. Throughout we will refer to volume differentials as dV and surface area differentials as $d\Omega$.

There are several different interpretations of Cosserat continuum. In the Cosserat and Cosserat (1909) couple stress theory, a differential material element admits not only normal and shear stresses but also couple stress components as shown in Fig. 5.12. For linear elastic behavior, the usual stress components are functions of the strains, and the couple stresses are functions of the strain gradients. Two distinct theories are identified in the Cosserat and Cosserat (1909) as described by Aero and Kuvshinsky (1961), Mindlin (1964), De Borst and Muhlhaus (1992), De Borst (1993), and Shu and Fleck (1999). As discussed earlier, first, there is a reduced Cosserat couple stress theory with the kinematic degrees of freedom being the displacement u_i and an associated material rotation θ_i tied to the displacements by a kinematic constraint given in Eq. (5.119) and with the definitions of strain and curvatures expressed in Eqs. (5.120) and (5.121).

According to this theory, the continuum is assumed to possess bending stiffness allowing for the introduction of additional stress measures in the form of moments per unit area. The presence of the couple stresses renders the Cauchy stress tensor asymmetric; however only the symmetric component generates work upon deformation. Second, there is a general couple stress theory in terms of a micro rotation $\boldsymbol{\omega}_i$ which is regarded an independent kinematic variable. The rotation and micro rotation are related by a relative rotation tensor $\alpha_{ij} = e_{ijk}\omega_k - e_{ijk}\theta_k$. For the particular choice of $\omega_k = \theta_k$, the general couple stress theory reduces to the more restrictive reduced couple stress theory. This general theory assumes that within the material point, there is also embedded a micro-volume giving rise to the micro rotation $\boldsymbol{\omega}_i$. Different theories have been postulated depending on the deformation properties assumed for the micro-volume; see Mindlin and Tiersten (1962), Toupin (1962), and Mindlin (1964, 1965) for a review of the different approaches. For instance, Fig. 5.14 shows the deformation state in a material point including the micro-volume for the case of pure shear in the solid proposed by Mindlin (1964). Figure 5.15 shows the particular case of reduced couple stress theory where the micro-volume is assumed rigid.

Newtonian mechanics equilibrium of the differential element shown in Fig. 5.12 (which is valid for both theories), after neglecting body forces and body couples, yields

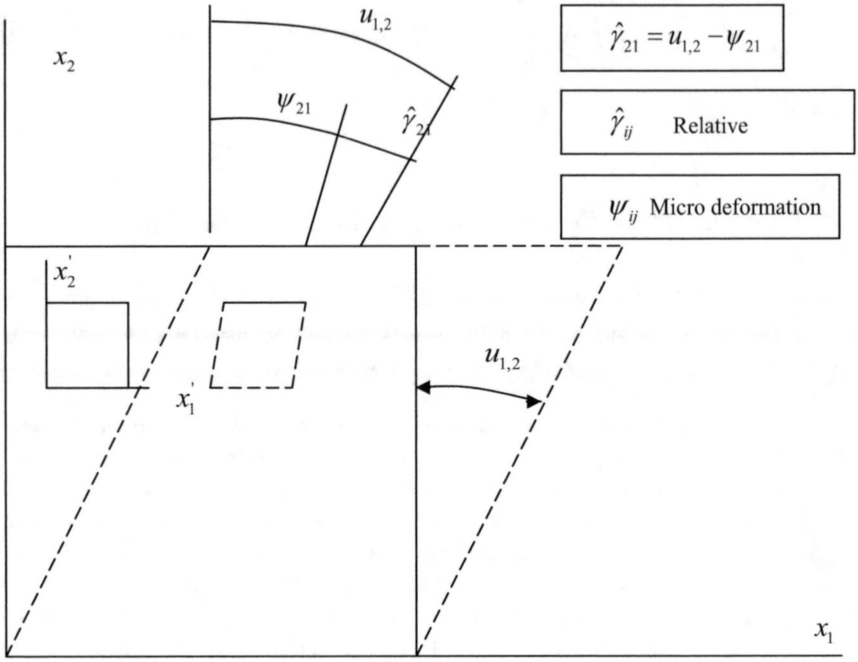

Fig. 5.14 Relative deformation in the couple stress theory by Mindlin (1964)

$$\sigma_{ji,j} + \tau_{ji,j} = 0 \tag{5.136}$$

$$\tau_{jk} + \frac{1}{2} e_{ijk} m_{pi,p} = 0 \tag{5.137}$$

where σ_{ij} and τ_{ij} are the symmetric and antisymmetric components of the Cauchy stress tensor, respectively, and m_{ij} is the couple stress tensor. Surface stress tractions and surface couple stress tractions are given by

$$\sigma^{(n)}{}_i = \left(\sigma_{ij} + \tau_{ij}\right) n_j \tag{5.138}$$

$$q_i = m_{ij} n_j \tag{5.139}$$

where $\sigma^{(n)}{}_i$ is the surface traction vector, q_i is the surface couple tractions vector, and n_i is a surface normal vector. In order to establish a general framework for elastic material behavior, Cosserat couple stress continuum can be obtained after postulating a strain energy density function dependent on strains and curvatures (rotation gradients). Toupin (1962) and Mindlin (1965) extended this theory to include also stretch gradients. In particular, they considered invariants of the strain and strain gradients into the strain energy density function in terms of the following generalized von Mises strain invariant which is given by (Fleck and Hutchinson 1997)

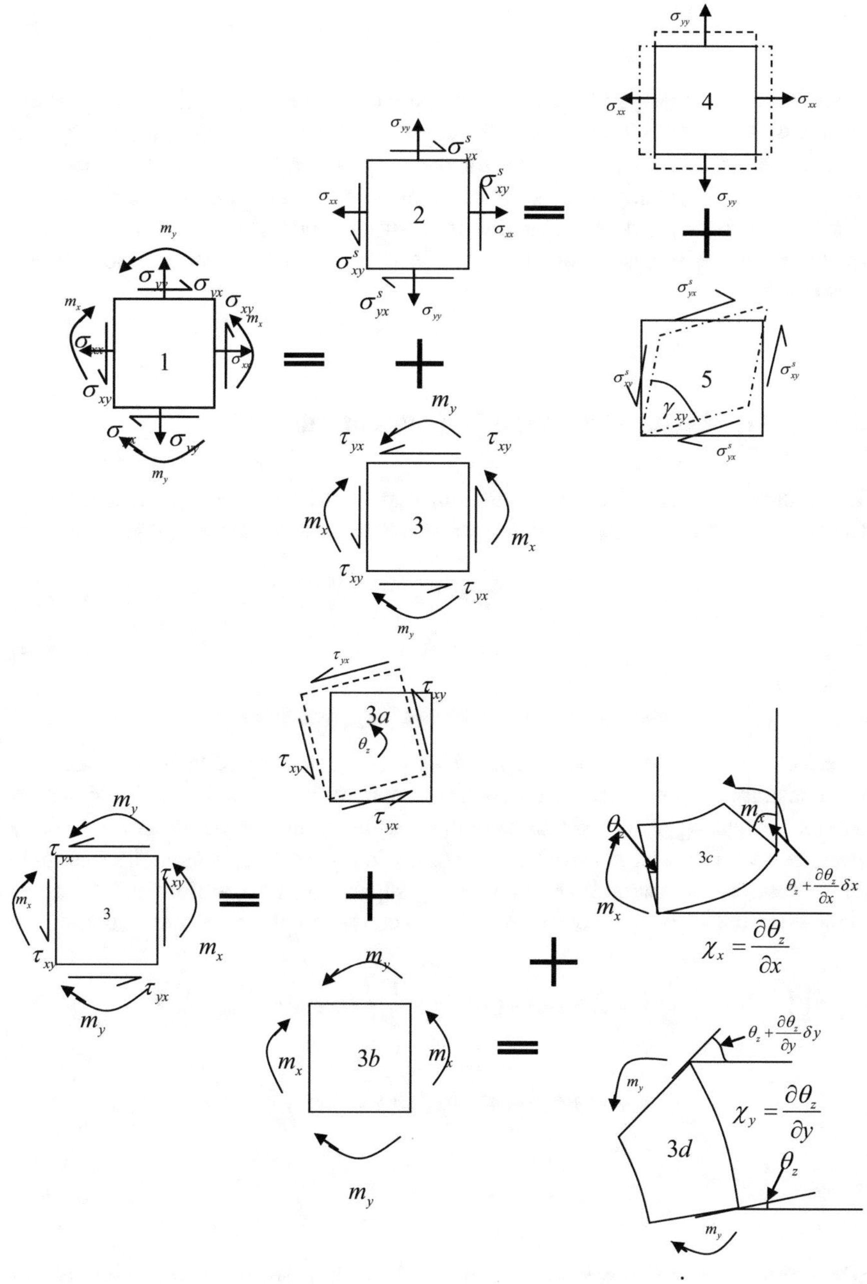

Fig. 5.15 Reduced Cosserat continuum stress point

$$E^2 = \frac{2}{3}\varepsilon'_{ij}\varepsilon'_{ij} + c_1\eta'_{iik}\eta'_{jjk} + c_2\eta'_{ijk}\eta'_{ijk} + c_3\eta'_{ijk}\eta'_{kij} \tag{5.140}$$

where the prime superscript denotes deviatoric component, $\eta_{ijk} = u_{k,\ ij}$ are the strain gradients, and c_i's are additional material constants with dimensions of length squared (L^2). Equation (5.140) is the basis for the most general strain gradient plasticity theory, and it essentially reveals the phenomenological coupling between the densities of statistically and geometrically stored dislocations. The reduced couple stress theory is just a special case of Eq. (5.140) where only rotation gradients are considered.

5.7 Finite Element Method Implementation

The following elastic constitutive relationships can be written for the symmetric Cauchy stress tensor, the asymmetric stress tensor, and the couple stress tensor:

$$\sigma_{ij} = C_{ijkl}\varepsilon_{kl} \tag{5.141}$$

$$\tau_{ij} = \overline{D}_{ijkl}\alpha_{kl} \tag{5.142}$$

$$l^{-1}m_{ij} = D_{ijkl}l\chi_{kl} \tag{5.143}$$

where C_{ijkl} is a tangential constitutive tensor relating strains to Cauchy stresses, D_{ijkl} is a constitutive tensor relating curvatures to couple stresses, and $\overline{D}_{ijkl}$ is a constitutive tensor relating relative rotations to the antisymmetric component of the Cauchy stress tensor. The total potential energy functional $\prod$ for the general elastic couple stress solid can be written by considering separately the contributions from the symmetric and antisymmetric stress tensors and the couple stress tensor as follows:

$$\begin{aligned}\prod(u_i, \omega_i) = &\frac{1}{2}\int\limits_V C_{ijkl}\varepsilon_{kl}(u_i)\varepsilon_{ij}(u_i)dV + \frac{1}{2}\int\limits_V D_{ijkl}\chi_{ij}(\omega_i)\chi_{kl}(\omega_i)dV \\ &+\frac{1}{2}\int\limits_V \overline{D}_{ijkl}\alpha_{ij}(u_i, \omega_i)\alpha_{kl}(u_i, \omega_i)dV - \int_\Omega \sigma_i^{(n)}u_i d\Omega \\ &- \int_\Omega q_i\omega_i d\Omega\end{aligned} \tag{5.144a}$$

where translational and rotational degrees of freedom are independent degrees of freedom. For the discussion that follows, it is convenient to write in the equivalent form of

$$\prod(u,\omega) = F(u) + G(\omega) + H(u,\omega)$$

where

$$F(u) = \frac{1}{2}a(u,u) - f(u)$$

$$G(\omega) = \frac{1}{2}b(\omega,\omega) - g(\omega)$$

$$H(u,\omega) = \frac{1}{2}\int_{\Omega} \overline{D}_{ijkl}\alpha_{ij}(u_i,\omega_i)\alpha_{kl}(u_i,\omega_i)d\Omega \tag{5.144b}$$

where $a(u,u) \equiv \int_{\Omega} C_{ijkl}\varepsilon_{ij}\varepsilon_{kl}d\Omega$ and $b(\omega,\omega) \equiv \int_{\Omega} D_{ijkl}\chi_{ij}\varepsilon_{kl}d\Omega$ are symmetric bilinear forms and $f(u)$ and $g(\omega)$ correspond to the boundary terms in Eq. (5.144a). For the particular case of the more restricted reduced couple stress theory where the stress tensor is symmetric and satisfies the constraint in Eq. (5.119), α_{ij} (or equivalently $H(u,\omega)$) vanishes, and Eq. (5.144b) reduces to Eq. (5.145a):

$$\overline{\Pi}(u_i) = \frac{1}{2}\int_{\Omega} C_{ijkl}\varepsilon_{kl}(u_i)\varepsilon_{ij}(u_i)d\Omega + \frac{1}{2}\int_{\Omega} D_{ijkl}\chi_{ij}(u_i)\chi_{kl}(u_i)d\Omega - \int_{\partial\Omega} t_i u_i d\Gamma - \int_{\partial\Omega} q_i\omega_i d\Gamma \tag{5.145a}$$

where now the strain energy contribution from the curvatures becomes a function of the translational degrees of freedom only. Note that Eq. (5.145a) is analogous to the total potential energy functional for the particular case of the so-called Timoshenko beam theory. Using the alternative notation, Eq. (5.145a) can be written as

$$\overline{\Pi}(u) = F(u) + G'(u) \tag{5.145b}$$

where $G'(u) = b'(u,u) - g'(u)$ with $b'(u,u) \equiv \int_{\Omega} D_{ijkl}\chi_{ij}(u_i)\chi_{kl}(u_i)d\Omega$ and $g'(u)$ corresponds to the boundary part associated to the rotation. The prime superscript notation $(\cdot)'$ has been introduced in order to clarify the fact that in the reduced theory, the curvatures are kinematically constrained to the displacements. Equations (5.144a) and (5.146a) are the basis for the finite element implementation of the theories described in this section. It is clear that the reduced theory continuum can also be formulated in terms of independent rotational degrees of freedom with the constraint to the translational degrees of freedom considered as the limit when the term $H(u,\omega) \to 0$ in the general couple stress continuum. This implies that Eq. (5.146a) is written in terms of additional independent rotational degrees of freedom but with the additional requirement imposed by the constraint between translational and rotational degrees of freedom. In this case the reduced theory can also be described by the functional $\widehat{\Pi}(u, \omega)$ as follows:

$$\widehat{\Pi}(u_i, \omega_i) = \frac{1}{2}\int_V C_{ijkl}\varepsilon_{kl}(u_i)\varepsilon_{ij}(u_i)dV + \frac{1}{2}\int_V D_{ijkl}x_{ij}(\omega_i)x_{kl}(\omega_i)dV$$

$$-\int_\Omega t_i u_i d\Omega - \int_\Omega q_i \omega_i d\Omega \quad (5.145c)$$

A comparison of Eqs. (5.144b) and (5.145c) reveals that $\Pi(u, \omega) \to \widehat{\Pi}(u, \omega)$ when $H(u, \omega) \to 0$. This means that the general couple stress theory approaches the reduced couple stress theory in the limit of vanishing relative rotation α_{ij}.

5.7.1 General Couple Stress Theory: Variational Formulation

Consider the case of the general couple stress continuum where there is no constraint set, but instead there is a space of admissible functions equipped with degrees of freedom having not only translational but also rotational components. We may think of the rotational degrees of freedom as independent elements belonging to the space $\overrightarrow{Q}$, or alternatively we can define displacements and rotations as elements of the single space $\vec{V} \times \vec{Q}$. In any case the product $\vec{V} \times \vec{Q}$ results in a third space with elements being all the ordered pairs of the form $\left(\vec{u}, \vec{\omega}\right) \in \vec{V} \times \vec{Q}$. In terms of the introduced notation, this is equivalent to the following variational problem; find $\left(\vec{u}, \vec{\omega}\right) \in \vec{V} \times \vec{Q}$ such that for any $\left(\vec{v}, \vec{\varphi}\right) \in \vec{V} \times \vec{Q}$, Π assumes its minimum value at $\left(\vec{u}, \vec{\omega}\right)$ where $\vec{V} \times \vec{Q}$ is now the corresponding space of admissible functions.

The generalized form of the principle of virtual displacements for the general couple stress theory can be given by

$$\int_V \sigma_{ij}\varepsilon_{ij}(v_i)dV + \int_V m_{ij}x_{ij}(\varphi_i)dV + \int_V \tau_{ij}\alpha_{ij}(v_i, \varphi_i)dV - \int_\Omega t_i v_i d\Omega$$

$$-\int_\Omega q_i \varphi_i d\Omega$$

$$= 0 \quad (5.146)$$

where v_i and φ_i denote virtual displacement and rotation, respectively. Displacement and rotation $\left(\vec{u}, \vec{\omega}\right)$ are regarded as independent kinematic degrees of freedom. This is consistent with the general couple stress theory, and the resulting variational problem is thus unconstrained.

5.7.2 *Reduced Couple Stress Theory: Variational Formulation*

The principle of virtual displacements for a reduced couple stress theory can be given by

$$\int_V \sigma_{ij}\varepsilon_{ij}(v_i)dV + \int_V m_{ij}\chi_{ij}(v_i)dV - \int_\Omega \sigma_i^{(n)} v_i d\Omega - \int_\Omega q_i\varphi_i(v_i)d\Omega = 0 \qquad (5.147)$$

In contrast to the case defined in Eq. (5.146), the present principle of virtual displacements shows that the only kinematic variable is $\vec{u}$ which at the same time completely defines the strains ε_{ij} and the curvatures χ_{ij}. This is consistent with the definition of the reduced couple stress theory. Notice that in this formulation there are no independent rotational degrees of freedom and there is no constraint to be enforced. However the displacement functions need to be C^1 continuous in a finite element implementation. However, when there is no kinematic constraint between displacements and rotations, mesh-dependent results are obtained in finite element analysis.

5.7.3 *Reduced Couple Stress Theory: Mixed Variational Principle*

The principle of virtual displacements with the kinematic constraint imposed in a weak sense leads to

$$\int_V \sigma_{ij}\varepsilon_{ij}(v_i)dV + \int_V m_{ij}\chi_{ji}(\varphi_i)dV + \int_V \tau_{ij}\alpha_{ij}(v_i,\varphi_i)dV$$

$$= \int_\Omega t_i v_i d\Omega + \int_\Omega q_i\varphi_i d\Omega \qquad (5.148a)$$

$$\int_V \rho_{ij}\alpha_{ij}(u_i,\omega_i)dV = 0 \qquad (5.148b)$$

The first two terms on the right-hand side of Eq. (5.148a) corresponds to the virtual work done by the stresses and couple stresses, respectively. The third term corresponds to the virtual work done by the asymmetric component of the stress tensor. The unique term in Eq. (5.148b) corresponds to the weak enforcement of the constraint condition of vanishing relative rotation. This approach was proposed by Xia and Hutchinson (1996) as an alternative method to solve the mesh dependency problem in the finite element method when using Eq. (5.147) and a pure displacement based finite element formulation.

5.7.3.1 Finite Element Method Implementation

This section describes the discretized versions of the equations derived above starting from their corresponding variational equations. In each case we will schematically show a typical finite element with its associated vector of nodal point parameters but without any restriction as to the number of nodes. The parameters may be translational degrees of freedom, translational and rotational degrees of freedom, or translational and rotational degrees of freedom with additional Lagrange multipliers. In order to distinguish between the values of the given parameter at any point within the element and its nodal point value, we will use the following notation. For instance, if displacement is being considered, the value at any point within the element will be denoted by the vector u, and its corresponding nodal point's vector representation will be denoted by $\widehat{u}$. In a given element, the value of a given parameter at any point within the element is obtained via interpolation from the known nodal point values. Strains and curvatures are usually calculated at Gauss integration points from the nodal point displacements and rotations values using interpolation functions. We will denote such a function by a subscripted symbol where the subscript indicates the variable that is being interpolated. For instance, we will denote the displacement-curvature interpolation matrix by B_χ where the curvature at any point within a given element is obtained out of the nodal point displacement vector as $\chi = B_\chi \widehat{u}$ where $\widehat{u}$ may have translational or translational and rotational degrees of freedom.

5.7.4 General Couple Stress Theory Implementation

In the case of the general couple stress theory, the continuum is free of the constraints; as a result, $\vec{u}, \vec{\omega}$ are independent degrees of freedom. In the finite element formulation, this fact implies that elements with C^0 continuity are enough to satisfy displacement compatibility requirements. The nodal point displacement vector for an n-nodded two-dimensional element has the following general form $\widehat{u}_e^T = [u_1 v_1 \omega_1 \ldots\ldots\ldots u_n v_n \omega_n]$ as shown in Fig. 5.16. In order to describe the finite

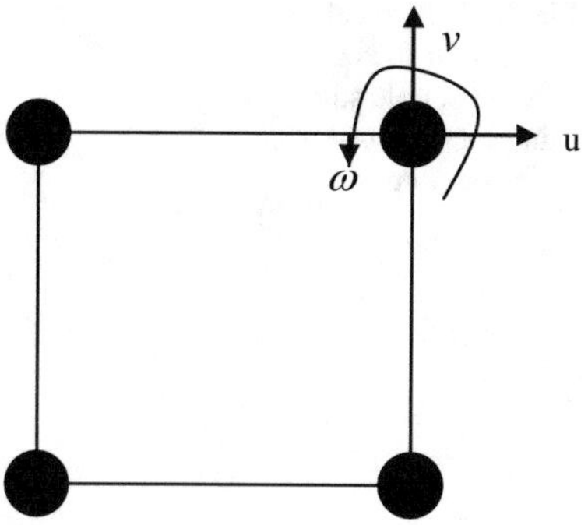

Fig. 5.16 2D finite element for the general couple stress theory

element discretization, we recall that for a general couple stress theory, the principle of virtual work was given by

$$\int_V \sigma_{ij}\varepsilon_{ij}(v_i)dV + \int_V m_{ij}\chi_{ij}(\varphi_i)dV + \int_V \tau_{ij}\alpha_{ij}(v_i,\varphi_i)dV - \int_\Omega \sigma_i^{(n)} v_i d\Omega - \int_\Omega q_i\varphi_i d\Omega = 0 \quad (5.149)$$

Letting $u = Nu_{i,}$, $\varepsilon = B_\varepsilon u_i$, $\chi = B_\chi u_i$, $\alpha = B_\alpha u_i$, and $\sigma_i^{(n)} = N\widehat{t}$ where $\widehat{t}^T = \left[t_1^u t_1^v t_1^\omega \ldots\ldots t_n^u t_n^v t_n^\omega\right]$ and using the constitutive relationships introduced in Eqs. (5.141)–(5.143), we can write the following matrix form after eliminating the virtual variables:

$$\left[\int_V \boldsymbol{B}_\varepsilon^T \boldsymbol{C} \boldsymbol{B}_\varepsilon dV + \int_V \boldsymbol{B}_\chi^T \boldsymbol{D} \boldsymbol{B}_\chi dV + \int_V \boldsymbol{B}_\alpha^T \overline{\boldsymbol{D}} \boldsymbol{B}_\alpha dV\right] u_i = \int_{\partial\Omega} \boldsymbol{N}^T \widehat{\boldsymbol{t}} d\Omega \quad (5.150)$$

or equivalently using stiffness matrices

$$\left[\boldsymbol{K}_\mathbf{e}^\varepsilon + \boldsymbol{K}_\mathbf{e}^\chi + \boldsymbol{K}_\mathbf{e}^\alpha\right]\mathbf{u}_i = \mathbf{F} \quad (5.151)$$

5.7.4.1 Reduced Couple Stress Theory: Pure Displacement Formulation

In the case of the reduced couple stress theory, $\vec{\boldsymbol{u}}, \vec{\boldsymbol{\omega}}$ are related trough the constraint relationship in Eqs. (5.119)–(5.121) and are not independent of each other. In terms of a formulation based on translational degrees of freedom only, this constraint

Fig. 5.17 Typical finite element for the case of reduced couple stress theory using translational degrees of freedom only

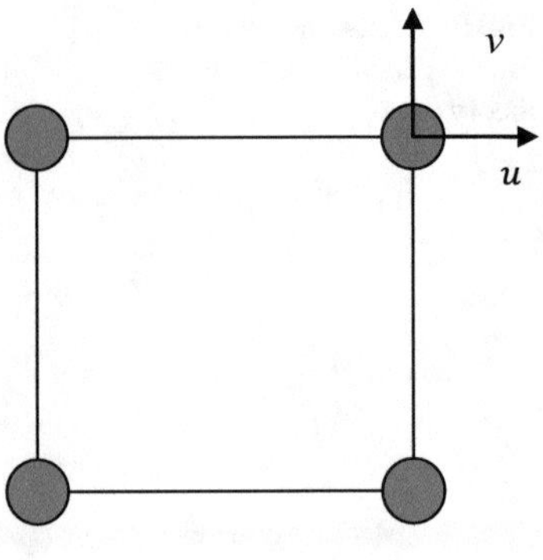

implies that elements with C^1 continuity are needed to satisfy displacement compatibility requirements. Assuming that such an element is readily available, the nodal point displacement vector for an n-nodded element has the following general form $\widehat{u}_e^T = [u_1 v_1 \ldots\ldots\ldots u_n v_n]$ as shown in Fig. 5.17.

In order to describe the finite element discretization for reduced couple stress theory, we start with the variational formulation:

$$\int_V \sigma_{ij}\varepsilon_{ij}(v_i)dV + \int_V m_{ij}\chi_{ij}(v_i)dV - \int_\Omega \sigma_i^{(n)} v_i d\Omega - \int_\Omega q_i\varphi_i(v_i)d\Omega = 0 \tag{5.152}$$

Letting $u = Nu_i$, $\varepsilon_e = B_\varepsilon u_i$, $\chi_e = B_\chi u_i$, and $q = N_q^T \widehat{t}$ and introducing the constitutive relationships introduced in Eqs. (5.141), (5.142), and (5.143), we can write

$$\left[\int_V \boldsymbol{B}_\varepsilon^T \boldsymbol{C} \boldsymbol{B}_\varepsilon dV + \int_V \boldsymbol{B}_\chi^T \boldsymbol{D} \boldsymbol{B}_\chi dV\right] u_i = \int_\Omega \left(\boldsymbol{N}^T + \boldsymbol{N}_q^T\right)\widehat{\boldsymbol{t}} d\Omega \tag{5.153}$$

Or equivalently using stiffness matrices we can write the following equilibrium equation:

$$\left[\boldsymbol{K}_\mathbf{e}^\varepsilon + \boldsymbol{K}_\mathbf{e}^\chi\right]\mathbf{u} = \mathbf{F} \tag{5.154}$$

5.7.4.2 Reduced Couple Stress Theory: Lagrange Multiplier Formulation

In the case of the reduced couple stress theory, $\vec{\boldsymbol{u}}, \vec{\boldsymbol{\omega}}$ are related through the constraint relationship in Eqs. (5.119)–(5.121) and are not independent of each other. Here we

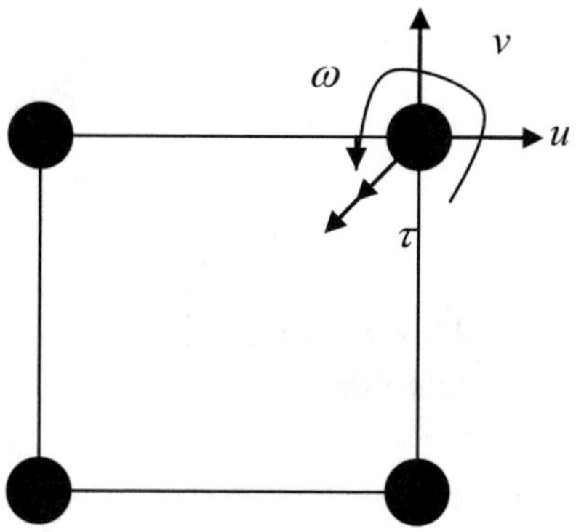

Fig. 5.18 Typical finite element for the case of reduced couple stress theory using Lagrange multipliers

present a formulation based on translational and rotational degrees of freedom with the kinematic constraint enforced via Lagrange multiplier approach. In this case the problem can be treated in terms of two vectors of nodal point parameters. The nodal degrees of freedom vector $\mathbf{u}_i^T = [u_1 v_1 \omega_1 \ldots\ldots\ldots u_n v_n \omega_n]$, and an additional nodal point vector of Lagrange multipliers $\hat{\tau}_e^T = [\tau_1 \ldots\ldots \tau_n]$ as shown in Fig. 5.18.

In order to describe the finite element discretization, we recall that reduced couple stress theory with enforced kinematic constraints via Lagrange multipliers is given by

$$\int_V \sigma_{ij}\varepsilon_{ij}(v_i)dV + \int_V m_{ij}\chi_{ji}(\varphi_i)dV + \int_V \tau_{ij}\alpha_{ij}(v_i,\varphi_i)dV = \int_\Omega t_i v_i d\Omega + \int_\Omega q_i\varphi_i d\Omega \tag{5.155a}$$

$$\int_\Omega \rho_{ij}\alpha_{ij}(u_i,\omega_i)d\Omega = 0 \tag{5.155b}$$

Using $u = N_u u_i$, $\tau_e = N_\tau \hat{\tau}_e$, $\varepsilon = B_\varepsilon u_i$, $\chi = B_\chi u_i$, $\alpha = B_\alpha u_i$, and $\bar{t} = N\hat{\bar{t}}$ where $\hat{\bar{t}}^T = [t_1^u t_1^v t_1^\omega \ldots\ldots t_n^u t_n^v t_n^\omega]$ and after introducing the constitutive relationships, we can write in the following matrix form:

$$\begin{bmatrix} \int_V \boldsymbol{B}_\varepsilon^T \boldsymbol{C}\boldsymbol{B}_\varepsilon dV + \int_V \boldsymbol{B}_\chi^T \boldsymbol{D}\boldsymbol{B}_\chi dV & \int_V \boldsymbol{B}_\alpha^T \boldsymbol{N}_\tau dV \\ \int_V \boldsymbol{N}_\tau^T \boldsymbol{B}_\alpha dV & 0 \end{bmatrix} \begin{bmatrix} \hat{\boldsymbol{u}} \\ \hat{\boldsymbol{\tau}} \end{bmatrix} = \begin{bmatrix} \int_\Omega \boldsymbol{N}^T \bar{t} d\Omega \\ 0 \end{bmatrix} \tag{5.155c}$$

or equivalently using stiffness matrices

$$\begin{bmatrix} \boldsymbol{K}_{uu} & \boldsymbol{K}_{u\tau} \\ \boldsymbol{K}_{u\tau}^{T} & \boldsymbol{0} \end{bmatrix} \begin{bmatrix} \widehat{u} \\ \widehat{\tau} \end{bmatrix} = \begin{bmatrix} \mathbf{F} \\ \mathbf{0} \end{bmatrix} \tag{5.155d}$$

5.8 Cosserat Continuum Implementation in Unified Mechanics Theory

Upto this point all our derivations have been based on Newtonian mechanics. In order to incorporate size effects, Cosserat continuum can be used. This implies the generalization of the stress space to include couple stresses. In Cosserat continuum the size effects are introduced by enhancing the definition of equivalent plastic strain with the addition of an equivalent plastic curvature. This approach leads to a straightforward extension of the classic flow theory (also known as theory of plasticity) which allows the treatment of cyclic loading and monotonic loading. First we will start with the formulation that presents the flow theory equations for the Newtonian mechanics rate-independent case. A key feature is the coupling between the Cauchy stress and couple stress components. This coupling is not present in the initial elastic material but progressively appears with the accumulation of plastic curvatures. In the case of degradation, additional terms appear in the expression for the Helmholtz free energy. On the other hand, consideration of rate-dependent effects assumes the validity of the same phenomenological description for the plastic curvatures as for the plastic strains. This assumption allows for using a single creep law for both components (strain and curvature) and is justified by the kinematic constraint present in the reduced couple stress continuum.

5.8.1 Rate-Independent Material Without Degradation

Recalling the relationship between the symmetric part of the Cauchy stress tensor and the elastic strains and the Couple stresses and the elastic curvatures can be written in rate form as follows:

$$\dot{\sigma}_{ij} = C_{ijkl}\dot{\varepsilon}_{kl}^{\mathrm{el}} \tag{5.156}$$

$$\ell^{-1}\dot{m}_{ij} = D_{ijkl}\ell\dot{\chi}_{kl}^{\mathrm{el}} \tag{5.157}$$

where $D_{ijkl} = \mu\delta_{ik}\delta_{jl}$. The strains and curvatures are decoupled into elastic and inelastic components which results in

$$\dot{\varepsilon}_{ij} = \dot{\varepsilon}_{ij}^{\mathrm{el}} + \dot{\varepsilon}_{ij}^{\mathrm{pl}} \tag{5.158}$$

$$\ell\dot{\chi}_{ij} = \ell\dot{\chi}_{ij}^{\mathrm{el}} + \ell\dot{\chi}_{ij}^{\mathrm{pl}} \tag{5.159}$$

Considering the following definition of generalized deviatoric stress norm $\|\Sigma\|$

$$|\Sigma'| = \left[S_{ij}S_{ij} + \ell^{-1}m_{ij}\ell^{-1}m_{ij}\right]^{1/2} \tag{5.160}$$

where S_{ij} is the deviatoric component of stress tensor σ_{ij} and m_{ij} is deviatoric in nature. (It does not have a spherical component). Similarly, the generalized strain tensor can be written as:

$$|E| = \left[\varepsilon_{ij}\varepsilon_{ij} + \ell\chi_{ij}\ell\chi_{ij}\right]^{1/2} \tag{5.161}$$

Next we introduce a yield surface separating the elastic and inelastic domains per classical plasticity model. In order to account for the Bauschinger effect observed in metals, a back-stress tensor β_{ij} and a couple back-stress tensor $\ell^{-1}\eta_{ij}$ defining the displacement of the yield surface in stress space can be defined. The difference $f_{ij} = S_{ij} - \beta_{ij}$ between the back-stress and the deviatoric component of the symmetric part of the Cauchy stress tensor is the relative stress tensor f_{ij}. In an analogous form for the couple back-stress, there follows that the relative couple back-stress $\ell^{-1}\widehat{C}_{ij}$ is defined by $\ell^{-1}\widehat{C}_{ij} = \ell^{-1}m_{ij} - \ell^{-1}\eta_{ij}$. The generalized relative stress $\|\xi\|$ can be described in terms of the relative stresses and given by

$$|\vec{\xi}| = \left[\, f_{ij}f_{ij} + \ell^{-2}\widehat{C}_{ij}\widehat{C}_{ij}\right]^{1/2} \tag{5.162}$$

We can now define a yield surface to distinguish between elastic domain and inelastic domain. In the classical theory of plasticity, the yield surface is defined in terms of a hardening parameter that can be shown to be proportional to the equivalent plastic strain. In the present couple stress-based strain gradient plasticity theory, this hardening parameter incorporates also the size effects via the equivalent plastic curvatures. The generalized yield surface can therefore be expressed as

$$F\left(\sigma, \ell^{-1}m, \alpha\right) = |\xi| - \sqrt{\frac{2}{3}}K(\alpha) \tag{5.163}$$

where α is the generalized hardening parameter and $K(\alpha)$ represents the radius of the yield surface which increases as material hardens. In order to complete the description of the flow theory representation of the constitutive model, it is necessary to define the flow rules (i.e., evolution equations for the plastic strains and curvatures) and hardening laws (i.e., evolution of the hardening parameter and back-stress components). Here it is assumed that the flow rule obeys associative plasticity.

Considering a more general stress space with normal stresses and couple stresses, the yield surface can be considered as a hypersphere in stress space with normal $\widehat{N}$ to the yield surface F defined as

$$\underline{\widehat{N}} = \frac{\partial F}{\partial \vec{\Sigma}} \equiv [\underline{\widehat{n}}, \underline{\widehat{v}}] \tag{5.164}$$

where $\underline{\widehat{n}} = \frac{\partial F}{\partial \underline{\sigma}} \equiv \frac{f_{ij}}{|\xi_{ij}|}$ and $\underline{\widehat{v}} = \frac{\partial F}{\partial \underline{l^{-1}m}} \equiv \ell^{-1} \frac{\widehat{C}_{ij}}{|\xi_{ij}|}$.

And the flow rules read

$$\dot{\varepsilon}_{ij}^{\mathrm{pl}} = \gamma \frac{\partial F}{\partial \sigma_{ij}} \equiv \gamma \frac{f_{ij}}{|\xi_{ij}|} \equiv \gamma \underline{\widehat{n}} \tag{5.165a}$$

$$\ell \dot{\chi}_{ij}^{\mathrm{pl}} = \gamma \frac{\partial F}{\partial \ell^{-1} m_{ij}} \equiv \gamma \ell^{-1} \frac{\widehat{C}_{ij}}{|\xi_{ij}|} \equiv \gamma \underline{\widehat{v}} \tag{5.165b}$$

γ is the consistency parameter which is defined from the loading/unloading conditions and is related to the evolution of the generalized equivalent plastic strain defined by the hardening law as follows:

$$\dot{\alpha} = \sqrt{\frac{2}{3}} \gamma \tag{5.166}$$

The constitutive model is completed with the evolution equations for the back-stresses:

$$\dot{\beta}_{ij} = \frac{2}{3} H' \gamma \underline{\widehat{n}}_{ij} \tag{5.167a}$$

$$\ell \dot{\eta}_{ij} = \frac{2}{3} H' \gamma \underline{\widehat{v}}_{ij} \tag{5.167b}$$

where H' represents a kinematic hardening modulus which may be a linear or a nonlinear function of the hardening parameter α. For instance, the assumption of a constant kinematic hardening modulus leads to the so-called Prager-Ziegler rule, (Fung and Tong 2001). Using Eqs. (5.165a) and (5.165b) into the generalized strain norm for the plastic quantities yields

$$|\dot{E}^{\mathrm{pl}}| = \left[\dot{\varepsilon}_{ij}^{\mathrm{pl}} \dot{\varepsilon}_{ij}^{\mathrm{pl}} + \ell^2 \dot{\chi}_{ij}^{\mathrm{pl}} \dot{\chi}_{ij}^{\mathrm{pl}} \right]^{1/2} \equiv \frac{\gamma}{|\xi_{ij}|} \left[f_{ij} f_{ij} + \ell^{-2} \widehat{C}_{ij} \widehat{C}_{ij} \right]^{1/2} \tag{5.168}$$

which implies $|\dot{E}^{\mathrm{pl}}| = \gamma$.

Using this result in Eq. (5.166) and integrating yields

$$\alpha(t) = \int_0^t \sqrt{\frac{2}{3}} |\dot{E}^{\text{pl}}(\tau)| d\tau \tag{5.169}$$

This is the generalized version of equivalent plastic strain (trajectory) but with the addition of the gradients of plastic strain. The evolution equations are complemented by the loading/unloading conditions which allow the determination of the consistency parameter. In terms of the yield function, the following loading/unloading condition must be satisfied:

$$\gamma \geq 0 \quad F\left(\sigma, \ell^{-1} m, \alpha\right) \leq 0 \tag{5.170a}$$

$$\gamma \geq 0 \quad \gamma F\left(\sigma, \ell^{-1} m, \alpha\right) = 0 \tag{5.170b}$$

And the consistency condition

$$\gamma \dot{F}\left(\boldsymbol{\sigma}, \ell^{-1} \boldsymbol{m}, \alpha\right) = 0 \tag{5.171}$$

5.8.1.1 Consistency Parameter Determination

Using the definition of the yield function, we have

$$\dot{F} = \frac{\partial F}{\partial \boldsymbol{\sigma}} : \dot{\boldsymbol{\sigma}} + \frac{\partial F}{\partial \boldsymbol{\ell}^{-1} m} : \boldsymbol{\ell}^{-1} \dot{\boldsymbol{m}} + \frac{\partial F}{\partial \boldsymbol{\varepsilon}^{p}} : \dot{\boldsymbol{\varepsilon}}^{\mathbf{pl}} + \frac{\partial F}{\partial \boldsymbol{\ell} \boldsymbol{\chi}^{pl}} : \dot{\boldsymbol{\chi}}^{\mathbf{pl}} \tag{5.172}$$

Using Hooke's law together with the flow rule yields

$$\begin{aligned} \dot{F} &= \left[\underline{\widehat{\boldsymbol{n}}} : \boldsymbol{C} : \dot{\boldsymbol{\varepsilon}} + \underline{\widehat{\boldsymbol{v}}} : \boldsymbol{D} : \boldsymbol{\ell} \dot{\boldsymbol{\chi}}\right] \\ &\quad - \gamma \left\{ \left[\underline{\widehat{\boldsymbol{n}}} : \boldsymbol{C} : \underline{\widehat{\boldsymbol{n}}} + \underline{\widehat{\boldsymbol{v}}} : \boldsymbol{D} : \underline{\widehat{\boldsymbol{v}}}\right] - \left(\frac{\partial F}{\partial \boldsymbol{\varepsilon}^{\mathbf{pl}}} : \underline{\widehat{\boldsymbol{n}}} + \frac{\partial F}{\partial \boldsymbol{\ell} \boldsymbol{\chi}^{\mathbf{pl}}} : \underline{\widehat{\boldsymbol{v}}} \right) \right\} \end{aligned} \tag{5.173}$$

And imposing the consistency condition yields the consistency parameter γ in the following form:

$$\gamma = \frac{\underline{\widehat{\boldsymbol{n}}} : \boldsymbol{C} : \dot{\boldsymbol{\varepsilon}} + \underline{\widehat{\boldsymbol{v}}} : \boldsymbol{D} : \ell \dot{\boldsymbol{\chi}}}{\underline{\widehat{\boldsymbol{n}}} : \boldsymbol{C} : \underline{\widehat{\boldsymbol{n}}} + \underline{\widehat{\boldsymbol{v}}} : \boldsymbol{D} : \underline{\widehat{\boldsymbol{v}}} - \left(\frac{\partial F}{\partial \boldsymbol{\varepsilon}^{\mathbf{pl}}} : \underline{\widehat{\boldsymbol{n}}} + \frac{\partial F}{\partial \ell \boldsymbol{\chi}^{\mathbf{pl}}} : \underline{\widehat{\boldsymbol{v}}} \right)} \tag{5.174}$$

Using

$$\frac{\partial F}{\partial \boldsymbol{\varepsilon}^{\mathbf{pl}}} = -\sqrt{\frac{2}{3}}K'\frac{2}{3}\frac{\boldsymbol{\varepsilon}^{\mathbf{pl}}}{\alpha} \quad \text{and} \quad \frac{\partial F}{\partial \ell \boldsymbol{\chi}^{\mathbf{pl}}} = -\sqrt{\frac{2}{3}}K'\frac{2}{3}\frac{\ell \boldsymbol{\chi}^{\mathbf{pl}}}{\alpha} \tag{5.175}$$

gives

$$\frac{\partial F}{\partial \boldsymbol{\varepsilon}^{\mathbf{pl}}} : \underline{\widehat{\boldsymbol{n}}} = -\sqrt{\frac{2}{3}}K'\frac{2}{3}\frac{\boldsymbol{\varepsilon}^{\mathbf{pl}}}{\alpha} : \frac{\boldsymbol{S}}{|\Sigma|} \quad \text{and} \quad \frac{\partial F}{\partial \ell \boldsymbol{\chi}^{\mathbf{pl}}} : \underline{\widehat{\boldsymbol{\nu}}} = -\sqrt{\frac{2}{3}}K'\frac{2}{3}\frac{\ell \boldsymbol{\chi}^{\mathbf{pl}}}{\alpha} : \frac{\ell^{-1}\boldsymbol{m}}{|\Sigma|} \tag{5.176}$$

Then the term within brackets in the denominator reduces to

$$\frac{\partial F}{\partial \boldsymbol{\varepsilon}^{\mathbf{pl}}} : \underline{\widehat{\boldsymbol{n}}} + \frac{\partial F}{\partial \ell \boldsymbol{\chi}^{\mathbf{pl}}} : \underline{\widehat{\boldsymbol{\nu}}} = -\sqrt{\frac{2}{3}}K'\frac{2}{3}\frac{1}{\alpha|\Sigma|}\left(\boldsymbol{\varepsilon}^{\mathbf{pl}} : \boldsymbol{S} + \ell \boldsymbol{\chi}^{\mathbf{pl}} : \ell^{-1}\boldsymbol{m}\right) = -\frac{2}{3}K' \tag{5.177}$$

after using $\boldsymbol{\varepsilon}^{\mathbf{pl}} : \boldsymbol{S} + \ell \boldsymbol{\chi}^{\mathbf{pl}} : \ell^{-1} m = K\alpha$.

Using the previous result, the consistency parameter γ reduces to

$$\gamma = \frac{\underline{\widehat{\boldsymbol{n}}} : \boldsymbol{C} : \dot{\boldsymbol{\varepsilon}} + \underline{\widehat{\boldsymbol{\nu}}} : \boldsymbol{D} : \ell \dot{\boldsymbol{\chi}}}{\underline{\widehat{\boldsymbol{n}}} : \boldsymbol{C} : \underline{\widehat{\boldsymbol{n}}} + \underline{\widehat{\boldsymbol{\nu}}} : \boldsymbol{D} : \underline{\widehat{\boldsymbol{\nu}}} + \frac{2}{3}K'} \tag{5.178}$$

Using

$$\underline{\widehat{\boldsymbol{n}}} : \boldsymbol{C} = 2\mu \underline{\widehat{\boldsymbol{n}}} \quad \text{and} \quad \underline{\widehat{\boldsymbol{n}}} : \boldsymbol{C} : \underline{\widehat{\boldsymbol{n}}} = 2\mu \frac{\boldsymbol{S} : \boldsymbol{S}}{\|\Sigma\|^2}$$

and

$$\underline{\widehat{\boldsymbol{\nu}}} : \boldsymbol{D} = 2\mu \underline{\widehat{\boldsymbol{\nu}}} \quad \text{and} \quad \underline{\widehat{\boldsymbol{\nu}}} : \boldsymbol{C} : \underline{\widehat{\boldsymbol{\nu}}} = 2\mu \frac{\ell^{-1}\boldsymbol{m} : \ell^{-1}\boldsymbol{m}}{\|\Sigma\|^2}$$

yields

$$\gamma = \frac{1}{|\Sigma|}\frac{S : \dot{\varepsilon} + \ell^{-1} : \ell \dot{\chi}}{\left(1 + \frac{K'}{3\mu}\right)} = \frac{1}{|\Sigma|}\frac{S : \dot{\varepsilon} + \ell^{-1} : \ell \dot{\chi}}{\widehat{K}} \tag{5.179}$$

Using this result into Hooke's law yields

$$\dot{\boldsymbol{\sigma}} = \boldsymbol{C} : \dot{\boldsymbol{\varepsilon}} - \frac{2\mu}{\widehat{K}}(\underline{\widehat{\boldsymbol{n}}} \otimes \underline{\widehat{\boldsymbol{n}}}) : \dot{\boldsymbol{\varepsilon}} - \frac{2\mu}{\widehat{K}}(\underline{\widehat{\boldsymbol{n}}} \otimes \underline{\widehat{\boldsymbol{\nu}}}) : \ell \dot{\boldsymbol{\chi}} \tag{5.180a}$$

$$\ell^{-1}\dot{\boldsymbol{m}} = \boldsymbol{D} : \ell \dot{\boldsymbol{\chi}} - \frac{2\mu}{\widehat{K}}(\underline{\widehat{\boldsymbol{n}}} \otimes \underline{\widehat{\boldsymbol{\nu}}}) : \dot{\boldsymbol{\varepsilon}} - \frac{2\mu}{\widehat{K}}(\underline{\widehat{\boldsymbol{\nu}}} \otimes \underline{\widehat{\boldsymbol{\nu}}}) : \ell \dot{\boldsymbol{\chi}} \tag{5.180b}$$

Using

$$\boldsymbol{C} = \kappa \mathbf{I} \otimes \mathbf{I} + 2\mu \left(\widehat{\prod} - \frac{1}{3} \mathbf{I} \otimes \mathbf{I} \right) \quad \text{and} \quad \boldsymbol{D} = 2\mu \widehat{\prod} \tag{5.181}$$

and after simplifying yields

$$\boldsymbol{M}^{\mathbf{ep}} = \begin{bmatrix} \kappa \mathbf{I} \otimes \mathbf{I} + 2\mu \left(\widehat{\prod} - \frac{1}{3} \mathbf{I} \otimes \mathbf{I} - \frac{\underline{\widehat{\boldsymbol{n}}} \otimes \underline{\widehat{\boldsymbol{n}}}}{\widehat{K}} \right) & -2\mu \frac{\underline{\widehat{\boldsymbol{n}}} \otimes \underline{\widehat{\boldsymbol{v}}}}{\widehat{K}} \\ -2\mu \frac{\underline{\widehat{\boldsymbol{n}}} \otimes \underline{\widehat{\boldsymbol{v}}}}{\widehat{K}} & 2\mu \left(\widehat{\prod} - \frac{\underline{\widehat{\boldsymbol{v}}} \otimes \underline{\boldsymbol{v}}}{\widehat{K}} \right) \end{bmatrix} \tag{5.182}$$

Alternatively it can be written as

$$\boldsymbol{M}^{\mathbf{ep}} = \begin{bmatrix} \boldsymbol{C} - \boldsymbol{C} : \frac{\underline{\widehat{\boldsymbol{n}}} \otimes \boldsymbol{C} : \underline{\widehat{\boldsymbol{n}}}}{\widehat{K}} & -\boldsymbol{C} : \frac{\underline{\widehat{\boldsymbol{n}}} \otimes \boldsymbol{D} : \underline{\widehat{\boldsymbol{v}}}}{\widehat{K}} \\ -\boldsymbol{D} : \frac{\underline{\widehat{\boldsymbol{v}}} \otimes \boldsymbol{C} : \underline{\widehat{\boldsymbol{n}}}}{\widehat{K}} & \boldsymbol{D} - \boldsymbol{D} : \frac{\underline{\widehat{\boldsymbol{v}}} \otimes \boldsymbol{D} : \underline{\widehat{\boldsymbol{v}}}}{\widehat{K}} \end{bmatrix} \tag{5.183}$$

The flow theory just described can be written in the following alternative form, which is useful in the numerical treatment of the problem. Using the generalized vector and matrix notation, we define for the simple case of isotropic hardening:

$$\boldsymbol{\Sigma} = \begin{bmatrix} \boldsymbol{\sigma} \\ \ell^{-1} \boldsymbol{m} \end{bmatrix}, \quad \boldsymbol{E} = \begin{bmatrix} \boldsymbol{\varepsilon} \\ \ell \chi \end{bmatrix}, \quad \boldsymbol{M} = \begin{bmatrix} \boldsymbol{C} & 0 \\ 0 & \boldsymbol{D} \end{bmatrix} \tag{5.184}$$

Then the constitutive model equations can be written as $\boldsymbol{\Sigma} = \boldsymbol{M} : \boldsymbol{E}^{\mathbf{el}}$. The yield surface and flow rule follow

$$F\left(\sigma, \ell^{-1} m, \alpha\right) = |\Sigma'| - \sqrt{\frac{2}{3}} K(\alpha) \tag{5.185}$$

$$\dot{\boldsymbol{E}}^{\mathbf{pl}} = \gamma \underline{\widehat{\boldsymbol{N}}} \tag{5.186}$$

then

$$\dot{F} = \frac{\partial F}{\partial \boldsymbol{\Sigma}} \cdot \dot{\boldsymbol{\Sigma}} + \frac{\partial F}{\partial \boldsymbol{E}^{\mathbf{pl}}} \cdot \dot{\boldsymbol{E}}^{\mathbf{pl}} \tag{5.187}$$

Using Hooke's law and the flow rule results in

$$\dot{F} = \underline{\widehat{\boldsymbol{N}}} : \boldsymbol{M} : \dot{\boldsymbol{E}} - \gamma \underline{\boldsymbol{N}} : \boldsymbol{M} : \underline{\widehat{\boldsymbol{N}}} + \frac{\partial F}{\partial \boldsymbol{E}^{\mathbf{pl}}} \cdot \left(\gamma \underline{\widehat{\boldsymbol{N}}} \right) \tag{5.188}$$

Utilizing the consistency conditions $\gamma \dot{F} = 0$ gives

$$\gamma = \frac{\underline{\boldsymbol{N}} : \boldsymbol{M} : \dot{\boldsymbol{E}}^{\mathbf{pl}}}{\underline{\widehat{\boldsymbol{N}}} : \boldsymbol{M} : \underline{\widehat{\boldsymbol{N}}} - \frac{\partial F}{\partial \boldsymbol{E}^{\mathbf{pl}}} \cdot \underline{\widehat{\boldsymbol{N}}}} \quad \text{or after using} \quad \frac{\partial F}{\partial \boldsymbol{E}^{\mathbf{pl}}} \cdot \underline{\widehat{\boldsymbol{N}}} = -\frac{2}{3}K' \tag{5.189}$$

we have

$$\gamma = \frac{\underline{\widehat{\boldsymbol{N}}} : \boldsymbol{M} : \dot{\boldsymbol{E}}^{\mathbf{pl}}}{\underline{\widehat{\boldsymbol{N}}} : \boldsymbol{M} : \underline{\widehat{\boldsymbol{N}}} + \frac{2}{3}K'} \tag{5.190}$$

Using

$$\underline{\widehat{\boldsymbol{N}}} : \boldsymbol{M} = 2\mu \underline{\widehat{\boldsymbol{N}}}^{\mathbf{T}}, \quad \underline{\widehat{\boldsymbol{N}}} : \boldsymbol{M} : \underline{\widehat{\boldsymbol{N}}} = 2\mu, \quad \underline{\widehat{\boldsymbol{N}}} : \boldsymbol{M} : \boldsymbol{E} = 2\mu \underline{\widehat{\boldsymbol{N}}} \cdot \boldsymbol{E}$$

yields

$$\gamma = \frac{\underline{\widehat{\boldsymbol{N}}} \cdot \dot{E}}{\widehat{K}} \tag{5.191}$$

Using this result in the generalized Hooke's law yields

$$\dot{\boldsymbol{\Sigma}} = \left[\boldsymbol{M} - \frac{2\mu}{\widehat{K}}\left(\underline{\widehat{\boldsymbol{N}}} \otimes \underline{\widehat{\boldsymbol{N}}}\right)\right] : \dot{\boldsymbol{E}} \tag{5.192}$$

The constitutive tensor given by Eq. (5.192) is equivalent to Eq. (5.183) where the coupling between strains and curvatures becomes evident in the off-diagonal terms in the generalized constitutive tensor.

5.8.2 Rate-Dependent Material Without Degradation

In classical continuum mechanics, in the case of a rate-dependent material, the conditions established by Eqs. (5.170a), (5.170b) and (5.171) are replaced by a constitutive equation of the form:

$$\gamma = \frac{\langle \varphi(F) \rangle}{\eta} \tag{5.193}$$

where η represents viscosity material parameter. In the case of a rate-independent material, F satisfies the conditions given by Eqs. (5.170a), (5.170b) and (5.171), and additionally stress states such $F(\boldsymbol{\sigma}, \ell^{-1}\boldsymbol{m}, \alpha) > 0$ are ruled out. In other words, the state of stress cannot be outside the yield surface. In the case of a rate-dependent material, on the other hand, the intensity of the viscoplastic flow is proportional to the distance of the state of stress to the yield surface defined by $F(\sigma, \ell^{-1}m, \alpha) = 0$.

Therefore, the yield condition can be written for a rate-dependent material in the following form:

$$F = \Theta(\gamma\eta) \tag{5.194}$$

whereas $\Theta(\gamma\eta) = \varphi^{-1}(\gamma\eta)$.

5.8.3 Thermodynamic State Index Coupling

In the formulation that is presented below, for the sake of simplicity in presentation, we will ignore derivatives with respect to entropy. In Cosserat continuum TSI is introduced into the formulation using the same approach as in the classical continuum. Accordingly, Hooke's law can be written as

$$\dot{\boldsymbol{\sigma}}_{ij} = (1 - \Phi)\boldsymbol{C}_{ijkl}\dot{\boldsymbol{\varepsilon}}^{\mathbf{e}}_{kl} \tag{5.195a}$$

$$\ell^{-1}\dot{\boldsymbol{m}}_{ij} = (1 - \Phi)\boldsymbol{D}_{ijkl}\boldsymbol{\ell}\dot{\boldsymbol{\chi}}^{\mathbf{e}}_{kl} \tag{5.195b}$$

And the yield function now becomes

$$\boldsymbol{F}\left(\boldsymbol{\sigma}, \boldsymbol{\ell}^{-1}\boldsymbol{m}, \alpha\right) = |\boldsymbol{\xi}^{\boldsymbol{\Phi}}| - \sqrt{\frac{\mathbf{2}}{\mathbf{3}}}K(\alpha) \tag{5.196}$$

The rate-dependent constitutive model equations are summarized in Table 5.5. The constitutive model can also be written in the following equivalent alternative form which is convenient for the numerical implementation of the algorithm using a return mapping scheme. Using the generalized stress and strain definitions, it follows that

Table 5.5 UMT rate-dependent-strain gradient formulation 1

Hooke's Law
$\dot{\boldsymbol{\sigma}}_{ij} = (\mathbf{1} - \boldsymbol{\Phi})\boldsymbol{C}_{ijkl}\dot{\boldsymbol{\varepsilon}}^{\mathbf{el}}_{kl}$
$\boldsymbol{\ell}^{-1}\dot{\boldsymbol{m}}_{ij} = (\mathbf{1} - \boldsymbol{\Phi})\boldsymbol{D}_{ijkl}\boldsymbol{\ell}\dot{\boldsymbol{\chi}}^{\mathbf{el}}_{kl}$
Yield function
$\boldsymbol{F}\left(\boldsymbol{\sigma}, \boldsymbol{\ell}^{-1}\boldsymbol{m}, \boldsymbol{\alpha}\right) = \lvert\boldsymbol{\xi}^{\boldsymbol{\Phi}}\rvert - \sqrt{\frac{2}{3}}\boldsymbol{K}(\boldsymbol{\alpha})$
Flow rule
$\dot{\boldsymbol{\varepsilon}}^{\mathbf{pl}}_{ij} = \gamma\frac{\partial F}{\partial \sigma_{ij}} \equiv \gamma\frac{f_{ij}}{\lvert\xi_{ij}\rvert} \equiv \gamma\underline{\widehat{n}}$ and $\ell\dot{\chi}^{\mathrm{pl}}_{ij} = \gamma\frac{\partial F}{\partial \ell^{-1}m_{ij}} \equiv \gamma\ell^{-1}\frac{C_{ij}}{\lvert\xi_{ij}\rvert} \equiv \gamma\underline{\widehat{\nu}}$
Hardening laws
$\dot{\alpha} = \sqrt{\frac{2}{3}}\gamma$
$\dot{\beta}^{\Phi}_{ij} = (1 - \Phi)\frac{2}{3}H'\gamma\underline{\widehat{n}}_{ij}$ and $\ell\dot{\eta}^{\Phi}_{ij} = (1 - \Phi)\frac{2}{3}H'\gamma\underline{\widehat{\nu}}_{ij}$.
Consistency parameter
$\gamma = \frac{\langle\varphi(F)\rangle}{\eta}$

$$\dot{\boldsymbol{\Sigma}} = (\mathbf{1} - \boldsymbol{\Phi})\boldsymbol{M}^{\mathrm{e}} : \left(\dot{\boldsymbol{E}} - \dot{\boldsymbol{E}}^{\mathrm{vp}} - \dot{\boldsymbol{E}}^{\theta}\right) \tag{5.197}$$

where $\dot{\boldsymbol{E}}^{\mathrm{vp}}$ is the viscoplastic strain rate and $\dot{\boldsymbol{E}}^{\theta}$ is the thermal strain rate. Then, yield function is written as

$$F(\boldsymbol{\Sigma}, \boldsymbol{\alpha}) = \sqrt{\boldsymbol{\xi}^{\boldsymbol{\Phi}^T} \boldsymbol{P} \boldsymbol{\xi}^{\boldsymbol{\Phi}}} - \sqrt{\frac{2}{3}} K(\alpha) \tag{5.198}$$

where $\boldsymbol{\xi}^{\boldsymbol{D\Phi}} = \boldsymbol{\Sigma}^{\boldsymbol{\Phi}} - \boldsymbol{X}^{\boldsymbol{\Phi}}$ is the generalized relative stress previously introduced with the generalized back-stress $\boldsymbol{X}^{\boldsymbol{\Phi}} \equiv \left[\boldsymbol{\beta}_{ij}^{\boldsymbol{\Phi}}, \boldsymbol{\ell}^{-1}\boldsymbol{\eta}_{ij}^{\boldsymbol{\Phi}}\right]$.where

$$\dot{\beta}_{ij}^{\Phi} = (1 - \Phi)\frac{2}{3} H' \gamma \underline{\widehat{\boldsymbol{n}}}_{ij} \tag{5.199}$$

$$\ell \dot{\eta}_{ij}^{\Phi} = (1 - \Phi)\frac{2}{3} H' \gamma \underline{\widehat{\boldsymbol{\nu}}}_{ij} \tag{5.200}$$

In Eq. (5.198) $\boldsymbol{P}$ is a constant matrix; hence $\boldsymbol{P}\boldsymbol{\xi}^{\boldsymbol{\Phi}}$ gives the deviatoric component of $\boldsymbol{\xi}^{\boldsymbol{\Phi}}$, and $\sqrt{\frac{2}{3}\boldsymbol{\xi}^{\boldsymbol{\Phi}^T}\boldsymbol{P}\boldsymbol{\xi}^{\boldsymbol{\Phi}}}$ is the von Mises equivalent stress for the generalized relative stress vector. In the same equation, $\boldsymbol{K}(\boldsymbol{\alpha})$ represents the size of the yield surface. The direction of plastic flow and the hardening laws are given for the associative plasticity case as

$$\dot{\boldsymbol{E}}^{\mathrm{vp}} = \gamma \frac{\boldsymbol{P}\boldsymbol{\xi}}{(1 - \Phi)} \tag{5.201}$$

$$\dot{\alpha} = \sqrt{\frac{2}{3}} \gamma (1 - \Phi) \sqrt{\boldsymbol{\xi}^{\boldsymbol{\Phi}^T} \boldsymbol{P} \boldsymbol{\xi}^{\boldsymbol{\Phi}}} \tag{5.202}$$

$$\dot{\boldsymbol{X}}^{\boldsymbol{\Phi}} = \gamma \frac{2}{3} H'(\alpha)(1 - \Phi)\underline{\widehat{N}} \tag{5.203}$$

Note that in Eq. (5.203) the hardening modulus $\boldsymbol{H}'$ can be a nonlinear function of the hardening parameter $\boldsymbol{\alpha}$. In this formulation Eq. (5.203) assumes that the couple back-stress evolves in the same manner as the symmetric stress tensor back-stress. However, this is not essential for the formulation. They can be different. The couple back-stress is needed to allow for the uniform movement of the yield surface in the enhanced stress space.

Hooke's law

$$\dot{\boldsymbol{\sigma}}_{ij} = (\mathbf{1} - \boldsymbol{\Phi})\boldsymbol{C}_{ijkl}\dot{\boldsymbol{\varepsilon}}_{kl}^{\mathrm{el}}$$

$$\boldsymbol{\ell}^{-1}\dot{\boldsymbol{m}}_{ij} = (\mathbf{1} - \boldsymbol{\Phi})\boldsymbol{D}_{ijkl}\boldsymbol{\ell}\dot{\boldsymbol{\chi}}_{kl}^{\mathrm{el}}$$

Yield function

$$\boldsymbol{F}\left(\boldsymbol{\sigma}, \boldsymbol{\ell}^{-1}\boldsymbol{m}, \boldsymbol{\alpha}\right) = |\boldsymbol{\xi}^{\boldsymbol{\Phi}}| - \sqrt{\frac{\boldsymbol{2}}{\boldsymbol{3}}}\boldsymbol{K}(\boldsymbol{\alpha})$$

Flow rule

$$\dot{\boldsymbol{\varepsilon}}_{ij}^{\mathbf{pl}} = \gamma \frac{\partial F}{\partial \sigma_{ij}} \equiv \gamma \frac{f_{ij}}{|\boldsymbol{\xi}_{ij}|} \equiv \gamma \underline{\widehat{n}} \quad \text{and} \quad \ell \dot{\chi}_{ij}^{\mathrm{pl}} = \gamma \frac{\partial F}{\partial \ell^{-1} m_{ij}} \equiv \gamma \ell^{-1} \frac{C_{ij}}{|\xi_{ij}|} \equiv \gamma \underline{\widehat{\nu}}$$

Hardening laws

$$\dot{\alpha} = \sqrt{\frac{2}{3}}\gamma$$

$$\dot{\beta}_{ij}^{\Phi} = (1 - \Phi)\frac{2}{3}H'\gamma \underline{\widehat{n}}_{ij} \quad \text{and} \quad \ell \dot{\eta}_{ij}^{\Phi} = (1 - \Phi)\frac{2}{3}H'\gamma \underline{\widehat{\nu}}_{ij}$$

Consistency parameter

$$\gamma = \frac{\langle \varphi(F) \rangle}{\eta}$$

5.8.3.1 Viscoplastic Creep Law

The evolution of the generalized viscoplastic strain is specified by the flow rule which in the enhanced theory specifies creep laws for the strains and the strain gradients. The following viscoplastic strain rate function is one of the generalized models. It accounts for multiple creep mechanisms in the polycrystalline materials as well as coarsening of the crystals:

$$\dot{\varepsilon}_{ij}^{\mathrm{vp}} = \frac{AD_0Eb}{k\theta}\left(\frac{\langle F \rangle}{E}\right)^n \left(\frac{b}{d}\right)^p \mathrm{e}^{-Q/R\theta} \frac{\partial F}{\partial \sigma_{ij}} \tag{5.204}$$

$$\ell \dot{\chi}_{ij}^{\mathrm{vp}} = \frac{AD_0Eb}{k\theta}\left(\frac{\langle F \rangle}{E}\right)^n \left(\frac{b}{d}\right)^p \mathrm{e}^{-Q/R\theta} \frac{\partial F}{\partial \ell^{-1} m_{ij}} \tag{5.205}$$

$$\dot{E}^{\mathrm{vp}} = \frac{AD_0Eb}{k\theta}\left(\frac{\langle F \rangle}{E}\right)^n \left(\frac{b}{d}\right)^p \mathrm{e}^{-Q/R\theta} \frac{\partial F}{\partial \Sigma} \tag{5.206}$$

The fluidity parameter η can be summarized as

$$\eta = \frac{k\theta}{AD_0E^{n-1}b}\left(\frac{d}{b}\right)^p e^{Q/R\theta} \tag{5.207}$$

5.8.4 Entropy Generation Rate in Cosserat Continuum

In the couple stress theory, the internal energy equation, which is an expression of the first law of thermodynamics, can be given by

$$\rho\frac{de}{dt} = \sigma_{ij}^{\mathrm{S}}D_{ij} + m_{ji}\,\chi_{ij} + \rho r - q_{i,i} \tag{5.208}$$

where σ_{ij}^{S} is the symmetric part of Cauchy stress tensor, m_{ji} is the couple stress tensor, and χ_{ij} are the corresponding curvatures. The rate of change of the Helmholtz free energy φ is written in terms of the symmetric part of the Cauchy stress tensor; thus

$$\rho\frac{d\varphi}{dt} = \sigma_{ij}^{\mathrm{S}}D_{ij}^{\mathrm{el}} \tag{5.209}$$

Combining Eqs. (5.175) and (5.176), the difference between the changes in the internal energy and the Helmholtz free energy with respect to a reference state in the presence of couple stresses is obtained:

$$\Delta e - \Delta\varphi = \frac{1}{\rho}\int_{t_1}^{t_2} \sigma_{ij}^{\mathrm{S}}D_{ij}^{\mathrm{pl}}\,dt + \frac{1}{\rho}\int_{t_1}^{t_2} m_{ji}\dot{\chi}_{ij}^{\mathrm{pl}}\,dt + \int_{t_1}^{t_2} r\,dt - \frac{1}{\rho}\int_{t_1}^{t_2} q_{i,i}dt \tag{5.210}$$

5.8.5 Integration Algorithms

Within the finite element method considering nonlinear material behavior, the solution is developed by a series of small increments with the solution at every increment found by a Newton method. During every increment the problem may be regarded as strain driven in the following sense. At the beginning of the time step, the total and viscoplastic strain fields and the internal state variables are considered to be known. Assuming that displacement increment is known, the basic problem is to update the field variables to their new values at the end of the time step in a manner consistent with the constitutive model. In what follows the integration, algorithms are presented following the same hierarchical framework that was used to introduce the model. First, the case of an undamaged rate-independent model using both a

Table 5.6 UMT rate-dependent model-strain gradient formulation 2

Hooke's Law	
$\dot{\sigma}_{ij} = (1-\Phi)C_{ijkl}\dot{\varepsilon}^{e}_{kl}$	$\ell^{-1}\dot{m}_{ij} = (1-\Phi)D_{ijkl}\ell\dot{\chi}^{e}_{kl}$
Yield function	
$F(\Sigma,\alpha) = \sqrt{\xi^{\Phi^T}P\xi^{\Phi}} - \sqrt{\frac{2}{3}}K(\alpha)$	
Flow rule	
$\dot{E}^{vp} = \gamma\frac{P\xi}{(1-\Phi)}$	
Hardening laws	
$\dot{\alpha} = \sqrt{\frac{2}{3}}\gamma(1-\Phi)\sqrt{\xi^{\Phi^T}P\xi^{\Phi}}$	$\dot{X}^{\Phi} = \gamma\frac{2}{3}H'(\alpha)(1-\Phi)\underline{\widehat{N}}$
Consistency parameter	
$\gamma = \frac{\langle\varphi(F)\rangle}{\eta}$	

radial return and return mapping scheme is presented. The next and final step treats the full damaged rate-dependent case and presents the return mapping algorithm which is based on the formulation presented in Table 5.6.

5.8.5.1 Radial Return Algorithm: Rate-Independent Model with Linear Isotropic Hardening

Using Eqs. (5.148a) and (5.148b) and backward Euler finite difference integration scheme, we can write

$$\begin{aligned} \boldsymbol{E}^{\mathrm{pl}}_{n+1} &= \boldsymbol{E}^{\mathrm{pl}}_{n} + \Delta\gamma\underline{\widehat{\boldsymbol{N}}}_{n+1} \\ \alpha_{n+1} &= \alpha_n + \sqrt{\frac{2}{3}}\Delta\gamma \end{aligned} \tag{5.211}$$

Now consider the following trial state obtained after freezing plastic flow

$$\begin{aligned} \boldsymbol{\Sigma}^{\mathrm{tr}}_{n+1} &= \boldsymbol{\Sigma}_n + 2\mu\Delta\boldsymbol{E}_{n+1} \\ \boldsymbol{\Sigma}_{n+1} &= \boldsymbol{\Sigma}^{\mathrm{tr}}_{n+1} - 2\mu\Delta\gamma\underline{\widehat{\boldsymbol{N}}}_{n+1} \end{aligned} \tag{5.212}$$

from which it is concluded that

$$|\boldsymbol{\Sigma}_{n+1}| + 2\mu\Delta\gamma = |\boldsymbol{\Sigma}^{\mathrm{tr}}_{n+1}| \tag{5.213a}$$

and

$$\underline{\widehat{\boldsymbol{N}}}_{n+1} = \frac{\boldsymbol{\Sigma}^{\mathrm{tr}}_{n+1}}{|\boldsymbol{\Sigma}^{\mathrm{tr}}_{n+1}|} \tag{5.213b}$$

Assuming linear isotropic hardening yields

$$|\boldsymbol{\Sigma}^{\mathrm{tr}}_{n+1}| - \sqrt{\frac{2}{3}}K(\alpha_n) - \frac{2}{3}K'\Delta\gamma - 2\mu\Delta\gamma = 0 \tag{5.214}$$

which is solved for the consistency parameter $\Delta\gamma$.

5.8.5.2 Linearization of Incremental Solution Process

Consider

$$\dot{\boldsymbol{\Sigma}}_{n+1} = \boldsymbol{M} : \dot{E}_{n+1} - 2\mu\Delta\gamma\underline{\widehat{N}}_{n+1} \tag{5.215}$$

Then it follows that

$$\frac{d\boldsymbol{\Sigma}_{n+1}}{d\boldsymbol{E}_{n+1}} = \boldsymbol{M} - 2\mu\left[\underline{\widehat{\boldsymbol{N}}}_{n+1} \otimes \frac{d\Delta\gamma}{d\boldsymbol{E}_{n+1}} + \Delta\gamma\frac{d\underline{\widehat{N}}_{n+1}}{d\boldsymbol{E}_{n+1}}\right] \tag{5.216}$$

using

$$\frac{d\Delta\gamma}{d\boldsymbol{E}_{n+1}} = \frac{1}{\widehat{K}}\underline{\widehat{\boldsymbol{N}}}_{n+1} \tag{5.217}$$

and

$$\frac{d\underline{\widehat{N}}_{n+1}}{d\boldsymbol{E}_{n+1}} = \frac{2\mu}{|\boldsymbol{\Sigma}^{\mathrm{tr}}_{n+1}|}\boldsymbol{H}_{n+1} \tag{5.218}$$

where

$$\boldsymbol{H}_{n+1} = \begin{bmatrix} \widehat{\amalg} - \frac{1}{3}\mathbf{I}\otimes\mathbf{I} - \underline{\widehat{\boldsymbol{n}}}_{n+1} \otimes \underline{\widehat{n}}_{n+1} & 0 \\ 0 & \widehat{\amalg} - \underline{\widehat{\boldsymbol{\nu}}}_{n+1} \otimes \underline{\widehat{\boldsymbol{\nu}}}_{n+1} \end{bmatrix} \tag{5.219}$$

thus

$$\frac{d\boldsymbol{\Sigma}_{n+1}}{d\boldsymbol{E}_{n+1}} = \boldsymbol{M} - 2\mu\left[\frac{\underline{\widehat{N}}_{n+1} \otimes \underline{\widehat{\boldsymbol{N}}}_{n+1}}{\widehat{K}} + \frac{\Delta\gamma 2\mu}{|\boldsymbol{\Sigma}^{\mathrm{tr}}_{n+1}|}\boldsymbol{H}_{n+1}\right] \tag{5.220}$$

Expanding Eq. (5.220) yields

$$M_{n+1}^{\text{ep}} = \begin{bmatrix} \kappa \mathbf{I} \otimes \mathbf{I} + 2\mu\delta_{n+1}\left(\widehat{\coprod} - \frac{1}{3}\mathbf{I} \otimes \mathbf{I}\right) - 2\mu\bar{\theta}_{n+1}\widehat{\underline{\boldsymbol{n}}}_{n+1} \otimes \widehat{\underline{\boldsymbol{n}}}_{n+1} & -\frac{2\mu}{\widehat{K}}\widehat{\underline{\boldsymbol{n}}}_{n+1} \otimes \widehat{\underline{\boldsymbol{v}}}_{n+1} \\ -\frac{2\mu}{\widehat{K}}\widehat{\underline{\boldsymbol{n}}}_{n+1} \otimes \widehat{\underline{\boldsymbol{v}}}_{n+1} & 2\mu\delta_{n+1}\widehat{\coprod} - 2\mu\bar{\theta}_{n+1}\underline{\boldsymbol{v}}_{n+1} \otimes \widehat{\underline{\boldsymbol{v}}}_{n+1} \end{bmatrix} \tag{5.221}$$

When $\Delta t \to 0$, Eq. (5.221) approaches to (5.183). The coupling between the curvature and strain in the tangent stiffness matrix is apparent.

5.8.5.3 Return Mapping Algorithm: Rate-Dependent Model-Combined Isotropic/Kinematic Hardening with Degradation

The constitutive model described in Table 5.6 is integrated using a return mapping algorithm as presented in Simo and Hughes (1997). A backward Euler integration scheme yields the following set of algorithmic equations:

$$\boldsymbol{E}_{n+1} = \boldsymbol{E}_n + \boldsymbol{\Delta E}_{n+1} \tag{5.222}$$

$$\alpha_{n+1} = \alpha_n + \Delta\gamma(1 - \Phi)\sqrt{\frac{2}{3}\boldsymbol{\xi}_{n+1}^T \boldsymbol{P}\boldsymbol{\xi}_{n+1}} \tag{5.223}$$

$$\boldsymbol{X}_{n+1} = \boldsymbol{X}_n + \Delta\gamma\frac{2}{3}\boldsymbol{H}'(1 - \Phi)\boldsymbol{\xi}_{n+1} \tag{5.224}$$

The standard operator split technique defines the following trial state:

$$\boldsymbol{\Sigma}_{n+1}^{\text{tr}} = \boldsymbol{\Sigma}_n + (1 - \Phi)\boldsymbol{M}\boldsymbol{\Delta E}_{n+1} \tag{5.225}$$

Using Eq. (5.225) and Hooke's law given by

$$\dot{\sigma}_{ij} = C_{ijkl}\,\dot{\epsilon}_{kl}^{\text{el}} \quad \text{and} \quad l^{-1}\dot{m}_{ij} = D_{ijkl} l \dot{\chi}_{kl}^{\text{el}}$$

we can write the following relations:

$$\boldsymbol{\Sigma}_{n+1} = \boldsymbol{\Sigma}_{n+1}^{\text{tr}} - \boldsymbol{M}\Delta\gamma\boldsymbol{P}\boldsymbol{\xi}_{n+1} \tag{5.226}$$

$$\boldsymbol{\xi}_{n+1}^{\text{tr}} = \boldsymbol{\Sigma}_{n+1}^{\text{tr}} - \boldsymbol{X}_n \tag{5.227}$$

From Eqs. (5.226) and (5.227), an updated relative stress can be obtained in terms of the algorithmic consistency parameter $\boldsymbol{\Delta\gamma}$:

$$\boldsymbol{\xi}_{n+1} = \Xi(\Delta\gamma)\frac{1}{1+\frac{2}{3}H'\Delta\gamma(1-\Phi)}\boldsymbol{M}^{-1}\boldsymbol{\xi}^{\mathbf{tr}}_{n+1} \tag{5.228}$$

where

$$\Xi(\Delta\gamma) = \left[\boldsymbol{M}^{-1} + \frac{\Delta\gamma P}{1+\frac{2}{3}H'(1-\Phi)\Delta\gamma}\right]^{-1} \tag{5.229}$$

To complete the above algorithm, it is still necessary to compute the consistency parameter $\boldsymbol{\Delta\gamma}$, which can be obtained from the yield condition and the corresponding constitutive model. Thus

$$F(\Delta\gamma) - \Theta\left(\frac{\eta\Delta\gamma}{\Delta t}\right) \equiv \widehat{f}_{n+1} - \sqrt{\frac{2}{3}}K(\alpha_{n+1}) - \Theta\left(\frac{\eta\Delta\gamma}{\Delta t}\right) = 0 \tag{5.230}$$

where $\widehat{\boldsymbol{f}}_{n+1} = \left[\boldsymbol{\xi}^T_{n+1}\boldsymbol{P}\boldsymbol{\xi}_{n+1}\right]^{1/2}$ and $\boldsymbol{\Theta}\left(\frac{\eta\Delta\gamma}{\Delta t}\right) = \varphi^{-1}\left(\frac{\eta\Delta\gamma}{\Delta t}\right)$ as defined in the previous section.

Equation (5.230) is a scalar nonlinear equation in the consistency parameter $\boldsymbol{\Delta\gamma}$, which can be solved by a local Newton iteration. In the Newton-Raphson iteration scheme, the elastoplastic tangent modulus consistent with the integration scheme is needed in order to preserve the convergence properties of the Newton algorithm. Linearizing the above set of algorithmic equations yields

$$\frac{d\boldsymbol{\Sigma}_{n+1}}{d\boldsymbol{E}_{n+1}} = (\mathbf{1}-\boldsymbol{\Phi})\left(\Xi(\boldsymbol{\Delta\gamma}) - \frac{1}{\left(1+\widetilde{\beta}\right)}\mathrm{N}\otimes\mathrm{N}\right) \tag{5.231}$$

where

$$\mathrm{N} = \frac{\Xi(\boldsymbol{\Delta\gamma})\boldsymbol{P}\boldsymbol{\xi}_{n+1}}{\sqrt{\boldsymbol{\xi}^T_{n+1}\boldsymbol{P}\Xi(\boldsymbol{\Delta\gamma})\boldsymbol{P}\boldsymbol{\xi}_{n+1}}} \tag{5.232}$$

$$\widetilde{\boldsymbol{\beta}} = \left[(\mathbf{1}-\boldsymbol{\Phi})\frac{\mathbf{2}\boldsymbol{\theta}_1}{\mathbf{3}\boldsymbol{\theta}_2}\widehat{\boldsymbol{f}}^{\,2}_{n+1}(\boldsymbol{K}'\boldsymbol{\theta}_1 + \boldsymbol{H}'\boldsymbol{\theta}_2) + \frac{\boldsymbol{\theta}_1^2}{\boldsymbol{\theta}_2}\frac{\boldsymbol{d\Theta}}{\boldsymbol{d\Delta\gamma}}\widehat{\boldsymbol{f}}_{n+1}\right]\frac{\mathbf{1}}{\boldsymbol{\xi}^{\mathbf{T}}_{n+1}\boldsymbol{P}\Xi(\boldsymbol{\Delta\gamma})\boldsymbol{P}\boldsymbol{\xi}_{n+1}} \tag{5.233}$$

$$\boldsymbol{\theta}_1 = \mathbf{1} + \frac{\mathbf{2}}{\mathbf{3}}\boldsymbol{H}'(\mathbf{1}-\boldsymbol{\Phi})\boldsymbol{\Delta\gamma}$$

$$\boldsymbol{\theta}_2 = \mathbf{1} - \frac{\mathbf{2}}{\mathbf{3}}\boldsymbol{K}'(\mathbf{1}-\boldsymbol{\Phi})\boldsymbol{\Delta\gamma}$$

5.8.5.4 Consistency Equation Solution

Recalling Eq. (5.198) and updating the formula given by Eq. (5.228) yields

$$F(\Delta\gamma) = \widehat{f}_{n+1} - \sqrt{\frac{2}{3}}K(\alpha_{n+1}) = 0 \tag{5.234}$$

$$\boldsymbol{\xi}_{n+1} = \Xi(\Delta\gamma)\frac{1}{1+\frac{2}{3}H'\Delta\gamma(1-\Phi)}\boldsymbol{M}^{-1}\boldsymbol{\xi}_{n+1}^{\mathbf{tr}} \tag{5.235}$$

In a Newton-Raphson iteration used to find the solution $\boldsymbol{x}$ of a nonlinear equation of the general form $\boldsymbol{F}(\boldsymbol{x}) = \mathbf{0}$, the ith correction to the solution is computed as $\boldsymbol{\Delta x}^i = \boldsymbol{\Delta x}^{i-1} - \frac{F(\boldsymbol{\Delta x}^{i-1})}{J(\boldsymbol{\Delta x}^{i-1})}$ where $\boldsymbol{J}(\boldsymbol{\Delta x}^{i-1}) = \frac{\partial \boldsymbol{F}}{\partial \boldsymbol{\Delta x}^{i-1}}$ represents the "Jacobian" of $\boldsymbol{F}$ $(\boldsymbol{x})$.

Solution of Eq. (5.230) with a local Newton iteration requires the "Jacobian" $\boldsymbol{J}$ $(\boldsymbol{\Delta\gamma})$ to be defined. In Newton-Raphson iteration, the Jacobian represents the derivative of the function; hence

$$J(\Delta\gamma) = \frac{dF}{d\Delta\gamma} - \frac{d\Theta}{d\Delta\gamma} \equiv \frac{d\widehat{f}_{n+1}}{d\Delta\gamma} - \sqrt{\frac{2}{3}}\frac{dK}{d\Delta\gamma} - \frac{d\Theta}{d\Delta\gamma} \tag{5.236}$$

Equation (5.235) is written as

$$\boldsymbol{\xi}_{n+1} = \left[I + \frac{2}{3}H'\Delta\gamma(1-\Phi)I + \Delta\gamma PM\right]^{-1}\boldsymbol{\xi}_{n+1}^{\mathbf{tr}} \tag{5.237}$$

To find $\frac{d\widehat{f}_{n+1}}{d\Delta\gamma}$ the symmetric matrices $\boldsymbol{P}$ and $\boldsymbol{M}$ are diagonalized. Since both matrices share the same characteristic space, the following relationships can be written

$$\boldsymbol{P} = \boldsymbol{Q}\boldsymbol{\Lambda}_P\boldsymbol{Q}^T \tag{5.238}$$

$$\boldsymbol{M} = \boldsymbol{Q}\boldsymbol{\Lambda}_M\boldsymbol{Q}^T \tag{5.239}$$

and then

$\boldsymbol{PM} = \boldsymbol{Q}\boldsymbol{\Lambda}_P\boldsymbol{\Lambda}_M\boldsymbol{Q}^T$ where $\boldsymbol{Q}^T = \boldsymbol{Q}^{-1}$. The matrices for a plane strain idealization are

$$\boldsymbol{Q} = \begin{bmatrix} \boldsymbol{Q}_1 & \mathbf{0} \\ \mathbf{0} & \boldsymbol{I}_{3\times 3} \end{bmatrix} \tag{5.240a}$$

with

$$Q_1 = \begin{bmatrix} 0 & 2/\sqrt{6} & 1/\sqrt{3} \\ -\sqrt{2}/2 & -1/\sqrt{6} & 1/\sqrt{3} \\ \sqrt{2}/2 & -1/\sqrt{6} & 1/\sqrt{3} \end{bmatrix} \tag{5.240b}$$

$$P = \begin{bmatrix} P_1 & 0 \\ 0 & P_2 \end{bmatrix} \tag{5.241a}$$

with

$$P_1 = \begin{bmatrix} 2/3 & -1/3 & -1/3 \\ -1/3 & 2/3 & -1/3 \\ -1/3 & -1/3 & 2/3 \end{bmatrix} \quad \text{and} \quad P_2 = \text{DIAG}[2, 1, 1] \tag{5.241b}$$

$$\Lambda_P = \text{DIAG}[\mathbf{1}, \mathbf{1}, \mathbf{0}, \mathbf{2}, \mathbf{1}, \mathbf{1}] \quad \textbf{and} \quad \Lambda_M = \textbf{DIAG}[\mathbf{2\mu}, \mathbf{2\mu}, \mathbf{3\lambda} + \mathbf{2\mu}, \boldsymbol{\mu}, \mathbf{2\mu}, \mathbf{2\mu}]$$

Now the update formula (5.237) can be written as

$$\boldsymbol{\xi}_{n+1} = \left[\left(1 + \frac{2}{3}\Delta\gamma H'\right)I + \Delta\gamma \boldsymbol{Q}\boldsymbol{\Lambda}_P\boldsymbol{\Lambda}_M\boldsymbol{Q}^{\mathbf{T}}\right]^{-1} \boldsymbol{\xi}_{n+1}^{\mathbf{tr}} \tag{5.242}$$

or after letting

$$\boldsymbol{\Gamma}(\boldsymbol{\Delta\gamma}) = \left[\left(\mathbf{1} + \frac{\mathbf{2}}{\mathbf{3}}\boldsymbol{\Delta\gamma H'(1-\Phi)}\right)\boldsymbol{I} + \boldsymbol{\Delta\gamma\Lambda_P\Lambda_M}\right]^{-1} \tag{5.243}$$

this expression becomes

$$\boldsymbol{\xi}_{n+1} = \boldsymbol{Q}\Gamma(\Delta\gamma)\boldsymbol{Q}^{\mathbf{T}}\boldsymbol{\xi}_{n+1}^{\mathbf{tr}} \tag{5.244}$$

where explicitly

$$\boldsymbol{\Gamma}(\boldsymbol{\Delta\gamma}) = \text{DIAG}\begin{bmatrix} \dfrac{1}{1+\left(\frac{2}{3}(1-\Phi)H'+2\mu\right)\Delta\gamma}, \dfrac{1}{1+\left(\frac{2}{3}(1-\Phi)H'+2\mu\right)\Delta\gamma}, \dfrac{1}{1+\frac{2}{3}(1-\Phi)H'\Delta\gamma}, \\ \dfrac{1}{1+\left(\frac{2}{3}(1-\Phi)H'+2\mu\right)\Delta\gamma}, \dfrac{1}{1+\left(\frac{2}{3}(1-\Phi)H'+2\mu\right)\Delta\gamma}, \dfrac{1}{1+\left(\frac{2}{3}(1-\Phi)H'+2\mu\right)\Delta\gamma} \end{bmatrix} \tag{5.245}$$

Making use of Eq. (5.214), $\widehat{\boldsymbol{f}}_{\boldsymbol{n+1}}$ becomes

$$\widehat{f}_{n+1} = \left[\xi_{n+1}^{\text{tr}^T} Q^T\Gamma(\Delta\gamma)QPQ^T\Gamma(\Delta\gamma)\xi_{n+1}^{\text{tr}}\right]^{1/2} \tag{5.246}$$

Using Eqs. (5.244)–(5.246), $\frac{\widehat{df_{n+1}}}{d\Delta\gamma}$ can be defined as

$$\frac{d\widehat{f}_{n+1}}{d\Delta\gamma} = \frac{\widehat{g}_{n+1}}{\widehat{f}_{n+1}} \tag{5.247}$$

where

$$\widehat{g}_{n+1} = \xi^{\mathrm{tr}T}_{n+1} Q G(\Delta\gamma) Q^T P Q \Gamma(\Delta\gamma) Q^T \xi^{\mathrm{tr}}_{n+1} \tag{5.248}$$

$$G(\Delta\gamma) = \frac{d\Gamma(\Delta\gamma)}{d\Delta\gamma} = \mathrm{DIAG}[G_{11}, G_{22}, G_{33}, G_{44}, G_{55}, G_{66}]$$

with

$$G_{11} = G_{22} = G_{44} = G_{55} = G_{66} = -\frac{\left[\frac{2}{3}(1-\Phi)H' + 2\mu\right]}{\left\{1 + \left[\frac{2}{3}(1-\Phi)H' + 2\mu\right]\Delta\gamma\right\}^2}$$

$$G_{33} = -\frac{\frac{2}{3}(1-\Phi)H'}{\left[1 + \frac{2}{3}(1-\Phi)H'\Delta\gamma\right]^2}$$

Using the definition of $K(\alpha)$, Eq. (5.247), and θ_2 from Eq. (5.231) yields the Jacobian needed for the local Newton iteration

Table 5.7 Return mapping integration algorithm-strain gradient model

Update strain	$E_{n+1} = E_n + \nabla \Delta u$
Compute trial state	$\Sigma^{tr}_{n+1} = (1-\Phi) M \left(E_{n+1} - E^{vp}_n - E^{\theta}_{n+1}\right)$
	$\xi^{tr}_{n+1} = \Sigma^{tr}_{n+1} - X_n$
Compute trial yield function	$F^{tr}_{n+1} = \sqrt{\xi^{trT}_{n+1} P \xi^{tr}_{n+1}} - \sqrt{\frac{2}{3}} K(\alpha_n)$
IF $F^{tr}_{n+1} > 0$ **THEN**	
Call Newton local and solve $f(\Delta\gamma) = 0$ for $\Delta\gamma$	
Compute	$\Xi(\Delta\gamma) = \left[M^{-1} + \frac{\Delta\gamma P}{1+\frac{2}{3}H'(1-\Phi)\Delta\gamma}\right]^{-1}$
Update	$\xi_{n+1} = \Xi(\Delta\gamma) \frac{1}{1+\frac{2}{3}H'(1-\Phi)\Delta\gamma} M^{-1} \xi^{tr}_{n+1}$
	$X_{n+1} = X_n + \Delta\gamma \frac{2}{3} H'(1-\Phi) \xi_{n+1}$
	$\Sigma_{n+1} = \xi_{n+1} + X_{n+1}$
	$\alpha_{n+1} = \alpha_n + \Delta\gamma (1-\Phi) \sqrt{\frac{2}{3} \xi^T_{n+1} P \xi_{n+1}}$
	$E^{vp}_{n+1} = E^{vp}_n + \Delta\gamma \frac{P\xi_{n+1}}{(1-\Phi)}$
	$E^{e}_{n+1} = E_{n+1} - E^{vp}_{n+1} - E^{\theta}_{n+1}$
Compute consistent Jacobian	
	$\frac{d\Sigma_{n+1}}{dE_{n+1}} = (1-\Phi)\left[\Xi(\Delta\gamma) - \frac{1}{(1+\beta)} \mathrm{N} \otimes \mathrm{N}\right]$
ELSE	
Elastic step (EXIT)	
END IF	
EXIT	

Table 5.8 Local Newton iteration to determine the consistency parameter-strain gradient theory

Let $\Delta\gamma^{(0)} \leftarrow 0$

$\alpha^{(0)}_{n+1} \leftarrow \alpha_n$

Start Iterations

DO_UNTIL $|F(\Delta\gamma| < tol$

$k \leftarrow k + 1$

Compute $\Delta\gamma^{(k+1)}$

$$F\left(\Delta\gamma^{(k)}\right) = \widehat{f}_{n+1}\left(\Delta\gamma^{(k)}\right) - \sqrt{2/3}K(\alpha_{n+1}) - \Theta\left(\frac{\eta\Delta\gamma^{(k)}}{\Delta t}\right)$$

$$J\left(\Delta\gamma^{(k)}\right) = \theta_2\frac{\widehat{g}_{n+1}\left(\Delta\gamma^{(k)}\right)}{\widehat{f}_{n+1}\left(\Delta\gamma^{(k)}\right)} - \frac{2}{3}K'(1-\Phi)\widehat{f}_{n+1}\left(\Delta\gamma^{(k)}\right) - \frac{d\Theta}{d\Delta\gamma}$$

$$\Delta\gamma^{(k+1)} \leftarrow \Delta\gamma^{(k)} - \frac{F\left(\Delta\gamma^{(k)}\right)}{J\left(\Delta\gamma^{(k)}\right)}$$

$$\alpha_{n+1}{}^{(k+1)} \leftarrow \alpha_{n+1}{}^{(k)} + \sqrt{2/3}\widehat{f}_{n+1}(1-\Phi)\left(\Delta\gamma^{(k)}\right)$$

END DO_UNITL

$$\frac{dF(\Delta\gamma)}{d\Delta\gamma} = \theta_2\frac{\widehat{g}_{n+1}}{\widehat{f}_{n+1}} - \frac{2}{3}K'\widehat{f}_{n+1} \tag{5.249}$$

The complete Newton-Raphson integration algorithm is shown in Table 5.7. The local Newton iteration for the solution of the consistency equation is shown in Table 5.8.

References

Aero, E., & Kuvshinsky, E. (1961). Fundamental equations of the theory of elastic media with rotationally interacting particles. *Soviet Physics Solid State, 2*, 1272.

Armstrong, P., & Frederick, C. (1966). *A mathematical representation of the multiaxial Bauschinger effect*. CEGB report RD/N731.

Basaran, C., Zhao, Y., Tang, H., & Gomez, J. (2005). A damage mechanics based unified constitutive model for Pb/Sn solder alloys. *ASME Journal of Electronic Packaging, 127*(3), 208–214.

Chaboche, J. (1989). Constitutive equations for cyclic plasticity and viscoplasticity. *International Journal of Plasticity, 3*, 247–302.

Cosserat, E., & Cosserat, F. (1909). *Theorie des Corps deformables*. Paris: A Hermann & Fils.

De Borst, R. (1993). A generalization of J2-flow theory for polar continua. *Computer Methods in Applied Mechanics and Engineering, 103*, 347–362.

De Borst, R., & Muhlhaus, H. (1992). Gradient dependent plasticity: Formulation and algorithmic aspects. *International Journal for Numerical Methods in Engineering, 35*, 521–539.

DeHoff, R. T. (1993). *Thermodynamics in materials science*. New York: McGraw-Hill.

Eringen, A. C. (1968). Theory of micropolar elasticity. In H. Leibowitz (Ed.), *Fracture, and advanced treatise* (pp. 621–729). New York: Academic Press.

Fleck, N., & Hutchinson, J. (1997). Strain gradient plasticity. In *Advances in Applied mechanics* (Vol. 33, pp. 295–361). New York: Academic Press.

Fung, Y. C., & Tong, P. (2001). *Classical and computational solid mechanics* (Advanced series in engineering science) (Vol. 1). Singapore: World Scientific.

Gomez, J. (2006, February). *A thermodynamics based damage mechanics framework for fatigue analysis of microelectronics solder joints with size effects*. PhD Dissertation, Submitted to Department of Civil, Structural and Environmental Engineering, University at Buffalo.

Kashyap, B., & Murty, G. (1981). Experimental constitutive relations for the high temperature deformation of a Pb-Sn eutectic alloy. *Materials Science and Engineering, 50*, 205–213.

Koiter, W. T. (1964). Couple stresses in the theory of elasticity. *Proceedings of the Koninklijke Nederlandse Akademie van Wetenschappen, Series B: Physical Sciences, 67*, 17–44.

Krajcinovic, D. (1996). *Damage mechanics* (North-Holland series in applied mathematics and mechanics). Amsterdam: Elsevier.

Lee, Y., & Basaran, C. (2011). A creep model for solder alloys. *Journal of Electronic Packaging, Transactions of the ASME, 133*(4), 044501.

Lemaitre, J. (1996). *A course on damage mechanics*. Berlin: Springer-Verlag.

Lemaitre, J., & Chaboche, J. L. (1990). *Mechanics of solid materials*. Cambridge: Cambridge University Press.

Li, K. W. (1989). *Applied thermodynamics: Availability method and energy conversion*. New York: Taylor & Francis.

Malvern, L. E. (1969). *Introduction to the mechanics of continuous medium*. Englewood Cliffs, NJ: Prentice-Hall.

Mazur, P., & De Groot, S. (1962). *Non-equilibrium thermodynamics*. New York: Dover Publications.

Mindlin, R. (1964). Micro-structure in linear elasticity. *Archive for Rational Mechanics and Analysis, 16*, 51–78.

Mindlin, R. (1965). Second gradient of strain and surface tension in linear elasticity. *International Journal of Solids and Structures, 1*, 417–438.

Mindlin, R., & Tiersten, H. F. (1962). Effects of couple-stresses in linear elasticity. Communicated by C. Truesdell. *Archive for Rational Mechanics and Analysis, 11*, 415–448.

Prager, W. (1955). The theory of plasticity: A survey of recent achievements (James Clayton Lecture). *Proceedings of the Institution of Mechanical Engineers, 169*, 41.

Rabotnov Y. N. (1969). Creep rupture. In: Hetényi M., & Vincenti W. G. (eds) *Applied mechanics. International Union of Theoretical and Applied Mechanics*. Berlin: Springer.

Shu, J. Y., & Fleck, N. A. (1999). Strain gradient plasticity: Size-dependent deformation of bicrystals. *Journal of the Mechanics and Physics of Solids, 47*, 297–324.

Simo, J., & Hughes, T. (1997). *Computational inelasticity* (Interdisciplinary applied mathematics). New York: Springer.

Toupin, R. (1962). Elastic materials with couple-stresses. *Archive for Rational Mechanics and Analysis, 11*, 385–414.

Xia, Z., & Hutchinson, J. (1996). Crack tip fields in strain gradient plasticity. *Journal of the Mechanics and Physics of Solids, 44*(10), 162–1648.

Chapter 6
Unified Micromechanics of Particulate Composites

6.1 Introduction

Modeling the macroscopic constitutive response of heterogeneous materials starting from local description of the microstructure is necessary. Composite materials are heterogeneous. However, continuum mechanics formulation requires homogenization of the medium. The homogenization from local level to macro level and the localization from macro-level quantities to the corresponding local micromechanics variables must be well defined.

Using Eshelby method (1957), Ju and Chen (1994a, b) formulated the effective mechanical properties of elastic multiphase composites containing many randomly dispersed ellipsoidal inhomogeneity with perfect bonding. Within the context of the representative volume element (REV), four governing micromechanical ensemble-volume averaged field equations are utilized to relate ensemble-volume averaged stresses, strains, volume fractions, eigenstrains, particle shapes and orientations, and elastic properties of constituent phases of the particulate composites. Ju and Tseng (1996) formulated combining a micromechanical interaction approach and the continuum plasticity to predict effective elastoplastic behavior of two-phase particulate composite containing many randomly dispersed elastic spherical inhomogeneity. Explicit pairwise interparticle interactions are considered in both the elastic and plastic responses. Furthermore, the ensemble-volume averaging procedure is employed and the formulation is of complete second order.

Let us consider a perfectly bonded two-phase composite consisting of an elastic matrix (phase 0) with bulk modulus k_0 and shear modulus μ_0 and randomly dispersed elastic spherical particles (phase 1) with bulk modulus k_1 and shear modulus μ_1. The effective bulk modulus k_* and effective shear modulus μ_* for this two-phase composite for the noninteracting solution, neglecting the interparticle interaction effects, was derived by Ju and Chen (1994a):

C. Basaran, *Introduction to Unified Mechanics Theory with Applications*,
https://doi.org/10.1007/978-3-030-57772-8_6

$$k_* = k_0\left\{1 + \frac{3(1-\nu_0)(k_1-k_0)\varphi}{3(1-\nu_0)k_0 + (1-\varphi)(1+\nu_0)(k_1-k_0)}\right\} \tag{6.1}$$

$$\mu_* = \mu_0\left\{1 + \frac{15(1-\nu_0)(\mu_1-\mu_0)\varphi}{15(1-\nu_0)\mu_0 + (1-\varphi)(8-10\nu_0)(\mu_1-\mu_0)}\right\} \tag{6.2}$$

where φ is the particle volume fraction and ν_0 is Poisson's ratio of the matrix.

If the effects due to interparticle interactions are included for the two-phase elastic composites with randomly located spherical particles, the solutions for the effective bulk modulus k_* and shear modulus μ_* are (Ju and Chen 1994b):

$$k_* = k_0\left\{1 + \frac{30(1-\nu_0)\varphi(3\gamma_1+2\gamma_2)}{3\alpha+2\beta-10(1+\nu_0)\varphi(3\gamma_1+2\gamma_2)}\right\} \tag{6.3}$$

$$\mu_* = \mu_0\left\{1 + \frac{30(1-\nu_0)\varphi\gamma_2}{\beta-4(4-5\nu_0)\varphi\gamma_2}\right\} \tag{6.4}$$

with

$$\alpha = 2(5\nu_0-1) + 10(1-\nu_0)\left(\frac{k_0}{k_1-k_0} - \frac{\mu_0}{\mu_1-\mu_0}\right) \tag{6.5}$$

$$\beta = 2(4-5\nu_0) + 15(1-\nu_0)\frac{\mu_0}{\mu_1-\mu_0} \tag{6.6}$$

and

$$\gamma_1 = \frac{5\varphi}{8\beta^2}\left\{(13-14\nu_0)\nu_0 - \frac{8\alpha}{3\alpha+2\beta}(1-2\nu_0)(1+\nu_0)\right\} \tag{6.7}$$

$$\gamma_2 = \frac{1}{2} + \frac{5\varphi}{16\beta^2}\left\{\left(25-34\nu_0+22\nu_0^2\right) - \frac{6\alpha}{3\alpha+2\beta}(1-2\nu_0)(1+\nu_0)\right\} \tag{6.8}$$

Alternatively, the effective Young's modulus E_* and Poisson's ratio ν_* of particulate composites are easily obtained through the following relationships:

$$E_* = \frac{9k_*\mu_*}{3k_*+\mu_*} \tag{6.9}$$

$$\nu_* = \frac{3k_*-2\mu_*}{6k_*+2\mu_*} \tag{6.10}$$

The experimental validation of Ju and Chen model was provided by Nie and Basaran (2005), by comparing the analytical predictions with the experimental data on the particulate composite prepared using lightly cross-linked poly-methyl methacrylate (PMMA) filled with alumina trihydrate (ATH). All filler particles are

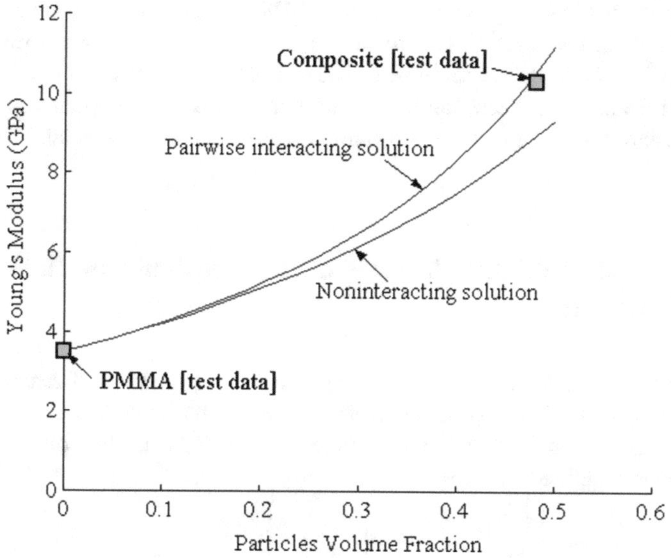

Fig. 6.1 Effective Young's modulus as a function of particle volume fraction (Nie 2005)

assumed spherical and both the matrix and the filler are isotropic elastic. The material properties involved are as follows:

ATH: $E_1 = 70$ GPa, $v_1 = 0.24$.
PMMA: $E_0 = 3.5$ GPa, $v_0 = 0.31$
Particle volume fraction: $\varphi = 0.48$

Experimental mean value of Young's modulus for this composite is about 10.2 GPa at room temperature. Figure 6.1 shows the comparisons among the analytical solution (including pairwise interacting solution and noninteracting solution) and experimental results. It is observed that agreement between the pairwise interacting prediction and experimental data is very good for the effective Young's modulus. Based on the foregoing preliminary analytical and experimental comparisons, it appears that Ju and Chen's analytical micromechanical approach offers a simple, approximate, yet sufficiently accurate framework for the prediction of effective elastic moduli of two-phase composites.

Particle-filled composites consist of bulk matrix, filler particles, and interfacial transition zone around particles, which often have very different properties such as coefficient of thermal expansion (CTE) and stiffness. Within the particulate composite microstructure, there are micro-stresses due to the coefficient of the thermal expansion (CTE) mismatch between the matrix and the filler particles. For the composite prepared using lightly cross-linked poly-methyl methacrylate (PMMA) filled with alumina trihydrate (ATH), these micro-stresses can be imaged using the fact that PMMA is optically birefringent. The micro-stresses imaged using this technique are indeed due to CTE mismatch as they dissipate at temperatures close to the glass transition temperature T_g of the PMMA. These stresses can be quantified by calculation or by direct measurement and are of the order between 15% and 75% of the tensile strength of the composite. Therefore, the thermal stress associated with the CTE mismatch between the matrix and particles is an important factor for the failure of particulate composites subjected to thermo-mechanical loads.

6.2 Ensemble-Volume Averaged Micromechanical Field Equations

In order to obtain effective constitutive equations and properties of random heterogeneous composites, one typically performs the ensemble-volume averaging process within a mesoscopic representative volume element (RVE). The volume-averaged stress tensor is defined as

$$\bar{\boldsymbol{\sigma}} = \frac{1}{V}\int_V \boldsymbol{\sigma}(\mathbf{x})d\mathbf{x} = \frac{1}{V}\left[\int_{V_0} \boldsymbol{\sigma}(\mathbf{x})d\mathbf{x} + \sum_{q=1}^{n}\int_{V_q} \boldsymbol{\sigma}(\mathbf{x})d\mathbf{x}\right] \tag{6.11}$$

where V is the volume of a RVE, V_0 is the volume of the matrix, V_q is the volume of the qth phase particles, and n denotes the number of particulate phases of different material properties (excluding the matrix).

Similarly, the volume-averaged strain tensor is defined as

$$\bar{\boldsymbol{\varepsilon}} = \frac{1}{V}\int_V \boldsymbol{\varepsilon}(\mathbf{x})d\mathbf{x} = \frac{1}{V}\left[\int_{V_0} \boldsymbol{\varepsilon}(\mathbf{x})d\mathbf{x} + \sum_{q=1}^{n}\int_{V_q} \boldsymbol{\varepsilon}(\mathbf{x})d\mathbf{x}\right]$$
$$\equiv \frac{1}{V}\left[V_0\bar{\boldsymbol{\varepsilon}}_0 + \sum_{q=1}^{n} V_q\bar{\boldsymbol{\varepsilon}}_q\right] \tag{6.12}$$

According to Eshelby's equivalence principle, the perturbed strain field $\boldsymbol{\varepsilon}'(\mathbf{x})$ induced by inhomogeneity can be related to specified eigenstrain $\boldsymbol{\varepsilon}^*(\mathbf{x})$ by replacing the inhomogeneities with the matrix material. Sometimes the inhomogeneities may also involve its own eigenstrain caused by, for example, phase transformation, precipitation, plastic deformation, or CTE mismatch between different constituents of the composites. However, it is not necessary to attribute the eigenstrain to any specific source. Then the total perturbed stress is the sum of the two parts, one caused by the inhomogeneity and the other one is the eigenstress caused by eigenstrain. For the domain of the qth-phase particles with elastic stiffness tensor $\mathbf{C}_q$, Mura (1987) provided the following relation:

$$\mathbf{C}_q : \left[\boldsymbol{\varepsilon}^0 + \boldsymbol{\varepsilon}'(\mathbf{x}) - \boldsymbol{\varepsilon}_q^T(\mathbf{x})\right] = \mathbf{C}_0 : \left[\boldsymbol{\varepsilon}^0 + \boldsymbol{\varepsilon}'(\mathbf{x}) - \boldsymbol{\varepsilon}_q^T(\mathbf{x}) - \boldsymbol{\varepsilon}_q^*(\mathbf{x})\right] \tag{6.13}$$

where $\mathbf{C}_0$ is the stiffness tensor of the matrix, $\boldsymbol{\varepsilon}^0$ is the uniform elastic strain field induced by far-field loads for a homogeneous matrix material only, $\boldsymbol{\varepsilon}_q^T$ is its own eigenstrain associated with the qth particle, $\boldsymbol{\varepsilon}_q^*$ is the fictitious equivalent eigenstrain by replacing the qth particles with the matrix material, $\boldsymbol{\varepsilon}'(\mathbf{x})$ is the perturbed strain due to distributed eigenstrain $\boldsymbol{\varepsilon}^T$, and $\boldsymbol{\varepsilon}^*$ is associated with all the particles in the RVE. The stress-strain relation is given in Newtonian mechanics formulation by

$$\boldsymbol{\sigma}^0 = \mathbf{C}_0 : \boldsymbol{\varepsilon}^0 \tag{6.14}$$

The strain at any point within an RVE is decomposed into two parts, the uniform strain and the perturbed strain, due to the distributed eigenstrain. It is emphasized that the eigenstrain $\boldsymbol{\varepsilon}^*$ and $\boldsymbol{\varepsilon}^T$ are nonzero in the particle domain and zero in the matrix domain, respectively. The perturbed strain field induced by all the distributed eigenstrains $\boldsymbol{\varepsilon}^*$ and $\boldsymbol{\varepsilon}^T$ can be expressed by (Mura 1987)

$$\boldsymbol{\varepsilon}'(\mathbf{x}) = \int_V \mathbf{G}(\mathbf{x} - \mathbf{x}') : \left[\boldsymbol{\varepsilon}^T(\mathbf{x}') + \boldsymbol{\varepsilon}^*(\mathbf{x}')\right] d\mathbf{x}' \tag{6.15}$$

where $\mathbf{x}, \mathbf{x}' \in V$, and $\mathbf{G}$ are Green's function in a linear elastic homogeneous matrix. For a linear elastic isotropic matrix, the fourth-rank tensor Green's function is (Ju and Chen 1994a)

$$G_{ijkl}(\mathbf{x} - \mathbf{x}') = \frac{1}{8\pi(1 - \nu_0)r^3} F_{ijkl}(-15, 3\nu_0, 3, 3 - 6\nu_0, -1 + 2\nu_0, 1 - 2\nu_0) \tag{6.16}$$

where $\mathbf{r} \equiv \mathbf{x} - \mathbf{x}'$, $r \equiv |\mathbf{r}|$ and ν_0 is Poisson's ratio of the matrix. The components of the fourth-rank tensor $\mathbf{F}$ are defined by

$$F_{ijkl}(B_m) = B_1 n_i n_j n_k n_l + B_2\left(\delta_{ik} n_j n_l + \delta_{il} n_j n_k + \delta_{jk} n_i n_l + \delta_{jl} n_i n_k\right) + B_3 \delta_{ij} n_k n_l + B_4 \delta_{kl} n_i n_j + B_5 \delta_{ij} \delta_{kl} + B_6\left(\delta_{ik}\delta_{jl} + \delta_{il}\delta_{jk}\right) \tag{6.17}$$

with the unit normal vector $\mathbf{n} \equiv \mathbf{r}/r$ and index $m = 1$ to 6.

From Eqs. (6.13) and (6.15), we arrive at

$$-\mathbf{A}_q : \boldsymbol{\varepsilon}_q^*(x) = \boldsymbol{\varepsilon}^0 - \boldsymbol{\varepsilon}_q^T(\mathbf{x}) + \int_V \mathbf{G}(\mathbf{x} - \mathbf{x}') : \left[\boldsymbol{\varepsilon}^T(\mathbf{x}') + \boldsymbol{\varepsilon}^*(\mathbf{x}')\right] d\mathbf{x}' \quad \mathbf{x}' \in V \tag{6.18}$$

where

$$\mathbf{A}_q = \left(\mathbf{C}_q - \mathbf{C}_0\right)^{-1} \cdot \mathbf{C}_0 \tag{6.19}$$

Furthermore, the total local strain field $\boldsymbol{\varepsilon}(\mathbf{x})$ can be expressed as

$$\boldsymbol{\varepsilon}(\mathbf{x}) = \boldsymbol{\varepsilon}^0 + \boldsymbol{\varepsilon}'(\mathbf{x}) = \boldsymbol{\varepsilon}^0 + \int_V \mathbf{G}(\mathbf{x}-\mathbf{x}') : \left[\boldsymbol{\varepsilon}^T(\mathbf{x}') + \boldsymbol{\varepsilon}^*(\mathbf{x}')\right] d\mathbf{x}' \tag{6.20}$$

Using the renormalization procedure of the volume-averaged strain tensor is given by Ju and Chen (1994a)

$$\overline{\boldsymbol{\varepsilon}} = \boldsymbol{\varepsilon}^0 + \frac{1}{V}\int_V \int_V \mathbf{G}(\mathbf{x}-\mathbf{x}') : \left[\boldsymbol{\varepsilon}^T(\mathbf{x}') + \boldsymbol{\varepsilon}^*(\mathbf{x}')\right] d\mathbf{x}' d\mathbf{x} = \boldsymbol{\varepsilon}^0 + \mathbf{s}$$
$$: \left[\sum_{q=1}^{n} \varphi_q \left(\overline{\boldsymbol{\varepsilon}}_q^T + \overline{\boldsymbol{\varepsilon}}_q^*\right)\right] \tag{6.21}$$

where $\mathbf{s}$ is a constant tensor for uni-directionally aligned and similarly ellipsoidal filler particles. If the linear elastic matrix material is isotropic and all inclusions are spherical, then the $\mathbf{s}$ takes the form of the Eshelby tensor $\mathbf{S}$:

$$S_{ijkl} = \frac{1}{15(1-\nu_0)}\left[(5\nu_0 - 1)\delta_{ij}\delta_{kl} + (4 - 5\nu_0)\left(\delta_{ik}\delta_{jl} + \delta_{il}\delta_{jk}\right)\right] \tag{6.22}$$

where δ_{ij} signifies the Kronecker delta.

Similarly, using Eqs. (6.11)–(6.13), the ensemble-volume averaged stress field can be recast as

$$\overline{\boldsymbol{\sigma}} = \frac{1}{V}\left[\int_{V_0} \mathbf{C}_0 : \boldsymbol{\varepsilon}(\mathbf{x}) d\mathbf{x} + \sum_{q=1}^{n} \int_{V_q} \mathbf{C}_0 : \left[\boldsymbol{\varepsilon}(\mathbf{x}) - \boldsymbol{\varepsilon}_q^T - \boldsymbol{\varepsilon}_q^*\right] d\mathbf{x}\right]$$
$$= \frac{1}{V}\left[V_0\mathbf{C}_0 : \overline{\boldsymbol{\varepsilon}}_0 + \sum_{q=1}^{n} V_q\mathbf{C}_0 : \left[\overline{\boldsymbol{\varepsilon}}_q - \overline{\boldsymbol{\varepsilon}}_q^T - \overline{\boldsymbol{\varepsilon}}_q^*\right]\right] = \mathbf{C}_0 : \left[\overline{\boldsymbol{\varepsilon}} - \sum_{q=1}^{n} \varphi_q\left(\overline{\boldsymbol{\varepsilon}}_q^T + \overline{\boldsymbol{\varepsilon}}_q^*\right)\right] \tag{6.23}$$

The effective elastic properties can be obtained, in principle, from Equations (6.18), (6.21), and (6.23) since the variables are $\overline{\boldsymbol{\sigma}}, \overline{\boldsymbol{\varepsilon}}, \boldsymbol{\varepsilon}^0, \overline{\boldsymbol{\varepsilon}}_q^*$. In essence, one needs to solve the relation between $\overline{\boldsymbol{\varepsilon}}$ and $\overline{\boldsymbol{\varepsilon}}_q^*$, which involves the solution of the integral equation (6.18). $\overline{\boldsymbol{\varepsilon}}_q^*$ depends on interparticle interactions, particle-matrix interactions, and microstructure (i.e., particle sizes, orientation, shapes, volume fractions, locations, configurations, and probability functions) of a composite system. For randomly dispersed particles, one needs to obtain the ensemble-volume averaged relation between $\bar{\boldsymbol{\varepsilon}}$ and $\bar{\boldsymbol{\varepsilon}}_q^*$ by averaging all possible solutions of the integral equation (6.18) for any particle configurations generated according to specified probability functions.

Taking the ensemble-volume average of Eq. (6.18) over all qth-phase particles, we obtain

$$-\mathbf{A}_q : \bar{\boldsymbol{\varepsilon}}_q^* = \boldsymbol{\varepsilon}^0 - \boldsymbol{\varepsilon}_q^T + \bar{\boldsymbol{\varepsilon}}_q{}' \tag{6.24}$$

where

$$\bar{\boldsymbol{\varepsilon}}_q' = \frac{1}{V_q} \int_{V_q} \int_{V} \mathbf{G}(\mathbf{x}-\mathbf{x}') : \left[\boldsymbol{\varepsilon}^T(\mathbf{x}') + \boldsymbol{\varepsilon}^*(\mathbf{x}')\right] d\mathbf{x}' d\mathbf{x} \tag{6.25}$$

If all particles in the qth-phase have the same ellipsoidal shape and orientation, then $\bar{\boldsymbol{\varepsilon}}_q{}'$ can be written as

$$\bar{\boldsymbol{\varepsilon}}_q' = \bar{\boldsymbol{\varepsilon}}_q'^p + \mathbf{S}_q : \left(\bar{\boldsymbol{\varepsilon}}_q^T + \bar{\boldsymbol{\varepsilon}}_q^*\right) \tag{6.26}$$

with

$$\bar{\boldsymbol{\varepsilon}}_q'^p = \frac{1}{V_q} \sum_{i=1}^{N_q} \int_{\Omega_q^i} \left\{ \int_{V-\Omega_q^i} \mathbf{G}(\mathbf{x}-\mathbf{x}') : \left[\boldsymbol{\varepsilon}^T(\mathbf{x}') + \boldsymbol{\varepsilon}^*(\mathbf{x}')\right] d\mathbf{x}' \right\} d\mathbf{x} \tag{6.27}$$

representing the interparticle interaction effects, where $\boldsymbol{\Omega}_q^i$ is the domain of the qth particle in the qth phase domain V_q, N_q is the number of the phase q particles dispersed in V, and $\mathbf{S}_q$ is the Eshelby tensor associated the qth particle.

From Eqs. (6.24) and (6.26), we arrive at

$$\left(-\mathbf{A}_q - \mathbf{S}_q\right) : \bar{\boldsymbol{\varepsilon}}_q^* = \boldsymbol{\varepsilon}^0 - \boldsymbol{\varepsilon}^T + \mathbf{S}_q : \boldsymbol{\varepsilon}^T + \bar{\boldsymbol{\varepsilon}}_q'^p \tag{6.28}$$

In summary, the three basic governing micromechanical ensemble-volume averaged field equations are recapitulated as follows:

$$\bar{\boldsymbol{\sigma}} = \mathbf{C}_0 : \left[\bar{\boldsymbol{\varepsilon}} - \sum_{q=1}^{n} \varphi_q \left(\bar{\boldsymbol{\varepsilon}}_q^* + \bar{\boldsymbol{\varepsilon}}_q^T\right)\right] \tag{6.29}$$

$$\bar{\boldsymbol{\varepsilon}} = \boldsymbol{\varepsilon}^0 + \sum_{q=1}^{n} \varphi_q \mathbf{S} : \left(\bar{\boldsymbol{\varepsilon}}_q^T + \bar{\boldsymbol{\varepsilon}}_q^*\right) \tag{6.30}$$

$$-(\mathbf{A}_q + \mathbf{S}_q) : \overline{\boldsymbol{\varepsilon}}_q^* = \boldsymbol{\varepsilon}^0 - \boldsymbol{\varepsilon}^T + \mathbf{S}_q : \overline{\boldsymbol{\varepsilon}}_q^T + \overline{\boldsymbol{\varepsilon}}_q^{\prime p} \tag{6.31}$$

In order to solve Eqs. (6.29)–(6.31) and obtain effective elastic properties of composite, it is essential to express the qth-phase average eigenstrain $\overline{\boldsymbol{\varepsilon}}_q^*$ in terms of the average strain $\overline{\boldsymbol{\varepsilon}}$. Namely, one has to solve the integral equation (4.18) exactly for each phase, which involves details of random microstructure.

6.3 Noninteracting Solution for Two-Phase Composites

Let us consider a perfectly bonded two-phase composite consisting of a viscoplastic matrix (phase 0) with elastic bulk modulus k_0 and elastic shear modulus μ_0 and randomly dispersed elastic spherical particles (phase 1) with bulk modulus k_1 and shear modulus μ_1. For the sake of simplicity, von Mises yield criterion is used for the matrix. Extension of the present framework to the general yield criterion and the general hardening law, however, is straightforward. Accordingly, at any matrix material point, the stress $\boldsymbol{\sigma}$ and the equivalent plastic strain e^{p} must satisfy the following yield function:

$$F(\boldsymbol{\sigma}, e^{\mathrm{p}}) = \sqrt{H(\boldsymbol{\sigma})} - K(e^{\mathrm{p}}) \tag{6.32}$$

where $K(e^{\mathrm{p}})$ is the isotropic hardening function of the matrix-only material. Furthermore, $H(\boldsymbol{\sigma})$ signifies the square of the deviatoric stress norm

$$H(\boldsymbol{\sigma}) = \boldsymbol{\sigma} : \mathbf{I}_{\mathrm{d}} : \boldsymbol{\sigma} \tag{6.33}$$

where $\mathbf{I}_{\mathrm{d}}$ denotes the deviatoric part of the fourth-rank identity tensor.

$$(I_{\mathrm{d}})_{ijkl} = -\frac{1}{3}\delta_{ij}\delta_{kl} + \frac{1}{2}(\delta_{ik}\delta_{jl} + \delta_{il}\delta_{jk}) \tag{6.34}$$

In order to solve the elastoplastic response exactly, the stress at any local point has to be known and used to determine the plastic response through the local yield criterion for all possible configurations. This approach is in general infeasible due to the complexity of the statistical and microstructural information. Therefore, a framework in which an ensemble-averaged yield criterion is constructed for the entire composite is used.

6.3.1 Average Stress Norm in Matrix

Total stress $\boldsymbol{\sigma}(\mathbf{x})$ at any point $\mathbf{x}$ in the matrix can be given by the superposition of the far-field stress $\boldsymbol{\sigma}_0$ and the perturbed stress $\boldsymbol{\sigma}'$ induced by the particles (Ju and Chen 1994a)

$$\boldsymbol{\sigma}(\mathbf{x}) = \boldsymbol{\sigma}_0 + \boldsymbol{\sigma}'(\mathbf{x}) \tag{6.35}$$

According the Eshelby theory, the perturbed stress $\boldsymbol{\sigma}'$ at any point in the matrix due to the presence of the particles can be written as

$$\boldsymbol{\sigma}'(\mathbf{x}) = \mathbf{C}_0 : \int_V \mathbf{G}(\mathbf{x} - \mathbf{x}') : \left[\boldsymbol{\varepsilon}^T + \boldsymbol{\varepsilon}^*(\mathbf{x}')\right] d\mathbf{x}' \tag{6.36}$$

where $\boldsymbol{\varepsilon}^*(\mathbf{x}')$ denotes the fictitious elastic eigenstrain in the particle induced by replacing the inhomogeneity with the matrix material, $\boldsymbol{\varepsilon}^T$ is its own eigenstrain associated with the inhomogeneity, assuming $\boldsymbol{\varepsilon}^T$ is uniform in the particles, and $\mathbf{G}(\mathbf{x} - \mathbf{x}')$ is the fourth-rank tensor of Green's function defined by Eq (6.16).

According the Eshelby theory, the eigenstrain $\boldsymbol{\varepsilon}^*(\mathbf{x}')$ due to a single ellipsoidal inclusion is uniform for the interior points of an isolated (noninteracting) inclusion. Therefore, the perturbed stress for any matrix point $\mathbf{x}$ due to a typical isolated inhomogeneity centered at $\mathbf{x}_1$ takes the form

$$\boldsymbol{\sigma}'(\mathbf{x}|\mathbf{x}_1) = \left[\mathbf{C}_0 \cdot \overline{\mathbf{G}}(\mathbf{x} - \mathbf{x}_1)\right] : \left(\boldsymbol{\varepsilon}^T + \boldsymbol{\varepsilon}^{*0}\right) \tag{6.37}$$

where

$$\overline{\mathbf{G}}(\mathbf{x} - \mathbf{x}_1) = \int_{\Omega_1} \mathbf{G}(\mathbf{x} - \mathbf{x}') d\mathbf{x}' \text{ For } \mathbf{x} \notin \boldsymbol{\Omega}_1 \tag{6.38}$$

Here $\boldsymbol{\Omega}_1$ is the particle domain centered at $\mathbf{x}_1$. Alternatively, we can derive

$$\overline{\mathbf{G}}(\mathbf{x} - \mathbf{x}_1) = \frac{\rho^3}{30(1 - \nu_0)}\left(\mathbf{H}^1 + \rho^2 \mathbf{H}^2\right) \tag{6.39}$$

where the components of $\mathbf{H}^1$ and $\mathbf{H}^2$ are given by

$$H^1_{ijkl}(\mathbf{r}) = 5F_{ijkl}(-15, 3\nu_0, 3, 3 - 6\nu_0, -1 + 2\nu_0, 1 - 2\nu_0) \tag{6.40}$$

$$H^2_{ijkl}(\mathbf{r}) = 3F_{ijkl}(35, -5, -5, -5, 1, 1) \tag{6.41}$$

where $\mathbf{r} \equiv \mathbf{x} - \mathbf{x}_1$, $r \equiv |\mathbf{r}|$, $\rho = a/r$, and a is the radius of a spherical particle. The components of the fourth-rank tensor $\mathbf{F}$ are given by Eq. (6.17). In Eq. (6.37), $\boldsymbol{\varepsilon}^{*0}$ denotes the solution of the eigenstrain $\boldsymbol{\varepsilon}^{*}$ for the single inclusion problem, which is given (from Eq. (6.31) when $\overline{\boldsymbol{\varepsilon}}_q^{\prime p}$ is dropped):

$$\overline{\boldsymbol{\varepsilon}}^{*0} = -(\mathbf{A}+\mathbf{S})^{-1} : \boldsymbol{\varepsilon}^0 + (\mathbf{A}+\mathbf{S})^{-1} \cdot (\mathbf{I}-\mathbf{S}) : \boldsymbol{\varepsilon}^T \tag{6.42}$$

where

$$\mathbf{A} = (\mathbf{C}_1 - \mathbf{C}_0)^{-1} \cdot \mathbf{C}_0 \tag{6.43}$$

We define $H(\mathbf{x}|\wp)$ as the square of the current deviatoric stress norm at the local point $\mathbf{x}$, which determines the plastic strain in a particulate composite for a given phase configuration $\wp$. Since there is no plastic strain in the elastic particles or voids, $H(\mathbf{x}|\wp)$ can be written as (Ju and Tseng 1997)

$$H(\mathbf{x}|\wp) = \begin{cases} \sigma(\mathbf{x}|\wp) : \mathbf{I}_{\mathrm{d}} : \boldsymbol{\sigma}(\mathbf{x}|\wp) & \mathbf{x} \text{ in the matrix} \\ 0 & \text{otherwise} \end{cases} \tag{6.44}$$

In addition, $\langle H \rangle_m(\mathbf{x})$ is defined as the ensemble average of $H(\mathbf{x}|\wp)$ over all possible points where point $\mathbf{x}$ is in the matrix phase. Matrix point receives the perturbations from particles. Therefore, the ensemble-average stress norm for any matrix point $\boldsymbol{x}$ can be evaluated by collecting and summing up all the current stress norm perturbations produced by any typical particle centered at $\mathbf{x}_1$ in the particle domain and averaging over all possible locations of $\mathbf{x}_1$, namely

$$\langle H \rangle_m(\mathbf{x}) = H^0 + \int_{r>a} \{H(\mathbf{x}|\mathbf{x}_1) - H^0\} P(\mathbf{x}_1) d\mathbf{x}_1 \tag{6.45}$$

for point $\mathbf{x}$ in the matrix. Here a is the radius of the particles, $P(\mathbf{x}_1)$ denotes the probability density functions for finding a particle centered at $\mathbf{x}_1$, and H_0 corresponds to the far-field stress norm in the matrix:

$$H^0 = \boldsymbol{\sigma}^0 : \mathbf{I}_{\mathrm{d}} : \boldsymbol{\sigma}^0 \tag{6.46}$$

where $\mathbf{I}_{\mathrm{d}}$ signifies the deviatoric part of the forth rank identity tensor.

Assuming that $P(\mathbf{x}_1)$ is statistically homogeneous, isotropic, and uniform, and $P(\mathbf{x}_1)$ takes the form

$$P(\mathbf{x}_1) = \frac{N}{V} \tag{6.47}$$

where N is the total number of particles dispersed in volume V. According to the assumption of statistical isotropy and uniformity, Eq. (4.45) can be recast into a more convenient form

$$\langle H\rangle_m(\mathbf{x}) \cong H^0 + \frac{N}{V}\int_{r>a} dr \int_{A(r)} \{H(\mathbf{r}) - H^0\} dA \tag{6.48}$$

with the help of Eqs. (6.35), (6.37), (6.44), and (6.46), plus using the following identities:

$$\int_{A(r)} n_i n_j dA = \frac{4\pi r^2}{3}\delta_{ij} \tag{6.49}$$

$$\int_{A(r)} n_i n_j n_k n_l dA = \frac{4\pi r^2}{15}\left(\delta_{ij}\delta_{kl} + \delta_{ik}\delta_{jl} + \delta_{il}\delta_{jk}\right) \tag{6.50}$$

$$\int_{r>a} dr \int_{A(r)} H^1(\mathbf{r}) dA = \mathbf{0} \tag{6.51}$$

$$\int_{r>a} dr \int_{A(r)} H^2(\mathbf{r}) dA = \mathbf{0} \tag{6.52}$$

$$\int_{r>a} dr \int_{A(r)} \overline{G}(\mathbf{r}) dA = \mathbf{0} \tag{6.53}$$

Equation (6.48) can be recast as

$$\langle H\rangle_m(\mathbf{x}) = H^0 + \frac{N}{V}\int_{r>a} dr \int_{A(r)} [\boldsymbol{\sigma}'(\mathbf{x}) : \mathbf{I}_{\mathrm{d}} : \boldsymbol{\sigma}'(\mathbf{x})] dA \tag{6.54}$$

After some lengthy yet straightforward derivations, one can obtain the ensemble-averaged current stress norm at any matrix point $\mathbf{x}$ as

$$\langle H\rangle_m(\mathbf{x}) = \boldsymbol{\sigma}^0 : \mathbf{T} : \boldsymbol{\sigma}^0 + \sigma^T : \mathbf{T}^* : \boldsymbol{\sigma}^T - 2\boldsymbol{\sigma}^0 : \mathbf{T}^* : \boldsymbol{\sigma}^T \tag{6.55}$$

where

$$\boldsymbol{\sigma}^T = \mathbf{AC}_1 : \boldsymbol{\varepsilon}^T \tag{6.56}$$

The components of the positive definite fourth-rank tensors $\mathbf{T}$ and $\mathbf{T}^*$ read

$$T_{ijkl} = T_1 \delta_{ij}\delta_{kl} + T_2(\delta_{ik}\delta_{jl} + \delta_{il}\delta_{jk}) \tag{6.57}$$

$$T^*_{ijkl} = T^*_1 \delta_{ij}\delta_{kl} + T^*_2(\delta_{ik}\delta_{jl} + \delta_{il}\delta_{jk}) \tag{6.58}$$

with

$$3T_1 + 2T_2 = 200(1 - 2\nu_0)^2 \frac{\varphi}{(3\alpha + 2\beta)^2} \tag{6.59}$$

$$T_2 = \frac{1}{2} + \left(23 - 50\nu_0 + 35\nu_0^2\right) \frac{\varphi}{\beta^2} \tag{6.60}$$

$$3T^*_1 + 2T^*_2 = 200(1 - 2\nu_0)^2 \frac{\varphi}{(3\alpha + 2\beta)^2} \tag{6.61}$$

$$T^*_2 = \left(23 - 50\nu_0 + 35\nu_0^2\right) \frac{\varphi}{\beta^2} \tag{6.62}$$

α and β are given by Eqs. (6.5) and (6.6) φ is the particle volume fraction.

The ensemble-averaged current stress norm at a matrix point must be established in terms of the macroscopic stress $\overline{\boldsymbol{\sigma}}$ in order to express the effective loading function in terms of the macroscopic stress. In the special case of uniform dispersions of identical elastic spheres in a homogeneous matrix, the macroscopic stress and the far-field stress take the form

$$\overline{\boldsymbol{\sigma}} = \mathbf{C}_0 : \left[\overline{\boldsymbol{\varepsilon}} - \varphi\left(\overline{\boldsymbol{\varepsilon}}^{*0} + \overline{\boldsymbol{\varepsilon}}^T\right)\right] \tag{6.63}$$

$$\boldsymbol{\sigma}^0 = \mathbf{C}_0 : \left[\overline{\boldsymbol{\varepsilon}} - \varphi \mathbf{S} : \left(\overline{\boldsymbol{\varepsilon}}^T + \overline{\boldsymbol{\varepsilon}}^{*0}\right)\right] \tag{6.64}$$

Using Eqs. (6.63), (6.64), and (6.42), the relation between the far-field stress $\boldsymbol{\sigma}^0$ and the macroscopic stress $\overline{\boldsymbol{\sigma}}$ takes the form

$$\overline{\boldsymbol{\sigma}} = \mathbf{P} : \boldsymbol{\sigma}^0 - \mathbf{Q} : \boldsymbol{\sigma}^T \tag{6.65}$$

where

$$\mathbf{P} = \mathbf{I} + \varphi(\mathbf{I} - \mathbf{S})(\mathbf{A} + \mathbf{S})^{-1} \tag{6.66}$$

$$\mathbf{Q} = \varphi(\mathbf{I} - \mathbf{S})(\mathbf{A} + \mathbf{S})^{-1} \tag{6.67}$$

with the components of **P** and **Q** are given by

$$P_{ijkl} = P_1 \delta_{ij}\delta_{kl} + P_2(\delta_{ik}\delta_{jl} + \delta_{il}\delta_{jk}) \tag{6.68}$$

$$Q_{ijkl} = Q_1 \delta_{ij}\delta_{kl} + Q_2(\delta_{ik}\delta_{jl} + \delta_{il}\delta_{jk}) \tag{6.69}$$

where

$$3P_1 + 2P_2 = a\varphi + 1 \tag{6.70}$$

$$P_2 = \frac{1}{2}(b\varphi + 1) \tag{6.71}$$

$$3Q_1 + 2Q_2 = a\varphi \tag{6.72}$$

$$Q_2 = \frac{1}{2}b\varphi \tag{6.73}$$

$$a = \frac{20(1 - 2\nu_0)}{3\alpha + 2\beta} \tag{6.74}$$

$$b = \frac{(7 - 5\nu_0)}{\beta} \tag{6.75}$$

With the help of Eq. (6.65), we arrive at the alternative expression for the ensemble-averaged current stress norm in a matrix point **x** as:

$$\langle H \rangle_m(\mathbf{x}) = \overline{\boldsymbol{\sigma}} : \overline{\mathbf{T}} : \overline{\boldsymbol{\sigma}} + 2\overline{\boldsymbol{\sigma}} : \overline{\mathbf{T}}^* : \overline{\boldsymbol{\sigma}}^T + \overline{\boldsymbol{\sigma}}^T : \overline{\mathbf{T}}^{**} : \overline{\boldsymbol{\sigma}}^T \tag{6.76}$$

where the positive definite fourth-rank tensor $\overline{\mathbf{T}}$, $\overline{\mathbf{T}}^*$, and $\overline{\mathbf{T}}^{**}$ are defined as

$$\overline{\mathbf{T}} = \left(\mathbf{P}^{-1}\right)^{\mathbf{T}} \cdot \mathbf{T} \cdot \mathbf{P}^{-1} \tag{6.77}$$

$$\overline{\mathbf{T}}^* = \left(\mathbf{P}^{-1}\right)^{\mathbf{T}} \cdot \mathbf{T} \cdot \mathbf{P}^{-1} \cdot \mathbf{Q} - \mathbf{T}^* \cdot \mathbf{P}^{-1} \tag{6.78}$$

$$\overline{\mathbf{T}}^{**} = \left(\mathbf{P}^{-1} \cdot \mathbf{Q}\right)^{\mathbf{T}} \cdot \mathbf{T} \cdot \left(\mathbf{P}^{-1} \cdot \mathbf{Q}\right) + \mathbf{T}^* - 2\mathbf{T}^* \cdot \left(\mathbf{P}^{-1} \cdot \mathbf{Q}\right) \tag{6.79}$$

where the components of $\overline{\mathbf{T}}, \overline{\mathbf{T}}^*$, and $\overline{\mathbf{T}}^{**}$ are

$$\overline{T}_{ijkl} = \overline{T}_1 \delta_{ij}\delta_{kl} + \overline{T}_2\left(\delta_{ik}\delta_{jl} + \delta_{il}\delta_{jk}\right) \tag{6.80}$$

$$\overline{T}_{ijkl}{}^* = \overline{T}_1{}^* \delta_{ij}\delta_{kl} + \overline{T}_2{}^*\left(\delta_{ik}\delta_{jl} + \delta_{il}\delta_{jk}\right) \tag{6.81}$$

$$\overline{T}_{ijkl}{}^{**} = \overline{T}_1{}^{**} \delta_{ij}\delta_{kl} + \overline{T}_2{}^{**}\left(\delta_{ik}\delta_{jl} + \delta_{il}\delta_{jk}\right) \tag{6.82}$$

with

$$3\overline{T}_1 + 2\overline{T}_2 = \frac{3T_1 + 2T_2}{(a\varphi + 1)^2} \tag{6.83}$$

$$\overline{T}_2 = \frac{T_2}{(b\varphi + 1)^2} \tag{6.84}$$

$$3\overline{T}_1^{\ *} + 2\overline{T}_2^{\ *} = -\frac{3T_1 + 2T_2}{(a\varphi + 1)^2} \tag{6.85}$$

$$\overline{T}_2^{\ *} = -\frac{T_2}{(b\varphi + 1)^2} + \frac{1}{2(b\varphi + 1)} \tag{6.86}$$

$$3\overline{T}_1^{\ **} + 2\overline{T}_2^{\ **} = \frac{3T_1 + 2T_2}{(a\varphi + 1)^2} \tag{6.87}$$

$$\overline{T}_2^{\ **} = \frac{T_2}{(b\varphi + 1)^2} + \frac{b\varphi - 1}{2(b\varphi + 1)} \tag{6.88}$$

Because $\boldsymbol{\sigma}^T = \mathbf{A}\mathbf{C}_1 : \boldsymbol{\varepsilon}^T$ is spherical stress, so Eq. (6.76) of the ensemble-averaged loading function can be simplified as

$$\langle H \rangle_m(\mathbf{x}) = \left(\overline{\boldsymbol{\sigma}} - \overline{\boldsymbol{\sigma}}^T\right) : \overline{\mathbf{T}} : \left(\overline{\boldsymbol{\sigma}} - \overline{\boldsymbol{\sigma}}^T\right) \tag{6.89}$$

It should be noted that the effective yield function is pressure dependent now and not of the von Mises type any more. Therefore, the particles have significant effects on the viscoplastic behavior of the matrix materials. Plastic yielding and plastic flow occur only in the matrix, because the filler particles are assumed elastic. The two-phase composite is in plastic deformation range when the ensemble-volume averaged current stress norm in the matrix reaches a critical level. The magnitude of the current equivalent stress norm is utilized to determine the possible viscoplastic strain for any point in the composite.

6.3.2 *Average Stress in Particles*

If the particle interaction for two-phase composite is ignored, Eq. (6.30) becomes

$$\overline{\boldsymbol{\varepsilon}} = \boldsymbol{\varepsilon}^0 + \varphi\mathbf{S} : \left(\overline{\boldsymbol{\varepsilon}}^T + \overline{\boldsymbol{\varepsilon}}^{*0}\right) \tag{6.90}$$

With the noninteracting solution $\overline{\boldsymbol{\varepsilon}}^{*0}$ of the eigenstrain given by Eq. (6.42), we arrive at

$$\overline{\boldsymbol{\varepsilon}} = \left[\mathbf{I} - \varphi\mathbf{S}(\mathbf{A} + \mathbf{S})^{-1}\right] : \boldsymbol{\varepsilon}^0 + \varphi\mathbf{S}(\mathbf{A} + \mathbf{S})^{-1}(\mathbf{A} + \mathbf{I}) : \overline{\boldsymbol{\varepsilon}}^T \tag{6.91}$$

The volume-averaged stress tensor for the particles is defined by

$$\overline{\boldsymbol{\sigma}}_1 = \frac{1}{V_1}\int_{V_1} \boldsymbol{\sigma}_1(\mathbf{x})d\mathbf{x} = \frac{1}{V_1}\int_{V_1} \mathrm{C}_1 : \left[\boldsymbol{\varepsilon}^0 + \boldsymbol{\varepsilon}'(\mathbf{x}) - \boldsymbol{\varepsilon}^T(\mathbf{x})\right]d\mathbf{x} = \mathbf{C}_1 : \left[\boldsymbol{\varepsilon}^0 + \overline{\boldsymbol{\varepsilon}}_1' - \overline{\boldsymbol{\varepsilon}}_1^T\right] \tag{6.92}$$

where $\overline{\boldsymbol{\varepsilon}}_1{}'$ can be recast as

$$\overline{\boldsymbol{\varepsilon}}_1' = \mathbf{S} : \left\{\frac{1}{V_1}\int_{V_1}\left[\boldsymbol{\varepsilon}^T(\mathbf{x}') + \boldsymbol{\varepsilon}^{*0}\right]d\mathbf{x}'\right\} = \mathbf{S} : \left(\overline{\boldsymbol{\varepsilon}}_1^T + \overline{\boldsymbol{\varepsilon}}_1^{*0}\right) \tag{6.93}$$

Therefore, we have

$$\overline{\sigma}_1 = \mathbf{C}_1\mathbf{A}(\mathbf{A}+\mathbf{S})^{-1} : \boldsymbol{\varepsilon}^0 - \mathbf{C}_1\mathbf{A}(\mathbf{I}-\mathbf{S})(\mathbf{A}+\mathbf{S})^{-1} : \overline{\boldsymbol{\varepsilon}}^T \tag{6.94}$$

As result of Eq. (6.91), the average internal stresses of particles can be expressed as:

$$\begin{aligned}\overline{\boldsymbol{\sigma}}_1 &= \mathbf{C}_1\mathbf{A}(\mathbf{A}+\mathbf{S})^{-1}\left[\mathbf{I} - \varphi\mathbf{S}(\mathbf{A}+\mathbf{S})^{-1}\right]^{-1} : \overline{\boldsymbol{\varepsilon}} \\ &- \mathbf{C}_1\mathbf{A}(\mathbf{A}+\mathbf{S})^{-1}\left\{\varphi\mathbf{S}(\mathbf{A}+\mathbf{S})^{-1}(\mathbf{A}+\mathbf{I})\left[\mathbf{I} - \varphi\mathbf{S}(\mathbf{A}+\mathbf{S})^{-1}\right]^{-1} + (\mathbf{I}-\mathbf{S})\right\} : \overline{\boldsymbol{\varepsilon}}^T \\ &= \mathbf{U} : \overline{\boldsymbol{\varepsilon}} - \mathbf{V} : \overline{\boldsymbol{\varepsilon}}^T\end{aligned} \tag{6.95}$$

where

$$\mathbf{U} = \mathbf{C}_1\mathbf{A}(\mathbf{A}+\mathbf{S})^{-1}\left[\mathbf{I} - \varphi\mathbf{S}(\mathbf{A}+\mathbf{S})^{-1}\right]^{-1} \tag{6.96}$$

$$\mathbf{V} = \mathbf{C}_1\mathbf{A}(\mathbf{A}+\mathbf{S})^{-1}\left\{\varphi\mathbf{S}(\mathbf{A}+\mathbf{S})^{-1}(\mathbf{A}+\mathbf{I})\left[\mathbf{I} - \varphi\mathbf{S}(\mathbf{A}+\mathbf{S})^{-1}\right]^{-1} + (\mathbf{I}-\mathbf{S})\right\} \tag{6.97}$$

By carrying out the lengthy algebra, the components of the positive definite fourth-rank tensor **U** and **V** are explicitly given by

$$U_{ijkl} = U_1\delta_{ij}\delta_{kl} + U_2\left(\delta_{ik}\delta_{jl} + \delta_{il}\delta_{jk}\right) \tag{6.98}$$

$$V_{ijkl} = V_1\delta_{ij}\delta_{kl} + V_2\left(\delta_{ik}\delta_{jl} + \delta_{il}\delta_{jk}\right) \tag{6.99}$$

where

$$3U_1 + 2U_2 = \frac{(3\alpha + 2\beta) - 10(1 + \nu_0)}{(3\alpha + 2\beta) - 10(1 + \nu_0)\varphi} \cdot 3k_1 \tag{6.100}$$

$$U_2 = \frac{\beta - (8 - 10\nu_0)}{\beta - (8 - 10\nu_0)\varphi} \cdot \mu_1 \tag{6.101}$$

$$3V_1 + 2V_2 = \frac{30(1 + \nu_0)k_1^2}{(k_1 - k_0)(3\alpha + 2\beta)} \left[\frac{(3\alpha + 2\beta) - 10(1 + \nu_0)}{(3\alpha + 2\beta) - 10(1 + \nu_0)\varphi} \cdot \varphi - 1 \right] + 3k_1 \tag{6.102}$$

$$V_2 = \frac{(8 - 10\nu_0)\mu_1^2}{(\mu_1 - \mu_0)\beta} \cdot \left[\frac{\beta - (8 - 10\nu_0)}{\beta - (8 - 10\nu_0)\varphi} \cdot \varphi - 1 \right] + \mu_1 \tag{6.103}$$

6.4 Pairwise Interacting Solution for Two-Phase Composites

In this section, we extend the noninteracting solution for two-phase composites developed in Sect. 6.3 to account for the interparticle interactions.

6.4.1 Approximate Solution of Two-Phase Interaction

If we neglect the interparticle interaction effects, then the ensemble-volume averaged perturbed strain $\bar{\boldsymbol{\varepsilon}}_q^{\prime p}$ can be dropped. The resulting noninteracting approximation for the particles becomes

$$-\mathbf{A} : \bar{\boldsymbol{\varepsilon}}^{*0} = \boldsymbol{\varepsilon}^0 - \boldsymbol{\varepsilon}^T + \mathbf{S} : \boldsymbol{\varepsilon}^T + \mathbf{S} : \bar{\boldsymbol{\varepsilon}}^{*0} \tag{6.104}$$

Within the present two-sphere context, the integral equation governing the distributed eigenstrain $\boldsymbol{\varepsilon}^*$ for a given particle configuration and remote strain field $\boldsymbol{\varepsilon}^0$ can be written as

$$-\mathbf{A} : \boldsymbol{\varepsilon}^{*(i)}(\mathbf{x}) = \boldsymbol{\varepsilon}^0 - \boldsymbol{\varepsilon}^T(\mathbf{x}) + \int_{\Omega_i} \mathbf{G}(\mathbf{x} - \mathbf{x}') : \left[\boldsymbol{\varepsilon}^T(\mathbf{x}') + \boldsymbol{\varepsilon}^{*(i)}(\mathbf{x}') \right] d\mathbf{x}' + \int_{\Omega_j} \mathbf{G}(\mathbf{x} - \mathbf{x}') : \left[\varepsilon^T(\mathbf{x}') + \boldsymbol{\varepsilon}^{*(j)}(\mathbf{x}') \right] d\mathbf{x}' \quad (i \neq j, \quad i,j = 1,2) \tag{6.105}$$

where assuming $\boldsymbol{\varepsilon}^T$ is uniform in the particles.

By subtracting the noninteracting solution Eq. (6.104), the effect of the interparticle interaction can be founded by solving the following integral equation:

$$-\mathbf{A} : \mathbf{d}^{*(i)}(\mathbf{x}) = \int_{\Omega_i} \mathbf{G}(\mathbf{x}-\mathbf{x}') : \mathbf{d}^{*(i)}(\mathbf{x}')d\mathbf{x}' + \int_{\Omega_j} \mathbf{G}(\mathbf{x}-\mathbf{x}') : \mathbf{d}^{*(j)}(\mathbf{x}')d\mathbf{x}'$$
$$+ \int_{\Omega_j} \mathbf{G}(\mathbf{x}-\mathbf{x}') : \left[\boldsymbol{\varepsilon}^T(\mathbf{x}') + \boldsymbol{\varepsilon}^{*0}\right]d\mathbf{x}' \quad (i \neq j, \quad i,j = 1,2) \tag{6.106}$$

where $\mathbf{d}^{*(i)}$ is given by

$$\mathbf{d}^{*(i)} = \boldsymbol{\varepsilon}^{*(i)} - \boldsymbol{\varepsilon}^{*0} \tag{6.107}$$

Following the procedure given by Ju and Chen (1994b), the approximate equations for $\overline{\mathbf{d}}^{*(i)}$ for the two-sphere interaction problem can be written as

$$-\mathbf{A} : \overline{\mathbf{d}}^{*(i)} = \mathbf{S} : \overline{\mathbf{d}}^{*(i)} + \mathbf{G}^1(\mathbf{x}_i - \mathbf{x}_j) : \overline{\mathbf{d}}^{*(j)} + \mathbf{G}^2(\mathbf{x}_i - \mathbf{x}_j)$$
$$: \left(\boldsymbol{\varepsilon}^T + \boldsymbol{\varepsilon}^{*0}\right) + \mathbf{0}\left(\rho^8\right) \tag{6.108}$$

where

$$\mathbf{G}^1(\mathbf{x}_i - \mathbf{x}_j) = \frac{\rho^3}{30(1-\nu_0)}\left(\mathbf{H}^1 + \rho^2\mathbf{H}^2\right) \tag{6.109}$$

$$\mathbf{G}^2(\mathbf{x}_i - \mathbf{x}_j) = \frac{\rho^3}{30(1-\nu_0)}\left(\mathbf{H}^1 + 2\rho^2\mathbf{H}^2\right) \tag{6.110}$$

In addition, $\mathbf{0}(\rho^6)$ denotes the terms, which are higher than the order of ρ^6.
where $\mathbf{r} \equiv \mathbf{x}_i - \mathbf{x}_j$, $r \equiv |\mathbf{r}|$, $\rho = a/r$, and a is the radius of a spherical particle. The components of $\mathbf{H}^1$ and $\mathbf{H}^2$ are given by Eqs. (6.40) and (6.41).

Furthermore, we observe that

$$\overline{\mathbf{d}}^{*(i)} = \overline{\mathbf{d}}^{*(j)} = \overline{\mathbf{d}}^* \tag{6.111}$$

Therefore, the solution of Eq. (6.107) is given by

$$\overline{\mathbf{d}}^* = -30(1-\nu_0)\mathbf{T}^{-1} \cdot \mathbf{G}^2 : \left(\boldsymbol{\varepsilon}^T + \boldsymbol{\varepsilon}^{*0}\right) + \mathbf{0}\left(\rho^8\right) \tag{6.112}$$

where

$$\mathbf{T}(\mathbf{x}_i - \mathbf{x}_j) = 30(1 - v_0)\left[\mathbf{A} + \mathbf{S} + \mathbf{G}^1(\mathbf{x}_i - \mathbf{x}_j)\right] \tag{6.113}$$

The corresponding expression to the order of $\mathbf{0}(\rho^3)$ is

$$\mathbf{T}^{-1} = \mathbf{K}^{-1} + \rho^3\mathbf{L} + \cdots \tag{6.114}$$

where

$$K_{ijkl} = F_{ijkl}(0, 0, 0, 0, \alpha, \beta) \tag{6.115}$$

$$L_{ijkl} = \frac{5}{4\beta^2} F_{ijkl}\left(-15, 3v_0, \frac{6\alpha(1 - 2v_0)}{3\alpha + 2\beta}, \frac{6\alpha(1 + v_0)}{3\alpha + 2\beta}, -\frac{2\alpha(2 - v_0)}{3\alpha + 2\beta}, 1 - 2v_0\right) \tag{6.116}$$

with α and β given in Eqs. (6.5) and (6.6).

The final expression for $\overline{\mathbf{d}}^*$ is

$$\begin{aligned} \overline{\mathbf{d}}^* &= -\rho^3\left[\mathbf{K}^{-1} \cdot \left(\mathbf{H}^1 + 2\rho^2\mathbf{H}^2\right)\right] : \left(\boldsymbol{\varepsilon}^T + \boldsymbol{\varepsilon}^{*0}\right) - \rho^6\left(\mathbf{L} \cdot \mathbf{H}^1\right) \\ &: \left(\boldsymbol{\varepsilon}^T + \boldsymbol{\varepsilon}^{*0}\right) + \mathbf{0}\left(\rho^8\right) \end{aligned} \tag{6.117}$$

In order to obtain the ensemble-average solution of $\overline{\mathbf{d}}^*$ within the context of approximate pairwise particle interaction, Eq. (6.117) must be integrated over all possible positions $\mathbf{x_2}$ of the second particle for a given location of the first particle $\mathbf{x_1}$. The ensemble-average process can be expressed as

$$\left\langle \overline{\mathbf{d}}^* \right\rangle = \int_{V-\Omega_1} \overline{\mathbf{d}}^*(\mathbf{x}_1 - \mathbf{x}_2) P(\mathbf{x}_2|\mathbf{x}_1) d\mathbf{x}_2 \tag{6.118}$$

It is often assumed that the two-point conditional probability function is statically isotropic and uniform and obeys the following form:

$$P(\mathbf{x}_2|\mathbf{x}_1) = \begin{cases} \dfrac{N}{V} & \text{if } |\mathbf{x}_2 - \mathbf{x}_1| \geq 2a \\ 0 & \text{otherwise} \end{cases} \tag{6.119}$$

where *N*/*V* is the number density of particles in a composite. Accordingly, the ensemble integration of Eq. (6.118) can be written as

$$\langle \overline{\mathbf{d}}^* \rangle = -\frac{N}{V}\left\{ \mathbf{K}^{-1} \cdot \left[\int_{2a}^{\infty} \rho^3 \int_A \mathbf{H}^1 dA dr + 2 \int_{2a}^{\infty} \rho^5 \int_A \mathbf{H}^2 dA dr \mathbf{H}^2 \right] \right\} : \left(\boldsymbol{\varepsilon}^T + \boldsymbol{\varepsilon}^{*0} \right)$$
$$- \frac{N}{V}\left[\int_{2a}^{\infty} \rho^6 \int_A \left(\mathbf{L} \cdot \mathbf{H}^1 \right) dA dr \right] : \left(\boldsymbol{\varepsilon}^T + \boldsymbol{\varepsilon}^{*0} \right) + \mathbf{0}\left(\rho^8 \right) \tag{6.120}$$

where A denotes the spherical surface with radius$r$. After lengthy but straightforward mathematical manipulation, the final expression takes the following form:

$$\langle \overline{\boldsymbol{\varepsilon}}^* \rangle = \boldsymbol{\Gamma} : \boldsymbol{\varepsilon}^{*\mathbf{0}} + \boldsymbol{\Gamma}^* : \boldsymbol{\varepsilon}^T \tag{6.121}$$

The components of the positive definite fourth-rank tensor $\boldsymbol{\Gamma}$ and $\boldsymbol{\Gamma}^*$ read

$$\Gamma_{ijkl} = \gamma_1 \delta_{ij} \delta_{kl} + \gamma_2 \left(\delta_{ik} \delta_{jl} + \delta_{il} \delta_{jk} \right) \tag{6.122}$$

$$\Gamma_{ijkl}{}^* = \gamma_1{}^* \delta_{ij} \delta_{kl} + \gamma_2{}^* \left(\delta_{ik} \delta_{jl} + \delta_{il} \delta_{jk} \right) \tag{6.123}$$

where

$$\gamma_1{}^* = \frac{5\varphi}{8\beta^2}\left\{ (13 - 14v_0)v_0 - \frac{8\alpha}{3\alpha + 2\beta}(1 - 2v_0)(1 + v_0) \right\} \tag{6.124}$$

$$\gamma_2{}^* = \frac{5\varphi}{16\beta^2}\left\{ \left(25 - 34v_0 + 22v_0^2\right) - \frac{6\alpha}{3\alpha + 2\beta}(1 - 2v_0)(1 + v_0) \right\} \tag{6.125}$$

With γ_1 and γ_2 are given in Eqs. (4.7) and (4.8).

6.4.2 *Ensemble-Average Stress Norm in the Matrix*

The total stress at any point $\mathbf{x}$ in the matrix is given by the superposition of the far-field stress $\boldsymbol{\sigma}_0$ and the perturbed stress $\boldsymbol{\sigma}'$ induced by the particles. Assuming the elastic eigenstrain in the particle $\boldsymbol{\varepsilon}^*$ is uniform, the perturbed stress for any matrix point $\mathbf{x}$ takes the form (Ju and Chen 1994a)

$$\boldsymbol{\sigma}'(\mathbf{x}|\mathbf{x}_1) = \left[\mathbf{C}_0 \cdot \overline{\mathbf{G}}(\mathbf{x} - \mathbf{x}_1) \right] : \left(\boldsymbol{\varepsilon}^T + \boldsymbol{\varepsilon}^* \right) \tag{6.126}$$

where $\boldsymbol{\varepsilon}^T$ is the eigenstrain caused by the CTE mismatch between the matrix and the particle. Using the ensemble-volume averaged eigenstrain given in Eq. (6.121), the stress perturbation can be given by the following relation:

$$\boldsymbol{\sigma}'(\mathbf{x}|\mathbf{x}_1) = \left[\mathbf{C}_0 \cdot \overline{\mathbf{G}}(\mathbf{x}-\mathbf{x}_1) \cdot \boldsymbol{\Gamma}\right] : \left(\boldsymbol{\varepsilon}^T + \boldsymbol{\varepsilon}^{*0}\right) \tag{6.127}$$

where $\boldsymbol{\varepsilon}^{*0}$ denotes the solution of the eigenstrain for the single inclusion problem, which is given by Eq. (6.42). Therefore

$$\begin{aligned}\boldsymbol{\sigma}'(\mathbf{x}|\mathbf{x}_1) &= -\left[\mathbf{C}_0 \cdot \overline{\mathbf{G}} \cdot \boldsymbol{\Gamma}(\mathbf{A}+\mathbf{S})^{-1}\right] \\ &: \boldsymbol{\varepsilon}^0 + \left[\mathbf{C}_0 \cdot \overline{\mathbf{G}} \cdot \boldsymbol{\Gamma}(\mathbf{A}+\mathbf{S})^{-1}(\mathbf{A}+\mathbf{I})\right] : \boldsymbol{\varepsilon}^T\end{aligned} \tag{6.128}$$

Following the same procedure used for the noninteracting solution, we obtain the ensemble-averaged current stress norm at any matrix point $\mathbf{x}$ as:

$$\langle H \rangle_m(\mathbf{x}) = \boldsymbol{\sigma}^0 : \mathbf{T} : \boldsymbol{\sigma}^0 + \boldsymbol{\sigma}^T : \mathbf{T}^* : \boldsymbol{\sigma}^T - 2\boldsymbol{\sigma}^0 : \mathbf{T}^* : \boldsymbol{\sigma}^T \tag{6.129}$$

where $\boldsymbol{\sigma}^T$ is given by Eq. (6.56) and the components of the positive definite fourth-rank tensor $\mathbf{T}$ and $\mathbf{T}^*$ read:

$$T_{ijkl} = T_1\delta_{ij}\delta_{kl} + T_2\left(\delta_{ik}\delta_{jl} + \delta_{il}\delta_{jk}\right) \tag{6.130}$$

$$T^*_{ijkl} = T^*_1\delta_{ij}\delta_{kl} + T^*_2\left(\delta_{ik}\delta_{jl} + \delta_{il}\delta_{jk}\right) \tag{6.131}$$

with

$$3T_1 + 2T_2 = 200(1-2\nu_0)^2 \frac{(3\gamma_1 + 2\gamma_2)^2}{(3\alpha + 2\beta)^2}\varphi \tag{6.132}$$

$$T_2 = \frac{1}{2} + \left(23 - 50\nu_0 + 35\nu_0^2\right)\frac{4\gamma_2^2}{\beta^2}\varphi \tag{6.133}$$

$$3T^*_1 + 2T^*_2 = 200(1-2\nu_0)^2 \frac{(3\gamma_1 + 2\gamma_2)^2}{(3\alpha + 2\beta)^2}\varphi \tag{6.134}$$

$$T^*_2 = \left(23 - 50\nu_0 + 35\nu_0^2\right)\frac{4\gamma_2^2}{\beta^2}\varphi \tag{6.135}$$

α and β are given in Eqs. (6.5) and (6.6).

The ensemble-averaged current stress norm at a matrix point can also be expressed in terms of the macroscopic stress $\overline{\boldsymbol{\sigma}}$. Following the same procedure as in former section, the relation between the far-field stress $\boldsymbol{\sigma}^0$ and the macroscopic stress $\overline{\boldsymbol{\sigma}}$ takes the form

$$\overline{\boldsymbol{\sigma}} = \mathbf{P} : \boldsymbol{\sigma}^0 - \mathbf{Q} : \boldsymbol{\sigma}^T \tag{6.136}$$

where

$$\mathbf{P} = \mathrm{I} + \varphi(\mathbf{I} - \mathbf{S})\boldsymbol{\Gamma}(\mathbf{A} + \mathbf{S})^{-1} \tag{6.137}$$

$$\mathbf{Q} = \varphi(\mathbf{I} - \mathbf{S})\boldsymbol{\Gamma}(\mathbf{A} + \mathbf{S})^{-1} \tag{6.138}$$

where the components of **P** and **Q** are given

$$P_{ijkl} = P_1\delta_{ij}\delta_{kl} + P_2\left(\delta_{ik}\delta_{jl} + \delta_{il}\delta_{jk}\right) \tag{6.139}$$

$$Q_{ijkl} = Q_1\delta_{ij}\delta_{kl} + Q_2\left(\delta_{ik}\delta_{jl} + \delta_{il}\delta_{jk}\right) \tag{6.140}$$

with

$$3P_1 + 2P_2 = a\varphi + 1 \tag{6.141}$$

$$P_2 = \frac{1}{2}(b\varphi + 1) \tag{6.142}$$

$$3Q_1 + 2Q_2 = a\varphi \tag{6.143}$$

$$Q_2 = \frac{1}{2}b\varphi \tag{6.144}$$

$$a = 20(1 - 2v_0)\frac{3\gamma_1 + 2\gamma_2}{3\alpha + 2\beta} \tag{6.145}$$

$$b = (7 - 5v_0)\frac{2\gamma_2}{\beta} \tag{6.146}$$

Using Eq. (6.136), we arrive at the alternative expression for the ensemble-averaged current stress norm in a matrix point**x**:

$$\langle H\rangle_m(\mathbf{x}) = \overline{\boldsymbol{\sigma}} : \overline{\mathbf{T}} : \overline{\boldsymbol{\sigma}} + 2\overline{\boldsymbol{\sigma}} : \overline{\mathbf{T}}^* : \overline{\boldsymbol{\sigma}}^T + \overline{\boldsymbol{\sigma}}^T : \overline{\mathbf{T}}^{**} : \overline{\boldsymbol{\sigma}}^T \tag{6.147}$$

where the positive definite fourth-rank tensor $\overline{\mathbf{T}}$, $\overline{\mathbf{T}}^*$ and $\overline{\mathbf{T}}^{**}$ are defined as

$$\overline{\mathbf{T}} = \left(\mathbf{P}^{-1}\right)^{\mathbf{T}} \cdot \mathbf{T} \cdot \mathbf{P}^{-1} \tag{6.148}$$

$$\overline{\mathbf{T}}^* = \left(\mathbf{P}^{-1}\right)^{\mathbf{T}} \cdot \mathbf{T} \cdot \mathbf{P}^{-1} \cdot \mathbf{Q} - \mathbf{T}^* \cdot \mathbf{P}^{-1} \tag{6.149}$$

$$\overline{\mathbf{T}}^{**} = \left(\mathbf{P}^{-1} \cdot \mathbf{Q}\right)^{\mathbf{T}} \cdot \mathbf{T} \cdot \left(\mathbf{P}^{-1} \cdot \mathbf{Q}\right) + \mathbf{T}^* - 2\mathbf{T}^* \cdot \left(\mathbf{P}^{-1} \cdot \mathbf{Q}\right) \tag{6.150}$$

where the components of $\overline{\mathbf{T}}$, $\overline{\mathbf{T}}^*$, and $\overline{\mathbf{T}}^{**}$ are

$$\overline{T}_{ijkl} = \overline{T}_1\delta_{ij}\delta_{kl} + \overline{T}_2\left(\delta_{ik}\delta_{jl} + \delta_{il}\delta_{jk}\right) \tag{6.151}$$

$$\overline{T}^*_{ijkl} = \overline{T}^*_1\delta_{ij}\delta_{kl} + \overline{T}^*_2\left(\delta_{ik}\delta_{jl} + \delta_{il}\delta_{jk}\right) \tag{6.152}$$

$$\overline{T}_{ijkl}^{**} = \overline{T}_1^{**}\delta_{ij}\delta_{kl} + \overline{T}_2^{**}\left(\delta_{ik}\delta_{jl} + \delta_{il}\delta_{jk}\right) \tag{6.153}$$

with

$$3\overline{T}_1 + 2\overline{T}_2 = \frac{3T_1 + 2T_2}{(a\varphi + 1)^2} \tag{6.154}$$

$$\overline{T}_2 = \frac{T_2}{(b\varphi + 1)^2} \tag{6.155}$$

$$3\overline{T}_1^* + 2\overline{T}_2^* = -\frac{3T_1 + 2T_2}{(a\varphi + 1)^2} \tag{6.156}$$

$$\overline{T}_2^* = -\frac{T_2}{(b\varphi + 1)^2} + \frac{1}{2(b\varphi + 1)} \tag{6.157}$$

$$3\overline{T}_1^{**} + 2\overline{T}_2^{**} = \frac{3T_1 + 2T_2}{(a\varphi + 1)^2} \tag{6.158}$$

$$\overline{T}_2^{**} = \frac{T_2}{(b\varphi + 1)^2} + \frac{b\varphi - 1}{2(b\varphi + 1)} \tag{6.159}$$

Because $\boldsymbol{\sigma}^T = \mathbf{AC}_1 : \boldsymbol{\varepsilon}^T$ is spherical stress, caused by CTE mismatch, Eq. (6.147) of the ensemble-averaged loading function can be simplified as

$$\langle H \rangle_m(\mathbf{x}) = \left(\overline{\boldsymbol{\sigma}} - \overline{\boldsymbol{\sigma}}^T\right) : \overline{\mathbf{T}} : \left(\overline{\boldsymbol{\sigma}} - \overline{\boldsymbol{\sigma}}^T\right) \tag{6.160}$$

6.4.3 Ensemble-Average Stress in the Filler Particles

The ensemble-volume averaged strain for two-phase composites can be given by

$$\overline{\boldsymbol{\varepsilon}} = \boldsymbol{\varepsilon}^0 + \varphi\mathbf{S} : \left(\overline{\boldsymbol{\varepsilon}}^T + \overline{\boldsymbol{\varepsilon}}^*\right) \tag{6.161}$$

Substituting Eq. (6.121)

$$\overline{\boldsymbol{\varepsilon}} = \boldsymbol{\varepsilon}^0 + \varphi\mathbf{S\Gamma} : \left(\overline{\boldsymbol{\varepsilon}}^T + \overline{\boldsymbol{\varepsilon}}^{*0}\right) \tag{6.162}$$

With the noninteracting solution $\overline{\boldsymbol{\varepsilon}}^{*0}$ of the eigenstrain given by Eq. (6.42), we arrive at

$$\overline{\boldsymbol{\varepsilon}} = \left[I - \varphi \mathbf{S\Gamma}(\mathbf{A}+\mathbf{S})^{-1}\right] : \boldsymbol{\varepsilon}^0 + \varphi\left[\mathbf{S\Gamma}(\mathbf{A}+\mathbf{S})^{-1}(\mathbf{A}+\mathbf{I})\right] : \overline{\boldsymbol{\varepsilon}}^T \tag{6.163}$$

The volume-averaged stress tensor for the particles is defined as

$$\overline{\boldsymbol{\sigma}_1} = \frac{1}{V_1}\int_{V_1} \boldsymbol{\sigma}_1(\mathbf{x})d\mathbf{x} = \frac{1}{V_1}\int_{V_1} \mathrm{C}_1 : \left[\boldsymbol{\varepsilon}^0 + \boldsymbol{\varepsilon}'(\mathbf{x}) - \boldsymbol{\varepsilon}^T(\mathbf{x})\right]d\mathbf{x} = \mathbf{C}_1 : \left[\boldsymbol{\varepsilon}^0 + \overline{\boldsymbol{\varepsilon}}' - \overline{\boldsymbol{\varepsilon}}^T\right] \tag{6.164}$$

where $\overline{\boldsymbol{\varepsilon}}'$ can be recast as

$$\overline{\boldsymbol{\varepsilon}}' = \mathbf{S} : \left\{\frac{1}{V_1}\int_{V_1}\left[\boldsymbol{\varepsilon}^T(\mathbf{x}') + \boldsymbol{\varepsilon}^*(\mathbf{x}')\right]d\mathbf{x}'\right\} = \mathbf{S\Gamma} : \left(\overline{\boldsymbol{\varepsilon}}^T + \overline{\boldsymbol{\varepsilon}}^{*0}\right) \tag{6.165}$$

Using Eq. (6.165), we have

$$\overline{\boldsymbol{\sigma}_1} = \mathbf{C}_1 : \boldsymbol{\varepsilon}^0 + \left[\mathbf{C}_1(\mathbf{S\Gamma} - \mathbf{I})\right] : \overline{\boldsymbol{\varepsilon}}^T + (\mathbf{C}_1\mathbf{S\Gamma}) : \overline{\boldsymbol{\varepsilon}}^{*0} \tag{6.166}$$

With the noninteracting solution $\overline{\boldsymbol{\varepsilon}}^{*0}$ of the eigenstrain, we arrive at

$$\overline{\boldsymbol{\sigma}_1} = \mathbf{C}_1\left[\mathbf{I} - \mathbf{S\Gamma}(\mathbf{A}+\mathbf{S})^{-1}\right] : \boldsymbol{\varepsilon}^0 + \mathbf{C}_1\left[\mathbf{S\Gamma}(\mathbf{A}+\mathbf{S})^{-1}(\mathbf{A}+\mathbf{I}) - \mathbf{I}\right] : \overline{\boldsymbol{\varepsilon}}^T \tag{6.167}$$

Combing with Eq. (6.163), the averaged internal stresses of particles can be expressed as

$$\begin{aligned}\overline{\boldsymbol{\sigma}_1} &= \mathbf{C}_1\left[\mathbf{I} - \mathbf{S\Gamma}(\mathbf{A}+\mathbf{S})^{-1}\right]\left[\mathbf{I} - \varphi\mathbf{S\Gamma}(\mathbf{A}+\mathbf{S})^{-1}\right]^{-1} : \overline{\boldsymbol{\varepsilon}} - \mathbf{C}_1\{\varphi\mathbf{S\Gamma}\left[\mathbf{I} - \mathbf{S\Gamma}(\mathbf{A}+\mathbf{S})^{-1}\right] \\ &\quad \left[\mathbf{I} - \varphi\mathbf{S\Gamma}(\mathbf{A}+\mathbf{S})^{-1}\right]^{-1}(\mathbf{A}+\mathbf{S})^{-1}(\mathbf{A}+\mathbf{I}) - \left[\mathbf{S\Gamma}(\mathbf{A}+\mathbf{S})^{-1}(\mathbf{A}+\mathbf{I}) - \mathbf{I}\right]\} : \overline{\boldsymbol{\varepsilon}}^T \\ &= \mathbf{U} : \overline{\boldsymbol{\varepsilon}} - \mathbf{V} : \overline{\boldsymbol{\varepsilon}}^T\end{aligned} \tag{6.168}$$

where

$$\mathbf{U} = \mathbf{C}_1\left[\mathbf{I} - \mathbf{S\Gamma}(\mathbf{A}+\mathbf{S})^{-1}\right]\left[\mathbf{I} - \varphi\mathbf{S\Gamma}(\mathbf{A}+\mathbf{S})^{-1}\right]^{-1} \tag{6.169}$$

$$\mathbf{V} = \mathbf{C}_1\{\varphi\mathbf{S}\boldsymbol{\Gamma}\left[\mathbf{I} - \mathbf{S}\boldsymbol{\Gamma}(\mathbf{A}+\mathbf{S})^{-1}\right]\left[\mathbf{I} - \varphi\mathbf{S}\boldsymbol{\Gamma}(\mathbf{A}+\mathbf{S})^{-1}\right]^{-1}(\mathbf{A}+\mathbf{S})^{-1}(\mathbf{A}+\mathbf{I}) - \left[\mathbf{S}\boldsymbol{\Gamma}(\mathbf{A}+\mathbf{S})^{-1}(\mathbf{A}+\mathbf{I}) - \mathbf{I}\right]\} \tag{6.170}$$

By carrying out the lengthy algebra, the components of the positive definite fourth-rank tensor **U** and **V** are explicitly given by

$$U_{ijkl} = U_1\delta_{ij}\delta_{kl} + U_2\left(\delta_{ik}\delta_{jl} + \delta_{il}\delta_{jk}\right) \tag{6.171}$$

$$V_{ijkl} = V_1\delta_{ij}\delta_{kl} + V_2\left(\delta_{ik}\delta_{jl} + \delta_{il}\delta_{jk}\right) \tag{6.172}$$

where

$$3U_1 + 2U_2 = \frac{(3\alpha+2\beta) - 10(1+\nu_0)(3\gamma_1+2\gamma_2)}{(3\alpha+2\beta) - 10(1+\nu_0)\varphi(3\gamma_1+2\gamma_2)} \cdot 3k_1 \tag{6.173}$$

$$U_2 = \frac{\beta - 4(4-5\nu_0)\gamma_2}{\beta - 4(4-5\nu_0)\varphi\gamma_2} \cdot \mu_1 \tag{6.174}$$

$$3V_1 + 2V_2 = \frac{30(1+\nu_0)(3\gamma_1+2\gamma_2)k_1^2}{(k_1-k_0)(3\alpha+2\beta)} \times \left[\frac{(3\alpha+2\beta) - 10(1+\nu_0)(3\gamma_1+2\gamma_2)}{(3\alpha+2\beta) - 10(1+\nu_0)\varphi(3\gamma_1+2\gamma_2)} \cdot \varphi - 1\right] + 3k_1 \tag{6.175}$$

$$V_2 = \frac{4(4-5\nu_0)\gamma_2\mu_1^2}{(\mu_1-\mu_0)\beta} \cdot \left[\frac{\beta - 4(4-5\nu_0)\gamma_2}{\beta - 4(4-5\nu_0)\gamma_2\varphi} \cdot \varphi - 1\right] + \mu_1 \tag{6.176}$$

6.5 Noninteracting Solution for Three-Phase Composites

Let us consider an initially perfectly bonded two-phase composite consisting of an elastic matrix (phase 0) with bulk modulus k_0 and shear modulus μ_0 and randomly dispersed elastic spherical particles (phase 1) with bulk modulus k_1 and shear modulus μ_1. Subsequently, as loadings or deformations are applied, some particles could experience complete interfacial debonding. These completely debonded particles can be regarded as spherical voids (phase 2).

6.5.1 Effective Elastic Modulus of Multiphase Composites

Effective elastic moduli of multiphase composites containing randomly located, uni-directionally aligned elastic ellipsoids were explicitly derived by Ju and Chen (1994a). For a multiphase composite, the effective (noninteracting) elasticity tensor $\mathbf{C}_*$ reads

$$\mathbf{C}_* = \mathbf{C}_0 \cdot \left[\mathbf{I} + \mathbf{B} \cdot (\mathbf{I} \cdot \mathbf{B})^{-1}\right] \tag{6.177}$$

where $\mathbf{B}$ takes the form

$$\mathbf{B} = \sum_{q=1}^{n} \varphi_q (\mathbf{S} + \mathbf{A_q})^{-1} \tag{6.178}$$

Since phase 2 contains spherical voids, we have $\mathbf{A_2} = -\mathbf{I}$. The effective bulk modulus k_* and shear modulus μ_* for the three-phase composite ($n = 2$) can be explicitly expressed as

$$k_* = k_0 \left(1 + \frac{30(1-\nu_0) \sum_{q=1}^{2} \frac{\varphi_q}{3\alpha_q + 2\beta_q}}{1 - 10(1+\nu_0) \sum_{q=1}^{2} \frac{\varphi_q}{3\alpha_q + 2\beta_q}} \right) \tag{6.179}$$

$$\mu^* = \mu_0 \left(1 + \frac{15(1-\nu_0) \sum_{q=1}^{2} \frac{\varphi_q}{\beta_q}}{1 - 2(4-5\nu_0) \sum_{q=1}^{2} \frac{\varphi_q}{\beta_q}} \right) \tag{6.180}$$

with

$$\alpha_q = 2(5\nu_0 - 1) + 10(1-\nu_0)\left(\frac{k_0}{k_q - k_0} - \frac{\mu_0}{\mu_q - \mu_0}\right) \tag{6.181}$$

$$\beta_q = 2(4 - 5\nu_0) + 15(1-\nu_0)\frac{\mu_0}{\mu_q - \mu_0} \tag{6.182}$$

6.5.2 Ensemble-Average Stress Norm in the Matrix

According to the Eshelby theory, the perturbed stress $\boldsymbol{\sigma}'$ at any point in the matrix due to the presence of the particles and voids can be written as

$$\boldsymbol{\sigma}'(\mathbf{x}) = \mathbf{C}_0 : \int_V \mathbf{G}(\mathbf{x}-\mathbf{x}') : (\boldsymbol{\varepsilon}_1^* + \boldsymbol{\varepsilon}_1^T) d\mathbf{x}' + \mathbf{C}_0 : \int_V \mathbf{G}(\mathbf{x}-\mathbf{x}') : (\boldsymbol{\varepsilon}_2^* + \boldsymbol{\varepsilon}_2^T) d\mathbf{x}' \tag{6.183}$$

The eigenstrain in a single ellipsoidal inclusion is uniform for the interior points of an isolated (no interacting) inclusion. Therefore, the perturbed stress for any matrix point $\mathbf{x}$ due to a typical isolated q-phase inhomogeneity centered at $\mathbf{x}_q$ takes the form

$$\boldsymbol{\sigma}'(\mathbf{x}|\mathbf{x}_q) = \left[\mathbf{C}_0 \cdot \overline{\mathbf{G}}(\mathbf{x}-\mathbf{x}_q)\right] : \left(\boldsymbol{\varepsilon}_q^{*0} + \boldsymbol{\varepsilon}_q^T\right) \tag{6.184}$$

where $\overline{\mathbf{G}}(\mathbf{x}-\mathbf{x}_q)$ is given by Eq. (6.39), and

$$\overline{\boldsymbol{\varepsilon}}_q^{*0} = -(\mathbf{A}_q + \mathbf{S})^{-1} : \boldsymbol{\varepsilon}^0 + (\mathbf{A}_q + \mathbf{S})^{-1} \cdot (\mathbf{I} - \mathbf{S}) : \boldsymbol{\varepsilon}_q^T \tag{6.185}$$

We denote by $H(\mathbf{x}|\wp)$ the square of the current stress norm at the local point $\mathbf{x}$, which determines the plastic strain for a given phase configuration $\wp$. Since there is no plastic strain in the elastic particles or voids, $H(\mathbf{x}|\wp)$ can be written as

$$H(\mathbf{x}|\wp) = \begin{cases} \boldsymbol{\sigma}(\mathbf{x}|\wp) : \mathbf{I}_\mathrm{d} : \boldsymbol{\sigma}(\mathbf{x}|\wp) & \mathbf{x} \text{ in the matrix} \\ 0 & \text{otherwise} \end{cases} \tag{6.186}$$

In addition, $\langle H \rangle_m(\mathbf{x})$ is defined as the ensemble average of $H(\mathbf{x}|\wp)$ over all possible points where $\mathbf{x}$ is in the matrix phase. Matrix point receives perturbations from particles and voids. Therefore, the ensemble-average stress norm for any matrix point $\mathbf{x}$ can be evaluated by collecting and summing up all the current stress norm perturbations produced by any typical particle centered at $\mathbf{x}_1$ in the particle domain and any typical void centered at $\mathbf{x}_2$ in the void domain, and averaging over all possible locations of $\mathbf{x}_1$ and $\mathbf{x}_2$

$$\langle H\rangle_m(\mathbf{x}) = H^0 + \int_{r_1>a} \{H(\mathbf{x}|\mathbf{x}_1) - H^0\}P(\mathbf{x}_1)d\mathbf{x}_1 + \int_{r_2>a} \{H(\mathbf{x}|\mathbf{x}_2) - H^0\}P(\mathbf{x}_2)d\mathbf{x}_2 \tag{6.187}$$

where $r_1 = |\mathbf{x} - \mathbf{x}_1|$, $r_2 = |\mathbf{x} - \mathbf{x}_2|$ $P(\mathbf{x}_1)$ and $P(\mathbf{x}_2)$ denote the probability density functions for finding a particle centered at $\mathbf{x}_1$ and a void centered at $\mathbf{x}_2$, respectively. For simplicity, $P(\mathbf{x}_1)$ and $P(\mathbf{x}_2)$ are assumed to be statistically homogeneous, isotropic, and uniform. Using the properties of the fourth-order tensor $\overline{\mathbf{G}}(\mathbf{x} - \mathbf{x}_q)$, we obtain the ensemble-averaged current stress norm at any matrix point $\mathbf{x}$ as

$$\langle H\rangle_m(\mathbf{x}) = H^0 + \frac{N_1}{V}\int_{r_1>a} dr_1 \int_{A(r_1)} (\boldsymbol{\sigma}'(\mathbf{x}|\mathbf{x}_1) : \mathbf{I}_\mathrm{d} : \boldsymbol{\sigma}'(\mathbf{x}|\mathbf{x}_1))dA + \frac{N_2}{V}\int_{r_2>a} dr_1 \int_{A(r_2)} (\boldsymbol{\sigma}'(\mathbf{x}|\mathbf{x}_2) : \mathbf{I}_\mathrm{d} : \boldsymbol{\sigma}'(\mathbf{x}|\mathbf{x}_2))dA \tag{6.188}$$

By carrying out the lengthy but straightforward algebra, we have

$$\langle H\rangle_m(\mathbf{x}) = \boldsymbol{\sigma}^0 : \mathbf{T} : \boldsymbol{\sigma}^0 + \boldsymbol{\sigma}^T : \mathbf{T}^* : \boldsymbol{\sigma}^T - 2\boldsymbol{\sigma}^0 : \mathbf{T}^* : \boldsymbol{\sigma}^T \tag{6.189}$$

where

$$\boldsymbol{\sigma}^T = \mathbf{A}_1\mathbf{C}_1 : \boldsymbol{\varepsilon}^T \tag{6.190}$$

The components of the positive definite fourth-rank tensor $\mathbf{T}$ and $\boldsymbol{T}^*$ read

$$T_{ijkl} = T_1\delta_{ij}\delta_{kl} + T_2(\delta_{ik}\delta_{jl} + \delta_{il}\delta_{jk}) \tag{6.191}$$

$$T^*_{ijkl} = T^*_1\delta_{ij}\delta_{kl} + T^*_2(\delta_{ik}\delta_{jl} + \delta_{il}\delta_{jk}) \tag{6.192}$$

with

$$3T_1 + 2T_2 = 200(1 - 2v_0)^2 \sum_{q=1}^{2} \frac{\varphi_q}{(3\alpha_q + 2\beta_q)^2} \tag{6.193}$$

$$T_2 = \frac{1}{2} + (23 - 50v_0 + 35v_0^2) \sum_{q=1}^{2} \frac{\varphi_q}{\beta_q^2} \tag{6.194}$$

$$3T_1^* + 2T_2^* = 200(1 - 2\nu_0)^2 \frac{\varphi_1}{(3\alpha_1 + 2\beta_1)^2} \tag{6.195}$$

$$T_2^* = \left(23 - 50\nu_0 + 35\nu_0^2\right) \frac{\varphi}{\beta_1^2} \tag{6.196}$$

The ensemble-averaged current stress norm at a matrix point must be established in terms of the macroscopic stress $\overline{\boldsymbol{\sigma}}$ in order to express the effective yield function in terms of the macroscopic stress. In the special case of uniform dispersions of identical elastic spheres in a homogeneous matrix, the macroscopic stress and the far-field stress take the form

$$\overline{\boldsymbol{\sigma}} = \boldsymbol{C}_0 : \left[\overline{\boldsymbol{\varepsilon}} - \sum_{q=1}^{2} \varphi_q \left(\overline{\boldsymbol{\varepsilon}}_q^{*0} + \overline{\boldsymbol{\varepsilon}}_q^T\right)\right] \tag{6.197}$$

$$\boldsymbol{\sigma}^0 = \mathbf{C}_0 : \left[\overline{\boldsymbol{\varepsilon}} - \sum_{q=1}^{2} \varphi_q \mathbf{S}\left(\overline{\boldsymbol{\varepsilon}}_q^{*0} + \overline{\boldsymbol{\varepsilon}}_q^T\right)\right] \tag{6.198}$$

Using Eqs. (6.197), (6.198), and (6.185), the relation between the far-field stress $\boldsymbol{\sigma}^0$ and the macroscopic stress $\overline{\boldsymbol{\sigma}}$ is given by

$$\overline{\boldsymbol{\sigma}} = \mathbf{P} : \boldsymbol{\sigma}^0 - \mathbf{Q} : \boldsymbol{\sigma}^T \tag{6.199}$$

where

$$\mathbf{P} = \mathbf{I} + \sum_{q=1}^{2} \varphi_q (\mathbf{I} - \mathbf{S})\left(\mathbf{A}_q + \mathbf{S}\right)^{-1} \tag{6.200}$$

$$\mathbf{Q} = \varphi_1 (\mathbf{I} - \mathbf{S})(\mathbf{A}_1 + \mathbf{S})^{-1} \tag{6.201}$$

with the components of $\mathbf{P}$ and $\mathbf{Q}$ are

$$P_{ijkl} = P_1 \delta_{ij}\delta_{kl} + P_2\left(\delta_{ik}\delta_{jl} + \delta_{il}\delta_{jk}\right) \tag{6.202}$$

$$Q_{ijkl} = Q_1 \delta_{ij}\delta_{kl} + Q_2\left(\delta_{ik}\delta_{jl} + \delta_{il}\delta_{jk}\right) \tag{6.203}$$

where

$$3P_1 + 2P_2 = \sum_{q=1}^{2} a_q \varphi_q + 1 \tag{6.204}$$

$$P_2 = \frac{1}{2}\left(\sum_{q=1}^{2} b_q\varphi_q + 1\right) \tag{6.205}$$

$$3Q_1 + 2Q_2 = a_1\varphi_1 \tag{6.206}$$

$$Q_2 = \frac{1}{2}b_1\varphi_1 \tag{6.207}$$

$$a_q = \frac{20(1 - 2\nu_0)}{3\alpha_q + 2\beta_q} \tag{6.208}$$

$$b_q = \frac{(7 - 5\nu_0)}{\beta_q} \tag{6.209}$$

Using Eqs. (6.199) and (6.189), we arrive at the alternative expression for the ensemble-averaged current stress norm in a matrix point $\mathbf{x}$ as

$$\langle H\rangle_m(\mathbf{x}) = \overline{\boldsymbol{\sigma}} : \overline{\mathbf{T}} : \overline{\boldsymbol{\sigma}} + 2\overline{\boldsymbol{\sigma}} : \overline{\mathbf{T}}^{*} : \overline{\boldsymbol{\sigma}}^{T} + \overline{\boldsymbol{\sigma}}^{T} : \overline{\mathbf{T}}^{**} : \overline{\boldsymbol{\sigma}}^{T} \tag{6.210}$$

where the positive definite fourth-rank tensor $\overline{\mathbf{T}}$**,** $\overline{\mathbf{T}}^{*}$ and $\overline{T}^{**}$ are defined as

$$\overline{\mathbf{T}} = \left(\mathbf{P}^{-1}\right)^{\mathbf{T}} \cdot \mathbf{T} \cdot \mathbf{P}^{-1} \tag{6.211}$$

$$\overline{\mathbf{T}}^{*} = \left(\mathbf{P}^{-1}\right)^{\mathbf{T}} \cdot \mathbf{T} \cdot \mathbf{P}^{-1} \cdot \mathbf{Q} - \mathbf{T}^{*} \cdot \mathbf{P}^{-1} \tag{6.212}$$

$$\overline{\mathbf{T}}^{**} = \left(\mathbf{P}^{-1} \cdot \mathbf{Q}\right)^{\mathbf{T}} \cdot \mathbf{T} \cdot \left(\mathbf{P}^{-1} \cdot \mathbf{Q}\right) + \mathbf{T}^{*} - 2\mathbf{T}^{*} \cdot \left(\mathbf{P}^{-1} \cdot \mathbf{Q}\right) \tag{6.213}$$

where the components of $\overline{\mathbf{T}}$, $\overline{\mathbf{T}}^{*}$ and $\overline{\mathbf{T}}^{**}$ are

$$\overline{T}_{ijkl} = \overline{T}_1\delta_{ij}\delta_{kl} + \overline{T}_2\left(\delta_{ik}\delta_{jl} + \delta_{il}\delta_{jk}\right) \tag{6.214}$$

$$\overline{T}^{*}_{ijkl} = \overline{T}^{*}_1\delta_{ij}\delta_{kl} + \overline{T}^{*}_2\left(\delta_{ik}\delta_{jl} + \delta_{il}\delta_{jk}\right) \tag{6.215}$$

$$\overline{T}^{**}_{ijkl} = \overline{T}^{**}_1\delta_{ij}\delta_{kl} + \overline{T}^{**}_2\left(\delta_{ik}\delta_{jl} + \delta_{il}\delta_{jk}\right) \tag{6.216}$$

with

$$3\overline{T}_1 + 2\overline{T}_2 = \frac{3T_1 + 2T_2}{\left(\sum_{q=1}^{2} a_q\varphi_q + 1\right)^2} \tag{6.217}$$

$$\overline{T}_2 = \frac{T_2}{\left(\sum_{q=1}^{2} b_q\varphi_q + 1\right)^2} \tag{6.218}$$

$$3\overline{T}_1^* + 2\overline{T}_2^* = -\frac{3T_1 + 2T_2}{\left(\sum_{q=1}^{2} a_q\varphi_q + 1\right)^2} \tag{6.219}$$

$$\overline{T}_2^* = -\frac{T_2}{\left(\sum_{q=1}^{2} b_q\varphi_q + 1\right)^2} + \frac{1}{2\left(\sum_{q=1}^{2} b_q\varphi_q + 1\right)} \tag{6.220}$$

$$3\overline{T}_1^{**} + 2\overline{T}_2^{**} = \frac{3T_1 + 2T_2}{\left(\sum_{q=1}^{2} a_q\varphi_q + 1\right)^2} \tag{6.221}$$

$$\overline{T}_2^{**} = \frac{T_2}{\left(\sum_{q=1}^{2} b_q\varphi_q + 1\right)^2} + \frac{\sum_{q=1}^{2} b_q\varphi_q - 1}{2\left(\sum_{q=1}^{2} b_q\varphi_q + 1\right)} \tag{6.222}$$

Because $\boldsymbol{\sigma}^T = \mathbf{AC}_1 : \boldsymbol{\varepsilon}^T$ caused by the CTE mismatch is spherical stress, so Eq. (6.147) of the ensemble-averaged loading function can be simplified as

$$\langle H \rangle_m(\mathbf{x}) = \left(\overline{\boldsymbol{\sigma}} - \overline{\boldsymbol{\sigma}}^T\right) : \overline{\mathbf{T}} : \left(\overline{\boldsymbol{\sigma}} - \overline{\boldsymbol{\sigma}}^T\right) \tag{6.223}$$

6.5.3 Ensemble-Average Stress in Filler Particles

The ensemble-volume averaged strain for three-phase composites takes the form

$$\overline{\boldsymbol{\varepsilon}} = \boldsymbol{\varepsilon}^0 + \sum_{q=1}^{2} \varphi_q \mathbf{S}\left(\overline{\boldsymbol{\varepsilon}}_q^{*0} + \overline{\boldsymbol{\varepsilon}}_q^T\right) \tag{6.224}$$

With the noninteracting solution $\overline{\boldsymbol{\varepsilon}}^{*0}$ of the eigenstrain given by Eq. (6.185), we arrive at

$$\bar{\boldsymbol{\varepsilon}} = \left[\mathbf{I} - \sum_{q=1}^{2} \varphi_q \mathbf{S}\left(\mathbf{A}_q + \mathbf{S}\right)^{-1}\right] : \boldsymbol{\varepsilon}^0 + \varphi_1 \mathbf{S}(\mathbf{A}_1 + \mathbf{S})^{-1}(\mathbf{A}_1 + \mathbf{I}) : \bar{\boldsymbol{\varepsilon}}_1^T \tag{6.225}$$

The volume-averaged stress tensor for the particles is defined as

$$\begin{aligned}\bar{\boldsymbol{\sigma}}_1 &= \frac{1}{V_1}\int_{V_1} \boldsymbol{\sigma}_1(\mathbf{x})d\mathbf{x} = \frac{1}{V_1}\int_{V_1} \mathbf{C}_1 : \left[\boldsymbol{\varepsilon}^0 + \boldsymbol{\varepsilon}'(\mathbf{x}) - \boldsymbol{\varepsilon}^T(\mathbf{x})\right]d\mathbf{x} \\ &= \mathbf{C}_1 : \left[\boldsymbol{\varepsilon}^0 + \bar{\boldsymbol{\varepsilon}}_1' - \bar{\boldsymbol{\varepsilon}}_1^T\right]\end{aligned} \tag{6.226}$$

where $\bar{\boldsymbol{\varepsilon}}_1{}'$ can be recast as

$$\bar{\boldsymbol{\varepsilon}}'_1 = \mathbf{S} : \left\{\frac{1}{V_1}\int_{V_1}\left[\boldsymbol{\varepsilon}^T(\mathbf{x}') + \boldsymbol{\varepsilon}^{*0}\right]d\mathbf{x}'\right\} = \mathbf{S} : \left(\bar{\boldsymbol{\varepsilon}}_1^T + \bar{\boldsymbol{\varepsilon}}_1^{*0}\right) \tag{6.227}$$

With the help of Eqs. (6.227) and (6.185), Eq. (6.226) can be rephrased as

$$\bar{\boldsymbol{\sigma}}_1 = \mathbf{C}_1\mathbf{A}_1(\mathbf{A}_1 + \mathbf{S})^{-1} : \boldsymbol{\varepsilon}^0 - \mathbf{C}_1\mathbf{A}_1(\mathbf{I} - \mathbf{S})(\mathbf{A}_1 + \mathbf{S})^{-1} : \bar{\boldsymbol{\varepsilon}}_1^T \tag{6.228}$$

As results of Eq. (6.225), the average internal stress of particles can be expressed as

$$\begin{aligned}\bar{\boldsymbol{\sigma}}_1 &= \mathbf{C}_1\mathbf{A}_1(\mathbf{A}_1 + \mathbf{S})^{-1}\left[\mathbf{I} - \sum_{q=1}^{2}\varphi_q\mathbf{S}\left(\mathbf{A}_q + \mathbf{S}\right)^{-1}\right]^{-1} : \bar{\boldsymbol{\varepsilon}} \\ &-\mathbf{C}_1\mathbf{A}_1(\mathbf{A}_1 + \mathbf{S})^{-1}\left\{\varphi_1\mathbf{S}(\mathbf{A}_1 + \mathbf{S})^{-1}(\mathbf{A}_1 + \mathbf{I})\left[\mathbf{I} - \sum_{q=1}^{2}\varphi_q\mathbf{S}\left(\mathbf{A}_q + \mathbf{S}\right)^{-1}\right]^{-1} + (\mathbf{I} - \mathbf{S})\right\} : \bar{\boldsymbol{\varepsilon}}_1^T \\ &= \mathbf{U} : \bar{\boldsymbol{\varepsilon}} - \mathbf{V} : \bar{\boldsymbol{\varepsilon}}_1^T\end{aligned} \tag{6.229}$$

where

$$\mathbf{U} = \mathbf{C}_1\mathbf{A}_1(\mathbf{A}_1 + \mathbf{S})^{-1}\left[\mathbf{I} - \sum_{q=1}^{2}\varphi_q\mathbf{S}\left(\mathbf{A}_q + \mathbf{S}\right)^{-1}\right]^{-1} \tag{6.230}$$

$$\mathbf{V} = \mathbf{C}_1\mathbf{A}_1(\mathbf{A}_1+\mathbf{S})^{-1}\left\{\varphi_1\mathbf{S}(\mathbf{A}_1+\mathbf{S})^{-1}(\mathbf{A}_1+\mathbf{I})\left[\mathbf{I}-\sum_{q=1}^{2}\varphi_q\mathbf{S}(\mathbf{A}_q+\mathbf{S})^{-1}\right]^{-1}+(\mathbf{I}-\mathbf{S})\right\} \tag{6.231}$$

By carrying out the lengthy algebra, the components of the positive definite fourth-rank tensor **U**and **V** are explicitly given by

$$U_{ijkl} = U_1\delta_{ij}\delta_{kl} + U_2\left(\delta_{ik}\delta_{jl} + \delta_{il}\delta_{jk}\right) \tag{6.232}$$

$$V_{ijkl} = V_1\delta_{ij}\delta_{kl} + V_2\left(\delta_{ik}\delta_{jl} + \delta_{il}\delta_{jk}\right) \tag{6.233}$$

where

$$3U_1 + 2U_2 = \frac{(3\alpha_1+2\beta_1)-10(1+v_0)}{(3\alpha_1+2\beta_1)\left[1-10(1+v_0)\sum_{q=1}^{2}\frac{\varphi_q}{3\alpha_q+2\beta_q}\right]}\cdot 3k_1 \tag{4.234}$$

$$U_2 = \frac{\beta_1-2(4-5v_0)}{\beta_1\left[1-2(4-5v_0)\sum_{q=1}^{2}\frac{\varphi_q}{\beta_q}\right]}\cdot\mu_1 \tag{6.235}$$

$$3V_1 + 2V_2 = \frac{(3\alpha_1+2\beta_1)-10(1+v_0)}{(3\alpha_1+2\beta_1)}$$

$$\left\{\frac{10(1+v_0)\varphi_1 k_1}{(3\alpha_1+2\beta_1)\left[1-10(1+v_0)\sum_{q=1}^{2}\frac{\varphi_q}{3\alpha_q+2\beta_q}\right](k_1-k_0)}+\frac{2(1-2v_0)}{3(1-v_0)}\right\}3k_1 \tag{6.236}$$

$$V_2 = \frac{\beta_1-2(4-5v_0)}{\beta_1}$$

$$\cdot\left[\frac{2(4-5v_0)\varphi_1\mu_1}{\beta_1\left[1-2(4-5v_0)\sum_{q=1}^{2}\frac{\varphi_q}{\beta_q}\right](\mu_1-\mu_0)}+\frac{7-5v_0}{15(1-v_0)}\right]\mu_1 \tag{6.237}$$

6.6 Effective Thermo-Mechanical Properties

In particulate composites, there is a thin layer of interfacial layer (interphase) between a particle and the matrix. The imperfect interface bond may be due to a very compliant thin interfacial layer that is assumed to have perfect boundary conditions with the matrix and the particle. This defines a three-phase composite that includes particles, thin interphase, and the matrix as shown in Fig. 6.2. Once the effective mechanical and thermal properties of the inner composite sphere assemblage (CSA), which consists of the particle and interphase layer of thickness δ, are found, the composite models used for perfect interface composite model can be readily applied to this case of composites with imperfect interface conditions.

The composite sphere assemblage (CSA) model was first proposed by Kerner (1956) and Van der Poel (1958) as shown in Fig. 6.3. Smith (1974, 1975), Christensen and Lo (1979), and Hashin and his co-workers (1962, 1968, 1990, 1991a, b; Hashin and Shtrikman 1963) improved on the CSA model. The CSA assumes that the particles are spherical and that the action on the particle is transmitted through a spherical interphase shell. The overall macro-behavior is assumed isotropic and is thus characterized by two effective moduli: the bulk modulus k and the shear modulus μ. In this section, we summarize the theoretical solution for effective thermo-mechanical properties of the CSA consisting of an elastic spherical particle and an elastic interphase layer with a thickness of δ. In the following formulae, k represents the bulk modulus, μ represents shear modulus, and α represents the coefficient of thermal expansion. The subscripts, f and m, refer to the interphase, filler, and matrix, respectively.

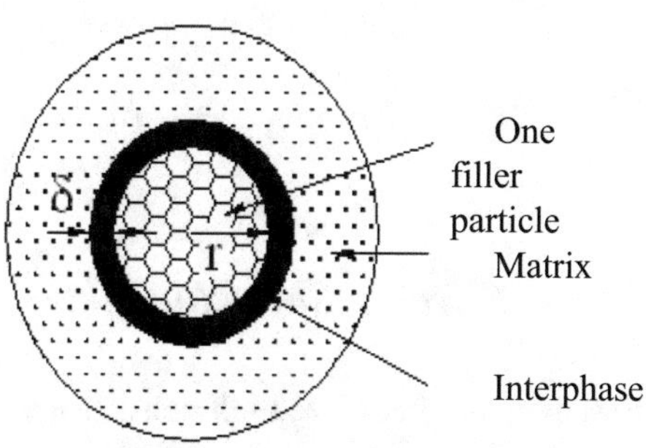

Fig. 6.2 Three-phase composite system. Note: filler particle is polycrystalline

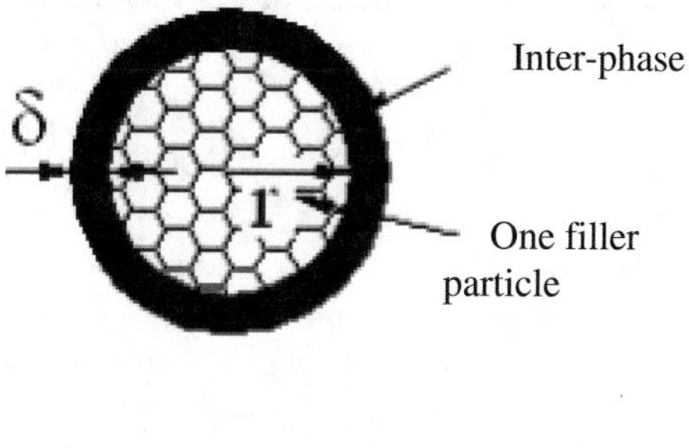

Fig. 6.3 Schematic illustration of composite spherical assemblage (CSA)

6.6.1 Effective Bulk Modulus

The effective bulk modulus k^* for the CSA as obtained by Hashin (1962) is given as

$$k^* = k_i + \frac{\varphi}{\frac{1}{k_f - k_i} + \frac{3(1-\varphi)}{3k_i + 4\mu_i}} \tag{6.238}$$

where

$$\varphi = \left(\frac{r}{r+\delta}\right)^3 \tag{6.239}$$

r is the radius of the filler particle
δ is the thickness of interphase

Hashin and Shtrikman's formulae (1963) and Walpole's formulae (1966) provide upper and lower bounds for the bulk modulus of the CSA. Equation (6.238) is same as the highest lower bound value.

6.6.2 Effective Coefficient of Thermal Expansion (ECTE)

The effective coefficient of thermal expansion α^* for the CSA as obtained by Levin (1967) is given as

$$\alpha^* = \alpha_i + \frac{\alpha_f - \alpha_i}{\frac{1}{k_f} - \frac{1}{k_i}}\left(\frac{1}{k^*} - \frac{1}{k_i}\right) \tag{6.240}$$

6.6.3 Effective Shear Modulus

Based on the generalized self-consistent scheme (GSCS) model proposed by Hashin (1962), Christensen and Lo (1979) have given the condition for determining the effective shear modulus of CSA as follows:

$$A\left(\frac{\mu^*}{\mu_i}\right)^2 + B\left(\frac{\mu^*}{\mu_i}\right) + D = 0 \tag{6.241}$$

where

$$A = 8[\mu_f/\mu_i - 1](4 - 5\nu_i)\eta_1\varphi^{10/3} - 2[63(\mu_f/\mu_i - 1)\eta_2 + 2\eta_1\eta_3]\varphi^{7/3}$$
$$+252[\mu_f/\mu_i - 1]\eta_2\varphi^{5/3} - 50\,[\mu_f/\mu_i - 1]\left(7 - 12\nu_i + 8\nu_i^2\right)\eta_2\varphi + 4(7 - 10\nu_i)\eta_2\eta_3 \tag{6.242}$$

$$B = -4[\mu_f/\mu_i - 1](1 - 5\nu_i)\eta_1\varphi^{10/3} + 4[63(\mu_f/\mu_i - 1)\eta_2 + 2\eta_1\eta_3]\varphi^{7/3}$$
$$-504[\mu_f/\mu_i - 1]\eta_2\varphi^{5/3} + 150\,[\mu_f/\mu_i - 1](3 - \nu_i)\nu_i\eta_2\varphi + 3(15\nu_i - 7)\eta_2\eta_3 \tag{6.243}$$

$$\begin{aligned} D = {} & 4[\mu_f/\mu_i - 1](5\nu_i - 7)\eta_1\varphi^{10/3} - 2[63(\mu_f/\mu_i - 1)\eta_2 + 2\eta_1\eta_3]\varphi^{7/3} \\ & +252[\mu_f/\mu_i - 1]\eta_2\varphi^{5/3} + 25\,[\mu_f/\mu_i - 1]\left(\nu_i^2 - 7\right)\eta_2\varphi - (7 + 5\nu_i)\eta_2\eta_3 \end{aligned} \tag{6.244}$$

with

$$\eta_1 = [\mu_f/\mu_i - 1](49 - 50\nu_f\nu_i) + 35(\mu_f/\mu_i)(\nu_f - 2\nu_i) + 35(2\nu_f - \nu_i) \tag{6.245}$$

$$\eta_2 = 5\nu_f[\mu_f/\mu_i - 8] + 7(\mu_f/\mu_i + 4) \tag{6.246}$$

$$\eta_3 = (\mu_f/\mu_i)(8 - 10\nu_i) + (7 - 5\nu_i) \tag{6.247}$$

φ is given in Eq. (6.239).

Based on the Van der Poel's formula for the shear modulus of a particulate composite, Smith (1974, 1975) also gave the condition for determining the effective shear modulus of the CSA as follows:

$$\alpha\left(\frac{\mu^*}{\mu_i} - 1\right)^2 + \beta\left(\frac{\mu^*}{\mu_i} - 1\right) + \gamma = 0 \tag{6.248}$$

where

$$\begin{aligned} \alpha = {} & \left[4P(7 - 10\nu_i) - S\varphi^{7/3}\right][Q - (8 - 10\nu_i)(M - 1)\varphi] \\ & - 126P(M - 1)\varphi\left(1 - \varphi^{2/3}\right)^2 \end{aligned} \tag{6.249}$$

$$\begin{aligned} \beta = {} & 35(1 - \nu_i)P[Q - (8 - 10\nu_i)(M - 1)\varphi] - 15(1 - \nu_i) \\ & \times \left[4P(7 - 10\nu_i) - S\varphi^{7/3}\right](M - 1)\varphi \end{aligned} \tag{6.250}$$

$$\gamma = -525P(1 - \nu_i)^2(M - 1)\varphi \tag{6.251}$$

with

$$M = \mu_f/\mu_i \tag{6.252}$$

$$P = (7 + 5\nu_f)M + 4(7 - 10\nu_f) \tag{6.253}$$

$$Q = (8 - 10\nu_i)M + (7 - 5\nu_i) \tag{6.254}$$

$$S = 35(7 + 5\nu_f)M(1 - \nu_i) - P(7 + 5\nu_i) \tag{6.255}$$

φ is given in Eq. (6.239).

Equations (6.241) and (6.248) are verified as equivalent in the determination of the effective shear modulus of the CSA. Solving the above equations, one can determine the exact solution for the effective shear modulus of the CSA model. One of the roots is negative and is extraneous. The positive root provides the value of the effective shear modulus.

6.6.4 Effective Young's Modulus and Effective Poisson's Ratio

Effective Young's modulus and Poisson's ratio for the CSA can be calculated from the well-known expressions

$$E^* = \frac{9k^*\mu^*}{3k^* + \mu^*} \tag{6.256}$$

$$\nu^* = \frac{3k^* - 2\mu^*}{6k^* + 2\mu^*} \tag{6.257}$$

6.6.5 Numerical Examples

In order to illustrate the effects of imperfect interface and interphase thickness on the overall effective mechanical and thermal properties of CSA, we consider a special case of CSA that consists of spherical alumina trihydrate (ATH) particle and interphase with the following properties:

For particles

$$E_f = 70\ \text{GPa}$$

$$\nu_f = 0.24$$

$$\alpha_f = 13 \times 10^{-6}/^\circ\text{C}$$

For interphase material

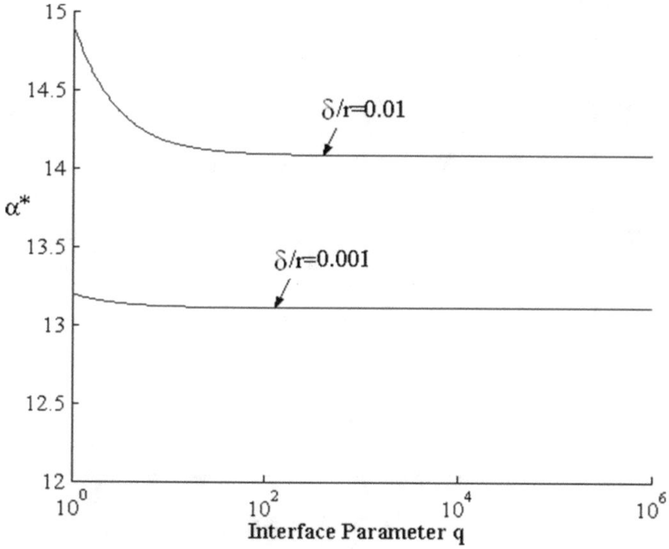

Fig. 6.4 Variation of effective CTE with the interface parameter q (Nie 2005)

$$\nu_i = 0.31$$

$$\alpha_i = 70 \times 10^{-6}/^0\mathrm{C}$$

$\delta/r = 0.01$ and $\delta/r = 0.001$

The nondimensional interface parameter is defined as $q = E_f/E_i$. A zero value of q implies that there are no interface displacement jumps and the perfectly bonded interface conditions exist. At the other extreme, infinite value of q implies that the interface tractions do not exist and the filler is debonded from the adjoining matrix media. Finite positive values for q define an imperfect interface, which lies between two extreme cases mentioned above. Figure 6.4 shows variation of effective Coefficient of Thermal Expansion with the interface parameter q.

Figures 6.5, 6.6, and 6.7 show the effects of the interphase bond modulus and thickness on the effective shear modulus, Young's modulus, and bulk modulus. The stiffness of the bond has a strong effect on the degradation of the shear modulus, Young's modulus, and bulk modulus. As the interface becomes thinner for the same E_i value, the effective shear modulus increases. On the other hand for the same interphase thickness, as the elastic modulus value E_i of the interphase decreases, the effective shear modulus decreases. This is also true for both the effective bulk modulus and Young's modulus. Figure 6.8 shows the effects of the interphase bond modulus and thickness on the effective Poisson's ratio. Effective Poisson's ratio seems to be independent of the interphase stiffness when $q \leq 10^2$ and $q \geq 10^4$.

Analytical and numerical evaluation of the effective elastic properties and the thermal expansion coefficients of the composite sphere assemblage show that the bulk and shear moduli are insensitive to the value of Poisson's ratio of interphase.

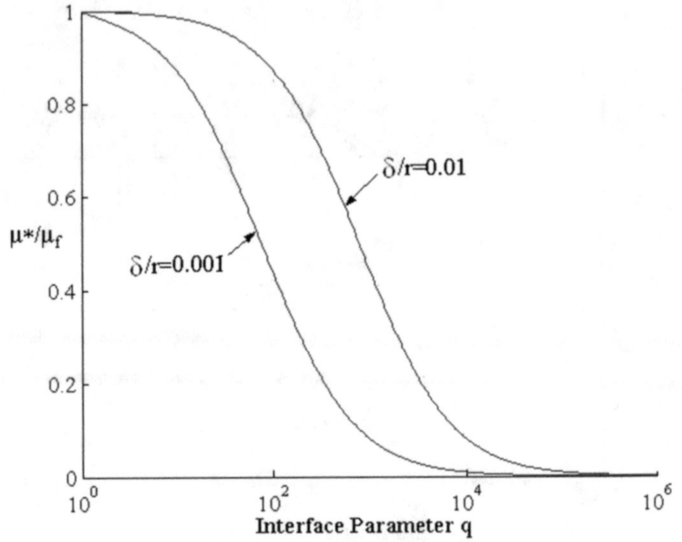

Fig. 6.5 Variation of effective shear modulus with the interface parameter q (Nie 2005)

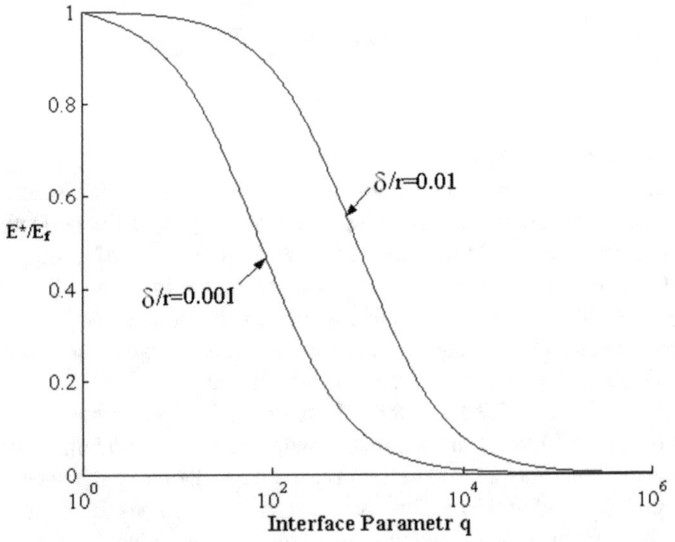

Fig. 6.6 Variation of effective Young's modulus with the interface parameter q (Nie 2005)

Numerical results also show that the nature of the interphase has significant effects on the stiffness and the thermal expansion coefficient of the CSA.

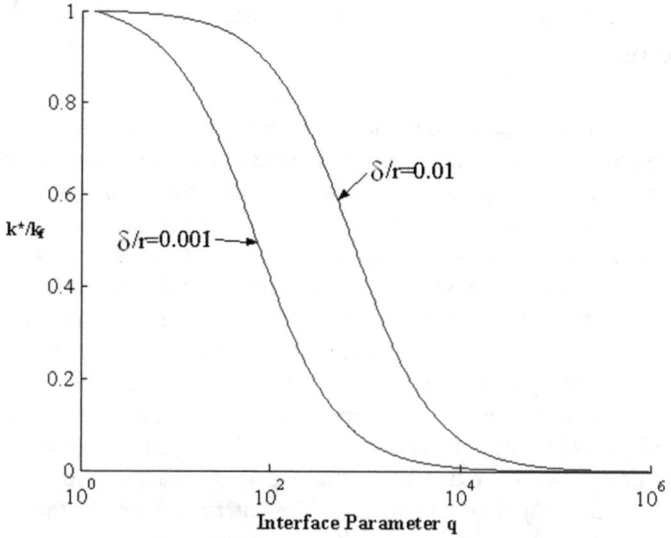

Fig. 6.7 Variation of effective bulk modulus with the interface parameter q (Nie 2005)

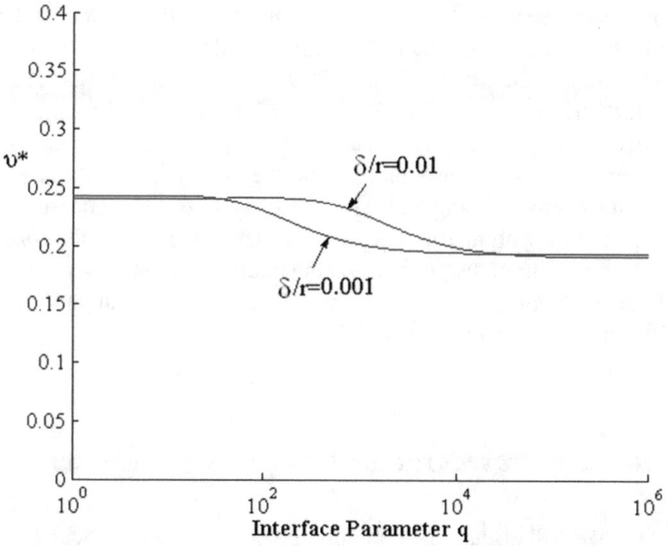

Fig. 6.8 Variation of effective Poisson's ratio with the interface parameter q (Nie 2005)

6.7 Micromechanical Constitutive Model of the Particulate Composite

The nature of the bond between particles and the matrix material has a significant effect on the mechanical behavior of particulate composites. Most analytical and numerical models assume that the bond between the filler and matrix is perfect and can be modeled using the continuity of tractions and displacements across a discrete interface. However, internal defects and imperfect interfaces are well-known to exist in composites, and the incorporation of such phenomena into the general theory requires modification and relaxation of the continuity of displacements between the constituents. The imperfect interface bond may be due to the compliant interfacial layer known as interphase or interface damage, which may have been created deliberately by coating the particles with a debonding agent. It may also develop during the manufacturing process due to chemical reactions between the particles and the matrix material or due to interface damage from cyclic loading. Most importantly, the strength of the bond at the interface controls the mechanical response and fatigue life of the composite (Basaran and Nie 2004). By controlling the stress-strain response of the interphase, it is possible to control overall behavior of the composite.

In the following section, a micromechanical model for particulate composites with imperfect interface between the filler particle and the matrix and CTE mismatch is presented, based on the work of Ju and Chen (1994a, b) and Ju and Tseng (1996), which was for perfectly bonded particle-matrix interfaces. In the micromechanical model, particulate composites are treated as three-phase composites consisting of agglomerate of particles, the bulk matrix, and the interfacial transition zone around the agglomerate as shown in Fig. 6.2. The interfacial transition zones are assumed to have perfect bonding with matrix and particles. The inner composite sphere assemblage (CSA), consisting of the particle and the interfacial transition zone, is regarded as an equivalent spherical particle with the effective mechanical and thermal properties derived in the previous section.

6.7.1 *Modeling Procedures for Particulate Composites*

In this section, we will discuss the modeling procedures for a particulate composite, including how the interface properties, CTE mismatch between the particle and the matrix, and an isotropic damage parameter are introduced.

The modeling process consists of four steps. The first step is the simplification of the real particulate composites, which is shown in Fig. 6.9. A is a representation of the real microstructure for a particulate composite, where particles have different sizes and shapes. In order to get an analytical expression for the effective behavior of the particulate composite, we must make some assumptions. For simplicity, we assume that the particles are uniform spherical. The interfacial layer is used to

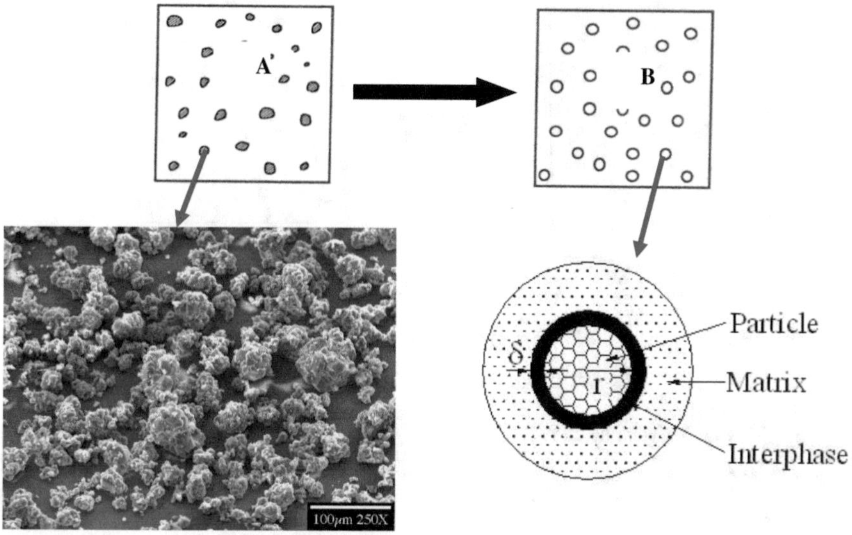

Fig. 6.9 Modeling procedures—step 1: simplification (Nie 2005)

model the imperfect interface condition. The displacement jumps are defined by the deformation of the interphase. The particulate composite is defined as a three-phase system that includes particles, the thin interphase, and the matrix. The simplified microstructure of a particulate composite is shown in Fig. 6.9.

The simplified microstructure for the particulate composites includes the particle, the interphase, and the matrix. For the sake of simplicity, the particle and the interphase are regarded as one, namely, a particle-interphase assemblage. The microstructure of composite is now simplified as C shown in Fig. 6.10. The microstructure of C includes the particle-interphase assemblage and the matrix. But the thermo-mechanical properties for the particle-interphase assemblage must be calculated for this two-phase assembly.

Finding the effective thermo-mechanical properties for the spherical particle-interphase assemblage is essential (Fig. 6.11).

The particulate composite is simplified as a two-phase system with perfect interfacial bonding between the particle-interphase assemblage and the matrix. This final modeling step is shown in Fig. 6.12.

6.7.2 Elastic Properties of Particulate Composites

Based on Ju and Chen's formulations (1994a, b), the noninteracting particles' solution for the effective properties of a two-phase composite with imperfect bonding can be modified by substituting the properties of the particle with the properties of composite sphere assemblage:

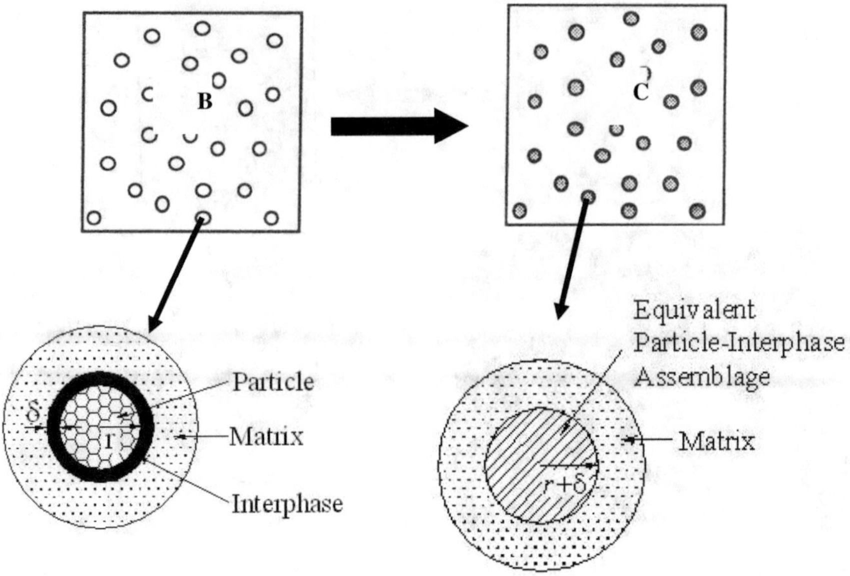

Fig. 6.10 Modeling procedures—step 2: equivalence (Nie 2005)

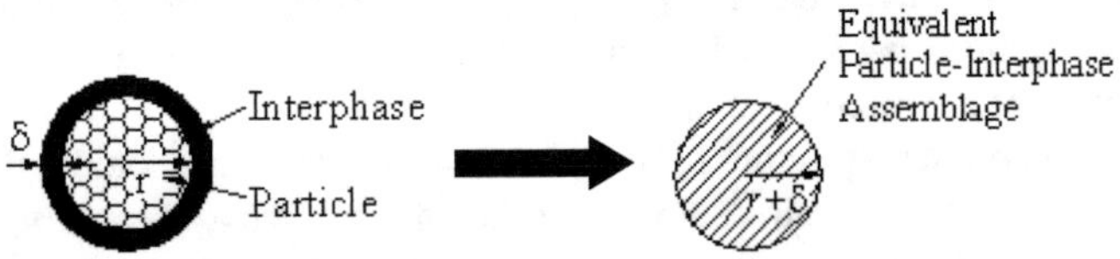

Fig. 6.11 Modeling procedures—step 3: thermo-mechanical properties (Nie 2005)

$$k = k_{\mathrm{m}}\left\{1 + \frac{3(1-v_{\mathrm{m}})(k^* - k_{\mathrm{m}})\varphi_{\mathrm{f}}}{3(1-v_{\mathrm{m}})k_{\mathrm{m}} + (1-\varphi_{\mathrm{f}})(1+v_{\mathrm{m}})(k^* - k_{\mathrm{m}})}\right\} \tag{6.258}$$

$$\mu = \mu_{\mathrm{m}}\left\{1 + \frac{15(1-v_{\mathrm{m}})(\mu^* - \mu_{\mathrm{m}})\varphi_{\mathrm{f}}}{15(1-v_{\mathrm{m}})\mu_{\mathrm{m}} + (1-\varphi_{\mathrm{f}})(8-10v_{\mathrm{m}})(\mu^* - \mu_{\mathrm{m}})}\right\} \tag{6.259}$$

Similarly, the pairwise interacting particles' solution for the effective properties of the two-phase composite with the imperfect bond can be modified as

$$k = k_{\mathrm{m}}\left\{1 + \frac{30(1-v_{\mathrm{m}})(3\gamma_1 + 2\gamma_2)\varphi_{\mathrm{f}}}{3\beta_1 + 2\beta_2 - 10(1+v_{\mathrm{m}})(3\gamma_1 + 2\gamma_2)\varphi_{\mathrm{f}}}\right\} \tag{6.260}$$

$$\mu = \mu_{\mathrm{m}}\left\{1 + \frac{30(1-v_{\mathrm{m}})\gamma_2\varphi_{\mathrm{f}}}{\beta_2 - 4(4-5v_{\mathrm{m}})\gamma_2\varphi_{\mathrm{f}}}\right\} \tag{6.261}$$

where

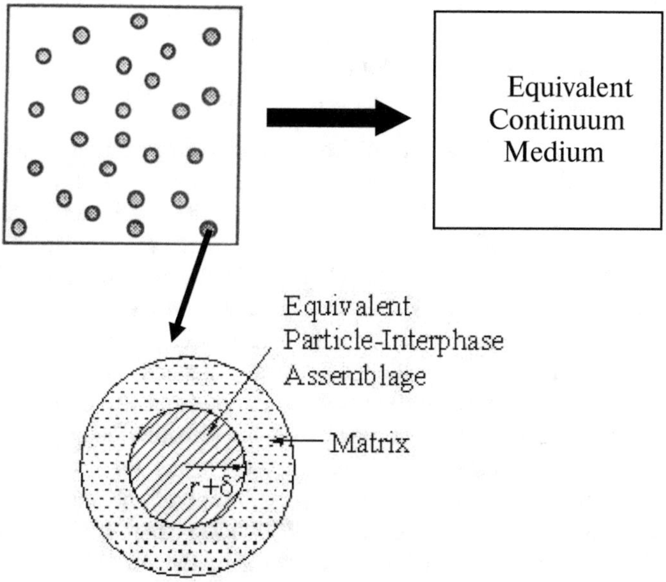

Fig. 6.12 Modeling procedures—step 4: micromechanics (Nie 2005)

$$\beta_1 = 2(5\nu_m - 1) + 10(1 - \nu_m)\left(\frac{k_m}{k^* - k_m} - \frac{\mu_m}{\mu^* - \mu_m}\right) \tag{6.262}$$

$$\beta_2 = 2(4 - 5\nu_m) + 15(1 - \nu_m)\frac{\mu_m}{\mu^* - \mu_m} \tag{6.263}$$

and

$$\gamma_1 = \frac{5\varphi_f}{8{\beta_2}^2}\left\{(13 - 14\nu_m)\nu_m - \frac{8\beta_1}{3\beta_1 + 2\beta_2}(1 - 2\nu_m)(1 + \nu_m)\right\} \tag{6.264}$$

$$\gamma_2 = \frac{1}{2} + \frac{5\varphi_f}{16{\beta_2}^2} \times \left\{\left(25 - 34\nu_m + 22\nu_m^2\right) - \frac{6\beta_1}{3\beta_1 + 2\beta_2}(1 - 2\nu_m)(1 + \nu_m)\right\} \tag{6.265}$$

where k^* and μ^* are the effective bulk modulus and shear modulus of the composite spherical filler assemblage, respectively, and φ_f is the particle volume fraction.

Example

Nie (2005) studied lightly cross-linked poly-methyl methacrylate (PMMA) filled with alumina trihydrate (ATH). The following phase properties were used for the numerical analysis:

Particle: $E_f = 70$ GPa, $\nu_f = 0.24$
Matrix: $E_m = 3.5$ GPa, $\nu_m = 0.31$
Particle volume fraction is $\varphi_f = 0.48$
Nondimensional interface parameter is defined as $q = E_f/E_i$.

Three kinds of composites with different interfacial properties were prepared. Composite A has the strongest interfacial adhesion due to the addition of a special adhesion promoting additives, where the value of the nondimensional interface parameter $q = E_f/E_i$ can be assumed one. The interfacial adhesion strength of composite C is the weakest due to the addition of debonding promoting additives. The value of the nondimensional interface parameter q for composite C can be assumed to be a very large number. The interfacial strength of the composite B is moderate. The value of nondimensional interface parameter q for composite B can be assumed to be some value between that for composite A and composite C.

The effects of the interphase properties on the effective Young's modulus of two-phase composite are shown in Figs. 6.13, 6.14, and 6.15, where the elastic properties of composites with different interphase properties obtained from test data were also plotted for comparison. These results indicate that the pairwise interacting particles' solution yields a better approximation of the overall elastic modulus. Observation of Figs. 6.13, 6.14, and 6.15 also indicates that as the thickness of the interphase gets larger, the effective elastic modulus value decreases for the same q value.

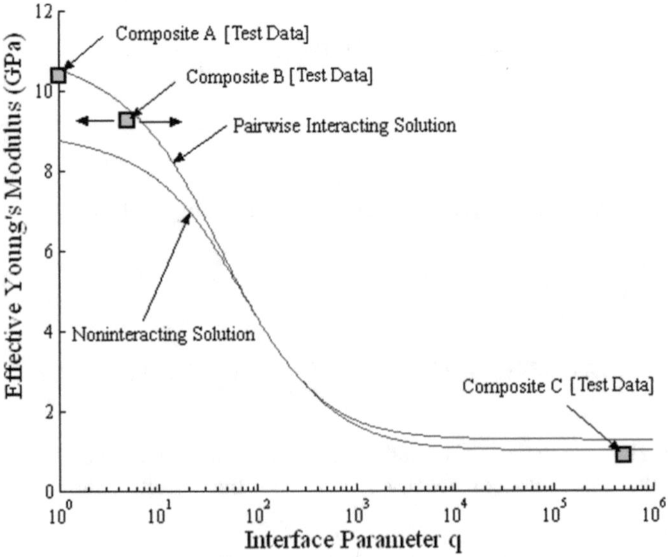

Fig. 6.13 Effective Young's modulus vs. interface parameter at volume fraction of 48% (the thickness of interphase is 1/10 of the diameter of particles) (Nie 2005)

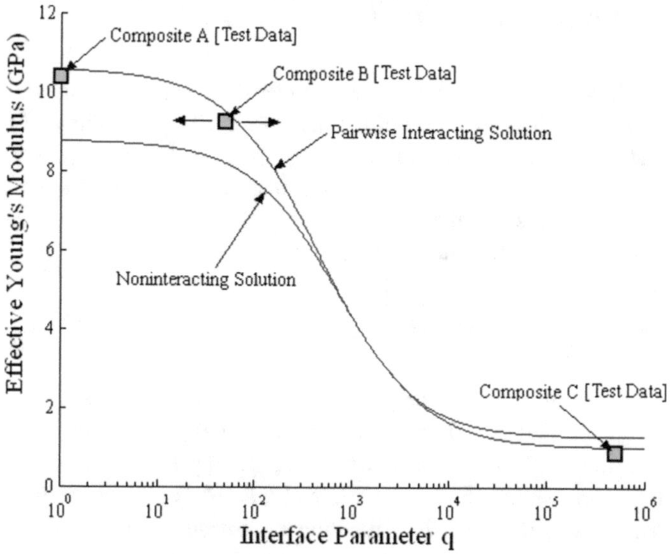

Fig. 6.14 Effective Young's modulus vs. interface parameter at volume fraction of 48% (the thickness of interphase is 1/100 of the diameter of particles) (Nie 2005)

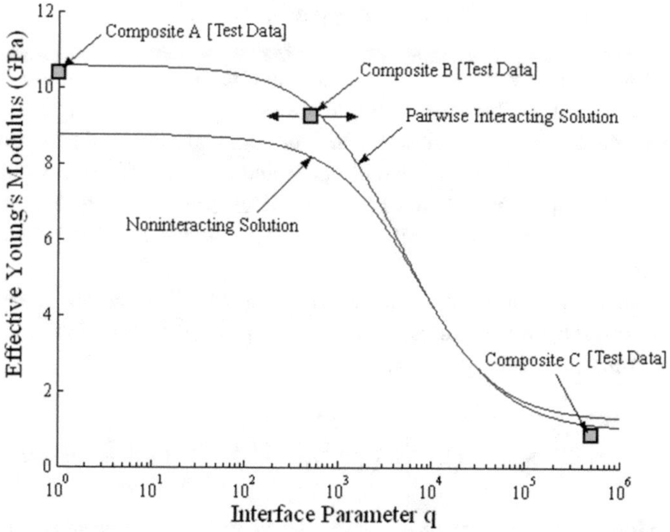

Fig. 6.15 Effective Young's modulus vs. interface parameter at volume fraction of 48% (the thickness of interphase is 1/1000 of the diameter of particles) (Nie 2005; Basaran et al. 2008a, b)

6.7.3 A Viscoplasticity Model

Viscoplastic flow occurs only in the matrix because filler particles have a very high melting temperature, and they are brittle linear elastic. The magnitude of the current equivalent stress norm of the matrix can be used to determine the viscoplastic behavior. Regarding the composite as viscoplastic overall, when the ensemble-volume averaged stress norm in the matrix reaches a certain level, the effective yield function for the composite can be given by

$$f(\overline{\boldsymbol{\sigma}}, \overline{\alpha}) = \sqrt{\left(\overline{\boldsymbol{\sigma}} - \overline{\boldsymbol{\sigma}}^T\right) : \overline{\mathbf{T}} : \left(\overline{\boldsymbol{\sigma}} - \overline{\boldsymbol{\sigma}}^T\right)} - \sqrt{T_1 + 2T_2}\overline{\sigma}_y(\overline{\alpha}) \tag{6.266}$$

where $\overline{\boldsymbol{\sigma}}$ is the average stress in the particulate composite; $\overline{\boldsymbol{\sigma}}^T$ is the stress caused by the CTE mismatch between the matrix and the particle which was derived earlier (6.190); $\mathbf{T}$ is the fourth-order tensor which is given by Eq. (6.211); T_1 and T_2 are given by Eqs. (6.217) and (6.218), respectively; $\overline{\sigma}_y(\overline{\alpha})$ is the current yield stress of the composite material, which is a function of isotropic hardening; and $\overline{\alpha}$ is the equivalent viscoplastic strain that defines isotropic hardening of the yield surface of the composites

$$\dot{\overline{\alpha}} = \sqrt{T_1 + 2T_2}\sqrt{\dot{\overline{\boldsymbol{\varepsilon}}}^{vp} : \overline{\mathbf{T}}^{-1} : \dot{\overline{\boldsymbol{\varepsilon}}}^{vp}} \tag{6.267}$$

The factors in the effective yield and the effective plastic strain increment equations are chosen so that the effective stress and effective plastic strain increments are equal to the uniaxial stress and uniaxial plastic strain increment in a uniaxial monotonic tensile test. It should be noted that the effective yield function is pressure dependent now and not of the von Mises type any more. So the particles and hydrostatic pressure have significant effects on the viscoplastic behavior of the matrix materials.

Unified mechanics theory provides a basic framework to introduce degradation evolution intrinsically. According to the strain equivalence principle, the effective yield stress function can be given by

$$f(\overline{\boldsymbol{\sigma}}, \overline{\alpha}) = \sqrt{\left(\frac{\overline{\boldsymbol{\sigma}}}{1-\Phi} - \overline{\boldsymbol{\sigma}}^T\right) : \overline{\mathbf{T}} : \left(\frac{\overline{\boldsymbol{\sigma}}}{1-\Phi} - \overline{\boldsymbol{\sigma}}^T\right)} - \sqrt{T_1 + 2T_2}\overline{\sigma}_y(\overline{\alpha}) \tag{6.268}$$

It is obvious that the damage increases the equivalent stress norm of the composite, which tend to amplify the viscoplastic behavior of composite.

If the kinematic hardening behavior is included, let $\overline{\boldsymbol{\beta}}$ define the center of the yield surface of the composite in the stress space, and the relative stress be defined as $\overline{\boldsymbol{\sigma}} - \overline{\boldsymbol{\beta}}$. Assume the particles have the same effects on the stress norm of the matrix as on the kinematic behavior of the matrix; the stress norm defining the viscoplastic behavior of the matrix can be updated as

$$\langle H\rangle_m(\mathbf{x}) = \left(\bar{\boldsymbol{\sigma}} - \bar{\boldsymbol{\beta}} - \bar{\boldsymbol{\sigma}}^T\right) : \bar{\mathbf{T}} : \left(\bar{\boldsymbol{\sigma}} - \bar{\boldsymbol{\beta}} - \bar{\boldsymbol{\sigma}}^T\right) \tag{6.269}$$

Therefore, the effective yield function for the composite can be given as

$$f(\bar{\boldsymbol{\sigma}}, \mathbf{q}) = \sqrt{\left(\frac{\bar{\boldsymbol{\sigma}}}{1-\Phi} - \bar{\boldsymbol{\beta}} - \bar{\boldsymbol{\sigma}}^T\right) : \bar{\mathbf{T}} : \left(\frac{\bar{\boldsymbol{\sigma}}}{1-\Phi} - \bar{\boldsymbol{\beta}} - \bar{\boldsymbol{\sigma}}^T\right)} - \sqrt{\bar{T}_1 + 2\bar{T}_2} \times \left[\sigma_y + K(\bar{\alpha})\right] \tag{6.270}$$

where $\mathbf{q} = \{\bar{\alpha}, \bar{\boldsymbol{\beta}}\}$ is chosen as the internal viscoplastic variable.

For simplicity, the Perzyna-type viscoplasticity model is employed to characterize rate (viscosity) effects in the matrix. Therefore, the effective ensemble-volume averaged plastic strain rate for the composite can be expressed as

$$\dot{\boldsymbol{\varepsilon}}^{vp} = \gamma \frac{\partial f}{\partial \boldsymbol{\sigma}} = \frac{1}{1-\Phi}\gamma \mathbf{n} \tag{6.271}$$

where

$$\mathbf{n} = \frac{\bar{\mathbf{T}} : \left(\frac{\bar{\boldsymbol{\sigma}}}{1-\Phi} - \bar{\boldsymbol{\beta}} - \bar{\boldsymbol{\sigma}}^T\right)}{\sqrt{\left(\frac{\bar{\boldsymbol{\sigma}}}{1-\Phi} - \bar{\boldsymbol{\beta}} - \bar{\boldsymbol{\sigma}}^T\right) : \bar{\mathbf{T}} : \left(\frac{\bar{\boldsymbol{\sigma}}}{1-\Phi} - \bar{\boldsymbol{\beta}} - \bar{\boldsymbol{\sigma}}^T\right)}} \tag{6.272}$$

and γ denotes the plastic consistency parameter

$$\gamma = \frac{\langle f\rangle}{\eta} = \frac{\langle f\rangle}{2\mu\tau} \tag{6.273}$$

where η is viscosity coefficient and τ is called relaxation time.

And the effective equivalent viscoplastic strain rate is defined as

$$\dot{\alpha} = \sqrt{\bar{T}_1 + 2\bar{T}_2}\frac{\gamma}{1-\Phi} \tag{6.274}$$

The evolution of the back stress, $\bar{\boldsymbol{\beta}}$, depends on the plastic strain or plastic work history. The earlier model for the evolution of the back stress is due to Prager (1956) and was subsequently modified by Ziegler (1959). The Prager hardening assumption is that during yielding, the back stress increment $d\bar{\boldsymbol{\beta}}$ is equal to the component of $d\bar{\boldsymbol{\sigma}}$ in the direction normal to the yield surface. Since the plastic strain increment is also normal to the yield surface, the increment $d\bar{\boldsymbol{\beta}}$ can be written as

$$d\bar{\boldsymbol{\beta}} = \left[d\bar{\boldsymbol{\sigma}} \cdot \frac{d\boldsymbol{\varepsilon}^p}{|d\boldsymbol{\varepsilon}^p|}\right]\frac{d\boldsymbol{\varepsilon}^p}{|d\boldsymbol{\varepsilon}^p|} \tag{6.275}$$

In the case of linear kinematic hardening, we have

$$d\bar{\boldsymbol{\beta}} = \frac{\left(\overline{T}_1 + 2\overline{T}_2\right)^{1.5}}{\left(\overline{T}_1 + 2\overline{T}_2\right)^2 + 2\overline{T}_1{}^2} H'^{\dot{\alpha}} \mathbf{n} \tag{6.276}$$

where H' is the kinematic hardening modulus. The factor in Eq. (6.276) is chosen so that the effective stress and effective plastic strain increments are equal to the uniaxial stress and uniaxial plastic strain increment in a uniaxial monotonic tensile test.

6.7.4 Thermodynamic State Index

The thermodynamic state index (TSI) is given by

$$\Phi = \Phi_{\text{cr}}\left(1 - \text{e}^{-\frac{m_s}{R}\Delta s}\right) \tag{6.277}$$

where Φ_{cr} is a temperature-dependent coefficient to map elastic modulus degradation onto TSI [however, this coefficient is not essential and can be take one if all entropy generation mechanisms are included in the fundamental equation and, especially, if the cyclic stress-strain data is not available], R is the gas constant, m_s is the molar mass, and Δs is the change in entropy and must be calculated from the entropy production rate at each Gauss integration point as follows:

$$\Delta s = \int_{t_0}^{t} \frac{\bar{\boldsymbol{\sigma}} : \dot{\bar{\boldsymbol{\varepsilon}}}^{vp}}{T\rho} dt + \int_{t}^{t_0} \left(\frac{k}{T^2 \rho} |\text{grad}T|^2 \right) dt + \int_{t_0}^{t} \frac{r}{T} dt \tag{6.278}$$

where ρ is the density, T is temperature, k is termed the thermal conductivity of the composite, and r is the distributed internal heat generation. Unfortunately, Eq. (6.278) ignores entropy generation due to aging in PMMA, microplasticity at particle corners and other long-term chemical reactions in the PMMA molecular chains. Of course, including all mechanisms in the entropy generation rate is more accurate.

6.7.5 Solution Algorithm

General return mapping algorithm is used to solve Eqs. (6.270)–(6.278). The general return mapping algorithm was proposed by Simo and Taylor (1985) and summarized in details by Simo and Hughes (1998). In order to minimize confusion, the symbol Δ is used to denote an increment over a time step, or an increment between successive iterations. For the rate-of-slip γ, we adhere to the conventions: $\Delta\gamma = \gamma\Delta t$ denotes the increment of γ over a time step, and $\Delta^2\gamma$ denotes the increment of $\Delta\gamma$ between iterations.

Let $\mathbf{C}$ be the elastic consistent tangent moduli of the particulate composite, then ignoring derivative of TSI with respect to displacement for simplicity, the overall stress-strain relation can be given by,

$$\boldsymbol{\sigma}_{n+1} = (1-\Phi)\mathbf{C} : \left(\boldsymbol{\varepsilon}_{n+1} - \boldsymbol{\varepsilon}_{n+1}^{vp} - \boldsymbol{\varepsilon}_{n+1}^{\text{th}}\right) \tag{6.279}$$

Since $\boldsymbol{\varepsilon}_{n+1}$ and $\boldsymbol{\varepsilon}_{n+1}^{\text{th}}$ are fixed during the return mapping stage, it follows that

$$\Delta\varepsilon^{vp} = -\frac{1}{1-\Phi}\mathbf{C}^{-1}\Delta\boldsymbol{\sigma} \tag{6.280}$$

From Eq. (6.277), we have

$$\dot{\Phi} = \frac{\Phi_{\text{cr}} m_{\text{s}}}{R} \exp\left(-\frac{m_{\text{s}}}{R}\Delta s\right)\Delta\dot{s} \tag{6.281}$$

assuming that irreversible entropy generation is only due to plastic work and heat generation. From Eq. (6.278) for isothermal process at each increment, we can write

$$\Delta\dot{s} = \frac{\boldsymbol{\sigma} : \dot{\boldsymbol{\varepsilon}}^{vp}}{T\rho} + \frac{r}{T} \tag{6.282}$$

Assuming that the heat generated within the system is negligible, so the distributed internal heat source of strength per unit mass is zero [which is not true]. Using Eqs. (6.280) and (6.282), Eq. (6.281) becomes

$$\Delta\Phi = -\frac{1}{1-\Phi}\frac{\Phi_{\text{cr}} m_{\text{s}}}{T\rho R} \exp\left(-\frac{m_{\text{s}}}{R}\Delta s\right)\boldsymbol{\sigma}\mathbf{C}^{-1}\Delta\boldsymbol{\sigma} \tag{6.283}$$

We should point out that in Eq. (6.282) we ignored many prominent irreversible entropy generation mechanisms such as relative motion between the filler and the matrix and the entropy generation in the filler polycrystals and aging in PMMA. Moreover, if the strain rate of the loading is very high, there will be additional entropy generation mechanism. However, our goal is to demonstrate the formulation as simple as possible.

From Eqs. (6.271), (6.274), and (6.276), we also have

$$\boldsymbol{\varepsilon}_{n+1}^{vp} = \boldsymbol{\varepsilon}_{n}^{vp} + \frac{1}{1-\Phi}\Delta\gamma\,\mathbf{n} \tag{6.284}$$

$$\alpha_{n+1} = \alpha_n + \frac{1}{1-\Phi}\sqrt{\overline{T}_1 + 2\overline{T}_2\Delta\gamma} \tag{6.285}$$

$$\overline{\boldsymbol{\beta}}_{n+1} = \overline{\boldsymbol{\beta}}_n + \frac{\left(\overline{T}_1 + 2\overline{T}_2\right)^{1.5}}{\left(\overline{T}_1 + 2\overline{T}_2\right)^2 + 2\overline{T}_1{}^2} \Delta H \mathbf{n} \tag{6.286}$$

Let

$$m_1 = \sqrt{\overline{T}_1 + 2\overline{T}_2} \tag{6.287}$$

$$m_2 = \frac{\left(\overline{T}_1 + 2\overline{T}_2\right)^{1.5}}{\left(\overline{T}_1 + 2\overline{T}_2\right)^2 + 2\overline{T}_1{}^2} \tag{6.288}$$

$$m_3 = -\frac{\Phi_{cr}\, m_s}{T\rho R} \exp\left(-\frac{m_s}{R}\Delta s\right) \tag{6.289}$$

Also we have

$$\frac{\partial f}{\partial \overline{\boldsymbol{\sigma}}} = \frac{1}{1-\Phi}\mathbf{n} \tag{6.290}$$

$$\frac{\partial^2 f}{\partial \overline{\boldsymbol{\sigma}}^2} = \frac{1}{(1-\Phi)^2}\mathbf{N} \tag{6.291}$$

$$\frac{\partial f}{\partial \overline{\boldsymbol{\beta}}} = -\mathbf{n} \tag{6.292}$$

$$\frac{\partial^2 f}{\partial \overline{\boldsymbol{\beta}}^2} = \mathbf{N} \tag{6.293}$$

$$\frac{\partial f}{\partial D} = \frac{\mathbf{n} : \boldsymbol{\sigma}}{(1-\Phi)^2} \tag{6.294}$$

$$\frac{\partial f}{\partial \overline{\alpha}} = -m_1 K' \tag{6.295}$$

$$\frac{\partial \mathbf{n}}{\partial \overline{\boldsymbol{\beta}}} = -\mathbf{N} \tag{6.296}$$

$$\frac{\partial \mathbf{n}}{\partial \overline{\boldsymbol{\sigma}}} = \frac{1}{1-\Phi}\mathbf{N} \tag{6.297}$$

$$\frac{\partial \mathbf{n}}{\partial D} = \frac{1}{(1-\Phi)^2}\mathbf{N} : \overline{\boldsymbol{\sigma}} \tag{6.298}$$

where $K' = \frac{\partial \overline{\sigma}_y}{\partial \overline{\alpha}}$ is the isotropic hardening modulus.

$$\mathbf{N} = \frac{\left(\bar{\mathrm{T}} - \mathbf{n} \otimes \mathbf{n}\right)}{\sqrt{\left(\frac{\bar{\boldsymbol{\sigma}}}{1-\Phi} - \bar{\boldsymbol{\beta}} - \bar{\boldsymbol{\sigma}}^T\right) : \bar{\mathbf{T}}\left(\frac{\bar{\boldsymbol{\sigma}}}{1-\Phi} - \bar{\boldsymbol{\beta}} - \bar{\boldsymbol{\sigma}}^T\right)}} \tag{6.299}$$

We define the following residual functions as follows:

$$R_\alpha = -\alpha_{n+1} + \alpha_n + \frac{m_1}{1-\Phi}\Delta\gamma \tag{6.300}$$

$$R_\gamma = \frac{2\mu\tau\Delta\gamma}{\Delta t} - \langle f \rangle \tag{6.301}$$

$$\mathbf{R}_\varepsilon = -\boldsymbol{\varepsilon}_{n+1}^{vp} + \boldsymbol{\varepsilon}_n^{vp} + \frac{1}{1-\Phi}\Delta\gamma\mathbf{n} \tag{6.302}$$

$$\mathbf{R}_\beta = -\bar{\boldsymbol{\beta}}_{n+1} + \bar{\boldsymbol{\beta}}_n + m_2\Delta H\mathbf{n} \tag{6.303}$$

Linearizing these residual functions yields

$$R_\alpha - \Delta\alpha_{n+1} + \frac{m_1}{1-\Phi}\Delta^2\gamma + \frac{m_1 m_3}{(1-\Phi)^3}\Delta\gamma\boldsymbol{\sigma}\mathbf{C}^{-1}\Delta\boldsymbol{\sigma} = 0 \tag{6.304}$$

$$n'\Delta\boldsymbol{\sigma} - \mathbf{n}\Delta\boldsymbol{\beta} = R_{\mathrm{f}} + m_1 K'^{R_\alpha} + \left(\frac{{m_1}^2 K'}{1-\Phi} + 2\mu\frac{\tau}{\Delta t}\right)\Delta^2\gamma \tag{6.305}$$

$$\mathbf{P} : \Delta\boldsymbol{\sigma} - \mathbf{Q} : \Delta\boldsymbol{\beta} + \frac{\Delta^2\gamma}{1-\Phi}\mathbf{n} + \mathbf{R}_\varepsilon = 0 \tag{6.306}$$

$$\mathbf{R} : \Delta\boldsymbol{\sigma} - \mathbf{S} : \Delta\boldsymbol{\beta} + \frac{m_1 m_2}{1-\Phi}H'^{\Delta^2}\gamma\mathbf{n} + m_2 H' R_\alpha\mathbf{n} + \mathbf{R}_\beta = 0 \tag{6.307}$$

where

$$\mathbf{P} = \frac{1}{1-\Phi}\left[\mathrm{C}^{-1} + \frac{\Delta\gamma}{1-\Phi}\mathbf{N} + \frac{\Delta\gamma m_3}{(1-\Phi)^2}\left(\frac{\mathbf{N}\boldsymbol{\sigma}}{1-\Phi} + \mathbf{n}\right) \otimes \left(\boldsymbol{\sigma}\mathrm{C}^{-1}\right)\right] \tag{6.308}$$

$$\mathbf{Q} = \frac{\Delta\gamma}{1-\Phi}\mathbf{N} \tag{6.309}$$

$$\mathbf{R} = \frac{m_2}{1-\Phi}\left[\Delta H\mathbf{N} + \frac{m_3}{(1-\Phi)^2}\left(\Delta H\mathbf{N}\boldsymbol{\sigma} + m_1 H'\Delta\gamma\mathbf{n}\right) \otimes \left(\boldsymbol{\sigma}\mathbf{C}^{-1}\right)\right] \tag{6.310}$$

$$\mathbf{S} = \mathbf{I} + m_2\Delta H\mathbf{N} \tag{6.311}$$

$$\mathbf{n}' = \frac{1}{1-\Phi}\left[\mathbf{n} + \frac{m_3}{(1-\Phi)^2}\left(\mathbf{n}\bar{\boldsymbol{\sigma}} - {m_1}^2 K'\Delta\gamma\right)\left(\bar{\boldsymbol{\sigma}}\mathbf{C}^{-1}\right)\right] \tag{6.312}$$

Combining Eqs. (6.306) and (6.307)

$$\begin{bmatrix} \mathbf{P} & \mathbf{Q} \\ \mathbf{R} & \mathbf{S} \end{bmatrix} \begin{Bmatrix} \Delta\overline{\boldsymbol{\sigma}} \\ -\Delta\overline{\boldsymbol{\beta}} \end{Bmatrix} + \frac{\Delta^2\gamma}{1-\Phi}\mathbf{n} \begin{Bmatrix} 1 \\ m_1 m_2 H' \end{Bmatrix} + \begin{Bmatrix} \mathbf{R}_\varepsilon \\ m_2 H' R_\alpha \mathbf{n} + \mathbf{R}_\beta \end{Bmatrix} = 0 \quad (6.313)$$

From Eqs. (6.313) and (6.305), we have

$$\Delta^2\gamma = \frac{-R_{\mathrm{f}} - m_1 K'^{R_\alpha} - [\mathbf{n}' \quad \mathbf{n}]\mathbf{A} \begin{Bmatrix} \mathbf{R}_\varepsilon \\ m_2 H' R_\alpha \mathbf{n} + \mathbf{R}_\beta \end{Bmatrix}}{\frac{m_1{}^2 K'}{1-\Phi} + 2\mu \frac{\tau}{\Delta t} + \frac{1}{1-\Phi}[\mathbf{n}' \quad \mathbf{n}]\mathbf{A} \begin{Bmatrix} \mathbf{n} \\ m_1 m_2 H' \mathbf{n} \end{Bmatrix}} \quad (6.314)$$

where

$$\mathbf{A}^{-1} = \begin{bmatrix} \mathbf{P} & \mathbf{Q} \\ \mathbf{R} & \mathbf{S} \end{bmatrix} \quad (6.315)$$

$$\mathbf{A} = \begin{bmatrix} \tilde{\mathbf{P}} & \tilde{\mathbf{Q}} \\ \tilde{\mathbf{R}} & \tilde{\mathbf{S}} \end{bmatrix} \quad (6.316)$$

$$\tilde{\mathbf{P}} = \left(\mathbf{P} - \mathbf{Q}\cdot\mathbf{S}^{-1}\cdot\mathbf{R}\right)^{-1} \quad (6.317)$$

$$\tilde{\mathbf{Q}} = -\left(\mathbf{P} - \mathbf{Q}\cdot\mathbf{S}^{-1}\cdot\mathbf{R}\right)^{-1}\cdot\left(\mathbf{Q}\cdot\mathbf{S}^{-1}\right) \quad (6.318)$$

$$\tilde{\mathbf{R}} = -\left(\mathbf{S}^{-1}\cdot\mathbf{R}\right)\cdot\left(\mathbf{P} - \mathbf{Q}\cdot\mathbf{S}^{-1}\cdot\mathbf{R}\right)^{-1} \quad (6.319)$$

$$\tilde{\mathbf{S}} = \mathbf{S}^{-1} + \left(\mathbf{S}^{-1}\cdot\mathbf{R}\right)\cdot\left(\mathbf{P} - \mathbf{Q}\cdot\mathbf{S}^{-1}\cdot\mathbf{R}\right)^{-1}\cdot\left(\mathbf{Q}\cdot\mathbf{S}^{-1}\right) \quad (6.320)$$

So we have

$$\begin{Bmatrix} \Delta\overline{\boldsymbol{\sigma}} \\ -\Delta\overline{\boldsymbol{\beta}} \end{Bmatrix} = -\mathbf{A} \begin{Bmatrix} \frac{\Delta^2\gamma}{1-\Phi}\mathbf{n} + \mathbf{R}_\varepsilon \\ \left(\frac{m_1 \Delta^2\gamma}{1-\Phi} + R_\alpha\right) m_2 H' \mathbf{n} + \mathbf{R}_\beta \end{Bmatrix} \quad (6.321)$$

$$\Delta\bar{\varepsilon}^{vp} = -\frac{1}{1-\Phi}\mathbf{C}^{-1}\Delta\overline{\boldsymbol{\sigma}} \quad (6.322)$$

$$\Delta^2 s = \frac{\overline{\boldsymbol{\sigma}}\Delta\overline{\boldsymbol{\varepsilon}}^{vp}}{T\rho} \quad (6.323)$$

$$\Delta s_{n+1} = \Delta s_n + \Delta^2 s \quad (6.324)$$

$$\Delta\Phi_{n+1} = -m_3\left(\overline{\boldsymbol{\sigma}}_{n+1}\Delta\overline{\boldsymbol{\varepsilon}}^{vp}\right) \quad (6.325)$$

$$\Delta\alpha_{n+1} = R_\alpha + \frac{m_1}{1-\Phi}\Delta^2\gamma - \frac{m_1 m_3}{(1-\Phi)^2}\Delta\gamma\boldsymbol{\sigma}\Delta\boldsymbol{\varepsilon}^{vp} \tag{6.326}$$

Then, we update the viscoplastic strain, consistency parameter, stress, TSI and entropy production, and effective equivalent viscoplastic strain at Gauss integration points and iterate until the norms of residual functions (6.300)–(6.303) are smaller than a predefined tolerance. Normally the convergence is achieved when the norms of these residual functions are all equal to or less than 1×10^{-5}. This tolerance is problem dependent.

The procedure summarized above is simply a systematic application of Newton's method to the system of Eqs. (6.270)–(6.278) that results in the computation of the closest point projection from the trial state onto the yield surface.

It should be noted that the general return mapping algorithm is unconditionally stable and the convergence of the algorithms toward the final value of the state variable is obtained at a quadratic rate. The further information on the general return mapping algorithm is given in Simo and Hughes (1998) and Ortiz and Martin (1989).

It is necessary to point out that formulation presented above calculates the TSI at Gauss integration points. In the finite element formulation, it is possible to define TSI at nodal points and solve for TSI and other nodal unknowns. However, this would significantly increase the computational cost. It is more cost-effective to calculate TSI at Gauss integration points from thermodynamic variables obtained in the previous step. Because the solution process is incremental, this simplification introduces very littler error.

6.7.6 Consistent Elastic-Viscoplastic Tangent Modulus

An important advantage of the algorithm lies in the fact that it can be exactly linearized in closed form. This leads to the notion of consistent elastic-viscoplastic tangent moduli.

Let **C** be the elastic consistent elastoviscoplastic tangent moduli, then the increment stress-strain relationship can be written as:

$$d\boldsymbol{\sigma} = (1-\Phi)\mathbf{C} : (d\boldsymbol{\varepsilon} - d\boldsymbol{\varepsilon}^{vp}) \tag{6.327}$$

Differentiating Eqs. (6.277) and (6.278), we have

$$d\Phi = -m_3\overline{\boldsymbol{\sigma}}d\overline{\boldsymbol{\varepsilon}}^{vp} \tag{6.328}$$

where m_3 is given by Eq. (6.289).

Differentiating Eq. (6.284), we have

$$d\boldsymbol{\varepsilon}^{vp} = \frac{\Delta\gamma}{(1-\Phi)^2}\mathbf{W}\mathbf{N}d\bar{\boldsymbol{\sigma}} - \frac{\Delta\gamma}{1-\Phi}\mathbf{W}\mathbf{N}d\bar{\boldsymbol{\beta}} + \frac{d\Delta\gamma}{1-\Phi}\mathbf{W}\mathbf{n} \tag{6.329}$$

where

$$\mathbf{W}^{-1} = \mathbf{I} + \frac{\Delta\gamma m_3}{(1-\Phi)^2}\left(\frac{\mathbf{N}\boldsymbol{\sigma}}{1-\Phi} + \mathbf{n}\right) \otimes \boldsymbol{\sigma} \tag{6.330}$$

Differentiating Eq. (4.285), we have

$$d\alpha = \frac{m_1}{1-\Phi}d\Delta\gamma - \frac{m_1 m_3}{(1-\Phi)^2}\Delta\gamma\bar{\boldsymbol{\sigma}}d\boldsymbol{\varepsilon}^{vp} \tag{6.331}$$

Differentiating Eq. (6.286), we have

$$d\bar{\boldsymbol{\beta}} = -\frac{m_2 m_3 \Delta H}{(1-\Phi)^2}(\mathbf{N}\bar{\boldsymbol{\sigma}} \otimes \bar{\boldsymbol{\sigma}})d\boldsymbol{\varepsilon}^{vp} + \frac{m_2\Delta H}{1-\Phi}\mathbf{N}d\bar{\boldsymbol{\sigma}} - m_2\Delta H\mathbf{N}d\boldsymbol{\beta} + m_2\mathbf{n}H'd\alpha \tag{6.332}$$

Differentiating Eq. (6.273), we have

$$\begin{aligned}
&\frac{1}{1-\Phi}\mathbf{n}d\bar{\boldsymbol{\sigma}} - \mathbf{n}d\bar{\boldsymbol{\beta}} - \frac{m_3}{(1-\Phi)^2}\left(\mathbf{n}\bar{\boldsymbol{\sigma}} - {m_1}^2K'\Delta\gamma\right)\bar{\boldsymbol{\sigma}}d\boldsymbol{\varepsilon}^{vp} \\
&= \left(2\mu\frac{\tau}{\Delta t} + \frac{{m_1}^2K'}{1-\Phi}\right)d\Delta\gamma
\end{aligned} \tag{6.333}$$

From Eqs. (6.327) and (6.329)

$$\mathbf{p} : d\bar{\boldsymbol{\sigma}} - q : d\bar{\boldsymbol{\beta}} + \frac{\mathbf{W}\mathbf{n}}{1-\Phi}d\Delta\gamma = d\boldsymbol{\varepsilon} \tag{6.334}$$

where

$$\mathbf{p} = \frac{1}{1-\Phi}\left(\mathbf{C}^{-1} + \frac{\Delta\gamma}{1-\Phi}\mathbf{W}\mathbf{N}\right) \tag{6.335}$$

$$\mathbf{q} = \frac{\Delta\gamma}{1-\Phi}\mathbf{W}\mathbf{N} \tag{6.336}$$

From Eqs. (6.329), (6.331), and (6.332), we have

$$\mathbf{r} : d\bar{\boldsymbol{\sigma}} - \mathbf{s} : d\bar{\boldsymbol{\beta}} + \frac{m_1 m_2 H'}{1-\Phi}\mathbf{W}\mathbf{n}d\Delta\gamma = 0 \tag{6.337}$$

where

$$\mathbf{r} = \frac{1}{1-\Phi}\left(m_2 \Delta H \mathbf{N} + \frac{\Delta\gamma}{1-\Phi}\mathbf{VWN}\right) \tag{6.338}$$

$$\mathbf{s} = \mathbf{I} + m_2 \Delta H \mathbf{N} + \frac{\Delta\gamma}{1-\Phi}\mathbf{VWN} \tag{6.339}$$

$$\mathbf{V} = -\frac{m_1 m_2 m_3}{(1-\Phi)^2} H' \Delta\gamma \left[\left(\frac{\mathbf{N}\boldsymbol{\sigma}}{1-\Phi} + \mathbf{n}\right) \otimes \boldsymbol{\sigma}\right] \tag{6.340}$$

Combining Eqs. (6.334) and (6.337)

$$\begin{bmatrix} \mathbf{p} & \mathbf{q} \\ \mathbf{r} & \mathbf{s} \end{bmatrix} \begin{Bmatrix} d\overline{\boldsymbol{\sigma}} \\ -d\overline{\boldsymbol{\beta}} \end{Bmatrix} + \frac{\mathbf{Wn}}{1-\Phi} \begin{Bmatrix} 1 \\ m_1 m_2 H' \end{Bmatrix} d\Delta\gamma = \begin{Bmatrix} d\boldsymbol{\varepsilon} \\ 0 \end{Bmatrix} \tag{6.341}$$

Combining Eqs. (6.329) and (6.333), we have

$$\frac{\mathbf{n}^*}{1-\Phi} d\overline{\boldsymbol{\sigma}} - \mathbf{n}^* d\overline{\boldsymbol{\beta}} = m_4 d\Delta\gamma \tag{6.342}$$

where

$$\mathbf{n}^* = \mathbf{n} - \frac{m_3 \Delta\gamma}{(1-\Phi)^3} \left(\mathbf{n}\overline{\boldsymbol{\sigma}} - {m_1}^2 K' \Delta\gamma\right)\left(\overline{\boldsymbol{\sigma}}\mathbf{WN}\right) \tag{6.343}$$

$$m_4 = \frac{m_3}{(1-\Phi)^3} \left(\mathbf{n}\overline{\boldsymbol{\sigma}} - {m_1}^2 K' \Delta\gamma\right)\left(\overline{\boldsymbol{\sigma}}\mathbf{Wn}\right) + 2\mu \frac{\tau}{\Delta t} + \frac{{m_1}^2 K'}{1-\Phi} \tag{6.344}$$

From Eqs. (6.341) and (6.342), we have

$$d\Delta\gamma = \frac{\mathbf{n}^* \left(\frac{1}{1-\Phi}\tilde{\mathbf{p}} + \tilde{\mathbf{r}}\right) d\boldsymbol{\varepsilon}}{m_4 + \frac{1}{1-\Phi} \begin{bmatrix} \frac{\mathbf{n}^*}{1-\Phi} & \mathbf{n}^* \end{bmatrix} \mathbf{a} \begin{Bmatrix} \mathbf{Wn} \\ m_1 m_2 H' \mathbf{Wn} \end{Bmatrix}} \tag{6.345}$$

where

$$\mathbf{a}^{-1} = \begin{bmatrix} \mathbf{p} & \mathbf{q} \\ \mathbf{r} & \mathbf{s} \end{bmatrix} \tag{6.346}$$

$$\mathbf{a} = \begin{bmatrix} \widetilde{\mathbf{p}} & \widetilde{\mathbf{q}} \\ \widetilde{\mathbf{r}} & \widetilde{\mathbf{s}} \end{bmatrix} \tag{6.347}$$

From Eqs. (4.334) and (4.337), we have

$$d\bar{\boldsymbol{\sigma}} = \tilde{\mathbf{p}}d\varepsilon - \left(\frac{1}{1-\Phi}\tilde{\mathbf{p}} + \tilde{\mathbf{r}}\right)\mathbf{W}\mathbf{n}d\Delta\gamma \tag{6.348}$$

From Eqs. (6.345) and (6.348), we have

$$\mathbf{C} = \frac{d\bar{\boldsymbol{\sigma}}}{d\varepsilon} = \tilde{\mathbf{p}} - \frac{\left[\left(\frac{1}{1-\Phi}\tilde{\mathbf{p}} + \tilde{\mathbf{r}}\right)\mathbf{W}\mathbf{n}\right] \otimes \left[\mathbf{n}^*\left(\frac{1}{1-\Phi}\tilde{\mathbf{p}} + \tilde{\mathbf{r}}\right)\right]}{m_4 + \frac{1}{1-\Phi}\left[\frac{\mathbf{n}^*}{1-\Phi} \quad \mathbf{n}^*\right] a \begin{Bmatrix} \mathbf{W}\mathbf{n} \\ m_1 m_2 H'\mathbf{W}\mathbf{n} \end{Bmatrix}} \tag{6.349}$$

6.8 Verification Examples

6.8.1 Material Properties of ATH

ATH (particle filler) is regarded as an isotropic elastic material. The thermo-mechanical properties of ATH as provided by the manufacturer are as follows:

Poisson ratio of ATH: $\nu_f = 0.24$
Elastic modulus of ATH: $E_f = 70$ GPa, CTE: $\alpha_f = 1.47 \times 10^{-6}/^\circ C$
The average diameter of a filler particle is 35 μm.

6.8.2 Properties of PMMA

The matrix PMMA is a very common polymer, which has been extensively studied. Young's modulus of PMMA as function of temperature is as shown in Fig. 6.16 (Cheng et al. 1990a, b, c):

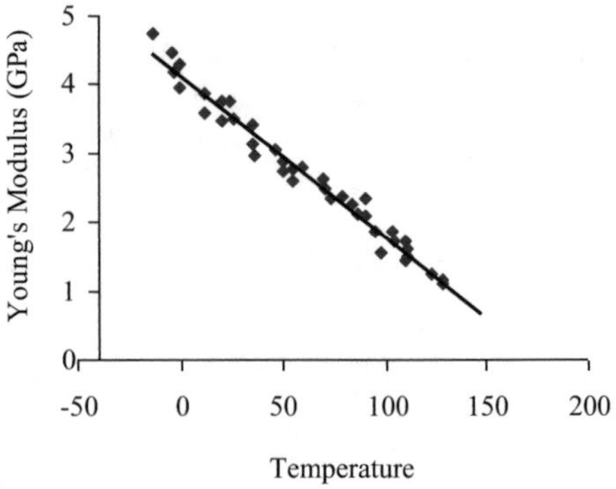

Fig. 6.16 Young's modulus of PMMA as a function of temperature (Cheng et al. 1990a, b, c)

$$E_m = -0.0234T + 4.124 \text{ (GPa)}$$

where T is the temperature in Celsius.

Poisson's ratio of PMMA is $\nu_m = 0.31$.

6.8.3 *Properties of Matrix-Filler Interphase*

The interphase around the particle is also regarded as an isotropic elastic material. Young's modulus and thickness of the interphase are adjustable parameters in the proposed model. In addition, it is reasonable to assume that Poisson's ratio and CTE of the interphase are the same as that of the matrix (PMMA); however, Young's modulus of the interphase is less than that of PMMA. For these simulations, the thickness of the interphase is taken as 1% of the diameter of filler particles; Young's modulus is half of that of PMMA except where specified differently.

6.8.3.1 Properties of Particulate Composites

The following thermo-mechanical properties of the particulate composite A are determined according to the data provided by the material supplier (Basaran and Nie 2007).

The average specific mass for the composite is $m_s = 85$ g/mol, and the density of the composite is $\rho = 1750$ kg/m^3. The volume fraction of particle in the composite is $\varphi = 0.48$

Poisson's ratio of composite A is given as a function of temperature as shown in Fig. 6.17:

$$\nu = 0.008T + 0.334$$

The coefficient of thermal expansion of composite A is given as function of temperature as shown in Fig. 6.18

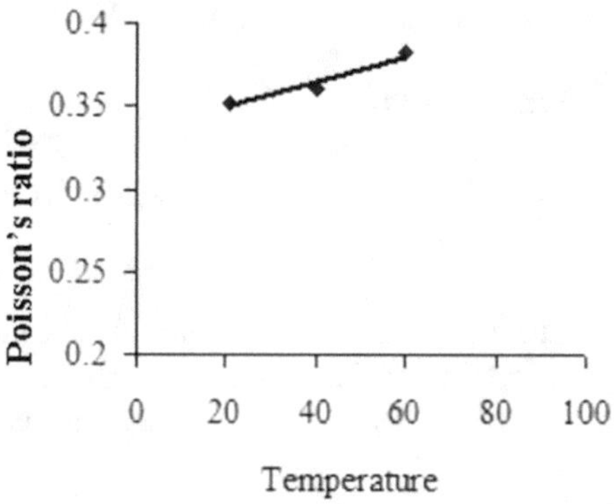

Fig. 6.17 Poisson's ratio of composite A as a function of temperature

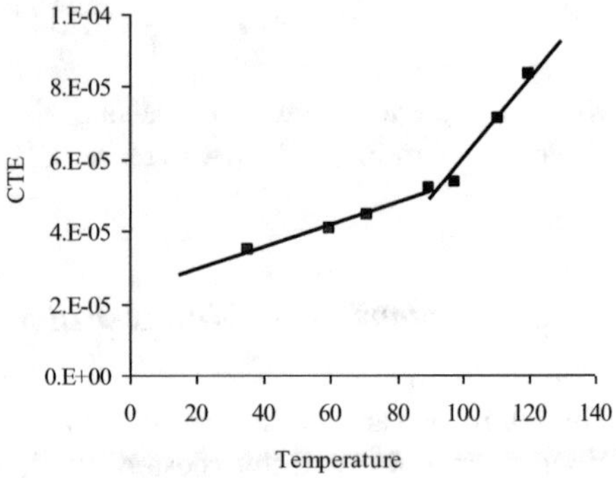

Fig. 6.18 Coefficient of thermal expansion (CTE) of composite A as a function of temperature

When $T \leq 90^{\circ}$C

$$\alpha = 3.035 \times 10^{-7}T + 2.347 \times 10^{-5}$$

When $T \geq 90^{\circ}$C

$$\alpha = 1.0992 \times 10^{-6}T - 5.012 \times 10^{-5}$$

Viscosity η is the ratio of the loss modulus to the angular frequency, which is determined during the forced harmonic oscillation test (Nielsen and Landel 1994). The viscosity relaxation time τ is defined as (Simo and Hughes 1998)

$$\tau = \frac{\eta}{2\mu}$$

where μ is the shear modulus. It is important to realize that the controlling factor in the relaxation process is the relative time t/τ. The absolute time t is regarded as short or long only when compared with τ. The concept of relaxation time is explained in greater details by Simo and Hughes (1998). From dynamic mechanical testing, the viscosity relaxation time τ is determined as shown in Fig. 6.19:

When $T \leq 90^{\circ}$C

$$\tau = 1.12406 \times 10^{-6}T^3 - 1.67823 \times 10^{-4}T^2 + 7.91134 \times 10^{-3}T + 7.35 \times 10^{-3}$$

When $T \geq 90^{\circ}$C

$$\tau = -2.6348 \times 10^{-6}T^4 1.08452 \times 10^{-3}T^3 - 1.64629 \times 10^{-1}T^2 + 10.9673T - 271.11$$

The elastic modulus of composite A can be determined from the properties of the particle, the matrix, and the interphase according to the micromechanical model presented earlier in this chapter. The elastic modulus of composite A is also

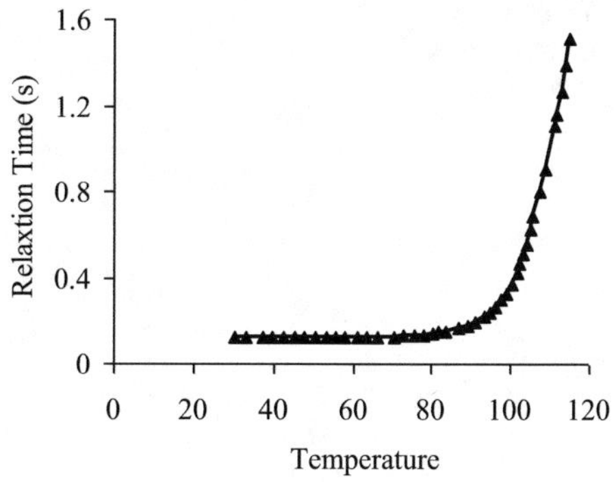

Fig. 6.19 Relaxation time of composite A as a function of temperature

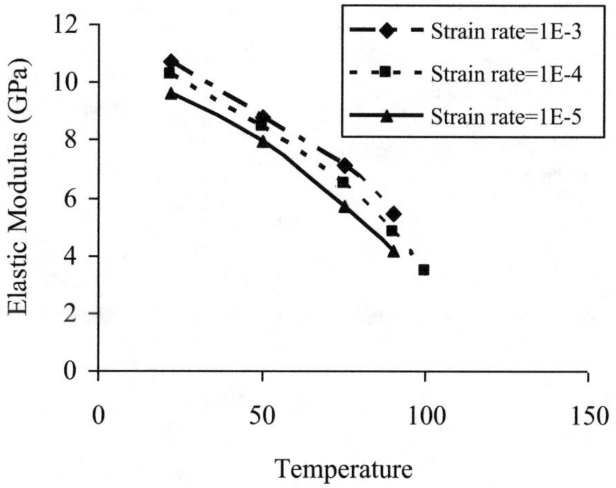

Fig. 6.20 Elastic modulus of composite A as a function of temperature (Nie 2005)

determined by uniaxial tensile tests with a different strain rate and temperature as shown in Fig. 6.20.

$$E_{\mathrm{m}} = -0.0005T^2 - 0.021T + 13.33 + 0.6\log(\dot{\varepsilon})(\mathrm{GPa})$$

The gas constant is $R = 8.3145$ J/mol/K.

The critical TSI as a function of temperature for composite A is determined from the uniaxial tensile tests at various temperatures as seen in Fig. 6.21.

$$\Phi_{\mathrm{cr}} = 0.1972T + 50.22 \quad T \le 100^{\circ}$$

Φ_{cr} ensures that Φ goes all the way to 1.00. If not used, nothing in the formulation changes.

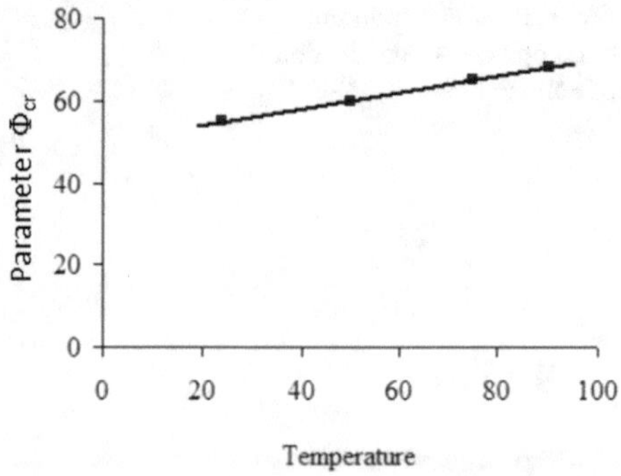

Fig. 6.21 Critical TSI of composite A as a function of temperature

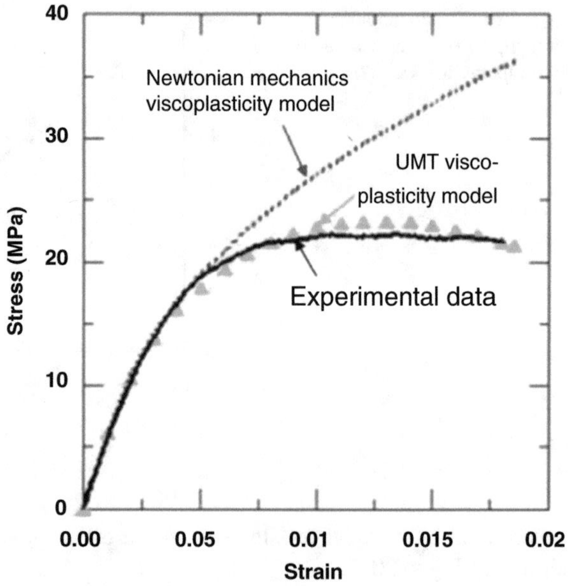

Fig. 6.22 Comparison of stress-strain relationship among viscoplastic model based on Newtonian mechanics, viscoplastic model based on UMT, and experiments at 75 °C

Where T is the temperature in Celsius and Φ_{cr} is determined using the statistical methods by matching the elastic modulus degradation determined from test data with the thermodynamic state index (TSI) evolution calculated from the internal irreversible entropy production during uniaxial tensile tests. This process allows us to map TSI onto elastic modulus degradation. If $\Phi_{cr} = 1$ used nothing changes in the formulation, TSI will still go from zero to one. Mapping elastic modulus degradation onto TSI space is not essential. System will degrade along TSI axis.

6.8.3.2 Finite Element Simulation Results

The stress-strain response obtained from the simulations was compared with experimental data as shown in Figs. 6.22 and 6.23 at 24 °C and 75 °C, respectively. These

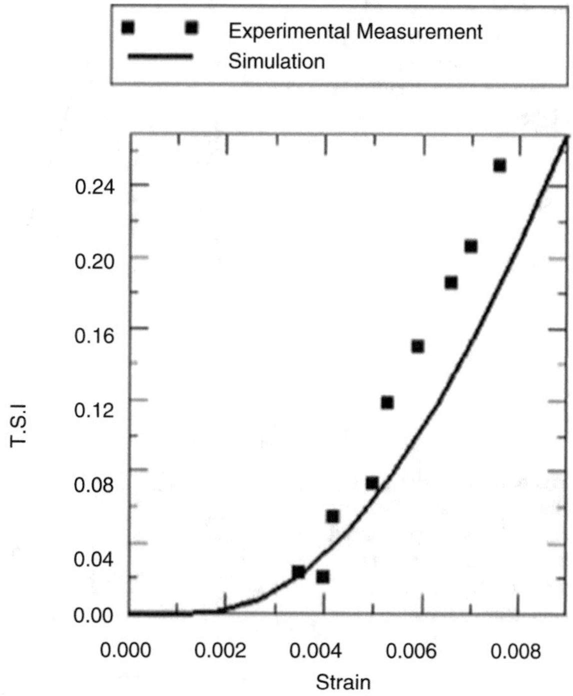

Fig. 6.23 Comparison of TSI obtained from simulations versus that measured in experiments in terms of elastic modulus degradation at 24 °C

simulation results are obtained based on the assumption that the thickness of interphase is 1% of the diameter of the particle and the elastic modulus of interphase is 50% of that of PMMA. Furthermore, the CTE mismatch effects are also included, where the temperature when residual stresses begin to build up is assumed to be 100 °C. It is seen that the degradation effects must be accounted for at relatively large strains.

The comparison of the TSI from the simulations with elastic modulus degradation that was measured during experiments is given in Fig. 6.22.

6.8.4 Cyclic Stress-Strain Response

The critical TSI Φ_{cr} is temperature dependent and can be determined from the tests data as follows: First, determine the elastic modulus degradation in cyclic tests. Second, calculate the dissipated plastic strain energy during cyclic testing using the stress-strain hysteresis loop, and then determine the TSI (Figs. 6.24, 6.25, 6.26, 6.27, 6.28, 6.29, 6.30, 6.31, and 6.32).

Finally, determine Φ_{cr} by mapping the elastic modulus degradation onto the TSI.

During experiments, Φ_{cr} was determined from the cyclic tests [at strain amplitude of 0.6%] as a function of temperature. Mapping elastic modulus degradation on to

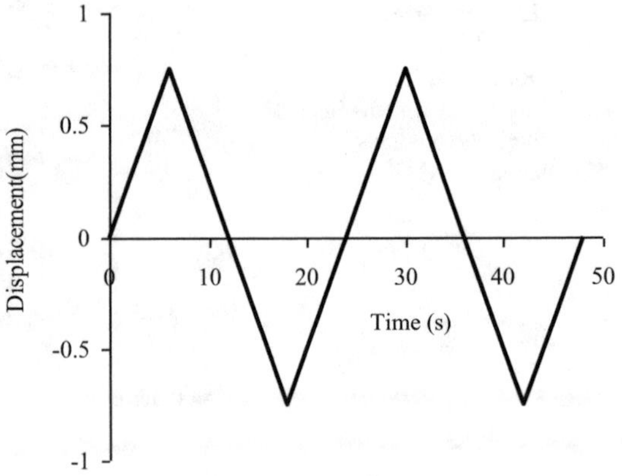

Fig. 6.24 Applied displacement profile for tension-compression fatigue test with a strain amplitude of 0.6%

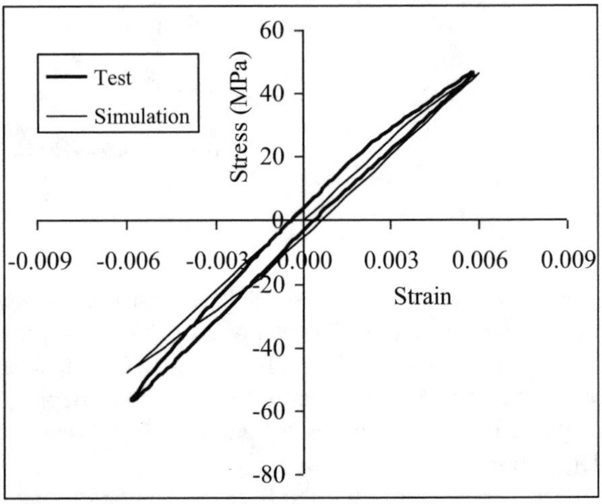

Fig. 6.25 Uniaxial stress-strain hysteresis loop from simulation vs. experimental data for cycle 5 at room temperature with strain amplitude of 0.6%

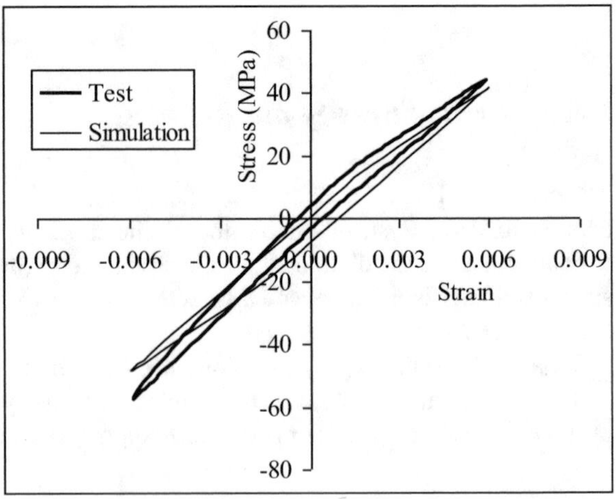

Fig. 6.26 Uniaxial stress-strain hysteresis loop from simulation vs. experimental data for cycle 50 at room temperature with strain amplitude of 0.6%

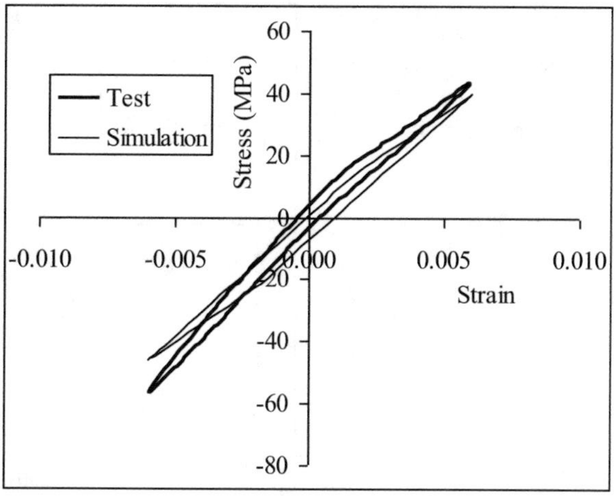

Fig. 6.27 Uniaxial stress-strain hysteresis loop from simulation vs. experimental data for cycle 104 at room temperature with strain amplitude of 0.6%

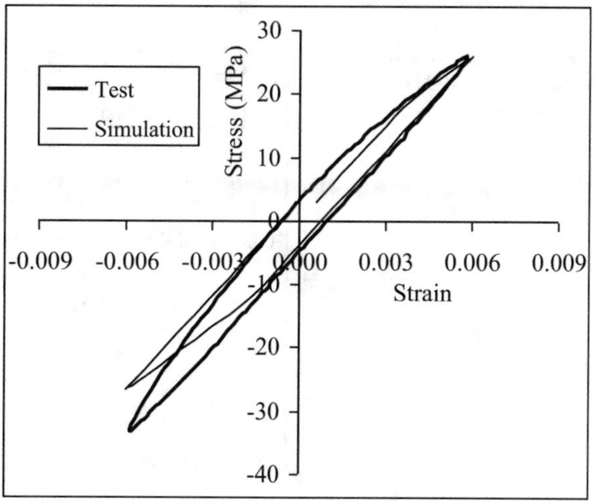

Fig. 6.28 Uniaxial stress-strain hysteresis loop from simulation vs. experimental data for cycle 5 at 75 °C with strain amplitude of 0.6%

TSI axis is necessary only if cyclic stress values are need to be predicted cycle to cycle. However, this mapping process is not needed if only final number of cycles to failure is needed. System will degrade along TSI axis.

$$\Phi_{cr} = 0.00545T + 0.659 \quad T \leq 100^{\circ}\mathrm{C}$$

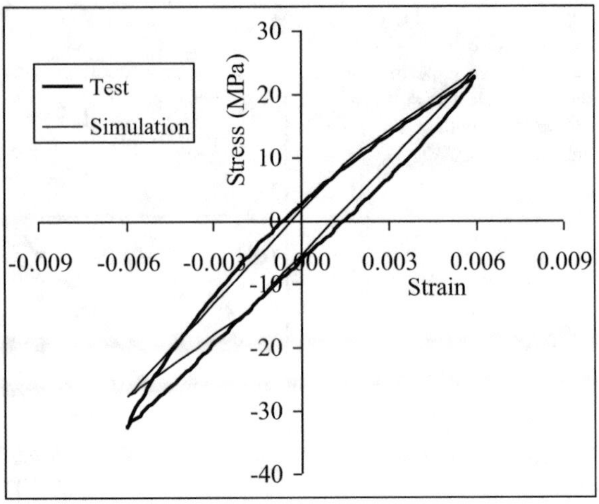

Fig. 6.29 Uniaxial stress-strain hysteresis loop from simulation vs. experimental data for cycle 100 at 75 °C with strain amplitude of 0.6%

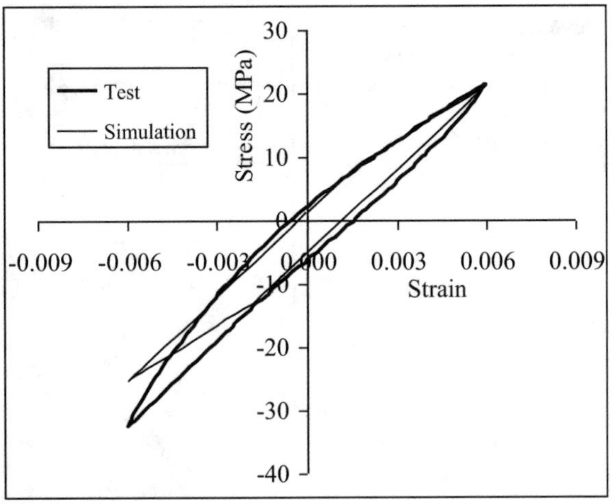

Fig. 6.30 Uniaxial stress-strain hysteresis loop from simulation vs. experimental data for cycle 495 at 75 °C with strain amplitude of 0.6%

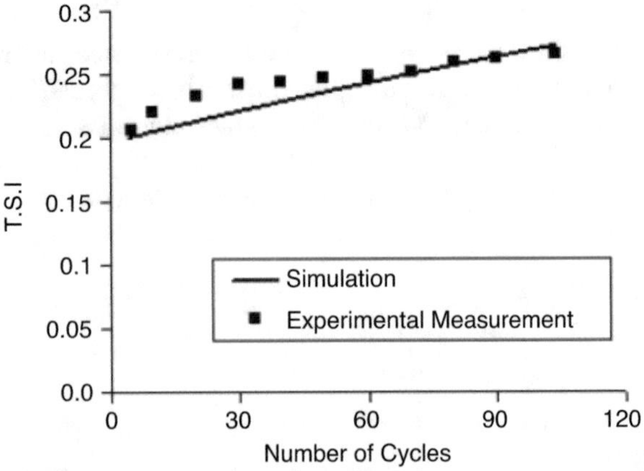

Fig. 6.31 Comparison of TSI at room temperature with strain amplitude of 0.6%

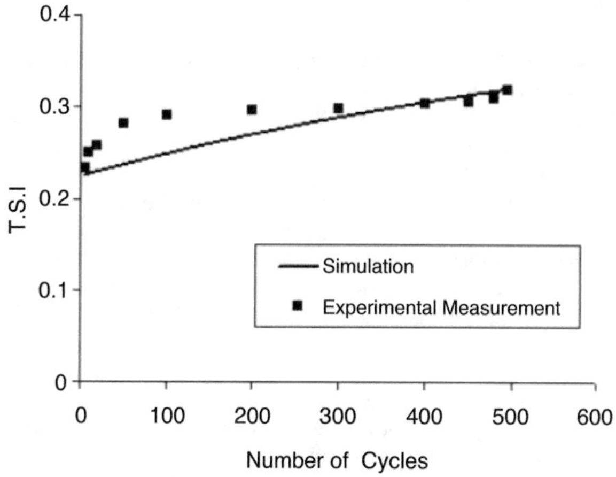

Fig. 6.32 Comparison of TSI at 750 °C with strain amplitude of 0.6%

References

Basaran, C., & Nie, S. (2004). An irreversible thermodynamic theory for damage mechanics of solids. *International Journal of Damage Mechanics, 13*(3), 205–224.

Basaran, C., & Nie, S. (2007). A thermodynamics based damage mechanics model for particulate composites. *International Journal of Solids and Structures, 44*, 1099–1114.

Basaran, C., Nie, S., Hutchins, S., & Ergun, H. (2008a). Influence of interfacial bond strength on fatigue life and mechanical behavior of a particulate composite: An experimental study. *International Journal of Damage Mechanics, 17*(2), 123–148.

Basaran, C., Nie, S., & Hutchins, C. (2008b). Time dependent behavior of a particle filled composite PMMA/ATH at elevated temperatures. *Journal of Composite Materials, 42*(19), 2003–2025.

Cheng, W.-M., Miller, G. A., Manson, J. A., Hertzberg, R. W., & Sperling, L. H. (1990a). Mechanical behavior of poly(methyl methacrylate). Part 1 Tensile strength and fracture toughness. *Journal of Materials Science, 25*, 1917–1923.

Cheng, W.-M., Miller, G. A., Manson, J. A., Hertzberg, R. W., & Sperling, L. H. (1990b). Mechanical behavior of poly(methyl methacrylate). Part 2 Temperature and frequency effects on the fatigue crack propagation behavior. *Journal of Materials Science, 25*, 1924–1930.

Cheng, W.-M., Miller, G. A., Manson, J. A., Hertzberg, R. W., & Sperling, L. H. (1990c). Mechanical behavior of poly(methyl methacrylate). Part 3 Activation processes for fracture mechanism. *Journal of Materials Science, 25*, 1931–1938.

Christensen, R. M., & Lo, K. H. (1979). Solutions for effective shear properties in three phase sphere and cylinder models. *Journal of the Mechanics and Physics of Solids, 27*, 315–330.

Eshelby, J. D. (1957). The deformation of the elastic field of an ellipsoidal inclusion, and related problems. *Proceedings of the Royal Society of London, Series A, Mathematical Science, 241* (1226), 376–396.

Hashin, Z. (1962). The elastic moduli of heterogeneous materials. *Transaction of the ASME, Journal of Applied Mechanics, 29*, 143–150.

Hashin, Z. (1968). Assessment of the self consistent scheme approximation: Conductivity of particulate composites. *Journal of Composite Materials, 2*(3), 284–301.

Hashin, Z. (1990). Thermoelastic properties and conductivity of carbon/carbon fiber composites. *Mechanics of Materials, 8*, 293–308.

Hashin, Z. (1991a). The spherical inclusion with imperfect interface. *Transaction of the ASME, Journal of Applied Mechanics, 58*, 444–449.

Hashin, Z. (1991b). Thermoelastic properties of particulate composites with imperfect interface. *Journal of the Mechanics and Physics of Solids, 39*(6), 745–762.

Hashin, Z., & Shtrikman, S. (1963). A variational approach to the theory of the elastic behavior of multiphase materials. *Journal of the Mechanics and Physics of Solids, 11*, 127–140.

Ju, J. W., & Chen, T. M. (1994a). Micromechanics and effective moduli of elastic composites containing randomly dispersed ellipsoidal inhomogeneities. *Acta Mechanica, 103*, 103–121.

Ju, J. W., & Chen, T. M. (1994b). Effective elastic moduli of two-phase composites containing randomly dispersed spherical inhomogeneities. *Acta Mechanica, 103*, 123–144.

Ju, J. W., & Tseng, K. H. (1996). Effective elastoplastic behavior of two-phase ductile matrix composites: A micromechanical framework. *International Journal of Solids & Structures, 33* (29), 4267–4291.

Ju, J. W., & Tseng, K. H. (1997). Effective elastoplastic algorithms for ductile matrix composites. *Journal of Engineering Mechanics, 123*(3), 260–266.

Kerner, E. H. (1956). The elastic and thermoelastic properties of composite media. *The Proceedings of Physical Society, 69B*, 808–813.

Levin, V. M. (1967). Thermal expansion coefficients of heterogeneous materials. *Mechanics of Solids, 2*(1), 58–94.

Mura, T. (1987). *Mechanics of elastic and inelastic solids: Micromechanics of defects in solids* (2nd ed.). Leiden: Martinus Nijhoff Publishers.

Nie, S. (2005). *A micromechanical study of the damage mechanics of acrylic particulate composites under thermomechanical loading.* PhD Dissertation, Submitted to Department of Civil, Structural and Environmental Engineering, University at Buffalo.

Nie, S., & Basaran, C. (2005). A micromechanical model for effective elastic properties of particulate composites with imperfect interfacial bonds. *International Journal of Solids & Structures, 42*(14), 4179–4191.

Nielsen, L. E., & Landel, R. F. (1994). *Mechanical properties of polymers and composites* (2nd ed.). New York: Marcel Dekker.

Ortiz, M., & Martin, J. E. (1989). Symmetry-preserving return-mapping algorithms and incrementally extremal paths: A unification of concepts. *International Journal for Numerical Methods in Engineering, 28*, 1839–1853.

Prager, W. (1956). A new method of analyzing stresses and strains in work-hardening plastic solids. *Journal of Applied Mechanics, 23*, 493–496.

Simo, J. C., & Hughes, T. J. R. (1998). *Interdisciplinary applied mathematics, mechanics and materials, computational inelasticity*. New York: Springer.

Simo, J. C., & Taylor, R. L. (1985). Consistent tangent operators for rate-independent elastoplasticity. *Computer Methods in Applied Mechanics and Engineering, 48*, 101–119.

Smith, J. C. (1974). Correction and extension of van der Poel's method for calculating the shear modulus of a particulate composite. *Journal of Research of the National Bureau of Standards-A. Physics and Chemistry, 78A*(3), 355–361.

Smith, J. C. (1975). Simplification of van der Poel's formula for the shear modulus of a particulate composite. *Journal of Research of the National Bureau of Standards-A. Physics and Chemistry, 79A*(2), 419–423.

Van der Poel, C. (1958). On the rheology of concentrated suspensions. *Rheologica Acta, 1*, 198.

Walpole, L. J. (1966). On bounds for the overall elastic moduli of inhomogeneous systems-I. *Journal of the Mechanics and Physics of Solids, 14*, 151.

Ziegler, H. (1959). A modification of Prager's hardening rule. *Quarterly of Applied Mathematics, 17*, 55–65.

Chapter 7
Unified Micromechanics of Finite Deformations

7.1 Introduction to Finite Deformations

In this chapter finite (large) deformation micromechanics is discussed. Formulation is more complicated than small deformation theory; therefore, it is easier to explain with an example. Micromechanical modeling of polymer is used as an example. A dual-micro-mechanism rate-dependent constitutive model is used to describe thermo-mechanical response of amorphous polymers below and above glass transition temperature. Material property definitions, evolution of internal state variables, and plastic flow rules are revisited to provide a smooth and continuous transition in material response around glass transition temperature, θ_g.

In order to formulate the large deformation mechanics, it is necessary to describe kinematics of constitutive model based on finite deformation tensors. Consider a body with a volume of V_o in undeformed (original or initial) configuration ($\boldsymbol{\Sigma}_\mathbf{o}$) at time t_o which deforms into a volume V in current (deformed) configuration ($\boldsymbol{\Sigma}$) at time t, as shown in Fig. 7.1, this body can be uniquely defined with a continuous one-to-one mapping (χ) of position vectors **r** and **x** in original configuration and current configuration, respectively. **r** is the material coordinate of a particle that is the coordinates of its location in the reference coordinate system. **x** is the spatial coordinates, defining location, at time t. Both material coordinates and spatial coordinates were discussed earlier in Chap. 2.

Gradient of deformation (**F**), which is the strain, describes transformation of a line element (**dr**) at position of r in the original configuration to a deformed line element (**dx**) at position of **x** in current configuration Eq. (7.1). Unique transformation ensures a non-singular, nonnegative determinant of deformation gradient or Jacobian (J) Eq. (7.3). Velocity gradient (**L**) is the gradient of velocity field (**v**) and relates deformation gradient to its material time derivative $\left(\dot{\mathbf{F}}\right)$. These relations can be given as follows,

C. Basaran, *Introduction to Unified Mechanics Theory with Applications*,
https://doi.org/10.1007/978-3-030-57772-8_7

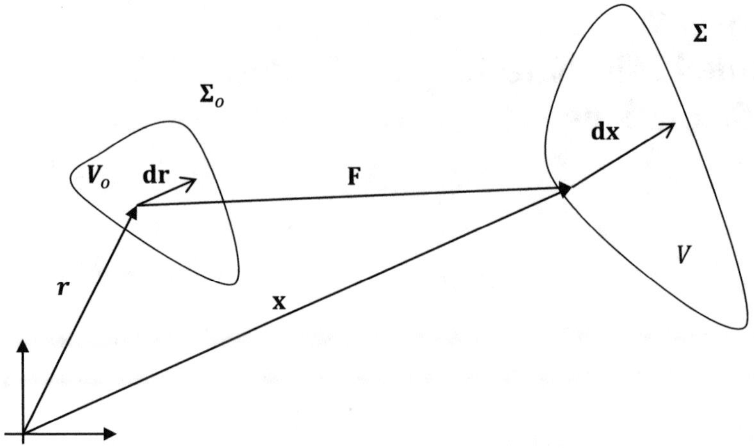

Fig. 7.1 Deformation from original configuration to current configuration

$$\mathbf{x} = \boldsymbol{\chi}(\mathbf{r}, t) \tag{7.1}$$

$$\mathbf{F} = \nabla_{\mathbf{X}}(\chi) = \frac{\partial \boldsymbol{\chi}}{\partial \mathbf{r}} \tag{7.2}$$

$$J = \det(\mathbf{F}) > 0 \tag{7.3}$$

$$\mathbf{v}(\mathbf{x}, t) = \dot{\mathbf{x}}(\mathbf{x}, t) \tag{7.4}$$

$$\mathbf{L} = \text{grad}(\mathbf{v}) = \frac{\partial \dot{\mathbf{x}}}{\partial \mathbf{r}} \tag{7.5}$$

$$\dot{\boldsymbol{F}} = \boldsymbol{L} \cdot \boldsymbol{F} \tag{7.6}$$

Deformation gradient (**F**) can be decomposed into elastic part ($\mathbf{F}^{e}$) and plastic part ($\mathbf{F}^{p}$), and information related to thermal deformation is included in the elastic part of deformation gradient. Plastic deformation gradient ($\mathbf{F}^{P}$) defines a deformation with respect to initial configuration ($\boldsymbol{\Sigma}_{\mathbf{o}}$) to an intermediate, “relaxed” or “natural” configuration ($\boldsymbol{\Sigma}'$) which is followed by an elastic transformation ($\mathbf{F}^{e}$) to the final configuration (**Σ**)**.** Symmetric part of velocity gradient (**L**) is defined as stretching rate tensor (**D**) and asymmetric part of velocity gradient is defined as spin tensor (**W**)

$$\mathbf{F} = \mathbf{F}^{e}\mathbf{F}^{p}; J = J^{e}J^{p} \tag{7.7}$$

$$J^{e} = \det(\mathbf{F}^{e}) > 0 \tag{7.8}$$

$$J^{p} = \det(\mathbf{F}^{p}) > 0 \tag{7.9}$$

$$\text{sym}(\mathbf{L}) = \mathbf{D}; \text{asym}(\mathbf{L}) = \mathbf{W}; \mathbf{L} = \mathbf{D} + \mathbf{W} \tag{7.10}$$

Using Eq. (7.7), elastic and plastic part of velocity gradient can be defined as

$$\mathbf{L} = \mathbf{L}^{\mathrm{e}} + \mathbf{F}^{\mathrm{e}}\mathbf{L}^{\mathrm{p}}\mathbf{F}^{\mathrm{e}^{-1}} \tag{7.11}$$

where

$$\mathbf{L}^{\mathrm{e}} = \dot{\mathbf{F}}^{\mathrm{e}}\mathbf{F}^{\mathrm{e}^{-1}} \tag{7.12}$$

$$\mathbf{L}^{\mathrm{p}} = \dot{\mathbf{F}}^{\mathrm{p}}\mathbf{F}^{\mathrm{p}^{-1}} \tag{7.13}$$

Stretching rate tensor (**D**) and spin tensor ***W*** for elastic and plastic parts can be derived similarly from Eq. (7.10), where superscripts "e" and "p" represent elastic and plastic parts of corresponding quantity, respectively

$$\mathsf{sym}(\mathbf{L}^{\mathrm{e}}) = \mathbf{D}^{\mathrm{e}}; \mathsf{asym}(\mathbf{L}^{\mathrm{e}}) = \mathbf{W}^{\mathrm{e}}; \mathbf{L}^{\mathrm{e}} = \mathbf{D}^{\mathrm{e}} + \mathbf{W}^{\mathrm{e}} \tag{7.14}$$

$$\mathsf{sym}(\mathbf{L}^{\mathrm{p}}) = \mathbf{D}^{\mathrm{p}}; \mathsf{asym}(\mathbf{L}^{\mathrm{p}}) = \mathbf{W}^{\mathrm{p}}; \mathbf{L}^{\mathrm{p}} = \mathbf{D}^{\mathrm{p}} + \mathbf{W}^{\mathrm{p}} \tag{7.15}$$

Right and left stretch tensors $(\mathbf{U}, \mathbf{V})$ and rotation tensor $(\mathbf{R})$ can be found from right and left polar decompositions of deformation gradient as follows:

$$\mathbf{F} = \mathbf{RU} \tag{7.16}$$

$$\mathbf{F} = \mathbf{VR} \tag{7.17}$$

U and **V** stretch tensors are positive definite symmetric tensors and **R** is a proper orthogonal tensor. Cauchy (**C**) and Almansi (**B**) tensors can be formulated as follows:

$$\mathbf{C} = \mathbf{F}^{\mathrm{T}}\mathbf{F} = \mathbf{UU} \tag{7.18}$$

$$\mathbf{B}^{-1} = \mathbf{F}^{-\mathrm{T}}\mathbf{F}^{-1} = \mathbf{V}^{-1}\mathbf{V}^{-1} \tag{7.19}$$

In the same fashion, the following relations can be written for elastic and plastic parts of the deformation gradient.

$$\mathbf{F}^{\mathrm{e}} = \mathbf{R}^{\mathrm{e}}\mathbf{U}^{\mathrm{e}} \tag{7.20}$$

$$\mathbf{F}^{\mathrm{e}} = \mathbf{V}^{\mathrm{e}}\mathbf{R}^{\mathrm{e}} \tag{7.21}$$

$$\mathbf{C}^{\mathrm{e}} = \mathbf{F}^{\mathrm{e}^{\mathrm{T}}}\mathbf{F}^{\mathrm{e}} = \mathbf{U}^{\mathrm{e}}\mathbf{U}^{\mathrm{e}} \tag{7.22}$$

$$\mathbf{B}^{\mathrm{e}^{-1}} = \mathbf{F}^{\mathrm{e}^{-\mathrm{T}}}\mathbf{F}^{\mathrm{e}^{-1}} = \boldsymbol{V}^{\mathrm{e}^{-1}}\boldsymbol{V}^{\mathrm{e}^{-1}} \tag{7.23}$$

$$\mathbf{F}^{\mathrm{p}} = \mathbf{R}^{\mathrm{p}}\mathbf{U}^{\mathrm{p}} \tag{7.24}$$

$$\mathbf{F}^{p} = \mathbf{V}^{p}\mathbf{R}^{p} \tag{7.25}$$

$$\mathbf{C}^{\mathrm{p}} = \mathbf{F}^{\mathrm{p}^{\mathrm{T}}}\mathbf{F}^{\mathrm{p}} = \mathbf{U}^{\mathrm{p}}\mathbf{U}^{\mathrm{p}} \tag{7.26}$$

$$\mathbf{B}^{\mathrm{p}^{-1}} = \mathbf{F}^{\mathrm{p}^{-\mathrm{T}}}\mathbf{F}^{\mathrm{p}^{-1}} = \mathbf{V}^{\mathrm{p}^{-1}}\mathbf{V}^{\mathrm{p}^{-1}} \tag{7.27}$$

Multiplicative split of deformation gradient tensor ($\mathbf{F}$) produces nonunique, locally defined intermediate (relaxed) configurations. One way to solve this problem is to introduce co-rotational rate definitions of elastic and plastic deformation gradients based on (director) orthonormal vectors which were proposed by Mandel (1972). Resulting elastic and plastic parts of rate deformation tensors are invariant upon any superposed rigid body rotations . However, arbitrariness of intermediate configuration can be also removed by setting plastic spin tensor equal to zero (Eq. (7.28)). In this case, elastic and plastic deformation gradients will include rotations, which can be handled with proper selection of stress and rate measures establishing a frame indifferent model.

$$\mathbf{W}^{\mathrm{p}} = 0 \tag{7.28}$$

$$J^{\mathrm{p}} = 1 \tag{7.29}$$

Assuming that plastic flow is irrotational, Eq. (7.28) and incompressible Eq. (7.29), Jacobian of total deformation gradient (J) and material time derivative of plastic deformation gradient $\left(\dot{\mathbf{F}}^{\mathrm{p}}\right)$ will become

$$J = J^{\mathrm{e}} \tag{7.30}$$

$$\dot{\mathbf{F}}^{\mathrm{p}} = \mathbf{D}^{\mathrm{p}}\mathbf{F}^{\mathrm{p}} \tag{7.31}$$

Because of incompressible plastic flow and irrotational plastic flow assumptions

$$\mathbf{L}^{\mathrm{p}} = \mathbf{D}^{\mathrm{p}} \tag{7.32}$$

$$\mathrm{tr}(\mathbf{L}^{\mathrm{p}}) = \mathrm{tr}(\mathbf{D}^{\mathrm{p}}) = 0 \tag{7.33}$$

$$\mathbf{L} = \mathbf{L}^{\mathrm{e}} + \mathbf{F}^{\mathrm{e}}\mathbf{D}^{\mathrm{p}}\mathbf{F}^{\mathrm{e}^{-1}} \tag{7.34}$$

System is assumed to be at rest initially ($t = 0$) which provides the following initial conditions:

$$\mathbf{F}^{\mathrm{p}}(\mathbf{r}, 0) = \mathbf{I} \tag{7.35}$$

$$\mathbf{F}^{\mathrm{e}}(\mathbf{r}, 0) = \mathbf{I} \tag{7.36}$$

7.2 Frame of Reference Indifference

Material constitutive model must be invariant with respect to frame of reference. Therefore, a quantity or an equation is frame indifferent (or objective) if it is invariant to changes in frame of reference.

Let rigid body motion of any material point $\boldsymbol{\chi}(\mathbf{r}, t)$ be defined by a proper orthogonal rotation tensor $\boldsymbol{\Omega}(t)$ and a vector $\mathbf{r}_o(t)$ for any time frame t as

$$\bar{\chi}(\mathbf{r}, t) = \boldsymbol{\Omega}(t)[\chi(\mathbf{r}, t) - \mathbf{O}] + \mathbf{r_o}(t) \tag{7.37}$$

where orthogonal characteristic of rotation tensor is defined as

$$\boldsymbol{\Omega}^{\mathrm{T}}\boldsymbol{\Omega} = \mathbf{I} \tag{7.38}$$

Here we define a stress tensor as objective if it does not depend on the frame of reference. In Eq. (7.37), $\boldsymbol{\Omega}(t)$ represents rigid body rotation and $\mathbf{r}_o(t)$ represents rigid body translation. In this context, frame indifference of a vector (**a**) and a second-order tensor (**A**) are defined in terms of transformation rules as follows:

$$\bar{\mathbf{a}} = \boldsymbol{\Omega}\mathbf{a} \tag{7.39}$$

$$\overline{\mathbf{A}} = \boldsymbol{\Omega}\mathbf{A}\boldsymbol{\Omega}^{\mathrm{T}} \tag{7.40}$$

Transformation of deformation gradient with respect to a change in frame of reference can be obtained from Eq. (7.31) as

$$\overline{\mathbf{F}} = \boldsymbol{\Omega}\mathbf{F} \tag{7.41}$$

Therefore, deformation gradient is not objective (it is not frame of reference indifferent) according to Eq. (7.40). Using Eq. (7.41) Cauchy tensor (**C**) in the new transformed configuration can be written as

$$\overline{\mathbf{C}} = \overline{\mathbf{F}}^{\mathrm{T}}\overline{\mathbf{F}} = \mathbf{F}^{\mathrm{T}}\boldsymbol{\Omega}^{\mathrm{T}}\boldsymbol{\Omega}\mathbf{F} = \mathbf{F}^{\mathrm{T}}\mathbf{F} = \mathbf{C} \tag{7.42}$$

Since Cauchy tensor is a Lagrangian tensor referring to original frame of reference, Cauchy tensor should not be expected to change with current coordinate system (frame of reference) according to Eq. (7.40). Therefore, Cauchy tensor (**C**) is not objective but still invariant with respect to changes in the current frame of reference in the deformed configuration. Using Eq. (7.41) and right polar decomposition rule in Eq. (7.16)

$$\overline{\mathbf{F}} = \overline{\mathbf{R}\mathbf{U}} = \boldsymbol{\Omega}\mathbf{F} = \boldsymbol{\Omega}\mathbf{R}\mathbf{U} \tag{7.43}$$

Since right polar decomposition is unique, Eq. (7.43) implies that

$$\overline{\mathbf{R}} = \boldsymbol{\Omega}\mathbf{R} \tag{7.44}$$

$$\overline{\mathbf{U}} = \mathbf{U} \tag{7.45}$$

Therefore, right stretch tensor (**U**) and rotation tensor (**R**) are not objective. However, similar to Cauchy tensor, right stretch tensor refers to original frame of reference and hence right stretch tensor is invariant to changes in current frame of reference. Similarly, using Eq. (7.19), Almansi tensor (**B**) in the transformed configuration we can write

$$\overline{\mathbf{B}} = \overline{\mathbf{F}}\overline{\mathbf{F}}^{\mathrm{T}} = \boldsymbol{\Omega}\mathbf{F}\mathbf{F}^{\mathrm{T}}\boldsymbol{\Omega}^{\mathrm{T}} = \boldsymbol{\Omega}\mathbf{B}\boldsymbol{\Omega}^{\mathrm{T}} \tag{7.46}$$

Therefore, Almansi tensor (**B**) is objective. Using Eq. (7.41) and left polar decomposition rule in Eq. (7.17) yields

$$\overline{\mathbf{F}} = \overline{\mathbf{V}\mathbf{R}} = \boldsymbol{\Omega}\mathbf{F} = \boldsymbol{\Omega}\mathbf{V}\mathbf{R} \tag{7.47}$$

According to transformation of rotation tensor in Eq. (7.44), Eq. (7.47) can be rewritten as

$$\overline{\mathbf{F}} = \overline{\mathbf{V}\mathbf{R}} = \boldsymbol{\Omega}\mathbf{V}\boldsymbol{\Omega}^{\mathrm{T}}\boldsymbol{\Omega}\mathbf{R} \tag{7.48}$$

Since left polar decomposition is also unique, Eq. (7.48) implies that

$$\overline{\mathbf{V}} = \boldsymbol{\Omega}\mathbf{V}\boldsymbol{\Omega}^{\mathrm{T}} \tag{7.49}$$

Therefore, left stretch tensor (**V**) is objective. Taking time derivative of Eq. (7.41), it can be shown that material time derivative of deformation gradient is not objective:

$$\dot{\overline{\mathbf{F}}} = \dot{\boldsymbol{\Omega}}\mathbf{F} + \boldsymbol{\Omega}\dot{\mathbf{F}} \tag{7.50}$$

Using Eq. (7.50) and the definition of velocity gradient in Eq. (7.6), we can write

$$\dot{\overline{\mathbf{F}}} = \overline{\mathbf{L}\mathbf{F}} = \dot{\boldsymbol{\Omega}}\mathbf{F} + \boldsymbol{\Omega}\dot{\mathbf{F}} = \left(\dot{\boldsymbol{\Omega}}\mathbf{F}\mathbf{F}^{-1}\boldsymbol{\Omega}^{\mathrm{T}} + \boldsymbol{\Omega}\dot{\mathbf{F}}\mathbf{F}^{-1}\boldsymbol{\Omega}^{\mathrm{T}}\right)\boldsymbol{\Omega}\mathbf{F} \tag{7.51}$$

Equation (7.51)implies that velocity gradient is not objective as shown below.

$$\overline{\mathbf{L}} = \boldsymbol{\Omega}\mathbf{L}\boldsymbol{\Omega}^{\mathrm{T}} + \dot{\boldsymbol{\Omega}}\boldsymbol{\Omega}^{\mathrm{T}} \tag{7.52}$$

Using definitions of stretch rate tensor and spin tensor given in Eq. (7.10)

$$\overline{\mathbf{L}} = \overline{\mathbf{D}} + \overline{\mathbf{W}} = \boldsymbol{\Omega}(\mathbf{D} + \mathbf{W})\boldsymbol{\Omega}^{\mathrm{T}} + \dot{\boldsymbol{\Omega}}\boldsymbol{\Omega}^{\mathrm{T}} \tag{7.53}$$

which implies that

$$\overline{\mathbf{D}} = \boldsymbol{\Omega}\mathbf{D}\boldsymbol{\Omega}^{\mathrm{T}} \tag{7.54}$$

$$\overline{\mathbf{W}} = \boldsymbol{\Omega}\mathbf{W}\boldsymbol{\Omega}^{\mathrm{T}} + \dot{\boldsymbol{\Omega}}\boldsymbol{\Omega}^{\mathrm{T}} \tag{7.55}$$

Therefore, stretch rate tensor (**D**) is objective, while spin tensor (**W**) is not objective. Using multiplicative decomposition definition of deformation gradient given in Eq. (7.7), we can write

$$\overline{F} = \overline{\mathbf{F}^{\mathrm{e}}\mathbf{F}^{\mathrm{p}}} = \boldsymbol{\Omega}\mathbf{F}^{\mathrm{e}}\mathbf{F}^{\mathrm{p}} \tag{7.56}$$

Hence we can write the following relations:

$$\overline{\mathbf{F}^{\mathrm{e}}} = \boldsymbol{\Omega}\mathbf{F}^{\mathrm{e}} \tag{7.57}$$

$$\overline{\mathbf{F}^{\mathrm{p}}} = \mathbf{F}^{\mathrm{p}} \tag{7.58}$$

Therefore, elastic and plastic deformation gradients are not objective. Since plastic deformation gradient refers to original and intermediate frame of reference, plastic deformation gradient is not objective but invariant to changes in current frame of reference. Using definitions for elastic and plastic rotation tensors, left stretch tensor, right stretch tensor, Cauchy tensor, and Almansi tensor in Eqs. (7.20)–(7.27), the following relations can be written as:

$$\overline{\mathbf{F}^{\mathrm{e}}} = \overline{\mathbf{R}^{\mathrm{e}}\mathbf{U}^{\mathrm{e}}} = \boldsymbol{\Omega}\mathbf{R}^{\mathrm{e}}\mathbf{U}^{\mathrm{e}} \tag{7.59}$$

$$\overline{\mathbf{R}^{\mathrm{e}}} = \boldsymbol{\Omega}\mathbf{R}^{\mathrm{e}} \tag{7.60}$$

$$\overline{\mathbf{U}^{\mathrm{e}}} = \mathbf{U}^{\mathrm{e}} \tag{7.61}$$

$$\overline{\mathbf{F}^{\mathrm{e}}} = \overline{\mathbf{V}^{\mathrm{e}}\mathbf{R}^{\mathrm{e}}} = \boldsymbol{\Omega}\mathbf{V}^{\mathrm{e}}\boldsymbol{\Omega}^{\mathrm{T}}\boldsymbol{\Omega}\mathbf{R}^{\mathrm{e}} \tag{7.62}$$

$$\overline{\mathbf{V}^{\mathrm{e}}} = \boldsymbol{\Omega}\mathbf{V}^{\mathrm{e}}\boldsymbol{\Omega}^{\mathrm{T}} \tag{7.63}$$

$$\overline{\mathbf{C}^{\mathrm{e}}} = \overline{\mathbf{U}^{\mathrm{e}}\mathbf{U}^{\mathrm{e}}} = \mathbf{U}^{\mathrm{e}}\mathbf{U}^{\mathrm{e}} = \mathbf{C}^{\mathrm{e}} \tag{7.64}$$

$$\overline{\mathbf{B}^{\mathrm{e}}} = \overline{\mathbf{V}^{\mathrm{e}}\mathbf{V}^{\mathrm{e}}} = \boldsymbol{\Omega}\mathbf{V}^{\mathrm{e}}\boldsymbol{\Omega}^{\mathrm{T}}\boldsymbol{\Omega}\mathbf{V}^{\mathrm{e}}\boldsymbol{\Omega}^{\mathrm{T}} = \boldsymbol{\Omega}\mathbf{B}^{\mathrm{e}}\boldsymbol{\Omega}^{\mathrm{T}} \tag{7.65}$$

$$\overline{\mathbf{F}^{\mathrm{p}}} = \overline{\mathbf{R}^{\mathrm{p}}\mathbf{U}^{\mathrm{p}}} = \boldsymbol{\Omega}\mathbf{R}^{\mathrm{p}}\mathbf{U}^{\mathrm{p}} \tag{7.66}$$

$$\overline{\mathbf{R}^{\mathrm{p}}} = \boldsymbol{\Omega}\mathbf{R}^{\mathrm{p}} \tag{7.67}$$

$$\overline{\mathbf{U}^{\mathrm{p}}} = \mathbf{U}^{\mathrm{p}} \tag{7.68}$$

$$\overline{\mathbf{F}^{\mathrm{p}}} = \overline{\mathbf{V}^{\mathrm{p}}\mathbf{R}^{\mathrm{p}}} = \boldsymbol{\Omega}\mathbf{V}^{\mathrm{p}}\boldsymbol{\Omega}^{\mathrm{T}}\boldsymbol{\Omega}\mathbf{R}^{\mathrm{p}} \tag{7.69}$$

$$\overline{\mathbf{V}^{\mathrm{p}}} = \boldsymbol{\Omega}\mathbf{V}^{\mathrm{p}}\boldsymbol{\Omega}^{\mathrm{T}} \tag{7.70}$$

$$\overline{\mathbf{C}^{\mathrm{p}}} = \overline{\mathbf{U}^{\mathrm{p}}\mathbf{U}^{\mathrm{p}}} = \mathbf{U}^{\mathrm{p}}\mathbf{U}^{\mathrm{p}} = \mathbf{C}^{\mathrm{p}} \tag{7.71}$$

$$\overline{\mathbf{B}^{\mathrm{p}}} = \overline{\mathbf{V}^{\mathrm{p}}\mathbf{V}^{\mathrm{p}}} = \boldsymbol{\Omega}\mathbf{V}^{\mathrm{p}}\boldsymbol{\Omega}^{\mathrm{T}}\boldsymbol{\Omega}\mathbf{V}^{\mathrm{p}}\boldsymbol{\Omega}^{\mathrm{T}} = \boldsymbol{\Omega}\mathbf{B}^{\mathrm{p}}\boldsymbol{\Omega}^{\mathrm{T}} \tag{7.72}$$

Therefore, elastic and plastic rotation tensors ($\mathbf{R}^{\mathrm{e}}, \mathbf{R}^{\mathrm{p}}$) are not objective Cauchy tensors. Elastic and plastic right stretch tensors and Cauchy stress tensors ($\mathbf{C}^{\mathrm{e}}, \mathbf{C}^{\mathrm{p}}, \mathbf{U}^{\mathrm{e}}, \mathbf{U}^{\mathrm{p}}$) are not objective but invariant to changes in current frame of reference. Elastic and plastic left stretch tensors and Almansi tensors ($\mathbf{B}^{\mathrm{e}}, \mathbf{B}^{\mathrm{p}}, \mathbf{V}^{\mathrm{e}}, \mathbf{V}^{\mathrm{p}}$) are objective. Using elastic and plastic velocity gradient definitions in Eqs. (7.12) and (7.13) with transformation rules for elastic and plastic deformation gradient in Eqs. (7.57) and (7.58) yields

$$\overline{\mathbf{L}^{\mathrm{e}}} = \dot{\overline{\mathbf{F}^{\mathrm{e}}}}\,\overline{\mathbf{F}^{\mathrm{e}}}^{-1} = \left(\boldsymbol{\Omega}\dot{\mathbf{F}}^{\mathrm{e}} + \dot{\boldsymbol{\Omega}}\mathbf{F}^{\mathrm{e}}\right)\left(\mathbf{F}^{\mathrm{e}^{-1}}\boldsymbol{\Omega}^{\mathrm{T}}\right) = \boldsymbol{\Omega}\mathbf{L}^{\mathrm{e}}\boldsymbol{\Omega}^{\mathrm{T}} + \dot{\boldsymbol{\Omega}}\boldsymbol{\Omega}^{\mathrm{T}} \tag{7.73}$$

$$\overline{\mathbf{L}^{\mathrm{p}}} = \dot{\overline{\mathbf{F}^{\mathrm{p}}}}\,\overline{\mathbf{F}^{\mathrm{p}}}^{-1} = \dot{\mathbf{F}}^{\mathrm{p}}\mathbf{F}^{\mathrm{p}^{-1}} = \mathbf{L}^{\mathrm{p}} \tag{7.74}$$

Equation (7.73) shows that elastic velocity gradient is not objective. According to Eq. (7.74), plastic velocity gradient is not objective but invariant to changes in current frame of reference, since plastic velocity gradient refers to original and intermediate reference frames, but does not refer to current frame of reference. Using definitions of elastic stretch rate tensor and spin tensor in Eq. (7.14), we can write

$$\overline{\mathbf{L}^{\mathrm{e}}} = \overline{\mathbf{D}^{\mathrm{e}}} + \overline{\mathbf{W}^{\mathrm{e}}} = \boldsymbol{\Omega}(\mathbf{D}^{\mathrm{e}} + \mathbf{W}^{\mathrm{e}})\boldsymbol{\Omega}^{\mathrm{T}} + \dot{\boldsymbol{\Omega}}\boldsymbol{\Omega}^{\mathrm{T}} \tag{7.75}$$

which implies that

$$\overline{\mathbf{D}^{\mathrm{e}}} = \boldsymbol{\Omega}\mathbf{D}^{\mathrm{e}}\boldsymbol{\Omega}^{\mathrm{T}} \tag{7.76}$$

$$\overline{\mathbf{W}^{\mathrm{e}}} = \boldsymbol{\Omega}\mathbf{W}^{\mathrm{e}}\boldsymbol{\Omega}^{\mathrm{T}} + \dot{\boldsymbol{\Omega}}\boldsymbol{\Omega}^{\mathrm{T}} \tag{7.77}$$

Therefore, elastic stretch rate tensor ($\mathbf{D}^{\mathrm{e}}$) is objective, while elastic spin tensor ($\mathbf{W}^{\mathrm{e}}$) is not objective. Similarly, using plastic stretch rate definition in Eq. (7.15) and applying irrotational plastic flow assumption in Eq. (7.28)

$$\overline{\mathbf{L}^{\mathrm{p}}} = \overline{\mathbf{D}^{\mathrm{p}}} = \mathbf{D}^{\mathrm{p}} \tag{7.78}$$

which implies that

$$\overline{\mathbf{D^p}} = \mathbf{D^p} \tag{7.79}$$

Therefore, plastic stretch rate tensor $\mathbf{D^p}$ is not objective but invariant to changes in current reference frame. To summarize, it has been shown that $\mathbf{F}, \mathbf{R}, \dot{\mathbf{F}}, \mathbf{L}, \mathbf{W}, \mathbf{F^e}, \mathbf{R^e}, \mathbf{R^p}, \mathbf{L^e}, \mathbf{W^e}$ are not objective, $\mathbf{C}$,$\mathbf{U}$, $\mathbf{F^p}$, $\mathbf{U^e}$, $\mathbf{C^e}$, $\mathbf{U^p}$, $\mathbf{C^p}$, $\mathbf{L^p}$, $\mathbf{D^p}$ are not objective but invariant to changes in current reference frame, and $\mathbf{B}$,$\mathbf{V}$,$\mathbf{D}$, $\mathbf{V^e}$, $\mathbf{B^e}$, $\mathbf{V^p}$, $\mathbf{B^p}$, $\mathbf{D^e}$ are objective.

7.3 Unified Mechanics Theory Formulation for Finite Strain

A thermodynamic potential is concave with respect to temperature (θ) and convex with respect to all other internal state variables. It provides a basis for satisfying conditions of thermodynamic stability imposed by Clausius-Duhem inequality. Specific Helmholtz free energy (ψ) forms such a basis. It is defined as the difference between specific internal energy (u) and product of absolute temperature (θ) and specific entropy (s)

$$\psi = u - \theta s \tag{7.80}$$

First law of thermodynamics (also known as conservation of energy law) states that energy can be transported, or converted from one form to another, but cannot be destroyed or created. Accordingly, internal energy of a system can be increased by heat flow into the system (δq), heat generated within the system due to internal scattering (r), mechanical work done on the system by external load (δw), or all other kinds of work done on the system ($\delta w'$) during any process. Considering only thermo-mechanical effects, rate of change in internal energy, $\dot{u}$ can be stated as

$$\dot{u} = Q + l + r \tag{7.81}$$

where Q is the rate of net heat flow into the system and l is rate of net work done on the system and r is the heat generated in the system, due to internal friction (scattering) chemical reactions etc. Second law of thermodynamic states that there exists a state function, entropy, which increases in universe for all type of processes due to entropy production. Unlike energy, entropy (s) is not only transferred across boundaries of system, but it may also be created in the system which is called as entropy production (S). Clausius-Duhem inequality describes second law of thermodynamics in terms of a nonnegative entropy production rate (γ) per unit volume for any kind of irreversible process which is defined as

$$\gamma = \rho \dot{s} + \operatorname{div}(\mathbf{J}_s) > 0 \tag{7.82}$$

$\mathbf{J}_s$ is net entropy flux into system. Heat flow into body can be defined in terms of heat flux as follows:

$$\rho Q = -\operatorname{div}\left(\mathbf{J}_q\right) \tag{7.83}$$

Substituting time derivative of Eq. (7.80) internal energy definition in Eq. (7.81) and heat flow equation in Eq. (7.83) into Eq. (7.82), internal entropy production density rate can be rewritten as

$$\gamma = \operatorname{div}(\mathbf{J}_s) - \frac{\operatorname{div}\left(\mathbf{J}_q\right)}{\theta} + \frac{\rho r}{\theta} + \frac{\rho}{\theta}\left(l - \dot{\psi} - \dot{\theta} s\right) > 0 \tag{7.84}$$

r is the internal heat source strength.

7.3.1 Thermodynamic Restrictions

The main premise of unified mechanics theory is the computation of the fundamental equation, which contains all entropy generation terms for all active mechanisms. For dual-mechanism rate-dependent material modeling, material response can be resolved into two components which necessitate multi-mechanism generalization of multiplicative decomposition in Eq. (7.7) and description of different Helmholtz free energy functions and associated fundamental equation assuming that linear addition is applicable for Eq. (7.80). Accordingly, Eqs. (7.12)–(7.15), (7.20), (7.27), and (7.31) hold true for each component of material resistance mechanism. Subscripts "I" and "M" will be used henceforth to designate the component of a quantity in **I**ntermolecular mechanism and **M**olecular network mechanism, respectively. For description of dissipation inequality, total Helmholtz free energy density in (original) reference configuration is written as a summation of defect energy (Ψ_{D}) and elastic energy stored in intermolecular structure (Ψ_{I}) and molecular network structure(Ψ_{M}):

$$\Psi\left(\mathbf{C}_{\mathrm{I}}^{\mathrm{e}}, \mathbf{C}_{\mathrm{M}}^{\mathrm{e}}, \mathbf{A}, \theta\right) = \Psi_{I}\left(\mathbf{E}_{\mathrm{I}}^{\mathrm{e}}, \theta\right) + \Psi_{\mathrm{M}}\left(\mathbf{C}_{\mathrm{M}}^{\mathrm{e}}, \theta\right) + \Psi_{\mathrm{D}}(\mathbf{A}, \theta) \tag{7.85}$$

Defect energy (Ψ_{D}) is assumed to depend on a stretch-like tensor ($\mathbf{A}$) and temperature (θ), elastic energy in intermolecular structure (Ψ_{I}) is assumed to depend on logarithmic elastic strain in intermolecular structure $\left(\mathbf{E}_{\mathrm{I}}^{\mathrm{e}}\right)$, and temperature ($\theta$) and elastic energy in molecular network structure (Ψ_{M}) are assumed to depend on elastic Cauchy tensor in molecular network structure $\left(\mathbf{C}_{\mathrm{M}}^{\mathrm{e}}\right)$ and temperature (θ). Assuming that a similar decomposition also holds for specific entropy and specific Helmholtz free energy, we can write

$$s\left(\mathbf{C}_{\mathrm{I}}^{\mathrm{e}}, \mathbf{C}_{\mathrm{M}}^{\mathrm{e}}, \mathbf{A}, \theta\right) = s_{\mathrm{I}}\left(\mathbf{E}_{\mathrm{I}}^{\mathrm{e}}, \theta\right) + s_{\mathrm{M}}\left(\mathbf{C}_{\mathrm{M}}^{\mathrm{e}}, \theta\right) + s_{\mathrm{D}}(\mathbf{A}, \theta) \tag{7.86}$$

$$\psi\left(\mathbf{C}_{\mathrm{I}}^{\mathrm{e}}, \mathbf{C}_{\mathrm{M}}^{\mathrm{e}}, \mathbf{A}, \theta\right) = \psi_{I}\left(\mathbf{E}_{\mathrm{I}}^{\mathrm{e}}, \theta\right) + \psi_{M}\left(\mathbf{C}_{\mathrm{M}}^{\mathrm{e}}, \theta\right) + \psi_{\mathrm{D}}(\mathbf{A}, \theta) \tag{7.87}$$

Helmholtz free energy density in reference configuration (Ψ) can be simply related to specific Helmholtz free energy function (ψ) through Eq. (7.88) and in rate form as follows:

$$\rho_o \psi\left(\mathbf{C}_{\mathrm{I}}^{\mathrm{e}}, \mathbf{C}_{\mathrm{M}}^{\mathrm{e}}, \mathbf{A}, \theta\right) = \Psi\left(\mathbf{C}_{\mathrm{I}}^{\mathrm{e}}, \mathbf{C}_{\mathrm{M}}^{\mathrm{e}}, \mathbf{A}, \theta\right) \tag{7.88}$$

$$\rho \dot{\psi}\left(\mathbf{C}_{\mathrm{I}}^{\mathrm{e}}, \mathbf{C}_{\mathrm{M}}^{\mathrm{e}}, \mathbf{A}, \theta\right) = J^{-1} \dot{\Psi}\left(\mathbf{C}_{\mathrm{I}}^{\mathrm{e}}, \mathbf{C}_{\mathrm{M}}^{\mathrm{e}}, \mathbf{A}, \theta\right) \tag{7.89}$$

where ρ_o and ρ are densities in reference configuration and deformed configuration, respectively. Note that, since eigenvalues of elastic Cauchy tensor $\left(\mathbf{C}_{\mathrm{I}}^{\mathrm{e}}\right)$ and logarithmic elastic strain tensor $\left(\mathbf{E}_{\mathrm{I}}^{\mathrm{e}}\right)$ corresponding to intermolecular structure are related through Eq. (7.90) and eigenvectors of these tensors are identical, it is possible to consider Helmholtz free energy density (Ψ_{I}) and specific entropy (s_{I}) associated with intermolecular structure as functions of temperature (θ) and elastic Cauchy tensor $\left(\mathbf{C}_{\mathrm{I}}^{\mathrm{e}}\right)$; hence we can write the following relations:

$$\mathrm{eigenval}\left(\mathbf{E}_{\mathrm{I}}^{\mathrm{e}}\right) = \frac{1}{2} \ln\left(\mathrm{eigenval}\left(\mathbf{C}_{\mathrm{I}}^{\mathrm{e}}\right)\right) \tag{7.90}$$

$$\Psi_{\mathrm{I}}\left(\mathbf{E}_{\mathrm{I}}^{\mathrm{e}}, \theta\right) \equiv \Psi_{\mathrm{I}}\left(\mathbf{C}_{\mathrm{I}}^{\mathrm{e}}, \theta\right) \tag{7.91}$$

$$s_{\mathrm{I}}\left(\mathbf{E}_{\mathrm{I}}^{\mathrm{e}}, \theta\right) \equiv s_{\mathrm{I}}\left(\mathbf{C}_{\mathrm{I}}^{\mathrm{e}}, \theta\right) \tag{7.92}$$

Time derivative of Helmholtz free energy can be formulated as follows:

$$\dot{\Psi}\left(\mathbf{C}_{\mathrm{I}}^{\mathrm{e}}, \mathbf{C}_{\mathrm{M}}^{\mathrm{e}}, \mathbf{A}, \theta\right) = \begin{bmatrix} \dfrac{\partial \Psi_{\mathrm{I}}\left(\mathbf{E}_{\mathrm{I}}^{\mathrm{e}}, \theta\right)}{\partial \mathbf{C}_{\mathrm{I}}^{\mathrm{e}}} : \dot{\mathbf{C}}_{\mathrm{I}}^{\mathrm{e}} + \dfrac{\partial \Psi_{\mathrm{M}}\left(\mathbf{C}_{\mathrm{M}}^{\mathrm{e}}, \theta\right)}{\partial C_{\mathrm{M}}^{\mathrm{e}}} : \dot{\mathbf{C}}_{\mathrm{M}}^{\mathrm{e}} \\ + \dfrac{\partial \Psi_{\mathrm{D}}(\mathbf{A}, \theta)}{\partial \mathbf{A}} : \dot{\mathbf{A}} + \dfrac{\partial \Psi_{\mathrm{I}}\left(\mathbf{E}_{\mathrm{I}}^{\mathrm{e}}, \theta\right)}{\partial \theta} : \dot{\theta} \\ + \dfrac{\partial \Psi_{\mathrm{M}}\left(\mathbf{C}_{\mathrm{M}}^{\mathrm{e}}, \theta\right)}{\partial \theta} : \dot{\theta} + \dfrac{\partial \Psi_{\mathrm{D}}(\mathbf{A}, \theta)}{\partial \theta} : \dot{\theta} \end{bmatrix} \tag{7.93}$$

According to principle of virtual work, rate of work done per unit volume of deformed body (external work power) is balanced with internal work, while total work done on the system is stored as elastic strain energy (represented with first two terms in Eq. (7.95)) and dissipated energy in plastic work (represented with the last two terms in Eq. (7.95))

$$\rho l = \dot{w}_{\mathrm{int}} \tag{7.94}$$

$$\dot{w}_{\mathrm{int}} = \boldsymbol{\Gamma}_{\mathrm{I}}^{\mathrm{e}} : \mathbf{L}_{\mathrm{I}}^{\mathrm{e}} + \boldsymbol{\Gamma}_{\mathrm{M}}^{\mathrm{e}} : \mathbf{L}_{\mathrm{M}}^{\mathrm{e}} + J^{-1}\boldsymbol{\Gamma}_{\mathrm{I}}^{\mathrm{p}} : \mathbf{L}_{\mathrm{I}}^{\mathrm{p}} + J^{-1}\boldsymbol{\Gamma}_{\mathrm{M}}^{\mathrm{p}} : \mathbf{L}_{\mathrm{M}}^{\mathrm{p}} \tag{7.95}$$

where, $\boldsymbol{\Gamma}_{\mathrm{I}}^{\mathrm{e}}$, $\boldsymbol{\Gamma}_{\mathrm{M}}^{\mathrm{e}}$, $\boldsymbol{\Gamma}_{\mathrm{I}}^{\mathrm{p}}$, and $\boldsymbol{\Gamma}_{\mathrm{M}}^{\mathrm{p}}$ are stress measures conjugate to rate of deformation measures $\mathbf{L}_{\mathrm{I}}^{\mathrm{e}}$, $\mathbf{L}_{\mathrm{M}}^{\mathrm{e}}$, $\mathbf{L}_{\mathrm{I}}^{\mathrm{p}}$, and $\mathbf{L}_{\mathrm{M}}^{\mathrm{p}}$, which were defined in Eqs. (7.12) and (7.13). Noting that $J^{-1} = J^{\mathrm{e}^{-1}}$, the J^{-1} multiplier in front of last two terms in Eq. (7.95) recovers work definitions from intermediate configuration to deformed configuration. Requirement of frame indifference for internal work can be given by the following relation:

$$\dot{w}_{\mathrm{int}} = \dot{\overline{w}}_{\mathrm{int}} \tag{7.96}$$

which implies that

$$\begin{pmatrix} \boldsymbol{\Gamma}_{\mathrm{I}}^{\mathrm{e}} : \mathbf{L}_{\mathrm{I}}^{\mathrm{e}} + \boldsymbol{\Gamma}_{\mathrm{M}}^{\mathrm{e}} : \mathbf{L}_{\mathrm{M}}^{\mathrm{e}} \\ +J^{-1}\boldsymbol{\Gamma}_{\mathrm{I}}^{\mathrm{p}} : \mathbf{L}_{\mathrm{I}}^{\mathrm{p}} + J^{-1}\boldsymbol{\Gamma}_{\mathrm{M}}^{\mathrm{p}} : \mathbf{L}_{\mathrm{M}}^{\mathrm{p}} \end{pmatrix} = \begin{pmatrix} \overline{\boldsymbol{\Gamma}}_{\mathrm{I}}^{\mathrm{e}} : \overline{\mathbf{L}}_{\mathrm{I}}^{\mathrm{e}} + \overline{\boldsymbol{\Gamma}}_{\mathrm{M}}^{\mathrm{e}} : \overline{\mathbf{L}}_{\mathrm{M}}^{\mathrm{e}} \\ +J^{-1}\overline{\boldsymbol{\Gamma}}_{\mathrm{I}}^{\mathrm{p}} : \overline{\mathbf{L}}_{\mathrm{I}}^{\mathrm{p}} + J^{-1}\overline{\boldsymbol{\Gamma}}_{\mathrm{M}}^{\mathrm{p}} : \overline{\mathbf{L}}_{\mathrm{M}}^{\mathrm{p}} \end{pmatrix} \tag{7.97}$$

Using transformation rules for elastic and plastic velocity gradients in Eqs. (7.73) and (7.74) we can write the following relations:

$$\begin{pmatrix} \boldsymbol{\Gamma}_{\mathrm{I}}^{\mathrm{e}} : \mathbf{L}_{\mathrm{I}}^{\mathrm{e}} + J^{-1}\boldsymbol{\Gamma}_{\mathrm{I}}^{\mathrm{p}} : \mathbf{L}_{\mathrm{I}}^{\mathrm{p}} \\ +\boldsymbol{\Gamma}_{\mathrm{M}}^{\mathrm{e}} : \mathbf{L}_{\mathrm{M}}^{\mathrm{e}} + J^{-1}\boldsymbol{\Gamma}_{\mathrm{M}}^{\mathrm{p}} : \mathbf{L}_{\mathrm{M}}^{\mathrm{p}} \end{pmatrix} = \begin{bmatrix} \overline{\boldsymbol{\Gamma}}_{\mathrm{I}}^{\mathrm{e}} : \left(\boldsymbol{\Omega}\mathbf{L}_{\mathrm{I}}^{\mathrm{e}}\boldsymbol{\Omega}^{\mathrm{T}} + \dot{\boldsymbol{\Omega}}\boldsymbol{\Omega}^{\mathrm{T}}\right) + J^{-1}\overline{\boldsymbol{\Gamma}}_{\mathrm{I}}^{\mathrm{p}} : \mathbf{L}_{\mathrm{I}}^{\mathrm{p}} \\ +\overline{\boldsymbol{\Gamma}}_{\mathrm{M}}^{\mathrm{e}} : \left(\boldsymbol{\Omega}\mathbf{L}_{\mathrm{M}}^{\mathrm{e}}\boldsymbol{\Omega}^{\mathrm{T}} + \dot{\boldsymbol{\Omega}}\boldsymbol{\Omega}^{\mathrm{T}}\right) + J^{-1}\overline{\boldsymbol{\Gamma}}_{\mathrm{M}}^{\mathrm{p}} : \mathbf{L}_{\mathrm{M}}^{\mathrm{p}} \end{bmatrix} \tag{7.98}$$

or

$$\begin{pmatrix} \boldsymbol{\Gamma}_{\mathrm{I}}^{\mathrm{e}} : \mathbf{L}_{\mathrm{I}}^{\mathrm{e}} + J^{-1}\boldsymbol{\Gamma}_{\mathrm{I}}^{\mathrm{p}} : \mathbf{L}_{\mathrm{I}}^{\mathrm{p}} \\ +\boldsymbol{\Gamma}_{\mathrm{M}}^{\mathrm{e}} : \mathbf{L}_{\mathrm{M}}^{\mathrm{e}} + J^{-1}\boldsymbol{\Gamma}_{\mathrm{M}}^{\mathrm{p}} : \mathbf{L}_{\mathrm{M}}^{\mathrm{p}} \end{pmatrix} = \begin{bmatrix} \left(\boldsymbol{\Omega}^{\mathrm{T}}\overline{\boldsymbol{\Gamma}}_{\mathrm{I}}^{\mathrm{e}}\boldsymbol{\Omega}\right) : \mathbf{L}_{\mathrm{I}}^{\mathrm{e}} + \left(\boldsymbol{\Omega}^{\mathrm{T}}\overline{\boldsymbol{\Gamma}}_{\mathrm{M}}^{\mathrm{e}}\boldsymbol{\Omega}\right) : \mathbf{L}_{\mathrm{M}}^{\mathrm{e}} \\ +J^{-1}\overline{\boldsymbol{\Gamma}}_{\mathrm{I}}^{\mathrm{p}} : \mathbf{L}_{\mathrm{I}}^{\mathrm{p}} + J^{-1}\overline{\boldsymbol{\Gamma}}_{\mathrm{M}}^{\mathrm{p}} : \mathbf{L}_{\mathrm{M}}^{\mathrm{p}} \\ +\overline{\boldsymbol{\Gamma}}_{\mathrm{I}}^{\mathrm{e}} : \dot{\boldsymbol{\Omega}}\boldsymbol{\Omega}^{\mathrm{T}} + \overline{\boldsymbol{\Gamma}}_{\mathrm{M}}^{\mathrm{e}} : \dot{\boldsymbol{\Omega}}\boldsymbol{\Omega}^{\mathrm{T}} \end{bmatrix} \tag{7.99}$$

The first two terms on the right-hand side of Eq. (7.99) indicate that stress measures corresponding elastic work $\left(\boldsymbol{\Gamma}_{\mathrm{I}}^{\mathrm{e}}, \boldsymbol{\Gamma}_{\mathrm{M}}^{\mathrm{e}}\right)$ are objective as shown in Eqs. (7.100) and (7.101). Since $\dot{\boldsymbol{\Omega}}\boldsymbol{\Omega}^{\mathrm{T}}$ is a skew symmetric tensor, Eqs. (7.55) and (7.77), the last two terms on the right-hand side of Eq. (7.99) imply that stress measures corresponding elastic work $\left(\boldsymbol{\Gamma}_{\mathrm{I}}^{\mathrm{e}}, \boldsymbol{\Gamma}_{\mathrm{M}}^{\mathrm{e}}\right)$ are symmetric (Eqs. (7.102) and (7.103)). The third and fourth terms on the right-hand side of Eq. (7.99) show that stress measures corresponding plastic work $(\boldsymbol{\Gamma}_{\mathrm{I}}^{\mathrm{p}}, \boldsymbol{\Gamma}_{\mathrm{M}}^{\mathrm{p}})$ are not objective but invariant to changes in current frame of reference as shown in Eqs. (7.104) and (7.105):

$$\overline{\boldsymbol{\Gamma}}_{\mathrm{I}}^{\mathrm{e}} = \boldsymbol{\Omega}\boldsymbol{\Gamma}_{\mathrm{I}}^{\mathrm{e}}\boldsymbol{\Omega}^{\mathrm{T}} \tag{7.100}$$

$$\overline{\boldsymbol{\Gamma}}_{\mathrm{M}}^{\mathrm{e}} = \boldsymbol{\Omega}\boldsymbol{\Gamma}_{\mathrm{M}}^{\mathrm{e}}\boldsymbol{\Omega}^{\mathrm{T}} \tag{7.101}$$

$$\overline{\boldsymbol{\Gamma}}_{\mathrm{I}}^{\mathrm{e}} = \overline{\boldsymbol{\Gamma}}_{\mathrm{I}}^{\mathrm{e}^{\mathrm{T}}}; \boldsymbol{\Gamma}_{\mathrm{I}}^{\mathrm{e}} = \boldsymbol{\Gamma}_{\mathrm{I}}^{\mathrm{e}^{\mathrm{T}}} \tag{7.102}$$

$$\overline{\boldsymbol{\Gamma}}_{\mathrm{M}}^{\mathrm{e}} = \overline{\boldsymbol{\Gamma}}_{\mathrm{M}}^{\mathrm{e}^{\mathrm{T}}}; \boldsymbol{\Gamma}_{\mathrm{M}}^{\mathrm{e}} = \boldsymbol{\Gamma}_{\mathrm{M}}^{\mathrm{e}^{\mathrm{T}}} \tag{7.103}$$

$$\overline{\boldsymbol{\Gamma}}_{\mathrm{I}}^{\mathrm{p}} = \boldsymbol{\Gamma}_{\mathrm{I}}^{\mathrm{p}} \tag{7.104}$$

$$\overline{\boldsymbol{\Gamma}}_{\mathrm{M}}^{\mathrm{p}} = \boldsymbol{\Gamma}_{\mathrm{M}}^{\mathrm{p}} \tag{7.105}$$

Using symmetry property of stress tensor in Eqs. (7.102) and (7.103), irrotational plastic flow definition in Eq. (7.28), total internal work over whole volume of a system can be written as

$$\begin{aligned} \dot{W}_{\mathrm{int}} &= \int_V \dot{w}_{\mathrm{int}} dV \\ &= \int_V \left[\boldsymbol{\Gamma}_{\mathrm{I}}^{\mathrm{e}} : \mathbf{D}_{\mathrm{I}}^{\mathrm{e}} + \boldsymbol{\Gamma}_{\mathrm{M}}^{\mathrm{e}} : \mathbf{D}_{\mathrm{M}}^{\mathrm{e}} + J^{-1}\boldsymbol{\Gamma}_{\mathrm{I}}^{\mathrm{p}} : \mathbf{D}_{\mathrm{I}}^{\mathrm{p}} + J^{-1}\boldsymbol{\Gamma}_{\mathrm{M}}^{\mathrm{p}} : \mathbf{D}_{\mathrm{M}}^{\mathrm{p}}\right] dV \end{aligned} \tag{7.106}$$

Total external work acting on the system can be described in terms of surface tractions on boundaries of the system and body force acting on the system as follows:

$$\dot{W}_{\mathrm{ext}} = \int_V \rho l dV = \int_\Omega \boldsymbol{\sigma}^{(n)} \bullet \dot{\chi}\, d\Omega + \int_V \underline{\boldsymbol{b}} \bullet \dot{\chi}\, dV \tag{7.107}$$

where Ω represents surface and V represents volume. Consider principal of virtual work for a special case defined from Eq. (7.34) as

$$\mathbf{L} = \mathrm{grad}(\dot{\chi}) = \mathbf{L}_{\mathrm{I}}^{\mathrm{e}} = \mathbf{L}_{\mathrm{M}}^{\mathrm{e}} \tag{7.108}$$

where it is assumed that

$$\mathbf{D}_{\mathrm{I}}^{\mathrm{p}} = \mathbf{D}_{\mathrm{M}}^{\mathrm{p}} = 0 \tag{7.109}$$

Principal of virtual work can then be rewritten for this special case from Eqs. (7.106) and (7.107) as

$$\int_{\Omega} \underline{\sigma^{(n)}} \bullet \dot{\chi} d\Omega + \int_{V} \underline{b} \bullet \dot{\chi} dV = \int_{V} \left[\mathbf{\Gamma}_{\mathrm{I}}^{\mathrm{e}} : \mathbf{D}_{\mathrm{I}}^{\mathrm{e}} + \mathbf{\Gamma}_{\mathrm{M}}^{\mathrm{e}} : \mathbf{D}_{\mathrm{M}}^{\mathrm{e}} \right] dV \tag{7.110}$$

$$\int_{\Omega} \sigma^{(n)} \bullet \dot{\chi}\, d\Omega + \int_{V} \underline{b} \bullet \dot{\chi} dV = \int_{V} \left[\left(\mathbf{\Gamma}_{\mathrm{I}}^{\mathrm{e}} + \mathbf{\Gamma}_{\mathrm{M}}^{\mathrm{e}} \right) : \mathrm{grad}(\dot{\chi}) \right] dV \tag{7.111}$$

$$\int_{\Omega} \boldsymbol{\sigma}^{(n)} \bullet \dot{\chi} d\Omega + \int_{V} \underline{b} \bullet \dot{\chi} dV = \int_{V} \mathrm{div}\left[\dot{\chi} \bullet \left(\mathbf{\Gamma}_{\mathrm{I}}^{\mathrm{e}} + \mathbf{\Gamma}_{\mathrm{M}}^{\mathrm{e}} \right) \right] dV - \int_{V} \mathrm{div}\left(\mathbf{\Gamma}_{\mathrm{I}}^{\mathrm{e}} + \mathbf{\Gamma}_{\mathrm{M}}^{\mathrm{e}} \right) \bullet \dot{\chi} dV \tag{7.112}$$

Since Eq. (7.112) is true for any choice of V and $\mathrm{grad}(\dot{\chi})$, from the first terms on the left- and right-hand side, we can write

$$\boldsymbol{\sigma}^{(n)} = \left(\mathbf{\Gamma}_{\mathrm{I}}^{\mathrm{e}} + \mathbf{\Gamma}_{\mathrm{M}}^{\mathrm{e}} \right) \underline{\boldsymbol{n}} \tag{7.113}$$

which is essentially Cauchy stress theorem describing the relation between stress tensor and surface tractions. From the second terms on the left- and right-hand side of Eq. (7.112), we can write

$$\mathrm{div}\left(\mathbf{\Gamma}_{\mathrm{I}}^{\mathrm{e}} + \mathbf{\Gamma}_{\mathrm{M}}^{\mathrm{e}} \right) + \underline{b} = 0 \tag{7.114}$$

which represents Cauchy's equation of motion for stationary systems. Therefore, stress measures in Eqs. (7.113) and (7.114) corresponds to Cauchy stress (**T**) components in intermolecular mechanism ($\mathbf{T}_{\mathrm{I}}$) and molecular network mechanism ($\mathbf{T}_{\mathrm{M}}$), respectively. Hence, we can write the following relations:

$$\mathbf{T} = \mathbf{\Gamma}_{\mathrm{I}}^{\mathrm{e}} + \mathbf{\Gamma}_{\mathrm{M}}^{\mathrm{e}} \tag{7.115}$$

$$\mathbf{\Gamma}_{\mathrm{I}}^{\mathrm{e}} = \mathbf{T}_{\mathrm{I}} = \mathbf{T}^{\mathrm{T}}{}_{\mathrm{I}} \tag{7.116}$$

$$\mathbf{\Gamma}_{\mathrm{M}}^{\mathrm{e}} = \mathbf{T}_{\mathrm{M}} = \mathbf{T}_{\mathrm{M}}{}^{\mathrm{T}} \tag{7.117}$$

Consider principal of virtual work for a second special case defined from Eq. (7.34) such that

$$L = \mathrm{grad}(\dot{\chi}) = \mathbf{L}_{\mathrm{I}}^{\mathrm{e}} + \mathbf{F}_{\mathrm{I}}^{\mathrm{e}} \mathbf{D}_{\mathrm{I}}^{\mathrm{p}} \mathbf{F}_{\mathrm{I}}^{\mathrm{e}^{-1}} = \mathbf{L}_{\mathrm{M}}^{\mathrm{e}} + \mathbf{F}_{\mathrm{M}}^{\mathrm{e}} \mathbf{D}_{\mathrm{M}}^{\mathrm{p}} \mathbf{F}_{\mathrm{M}}^{\mathrm{e}^{-1}} = 0 \tag{7.118}$$

or

$$\mathbf{L}_{\mathrm{I}}^{\mathrm{e}} = -\mathbf{F}_{\mathrm{I}}^{\mathrm{e}}\mathbf{D}_{\mathrm{I}}^{\mathrm{p}}\mathbf{F}_{\mathrm{I}}^{\mathrm{e}^{-1}} \tag{7.119}$$

$$\mathbf{L}_{\mathrm{M}}^{\mathrm{e}} = -\mathbf{F}_{\mathrm{M}}^{\mathrm{e}}\mathbf{D}_{\mathrm{M}}^{\mathrm{p}}\mathbf{F}_{\mathrm{M}}^{\mathrm{e}^{-1}} \tag{7.120}$$

Accordingly, principal of virtual work can be rewritten for this special case using Eqs. (7.106) and (7.197) as

$$\dot{W}_{\mathrm{int}} = \int_V \begin{bmatrix} \boldsymbol{\Gamma}_{\mathrm{I}}^{\mathrm{e}} : \left(-\mathbf{F}_{\mathrm{I}}^{\mathrm{e}}\mathbf{D}_{\mathrm{I}}^{\mathrm{p}}\mathbf{F}_{\mathrm{I}}^{\mathrm{e}^{-1}}\right) + \boldsymbol{J}^{-1}\boldsymbol{\Gamma}_{\mathrm{I}}^{\mathrm{p}} : \mathbf{D}_{\mathrm{I}}^{\mathrm{p}} \\ +\boldsymbol{\Gamma}_{\mathrm{M}}^{\mathrm{e}} : \left(-\mathbf{F}_{\mathrm{M}}^{\mathrm{e}}\mathbf{D}_{\mathrm{M}}^{\mathrm{p}}\mathbf{F}_{\mathrm{M}}^{\mathrm{e}^{-1}}\right) + \boldsymbol{J}^{-1}\boldsymbol{\Gamma}_{\mathrm{M}}^{\mathrm{p}} : \mathbf{D}_{\mathrm{M}}^{\mathrm{p}} \end{bmatrix} dV \tag{7.121}$$

or

$$\dot{W}_{\mathrm{int}} = \int_V \left[\left(\boldsymbol{J}^{-1}\boldsymbol{\Gamma}_{\mathrm{I}}^{\mathrm{p}} - \mathbf{F}_{\mathrm{I}}^{\mathrm{e}^{\mathrm{T}}}\boldsymbol{\Gamma}_{\mathrm{I}}^{\mathrm{e}}\mathbf{F}_{\mathrm{I}}^{\mathrm{e}^{-\mathrm{T}}}\right) : \mathbf{D}_{\mathrm{I}}^{\mathrm{p}} + \left(\boldsymbol{J}^{-1}\boldsymbol{\Gamma}_{\mathrm{M}}^{\mathrm{p}} - \mathbf{F}_{\mathrm{M}}^{\mathrm{e}^{\mathrm{T}}}\boldsymbol{\Gamma}_{\mathrm{M}}^{\mathrm{e}}\mathbf{F}_{\mathrm{M}}^{\mathrm{e}^{-\mathrm{T}}}\right) : \mathbf{D}_{\mathrm{M}}^{\mathrm{p}}\right] dV \tag{7.122}$$

Since velocity gradient is assumed to be zero for this special case, velocity field will be also equal to zero. Therefore, external work will be equal to zero $\left(\dot{W}_{\mathrm{ext}} = 0\right)$. As a result, individual terms inside parenthesis in Eq. (7.122) should be equal to zero for arbitrary selection of V, $\mathbf{D}_{\mathrm{I}}^{\mathrm{p}}$, and $\mathbf{D}_{\mathrm{M}}^{\mathrm{p}}$.

$$\boldsymbol{J}^{-1}\boldsymbol{\Gamma}_{\mathrm{I}}^{\mathrm{p}} - \mathbf{F}_{\mathrm{I}}^{\mathrm{e}^{\mathrm{T}}}\boldsymbol{\Gamma}_{\mathrm{I}}^{\mathrm{e}}\mathbf{F}_{\mathrm{I}}^{\mathrm{e}^{-\mathrm{T}}} = 0 \tag{7.123}$$

$$\boldsymbol{J}^{-1}\boldsymbol{\Gamma}_{\mathrm{M}}^{\mathrm{p}} - \mathbf{F}_{\mathrm{M}}^{\mathrm{e}^{\mathrm{T}}}\boldsymbol{\Gamma}_{\mathrm{M}}^{\mathrm{e}}\mathbf{F}_{\mathrm{M}}^{\mathrm{e}^{-\mathrm{T}}} = 0 \tag{7.124}$$

Using definitions of stress measures in Eqs. (7.116) and (7.117), it can be shown that

$$\boldsymbol{\Gamma}_{\mathrm{I}}^{\mathrm{p}} = \mathbf{J}\mathbf{F}_{\mathrm{I}}^{\mathrm{e}^{\mathrm{T}}}\mathbf{T}_{\mathrm{I}}\mathbf{F}_{\mathrm{I}}^{\mathrm{e}^{-\mathrm{T}}} \tag{7.125}$$

$$\boldsymbol{\Gamma}_{\mathrm{M}}^{\mathrm{p}} = \mathbf{J}\mathbf{F}_{\mathrm{M}}^{\mathrm{e}^{\mathrm{T}}}\mathbf{T}_{\mathrm{M}}\mathbf{F}_{\mathrm{M}}^{\mathrm{e}^{-\mathrm{T}}} \tag{7.126}$$

which form definition of elastic symmetric Mandel stress in intermolecular structure and molecular network structure as

$$\mathbf{M}_{\mathrm{I}}^{\mathrm{e}} = J\mathbf{F}_{\mathrm{I}}^{\mathrm{e}^{\mathrm{T}}}\mathbf{T}_{\mathrm{I}}\mathbf{F}_{\mathrm{I}}^{\mathrm{e}^{-\mathrm{T}}} \tag{7.127}$$

$$\mathbf{M}_{\mathrm{M}}^{\mathrm{e}} = J\mathbf{F}_{\mathrm{M}}^{\mathrm{e}^{\mathrm{T}}}\mathbf{T}_{\mathrm{M}}\mathbf{F}_{\mathrm{M}}^{\mathrm{e}^{-\mathrm{T}}} \tag{7.128}$$

whereas since trace of plastic stretch rate is equal to zero due to incompressible plastic flow, assumption stress conjugate to plastic stretch rate should be a deviatoric tensor. Therefore, we can write

$$\boldsymbol{\Gamma}_{\mathrm{I}}^{\mathrm{p}} = \operatorname{dev}\left(\mathbf{M}_{\mathrm{I}}^{\mathrm{e}}\right) \tag{7.129}$$

$$\boldsymbol{\Gamma}_{\mathrm{M}}^{\mathrm{p}} = \operatorname{dev}\left(\mathbf{M}_{\mathrm{M}}^{\mathrm{e}}\right) \tag{7.130}$$

Finally, elastic second Piola-Kirchhoff stress tensor in intermolecular structure and molecular network structure can be defined as follows:

$$\mathbf{S}_{\mathrm{I}}^{\mathrm{e}} = J\mathbf{F}_{\mathrm{I}}^{\mathrm{e}^{-1}}\mathbf{T}_{\mathrm{I}}\mathbf{F}_{\mathrm{I}}^{\mathrm{e}^{-\mathrm{T}}} \tag{7.131}$$

$$\mathbf{S}_{\mathrm{M}}^{\mathrm{e}} = J\mathbf{F}_{\mathrm{M}}^{\mathrm{e}^{-1}}\mathbf{T}_{\mathrm{M}}\mathbf{F}_{\mathrm{M}}^{\mathrm{e}^{-\mathrm{T}}} \tag{7.132}$$

Using definitions of elastic Mandel stress (Eqs. (7.127) and (7.128)) and symmetric second Piola-Kirchhoff stress (Eqs. (7.131) and (7.132))

$$\mathbf{M}_{\mathrm{I}}^{\mathrm{e}} = \mathbf{C}_{\mathrm{I}}^{\mathrm{e}}\mathbf{S}_{\mathrm{I}}^{\mathrm{e}} \tag{7.133}$$

$$\mathbf{M}_{\mathrm{M}}^{\mathrm{e}} = \mathbf{C}_{\mathrm{M}}^{\mathrm{e}}\mathbf{S}_{\mathrm{M}}^{\mathrm{e}} \tag{7.134}$$

Therefore, it has been shown that stress measures $\left(\boldsymbol{\Gamma}_{\mathrm{I}}^{\mathrm{e}}, \boldsymbol{\Gamma}_{\mathrm{M}}^{\mathrm{e}}, \boldsymbol{\Gamma}_{\mathrm{I}}^{\mathrm{p}}, \boldsymbol{\Gamma}_{\mathrm{M}}^{\mathrm{p}}\right)$ or $\left(\mathbf{T}_{\mathrm{I}}, \mathbf{T}_{\mathrm{M}}, \mathbf{M}_{\mathrm{I}}^{\mathrm{e}}, \mathbf{M}_{\mathrm{M}}^{\mathrm{e}}\right)$ with deformation rate conjugates of $\left(\mathbf{D}_{\mathrm{I}}^{\mathrm{e}}, \mathbf{D}_{\mathrm{M}}^{\mathrm{e}}, \mathbf{D}_{\mathrm{I}}^{\mathrm{p}}, \mathbf{D}_{\mathrm{M}}^{\mathrm{p}}\right)$ form a frame indifferent framework for dual-mechanism elastic-viscoplastic constitutive model. The thermodynamic restrictions on constitutive relations can be obtained by substituting Eqs. (7.86), (7.89), and (7.93) and into Eq. (7.84), as follows:

$$\operatorname{div}(\mathbf{J}_s) - \operatorname{div}\left(\frac{\mathbf{J}_q}{\theta}\right) = 0 \tag{7.135}$$

$$\mathbf{J}_s = \frac{\mathbf{J}_q}{\theta} \tag{7.136}$$

$$\gamma_{\text{ther}} = -\frac{1}{\theta^2}\operatorname{div}\left(\mathbf{J}_q\right) \bullet \nabla_x(\theta) + \frac{\rho r}{\theta} > 0 \tag{7.137}$$

$$\left[J^{-1}\left(\frac{\partial \Psi_{\mathrm{I}}\left(\mathbf{E}_{\mathrm{I}}^{\mathrm{e}}, \theta\right)}{\partial \theta} + \frac{\partial \Psi_{\mathrm{M}}\left(\mathbf{C}_{\mathrm{M}}^{\mathrm{e}}, \theta\right)}{\partial \theta} + \frac{\partial \Psi_{\mathrm{D}}(\mathbf{A}, \theta)}{\partial \theta}\right) + \rho s\right]\dot{\theta} = 0 \tag{7.138}$$

$$\rho s\left(\mathbf{E}_{\mathrm{I}}^{\mathrm{e}}, \mathbf{C}_{\mathrm{M}}^{\mathrm{e}}, \mathbf{A}, \theta\right) = -J^{-1}\frac{\partial \Psi\left(\mathbf{E}_{\mathrm{I}}^{\mathrm{e}}, \mathbf{C}_{\mathrm{M}}^{\mathrm{e}}, \mathbf{A}, \theta\right)}{\partial \theta} \tag{7.139}$$

$$s\left(\mathbf{E}_{\mathrm{I}}^{\mathrm{e}}, \mathbf{C}_{\mathrm{M}}^{\mathrm{e}}, \mathbf{A}, \theta\right) = -\frac{\partial \psi\left(\mathbf{E}_{\mathrm{I}}^{\mathrm{e}}, \mathbf{C}_{\mathrm{M}}^{\mathrm{e}}, \mathbf{A}, \theta\right)}{\partial \theta} \tag{7.140}$$

$$\frac{1}{\theta}\left(\boldsymbol{\Gamma}_{\mathrm{I}}^{\mathrm{e}} : \mathbf{L}_{\mathrm{I}}^{\mathrm{e}} - J^{-1}\frac{\partial \Psi_{\mathrm{I}}\left(\mathbf{E}_{\mathrm{I}}^{\mathrm{e}}, \theta\right)}{\partial \mathbf{C}_{\mathrm{I}}^{\mathrm{e}}} : \dot{\mathbf{C}}_{\mathrm{I}}^{\mathrm{e}}\right) = \tag{7.141}$$

$$\boldsymbol{\Gamma}_{\mathrm{I}}^{\mathrm{e}} : \mathbf{L}_{\mathrm{I}}^{\mathrm{e}} = \mathbf{T}_{\mathrm{I}} : \mathbf{L}_{\mathrm{I}}^{\mathrm{e}} = \mathbf{T}_{\mathrm{I}} : \mathbf{D}_{\mathrm{I}}^{\mathrm{e}} = J^{-1}\frac{\partial \Psi_{\mathrm{I}}(\mathbf{E}_{\mathrm{I}}^{\mathrm{e}}, \theta)}{\partial \mathbf{C}_{\mathrm{I}}^{\mathrm{e}}} : \dot{\mathbf{C}}_{\mathrm{I}}^{\mathrm{e}} \tag{7.142}$$

$$\mathbf{T}_{\mathrm{I}} : \mathbf{D}_{\mathrm{I}}^{\mathrm{e}} = \frac{1}{2}\left(\mathbf{F}_{\mathrm{I}}^{\mathrm{e}^{-1}}\mathbf{T}_{\mathrm{I}}\mathbf{F}_{\mathrm{I}}^{\mathrm{e}^{-\mathrm{T}}}\right) : \dot{\mathbf{C}}_{\mathrm{I}}^{\mathrm{e}} = J^{-1}\frac{\partial \Psi_{\mathrm{I}}(\mathbf{E}_{\mathrm{I}}^{\mathrm{e}}, \theta)}{\partial \mathbf{C}_{\mathrm{I}}^{\mathrm{e}}} : \dot{\mathbf{C}}_{\mathrm{I}}^{\mathrm{e}} \tag{7.143}$$

$$J\left(\mathbf{F}_{\mathrm{I}}^{\mathrm{e}^{-1}}\mathbf{T}_{\mathrm{I}}\mathbf{F}_{\mathrm{I}}^{\mathrm{e}^{-\mathrm{T}}}\right) = \mathbf{S}_{\mathrm{I}}^{\mathrm{e}} = 2\frac{\partial \Psi_{\mathrm{I}}(\mathbf{E}_{\mathrm{I}}^{\mathrm{e}}, \theta)}{\partial \mathbf{C}_{\mathrm{I}}^{\mathrm{e}}} \tag{7.144}$$

$$\frac{1}{\theta}\left(\boldsymbol{\Gamma}_{\mathrm{M}}^{\mathrm{e}} : \mathbf{L}_{\mathrm{M}}^{\mathrm{e}} - J^{-1}\frac{\partial \Psi_{\mathrm{M}}(\mathbf{C}_{\mathrm{M}}^{\mathrm{e}}, \theta)}{\partial \mathbf{C}_{\mathrm{M}}^{\mathrm{e}}} : \dot{\mathbf{C}}_{\mathrm{M}}^{\mathrm{e}}\right) = \tag{7.145}$$

$$\boldsymbol{\Gamma}_{\mathrm{M}}^{\mathrm{e}} : \mathbf{L}_{\mathrm{M}}^{\mathrm{e}} = \mathbf{T}_{\mathrm{M}} : \mathbf{L}_{\mathrm{M}}^{\mathrm{e}} = \mathbf{T}_{\mathrm{M}} : \mathbf{D}_{\mathrm{M}}^{\mathrm{e}} = J^{-1}\frac{\partial \Psi_{\mathrm{M}}(\mathbf{C}_{\mathrm{M}}^{\mathrm{e}}, \theta)}{\partial \mathbf{C}_{\mathrm{M}}^{\mathrm{e}}} : \dot{\mathbf{C}}_{\mathrm{M}}^{\mathrm{e}} \tag{7.146}$$

$$\mathbf{T}_{\mathrm{M}} : \mathbf{D}_{\mathrm{M}}^{\mathrm{e}} = \frac{1}{2}\left(\mathbf{F}_{\mathrm{M}}^{\mathrm{e}}\mathbf{T}_{\mathrm{M}}\mathbf{F}_{\mathrm{M}}^{\mathrm{e}}\right) : \dot{\mathbf{C}}_{\mathrm{M}}^{\mathrm{e}} = J^{-1}\frac{\partial \Psi_{\mathrm{M}}(\mathbf{C}_{\mathrm{M}}^{\mathrm{e}}, \theta)}{\partial \mathbf{C}_{\mathrm{M}}^{\mathrm{e}}} : \dot{\mathbf{C}}_{\mathrm{M}}^{\mathrm{e}} \tag{7.147}$$

$$J\left(\mathbf{F}_{\mathrm{M}}^{\mathrm{e}^{-1}}\mathbf{T}_{\mathrm{M}}\mathbf{F}_{\mathrm{M}}^{\mathrm{e}^{-\mathrm{T}}}\right) = \mathbf{S}_{\mathrm{M}}^{\mathrm{e}} = 2\frac{\partial \Psi_{\mathrm{M}}(\mathbf{C}_{\mathrm{M}}^{\mathrm{e}}, \theta)}{\partial \mathbf{C}_{\mathrm{M}}^{\mathrm{e}}} \tag{7.148}$$

$$\gamma_{\mathrm{mech}} = \frac{1}{\theta}\left(J^{-1}\boldsymbol{\Gamma}_{\mathrm{I}}^{\mathrm{p}} : \mathbf{L}_{\mathrm{I}}^{\mathrm{p}} + J^{-1}\boldsymbol{\Gamma}_{\mathrm{M}}^{\mathrm{p}} : \mathbf{L}_{\mathrm{M}}^{\mathrm{p}} - J^{-1}\frac{\partial \Psi_{\mathrm{D}}(\mathbf{A}, \theta)}{\partial \mathbf{A}} : \dot{\mathbf{A}}\right) > 0 \tag{7.149}$$

Equation (7.136) relates entropy flux to heat flux, and Eq. (7.137) defines irreversible entropy production per unit volume in deformed configuration associated with heat conduction which is always positive according to Fourier's law (Eq. (7.150)). Equation (7.140) provides the relation between specific entropy and specific Helmholtz energy, while Eqs. (7.144) and (7.148) describe stress measures derived from Helmholtz free energy functions. Equation (7.149) defines irreversible entropy production due to mechanical dissipation per unit volume in deformed configuration. The reason for appearance of $\boldsymbol{J}^{-1}$ term in front of Eq. (7.149) is that all stress measures and their conjugate rate measures refer to intermediate (relaxed configuration), while irreversible entropy production rate density refers to deformed configuration. Heat flux is given by

$$\mathbf{J}_q = -k\nabla_x(\theta) \tag{7.150}$$

where k is temperature-dependent thermal conductivity of material. Specific heat can be expressed in terms of specific entropy (Eq. 7.151) and in terms of Helmholtz free energy (Eq. 7.153) by exploiting the relation in Eq. (7.139). Using linear decomposition assumption in Eqs. (7.85) and (7.86), specific heat can be rewritten in terms of separate component of specific entropy and Helmholtz free energy Eq. (7.154):

$$c = \theta \frac{\partial s\left(\mathbf{E}_{\mathrm{I}}^{\mathrm{e}}, \mathbf{C}_{\mathrm{M}}^{\mathrm{e}}, \mathbf{A}, \theta\right)}{\partial \theta} \tag{7.151}$$

$$c = \theta\left(\frac{\partial s_{\mathrm{I}}\left(\mathbf{E}_{\mathrm{I}}^{\mathrm{e}}, \theta\right)}{\partial \theta} + \frac{\partial s_{\mathrm{M}}\left(\mathbf{C}_{\mathrm{M}}^{\mathrm{e}}, \theta\right)}{\partial \theta} + \frac{\partial s_{\mathrm{D}}(\mathbf{A}, \theta)}{\partial \theta}\right) \tag{7.152}$$

$$c = -\theta \frac{\partial}{\partial \theta}\left[\frac{J^{-1}}{\rho} \frac{\partial \Psi\left(\mathbf{E}_{\mathrm{I}}^{\mathrm{e}}, \mathbf{C}_{\mathrm{M}}^{\mathrm{e}}, \mathbf{A}, \theta\right)}{\partial \theta}\right] \tag{7.153}$$

$$c = -\theta \frac{\partial}{\partial \theta}\left[\frac{J^{-1}}{\rho}\left(\frac{\partial \Psi_{\mathrm{I}}\left(\mathbf{E}_{\mathrm{I}}^{\mathrm{e}}, \theta\right)}{\partial \theta} + \frac{\partial \Psi_{\mathrm{M}}\left(\mathbf{C}_{\mathrm{M}}^{\mathrm{e}}, \theta\right)}{\partial \theta} + \frac{\partial \Psi_{\mathrm{D}}(\mathbf{A}, \theta)}{\partial \theta}\right)\right] \tag{7.154}$$

7.3.2 Constitutive Relations

In viscoplastic constitutive modeling of amorphous polymers for large deformations, dual decomposition of material response into two parallel working mechanisms of intermolecular structure and molecular network structure is widely used, Gunel and Basaran (2009, 2011a, 2011b). More recently, a trial mechanism has been also proposed to include a secondary mechanism for molecular network structure (Srivastava and Anand 2010). Both dual- and triple-mechanism models are proven to be successful in describing large deformation behavior of amorphous polymers at different isothermal test conditions. In order to extend applicability of such models to non-isothermal conditions, several refinements on material property definitions and viscoplastic flow rule definitions are necessary.

In dual-mechanism constitutive models, material response is assumed to be controlled by states of two parallel working mechanisms (intermolecular structure and molecular network structure), as depicted in Fig. 7.2. Intermolecular (I) and molecular network mechanisms (M) work in parallel, deformation in both mechanisms are equal to each other and equal to total deformation (Eq. (7.155)), while total stress is the summation of stresses due to intermolecular interactions (I) and molecular network interactions (M) (Eq. (7.156)). Subscripts "I" and "M" represent

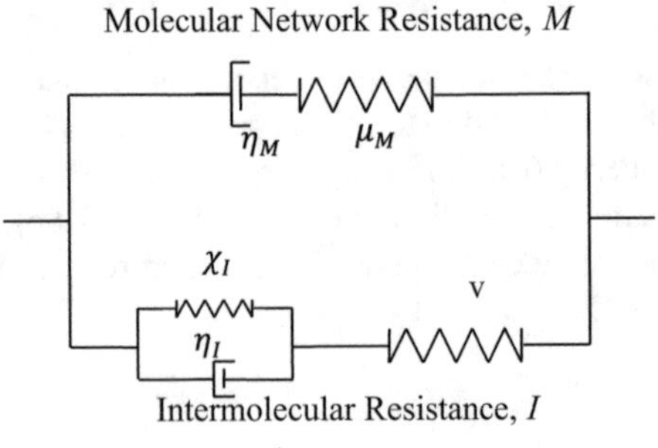

Fig. 7.2 Schematic representation of material model

intermolecular and molecular network components of associated quantity, respectively.

$$\mathbf{F} = \mathbf{F}_{\mathrm{I}} = \mathbf{F}_{\mathrm{M}} \tag{7.155}$$

$$\mathbf{T} = \mathbf{T}_{\mathrm{I}} + \mathbf{T}_{\mathrm{M}} \tag{7.156}$$

7.3.2.1 Intermolecular Resistance (I)

In most polymers, initial elastic response due to intermolecular resistance is governed by van der Waals bonds with surrounding molecules. A Helmholtz free energy per unit volume in reference configuration is considered for constitutive relation describing intermolecular resistance which was developed by Anand (Anand and On 1979; Anand 1986):

$$\Psi_I\left(\mathbf{E}_{\mathrm{I}}^{\mathrm{e}}, \theta\right) = \left\{ \begin{array}{r} G\left|\mathrm{dev}\left(\mathbf{E}_{\mathrm{I}}^{\mathrm{e}}\right)\right|^2 + \frac{1}{2}\left(K - \frac{2}{3}\mathrm{G}\left[\mathrm{tr}\left(\mathbf{E}_{\mathrm{I}}^{\mathrm{e}}\right)\right]^2\right) \\ -3K\alpha(\theta - \theta_o)\mathrm{tr}\left(\mathbf{E}_{\mathrm{I}}^{\mathrm{e}}\right) \end{array} \right\} \tag{7.157}$$

where θ_o is initial temperature, $\theta_o = \theta(\mathbf{r}, t_o)$, and G, K, α are temperature-dependent shear modulus, bulk modulus, and coefficient of thermal expansion, respectively. Elastic logarithmic strain $\left(\mathbf{E}_{\mathrm{I}}^{\mathrm{e}}\right)$ is related to right elastic stretch tensor $\left(\mathbf{C}_{\mathrm{I}}^{\mathrm{e}}\right)$ and elastic deformation gradient $\left(\mathbf{F}_{\mathrm{I}}^{\mathrm{e}}\right)$ through the following relations:

$$\mathbf{E}_{\mathrm{I}}^{\mathrm{e}} = \frac{1}{2} \ln \mathbf{C}_{\mathrm{I}}^{\mathrm{e}} \tag{7.158}$$

$$\mathbf{E}_{\mathrm{I}}^{\mathrm{e}} = \frac{1}{2} \ln \left(\mathbf{F}_{\mathrm{I}}^{\mathrm{e}^{\mathrm{T}}} \mathbf{F}_{\mathrm{I}}^{\mathrm{e}}\right) \tag{7.159}$$

Utilizing (Eq. (7.144)) symmetric second Piola-Kirchhoff stress tensor $\mathbf{S}_{\mathrm{I}}^{\mathrm{e}}$ and Cauchy stress tensor $(\mathbf{T}_{\mathrm{I}})$ can be obtained from Helmholtz free energy density function corresponding to intermolecular resistance as

$$\mathbf{S}_{\mathrm{I}}^{\mathrm{e}} = 2\frac{\partial \Psi_{\mathrm{I}}\left(\mathbf{E}_{\mathrm{I}}^{\mathrm{e}}, \theta\right)}{\partial \mathbf{C}_{\mathrm{I}}^{\mathrm{e}}} \tag{7.160}$$

$$\mathbf{T}_{\mathrm{I}} = \boldsymbol{J}^{-1}\mathbf{F}_{\mathrm{I}}^{\mathrm{e}}\mathbf{S}_{\mathrm{I}}^{\mathrm{e}}\mathbf{F}_{\mathrm{I}}^{\mathrm{e}^{\mathrm{T}}} \tag{7.161}$$

Since Helmholtz free energy density corresponding to macroscopic elastic energy stored (Ψ_{I}) is an isotropic function of elastic right Cauchy tensor $\left(\mathbf{C}_{\mathrm{I}}^{\mathrm{e}}\right)$, $\mathbf{C}_{\mathrm{I}}^{\mathrm{e}}$ and $\partial\Psi_{\mathrm{I}}/\partial\mathbf{C}_{\mathrm{I}}^{\mathrm{e}}$ are coaxial, and their product is a symmetric tensor; elastic Mandel stress $\left(\mathbf{M}_{\mathrm{I}}^{\mathrm{e}}\right)$ is given by

$$\mathbf{M}_{\mathrm{I}}^{\mathrm{e}} = \mathbf{C}_{\mathrm{I}}^{\mathrm{e}}\mathbf{S}_{\mathrm{I}}^{\mathrm{e}} \tag{7.162}$$

Relation between elastic Mandel stress $\left(\mathbf{M}_{\mathrm{I}}^{\mathrm{e}}\right)$ and elastic logarithmic strain $\left(\mathbf{E}_{\mathrm{I}}^{\mathrm{e}}\right)$ can be obtained from Helmholtz free energy function Eq. (7.157) and second Piola-Kirchhoff stress tensor $\left(\mathbf{S}_{\mathrm{I}}^{\mathrm{e}}\right)$ definition Eq. (7.160) as follows:

$$\mathbf{M}_{\mathrm{I}}^{\mathrm{e}} = 2\mathrm{G}\,\mathrm{dev}\left(\mathbf{E}_{\mathrm{I}}^{\mathrm{e}}\right) + K\left[\mathrm{tr}\left(\mathbf{E}_{\mathrm{I}}^{\mathrm{e}}\right) - 3\alpha(\theta - \theta_o)\right]\mathbf{I} \tag{7.163}$$

Kinematic hardening in intermolecular structure is modeled by a defect energy function per unit volume in intermediate (relaxed) configuration (Anand et al. 2009):

$$\Psi_{\mathrm{D}}(\mathbf{A}, \theta) = \frac{1}{4}B\left[\ln(a_1)^2 + \ln(a_2)^2 + \ln(a_3)^2\right] \tag{7.164}$$

where a_i represents eigenvalues of a stretch-like internal variable ($\mathbf{A}$) which is a symmetric unimodular tensor, $\det(\mathbf{A}(x,t)) = 1$. Since defect energy ($\Psi_{\mathrm{D}}$) is an isotropic function of symmetric unimodular stretch-like tensor ($\mathbf{A}$), $\mathbf{A}$ and $\partial\Psi_{\mathrm{D}}/\partial\mathbf{A}$ are coaxial, and their product is symmetric deviatoric back stress tensor ($\mathbf{M}_{\mathrm{back}}$). Back stress (Eq. (7.165)) and evolution equation for ($\mathbf{A}$) (Eq. (7.166) with initial condition in Eq. (7.167)) are defined as

$$\mathbf{M}_{\mathrm{back}} = 2\mathrm{dev}\left(\frac{\partial\Psi(\mathbf{A},\theta)}{\partial\mathbf{A}}\mathbf{A}\right) = \mathrm{B}\ln(\mathbf{A}) \tag{7.165}$$

$$\dot{\mathbf{A}} = \mathbf{D}_{\mathrm{I}}^{\mathrm{p}}\mathbf{A} + \mathbf{A}\mathbf{D}_{\mathrm{I}}^{\mathrm{p}} - \gamma\mathbf{A}\ln(\mathbf{A})\nu_{\mathrm{I}}^{\mathrm{p}} \tag{7.166}$$

$$\mathbf{A}(\mathbf{X}, 0) = \mathbf{I} \tag{7.167}$$

where γ represent the dynamic recovery, B is temperature-dependent back stress modulus, and $\nu_{\mathrm{I}}^{\mathrm{p}}$ is equivalent to plastic stretch rate in intermolecular structure. Driving stress for plastic flow in intermolecular structure is defined as

$$\mathbf{M}_{\mathrm{eff}} = \mathrm{dev}\left(\mathbf{M}_{\mathrm{I}}^{\mathrm{e}} - \mathbf{M}_{\mathrm{back}}\right) \tag{7.168}$$

Equivalent plastic stretch rate (Eq. (7.169)), effective equivalent shear stress (Eq. (7.170)), and mean normal pressure (Eq. (7.171)) are defined in terms of tensorial variables as follows:

$$\nu_{\mathrm{I}}^{\mathrm{p}} = \sqrt{2}\left|\mathbf{D}_{\mathrm{I}}^{\mathrm{p}}\right| \tag{7.169}$$

$$\tau_{\mathrm{I}} = \frac{1}{\sqrt{2}}\left|\mathbf{M}_{\mathrm{eff}}\right| \tag{7.170}$$

$$p_{\mathrm{I}} = -\frac{1}{3}\mathrm{tr}\left(\mathbf{M}_{\mathrm{I}}^{\mathrm{e}}\right) \tag{7.171}$$

Evolution of plastic deformation gradient in intermolecular mechanism can be rewritten from Eq. (7.31) as

$$\dot{\mathbf{F}}_{\mathrm{I}}^{\mathrm{p}} = \mathbf{D}_{\mathrm{I}}^{\mathrm{p}}\mathbf{F}_{\mathrm{I}}^{\mathrm{p}} \tag{7.172}$$

$$\mathbf{F}_{\mathrm{I}}^{\mathrm{p}}(\mathbf{r}, 0) = \mathbf{I} \tag{7.173}$$

Only the effective equivalent shear stress drives the plastic flow, and it is the source of plastic dissipation. Some part of plastic work is stored as energy associated with back stress. Once effective shear stress level reaches a critical level, so that energy barrier to molecular chain segment rotation is exceeded, plastic flow takes place.

It is important to point out that the concept of dislocation and dislocation motion-induced metal plasticity model is not applicable to amorphous polymers because the concept of dislocation cannot be justified in amorphous polymers.

According to cooperative model, viscous flow in a solid amorphous polymer may take place only when a number of polymer segments move cooperatively which also account for the significance of activation volume during yield process. The flow rule for amorphous polymers is essentially based on the energy distribution statistics of individual segments (Fotheringham and Cherry 1978). In simple terms, cooperative model flow rule is based on average probability of simultaneous occurrence of n thermally activated transitions across an energy barrier (activation energy, Q) inducing a macroscopic strain increment of ν_o (Fotheringham et al. 1976; Fotheringham and Cherry 1978). Yield characteristics of amorphous polymers are strongly temperature and rate dependent. According to strain rate-temperature superposition principle, an increase in temperature will have the same effect on the yield stress as a decrease in strain rate (Francisco et al. 1996). Equivalence of time and temperature describes that yielding of amorphous polymers at low temperatures is comparable to that at high strain rates. Therefore, Eyring plots (yield stress-temperature ratio versus plastic strain rate curves) for various temperatures can be shifted vertically and horizontally with respect to a reference temperature (θ_{ref}) in order to obtain a master curve describing yield stress behavior over a wide range of temperatures and strain rates.

Richeton et al. (2006) has proposed that both horizontal shift (ΔH_{h}) and vertical shift (ΔH_{v}) should follow Arrhenius-type temperature dependence. Resulting yield stress definition relates yield behavior of polymer with β mechanical loss peak at temperatures below glass transition temperature θ_{g} through introducing activation energy at β-transition temperature, i.e., yield behavior is controlled by segmental motions of polymer chains and reference state for yielding is chosen as β-transition. Increase in yield stress due to an increase in strain rate is attributed to decrease in molecular mobility of molecular chains, while a slow deformation rate allows polymer chains to slip past each other, resulting in a lower resistance to flow. At low temperatures near secondary transition temperature (θ_β), secondary molecular motions are restricted and chains become stiffer which also increase yield stress, while increase in temperature provides more energy to polymer chains facilitating relative motion between polymer molecular chains. For temperatures above glass transition temperature θ_g, characteristic plastic strain rate equation was modified by Williams-Landel-Ferry (WLF) parameters (c_1, c_2). Although characteristic plastic strain rate definitions at temperatures below and above θ_g are continuous functions of temperature in separate domains, piece-wise definition with respect to glass transition (θ_g) results in unrealistic change in plastic flow behavior around glass transition, i.e., derivative of plastic strain rate equation is discontinuous at θ_g. Srivastava and Anand (2010) proposed a modified version of flow rule in intermolecular structure which incorporates different values of activation energy for glassy region and rubbery region but still abrupt change in activation energies at θ_g which creates problem in material response. In order to provide a smoother transition in flow characteristics around θ_g, characteristic plastic strain rate (Eq. (7.174)) and equivalent shear plastic stretch rate (Eq. (7.175)) can be given in the following forms:

$$\nu^* = \nu_{\mathrm{I}}^{\mathrm{o}} \exp\left(-\frac{Q_{\mathrm{I}}}{k_{\mathrm{B}}\theta}\right)\left[1 + \exp\left(\frac{\ln(10)c_1(\theta - \theta_g)}{c_2 + (\theta - \theta_g)}\right)\right] \tag{7.174}$$

$$\nu_{\mathrm{I}}^{\mathrm{p}} = \nu^*\left[\sinh\left(\frac{\overline{\tau}_{\mathrm{I}} V}{2k_{\mathrm{B}}\theta}\right)\right]^{n_{\mathrm{I}}} \tag{7.175}$$

where $\nu_{\mathrm{I}}^{\mathrm{o}}$ is the pre-exponential factor, Q_{I} is the activation energy for plastic flow in intermolecular structure, k_{B} is the Boltzmann's constant, c_1 and c_2 are WLF parameters, n_I is the number of thermally activated transitions necessary for plastic flow, V is the activation volume, and $\overline{\tau}_I$ is the net effective stress which is defined as

$$\overline{\tau}_{\mathrm{I}} = \tau_{\mathrm{I}} - S_{\mathrm{I}} - \alpha_{\mathrm{P}} p_{\mathrm{I}} \tag{7.176}$$

where α_{p} is hydrostatic pressure sensitivity parameter and S_{I} is plastic flow resistance in intermolecular structure. Evolution of intermolecular resistance to plastic flow can be given by

$$\dot{S}_{\mathrm{I}} = h_{\mathrm{I}}\left(S_{\mathrm{I}}^* - S_{\mathrm{I}}\right)\nu_{\mathrm{I}}^{\mathrm{p}} \tag{7.177}$$

with initial condition given by

$$S_{\mathrm{I}}^{\mathrm{o}} = S_{\mathrm{I}}(\mathbf{r}, 0) \tag{7.178}$$

where h_{I} is a parameter characterizing hardening-softening and S_{I}^* is the saturation value for plastic flow resistance in intermolecular structure which can be defined as follows:

$$S_{\mathrm{I}}^{*} = b(\phi^{*} - \phi) \tag{7.179}$$

where b is a temperature and rate-dependent parameter which relates saturation value of plastic flow resistance to an order function $(\phi^{*} - \phi)$. Resistance to plastic flow (S_{I}) increases with disorder in the material and becomes constant $\left(S_{\mathrm{I}}^{*}\right)$ when order parameter (ϕ) reaches a critical value (ϕ^{*}) which is also a temperature and rate dependent variable. When intermolecular resistance reaches saturation value, steady-state plastic flow occurs, and plastic flow rate becomes equal to applied strain rate. Evolution equation for order parameter is defined as

$$\dot{\phi} = \mathrm{g}(\phi^{*} - \phi)\nu_{\mathrm{I}}^{\mathrm{p}} \tag{7.180}$$

$$\phi_{\mathrm{o}} = \phi(\mathbf{r}, 0) \tag{7.181}$$

where g is a temperature-dependent parameter. Evolution equation for plastic deformation gradient in Eqs. (7.172) and (7.173) completes definition of material behavior in intermolecular structure. Strain hardening becomes insignificant as temperatures approach θ_{g} and completely vanishes above θ_{g} (Richeton et al. 2006). Definition of a vanishing internal resistance right at θ_{g} causes also discontinuity in yield behavior of polymer. Since annealing at high temperatures well above θ_{g} clears past thermo-mechanical history of materials by providing an alternative stationary molecular configuration at a higher energy level, internal resistance is bound to vanish above or around θ_{g}. Therefore, underlying problem is essentially on the assumption that glass transition takes place at a single temperature and internal resistance becomes zero abruptly at θ_{g}. Similarly, variables $\left(b, \mathrm{g}, S_{\mathrm{I}}^{*}, \phi^{*}, h_{\mathrm{I}}\right)$ that characterize hardening-softening behavior in post-yield region shall also provide a smooth transition from temperatures below θ_{g} to temperatures above θ_{g}.

It should be noted that viscoplastic models currently available in literature are all phenomenological and serve as a mathematical tool to fit experimentally observed behavior into a curve. These models can provide reasonably accurate predictions for yield characteristics of amorphous polymers for only isothermal cases. In the case of non-isothermal loading, which include temperature change in the material concurrently with loading, most material models available in literature predict unrealistic results. A comparison of viscoplastic models for amorphous polymers from literature and improved version of dual-mechanism model are presented in Fig. 7.3.

Temperature variations of characteristic viscoplastic shear strain rates in different models are presented by normalizing with respect to characteristic viscoplastic strain rate at reference glass transition temperature for PMMA (387K). Material properties are taken from Srivastava and Anand (2010), while WLF parameters in the model discussed in this chapter and in Richeton et al. (2006) are taken as their original values. Viscoplastic model presented in this chapter are applicable for temperatures both above and below glass transition. In the model presented earlier in this chapter, temperature dependence of viscoplastic stretch rate is directly employed utilizing physically motivated Williams-Landel-Ferry parameters in a completely new form of expression as presented in Eq. (7.174). It is clear that temperature-dependent

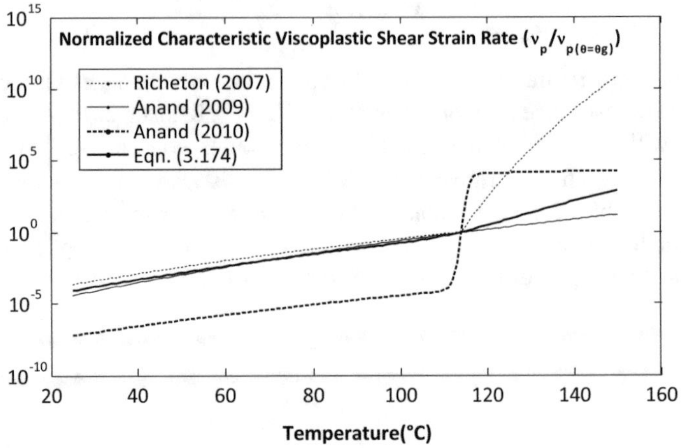

Fig. 7.3 Comparison of different viscoplastic models in literature in terms of temperature dependence of normalized characteristic viscoplastic shear strain rate $\left(\nu_p/\nu_{p(\theta=\theta_g)}\right)$ (Gunel and Basaran 2010)

activation energy approach in Anand's model predicts not a gradual change in behavior in intermolecular mechanism but an abrupt increase in viscoplastic strain rate over a relatively short temperature range. Therefore, according to Anand model, there is actually no rubbery region, but material response is liquid like as viscoplastic strain rate increases by six orders in a narrow temperature interval (2 °C) around glass transition temperature. On the other hand, in Richeton's model [142], there is also a remarkable change in viscoplastic strain rate due to the piece-wise definition with respect to glass transition temperature, but it also causes discontinuous derivative of viscoplastic rate at glass transition. Accordingly, material response predicted by Anand's or Richeton's viscoplastic models will present a significant (abrupt) change in stress at the glass transition temperature. Improved version of dual-mechanism model presented in this chapter predicts definitely a more gradual transition in material response with respect to temperature around glass transition. Though Anand's model has a continuous definition of activation energy in temperature domain, remarkable difference between activation energies in glassy and rubbery region still produce an abrupt change in response. According to Fig. 7.3 Anand's viscoplastic model is invariant of temperature above glass transition temperature, while Richeton's model provides relatively gentler transition in response. As a result of this rapid change in viscoplastic response in Anand's model, it cannot predict material behavior accurately under non-isothermal conditions. These models were further studied by Gunel and Basaran (2010) in terms of predictions for creep strain rate in a creep test with a stress level of 0.6 MPa which is conducted at different temperatures as depicted in Fig. 7.4.

Figure 7.4 supports previous observations just discussed. All models yield same predictions for creep strain rate at temperatures below glass transition. For assurance of accurate and realistic modeling of material response around glass transition, every

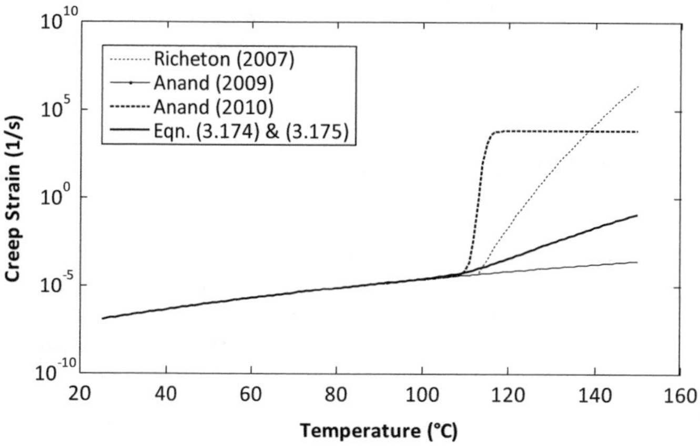

Fig. 7.4 Comparison of different viscoplastic models in literature in terms of creep strain rates at different temperatures in response to an applied stress of 0.6 MPa

aspect of material property definitions that describe hardening-softening behavior and flow characteristics should be continuous in temperature domain and should have continuous (at least) first derivatives with respect to temperature.

7.3.2.2 Molecular Network Resistance (M)

Resistance in molecular network to deformation is based on molecular orientation and relaxation process. If there is enough stretch in polymer chains, network resists relaxation and resistance increases with increasing stretch. In literature, there is a general consensus on modeling of plastic flow behavior in intermolecular structure based on Eyring cooperative model, but there is still some debate on modeling of molecular network structure. Arruda and Boyce (1993) modeled molecular network resistance with rubber elasticity model based on eight-chain network of non-Gaussian chains similar to transient response of elastomers which represents nonlinear rate-dependent deviation from equilibrium state Bergström and Boyce (1998). However, network resistance described by eight-chain model is not accurate enough to be a physically consistent model describing orientational hardening behavior in amorphous polymers, since temperature dependence of rubbery modulus and number of rigid links between polymer chain segments which essentially control response does not match experimental observations. Therefore, molecular network description based on eight-chain model (Richeton et al. 2007; Arruda et al. 1995; Palm et al. 2006; Boyce et al. 2000) becomes merely a numerical tool to match experimentally observed stress-strain response. Instead, a simpler two constant constitutive relation for rubber networks developed by Gent (1996) was shown to describe strain hardening due to polymer chain stretching better than statistical-mechanical entropic rubber elasticity models (eight-chain model) while resulting in a similar stress-strain response (Ames et al. 2009; Srivastava and Anand 2010). Gent

free energy per unit volume in reference configuration describes elastic energy stored in molecular network structure in terms of first invariant of stretch in polymer chains as

$$\Psi_{\mathrm{M}}\left(\mathbf{C}_{\mathrm{M}}^{\mathrm{e}},\theta\right) = -\frac{1}{2}\mu_{\mathrm{M}} I_{\mathrm{M}} \ln\left(1-\frac{I_1-3}{I_m}\right) \tag{7.182}$$

where μ_{M} and I_{M} are temperature-dependent rubbery shear modulus and limit on extensibility of polymer chains, respectively. Since volume change in material is considered in elastic deformation gradient associated with intermolecular structure, it is essential to define Gent free energy in terms of distortional elastic deformation gradient in network structure, $\left(\mathbf{F}_{\mathrm{M}}^{\mathrm{e}}\right)_d$ (Eq. (7.183)), which produces no change in volume. I_1 is first invariant of elastic distortional Cauchy tensor in network structure $\left(\mathbf{C}_{\mathrm{M}}^{\mathrm{e}}\right)_d$

$$\left(\mathbf{F}_{\mathrm{M}}^{\mathrm{e}}\right)_d = \boldsymbol{J}^{-1/3}\mathbf{F}_{\mathrm{M}}^{\mathrm{e}} \tag{7.183}$$

$$\det\left(\left(\mathbf{F}_{\mathrm{M}}^{\mathrm{e}}\right)_d\right) = 1 \tag{7.184}$$

$$\left(\mathbf{C}_{\mathrm{M}}^{\mathrm{e}}\right)_d = \left(\mathbf{F}_{\mathrm{M}}^{\mathrm{e}}\right)_d^T\left(\mathbf{F}_{\mathrm{M}}^{\mathrm{e}}\right)_d \tag{7.185}$$

$$I_1 = \mathrm{tr}\left[\left(\mathbf{C}_{\mathrm{M}}^{\mathrm{e}}\right)_d\right] \tag{7.186}$$

Second Piola-Kirchhoff stress $\left(\mathbf{S}_{\mathrm{M}}^{\mathrm{e}}\right)$ and Cauchy stress ($\mathbf{T}_{\mathrm{M}}$)can be derived from Gent free energy as follows:

$$\mathbf{S}_{\mathrm{M}}^{\mathrm{e}} = 2\frac{\partial\Psi_{\mathrm{M}}\left(\mathbf{C}_{\mathrm{M}}^{\mathrm{e}},\theta\right)}{\partial\mathbf{C}_{\mathrm{M}}^{\mathrm{e}}} \tag{7.187}$$

$$\mathbf{T}_{\mathrm{M}} = J^{-1}\mathbf{F}_{\mathrm{M}}^{\mathrm{e}}\mathbf{S}_{\mathrm{M}}^{\mathrm{e}}\mathbf{F}_{\mathrm{M}}^{\mathrm{e}^{\mathrm{T}}} \tag{7.188}$$

Using Gent free energy definition in and stress definitions in

$$\mathbf{S}_{\mathrm{M}}^{\mathrm{e}} = \boldsymbol{J}^{-2/3}\mu_{\mathrm{M}}\left(1-\frac{I_1-3}{I_{\mathrm{M}}}\right)^{-1}\left[\mathbf{I}-\frac{1}{3}\mathrm{tr}\left(\left(\mathbf{C}_{\mathrm{M}}^{\mathrm{e}}\right)_{\mathrm{d}}\right)\left(\mathbf{C}_{\mathrm{M}}^{\mathrm{e}}\right)_{\mathrm{d}}^{-1}\right] \tag{7.189}$$

$$\mathbf{T}_{\mathrm{M}} = \boldsymbol{J}^{-1}\mu_{\mathrm{M}}\left(1-\frac{I_1-3}{I_{\mathrm{M}}}\right)^{-1}\mathrm{dev}\left(\left(\mathbf{B}_{\mathrm{M}}^{\mathrm{e}}\right)_{\mathrm{d}}\right) \tag{7.190}$$

Elastic distortional Almansi tensor, $\left(\mathbf{B}_{\mathrm{M}}^{\mathrm{e}}\right)_{\mathrm{d}}$, is defined in terms of distortional elastic deformation gradient $\left(\left(\mathbf{F}_{\mathrm{M}}^{\mathrm{e}}\right)_{\mathrm{d}}\right)$ as

$$\left(\mathbf{B}_{\mathrm{M}}^{\mathrm{e}}\right)_{\mathrm{d}} = \left(\mathbf{F}_{\mathrm{M}}^{\mathrm{e}}\right)_{\mathrm{d}}\left(\mathbf{F}_{\mathrm{M}}^{\mathrm{e}}\right)_{\mathrm{d}}^{\mathrm{T}} \tag{7.191}$$

Elastic Mandel stress $\mathbf{M}_{\mathrm{M}}^{\mathrm{e}}$ and equivalent shear stress τ_{M} and equivalent plastic shear strain rate $\nu_{\mathrm{M}}^{\mathrm{p}}$ for molecular relaxation in network structure can be found as

$$\mathbf{M}_{\mathrm{M}}^{\mathrm{e}} = \mathbf{C}_{\mathrm{M}}^{\mathrm{e}}\mathbf{S}_{\mathrm{M}}^{\mathrm{e}} \tag{7.192}$$

$$\mathbf{M}_{\mathrm{M}}^{\mathrm{e}} = \mu_{\mathrm{M}}\left(1 - \frac{I_1 - 3}{I_{\mathrm{M}}}\right)^{-1} \mathrm{dev}\left(\left(\mathbf{C}_{\mathrm{M}}^{\mathrm{e}}\right)_{\mathrm{d}}\right) \tag{7.193}$$

$$\tau_{\mathrm{M}} = \frac{1}{\sqrt{2}}\left|\mathbf{M}_{\mathrm{M}}^{\mathrm{e}}\right| \tag{7.194}$$

$$\nu_{\mathrm{M}}^{\mathrm{p}} = \sqrt{2}\left|\mathbf{D}_{\mathrm{M}}^{\mathrm{p}}\right| \tag{7.195}$$

Evolution of plastic deformation gradient in molecular network mechanism can be rewritten from Eq. (7.31) as

$$\dot{\mathbf{F}}_{\mathrm{M}}^{\mathrm{p}} = \mathbf{D}_{\mathrm{M}}^{\mathrm{p}}\mathbf{F}_{\mathrm{M}}^{\mathrm{p}} \tag{7.196}$$

$$\mathbf{F}_{\mathrm{M}}^{\mathrm{p}}(\mathbf{r}, 0) = \mathbf{I} \tag{7.197}$$

Molecular network is responsible for resistance to chain alignment which resists relaxation as stretch in network increases (Gunel 2010). A similar observation for elastomers was also found that plastic molecular chain stretch is inversely proportional to effective creep rate (Bergström and Boyce 1998). In experimental and numerical studies, it was observed that in post-yield region (at large deformations), controlling mechanism is molecular network mechanism. Molecular network mechanism has a dominant contribution to stress change in post-yield region while the amount of elastic recovery upon unloading is associated with plastic strain in molecular network mechanism. Temperature dependence of molecular relaxation in network structure is characterized with a classical Arrhenius term, and flow rule is completed with a simple power law as follows:

$$\nu_{\mathrm{M}}^{\mathrm{p}} = \nu_{\mathrm{M}}^{\mathrm{o}} \exp\left(-\frac{Q_{\mathrm{M}}}{k_{\mathrm{B}}\theta}\right)\left(\frac{\tau_{\mathrm{M}}}{S_{\mathrm{M}}}\right)^{n_{\mathrm{M}}} \tag{7.198}$$

where $\nu_{\mathrm{M}}^{\mathrm{o}}$ is the pre-exponential factor, Q_{M} is the activation energy for molecular relaxation in network structure, n_{M} is a strain-rate sensitivity parameter, and S_{M} is a stress measure describing resistance of network structure to relaxation which increases with increasing plastic stretch rate as defined as follows:

$$\dot{S}_{\mathrm{M}} = h_{\mathrm{M}}\left(\lambda_{\mathrm{M}}^{\mathrm{p}} - 1\right)\left(S_{\mathrm{M}}^{*} - S_{\mathrm{M}}\right)\nu_{\mathrm{M}}^{\mathrm{p}} \tag{7.199}$$

with initial condition

$$S_{\mathrm{M}}^{\mathrm{o}} = S_{\mathrm{M}}(\mathbf{r}, 0) \tag{7.200}$$

where h_{M} is a parameter characterizing molecular relaxation in material, S_{M}^{*} is the temperature-rate dependent saturation value for network resistance, and $\lambda_{\mathrm{M}}^{\mathrm{p}}$ is the plastic stretch which is related to plastic Almansi tensor in network structure $(\mathbf{B}_{\mathrm{M}}^{\mathrm{p}})$ as follows:

$$\lambda_{\mathrm{M}} = \sqrt{\frac{\mathrm{tr}(\mathbf{B}_{\mathrm{M}}^{\mathrm{p}})}{3}} \tag{7.201}$$

$$\mathbf{B}_{\mathrm{M}}^{\mathrm{p}} = \mathbf{F}_{\mathrm{M}}^{\mathrm{p}} \mathbf{F}_{\mathrm{M}}^{\mathrm{p}^{\mathrm{T}}} \tag{7.202}$$

Plastic deformation gradient evolution given by Eqs. (7.196) and (7.197) completes the definition of material behavior in molecular network structure. According to Eq. (7.199), network resistance will increase continuously as plastic stretch $(\lambda_{\mathrm{M}}^{\mathrm{p}})$ in polymer chains increases and reaches a constant value $\left(S_{\mathrm{M}}^{*}\right)$ which depends on temperature and stretch rate. Plastic stretch-dependent evolution of resistance to plastic flow also ensures correct prediction of elastic recovery in unloading path. In Anand's model, net driving stress for plastic flow in intermolecular mechanism includes an additional resistance term accounting for dissipative resistance to plastic flow $(S_b \,\text{or}\, S_2)$ taking place at large deformations. According to Anand's model, this dissipative resistance evolves with plastic stretch in intermolecular mechanism or total stretch in intermolecular mechanism. In the model derived in this chapter, dissipative resistance (S_{M}) at large deformations in molecular network mechanism evolves with plastic stretch in molecular network branch which actually controls material response at large deformations (post-yield region). A third mechanism must be added for true prediction of elastic recovery during unloading and cooling of a pre-heated sample. It was argued that a second mechanism for molecular network structure was introduced due to a necessity driven by an experimentally observed complex response. However, constitutive model framework presented in this chapter involves non-isothermal condition without introducing the additional third mechanism.

Finally, from relation given by Eq. (7.196) and specific Helmholtz free energy definitions in Eqs. (7.157), (7.164), and (7.182), temperature-dependent governing equation can be given as

$$\begin{aligned}
\rho c\dot{\theta} &= \nabla_x(k\nabla_x(\theta)) + r \\
&+ J^{-1}\left(\left(\tau_I + \frac{1}{2}\gamma \mathbf{B}|\ln(\mathbf{A})|^2\right)\nu_{\mathrm{I}}^{\mathrm{p}} + \tau_{\mathrm{M}}\nu_{\mathrm{M}}^{\mathrm{p}}\right) \\
&+ J^{-1}\theta\left(\frac{1}{2}\frac{\partial \mathbf{S}_{\mathrm{I}}^{\mathrm{e}}}{\partial\theta} : \dot{\mathbf{C}}_{\mathrm{I}}^{\mathrm{e}} + \frac{1}{2}\frac{\partial \mathbf{S}_{\mathrm{M}}^{\mathrm{e}}}{\partial\theta} : \dot{C}_{\mathrm{M}}^{\mathrm{e}} + \frac{1}{2}\frac{\partial\left(\mathbf{M}_{\mathrm{back}}\mathbf{A}^{-1}\right)}{\partial\theta} : \dot{\mathbf{A}}\right)
\end{aligned} \tag{7.203}$$

where the first two terms are heat conduction representing heat transfer within material during transient state and heat source due to passive heating (external heating) or active heating (internally generated heat). The last two terms represent heat induced due to intrinsic dissipation and thermo-elastic effect representing conversion between mechanical and thermal energy in elastic range. In the case of coupled thermo-mechanical loading, temperature increase due to mechanical work and temperature change due to heat transfer between material and surroundings are mixed together. Based on descriptions of stress components and strain rate measures, irreversible entropy production due to mechanical dissipation per unit volume in deformed configuration can be rewritten from Eq. (7.149) as

$$\gamma_{\text{mech}} = \frac{J^{-1}}{\theta}\left[\begin{array}{c}\left(\text{dev}\left(\mathbf{M}_{\text{I}}^{\text{e}}\right) - 2\text{dev}\left(\frac{\partial\Psi_{\text{D}}(\mathbf{A},\theta)}{\partial\mathbf{A}}\mathbf{A}\right)\right) : \mathbf{D}_{\text{I}}^{\text{p}} \\ +\gamma\left(\frac{\partial\Psi_{\text{D}}(\mathbf{A},\theta)}{\partial\mathbf{A}}\mathbf{A}\right) : \ln(\mathbf{A})\nu_{\text{I}}^{\text{p}} + \text{dev}\left(\mathbf{M}_{\text{M}}^{\text{e}}\right) : \mathbf{D}_{\text{M}}^{\text{p}}\end{array}\right] > 0 \quad (7.204)$$

$$\gamma_{\text{mech}} = \frac{J^{-1}}{\theta}\left[\left(\tau_I + \frac{1}{2}\gamma B|\ln(\mathbf{A})|^2\right)\nu_{\text{I}}^{\text{p}} + \tau_{\text{M}}\nu_{\text{M}}^{p}\right] > 0 \quad (7.205)$$

Due to associated plastic flow assumption, irreversible mechanical entropy production (γ_{mech}) is always positive:

$$\mathbf{N}_{\text{I}}^{\text{p}} = \frac{\mathbf{D}_{\text{I}}^{\text{p}}}{\left|\mathbf{D}_{\text{I}}^{\text{p}}\right|} = \frac{\mathbf{M}_{\text{eff}}}{\left|\mathbf{M}_{\text{eff}}\right|} \quad (7.206)$$

$$\mathbf{N}_{\text{M}}^{\text{p}} = \frac{\mathbf{D}_{\text{M}}^{\text{p}}}{\left|\mathbf{D}_{\text{M}}^{\text{p}}\right|} = \frac{\mathbf{M}_{\text{M}}^{\text{e}}}{\left|\mathbf{M}_{\text{M}}^{\text{e}}\right|} \quad (7.207)$$

7.4 Thermodynamic State Index

Entropy production can be decomposed into thermal dissipation (Eq. (7.137)) due to heat exchange between system and surroundings and mechanical dissipation (Eq. (7.205)) as a result of permanent changes in material molecular structure. Critical entropy production (S_{cr}) [also called fatigue fracture entropy] is a characteristic value of a material which is independent of loading rate, boundary conditions, and geometry of structure. Link between irreversible material degradation and amount of heat generated due to some non-conservative forces (plastic dissipation, friction forces, chemical reactions, etc.) or entropy production due to mechanical dissipation is well established. Heat conduction within a metal will be very fast, leading to negligible thermal gradients and thermal dissipation due to high thermal diffusivity of metals. In the case of materials with low thermal diffusivity such as polymers, heat transfer may take place over a prolonged period of times with significant thermal gradients. However, thermal gradients cannot be responsible for failure of chemical bonds since material degradation or damage is a consequence of formation of small voids or cracks at microscale by breakage of chemical bonds between molecules. Thermal dissipation may result in deterioration of material properties which is usually insignificant compared to degradation by mechanical dissipation and other entropy generation mechanisms.

Thermodynamic state index (TSI) is given by

$$\Phi = \Phi_{\mathrm{cr}}\left[1 - e^{-m_s \frac{\Delta S}{R}}\right] \tag{7.208}$$

When TSI value reaches a critical level that can be defined as failure or $\Phi = 1$ depending on the application. Critical TSI depends on critical entropy level (S_{cr}), and it can also be a characteristic of material and should be calculated or measured for different materials separately.

$$\Phi_{\mathrm{cr}} = \left[1 - \mathrm{e}^{-m_s \frac{[S_{\mathrm{cr}} - S_{\mathrm{o}}]}{R}}\right] \tag{7.209}$$

where S_o is initial internal entropy value, which can be taken as zero. During any irreversible process inducing changes, degradation in microstructure, internal entropy production increases according to second law of thermodynamics. Total entropy production due to mechanical and thermal dissipation can be calculated at any time step as follows:

$$S_{\mathrm{mech}} = S_{\mathrm{mech}}|_{t=t_{\mathrm{o}}} + \int_{t_{\mathrm{o}}}^{t} \gamma_{\mathrm{mech}} dt \tag{7.210}$$

$$S_{\mathrm{mech}} = S_{\mathrm{mech}}|_{t=t_{\mathrm{o}}} + \int_{t_{\mathrm{o}}}^{t} \left\{ \frac{\boldsymbol{J}^{-1}}{\theta} \left[\left(\tau_{\mathrm{I}} + \frac{1}{2}\gamma B |\ln(\mathbf{A})|^2 \right) \boldsymbol{\nu}_{\mathrm{I}}^{\mathrm{p}} + \boldsymbol{\tau}_{\mathrm{M}} \boldsymbol{\nu}_{\mathrm{M}}^{\mathrm{p}} \right] \right\} dt \tag{7.211}$$

$$S_{\mathrm{ther}} = S_{\mathrm{ther}}|_{t=t_{\mathrm{o}}} + \int_{t_{\mathrm{o}}}^{t} \gamma_{\mathrm{ther}} dt \tag{7.212}$$

$$S_{\mathrm{ther}} = S_{\mathrm{ther}}|_{t=t_{\mathrm{o}}} + \int_{t_{\mathrm{o}}}^{t} \left\{ -\frac{1}{\theta^2} \mathrm{div}\left(\mathbf{J}_q\right) \bullet \nabla_x(\theta) + \frac{\rho r}{\theta} \right\} dt \tag{7.213}$$

$$S_{\mathrm{mech}}|_{t=t_{\mathrm{o}}} = S_{\mathrm{ther}}|_{t=t_o} = 0 \tag{7.214}$$

At failure internal entropy production reaches a critical value (S_{cr}) which is temperature dependent. Unlike metals, temperature dependence of critical entropy is essential for the case of stretching of polymers due to change in failure mode of amorphous polymer chains. Amorphous polymers display a brittle failure at low temperatures ($\theta < < \theta_g$) without any significant plastic dissipation, while at high temperatures ($\theta > \theta_g$) ductile failure occurs after significant amount of plastic work. Critical TSI parameter is defined in such a way that at constant temperatures as $S \rightarrow S_{cr}(\theta)$, $\Phi \rightarrow 1$. Nonnegative entropy production assures that $\Phi \geq 0$, while for an pristine material ($S = 0$), damage is assumed to be zero:

$$\Phi(\theta) = \left[1 - \exp\left(-\frac{m_s}{R} S_{cr}(\theta)\right)\right] \tag{7.215}$$

According to incremental form of TSI evolution, degradation will increase at a much faster rate at low temperatures (sudden brittle failure), while increase in degradation will be relatively smaller at high temperatures (prolonged ductile failure). TSI evolution merely depends on the fundamental equation which incorporates all micro mechanisms responsible for entropy generation due to thermo-mechanical loading. Derivative of TSI with respect to entropy can be given by

$$\frac{\Delta \Phi}{\Delta S} = \Phi_{cr} \frac{m_s}{R} \exp\left(-\frac{m_s}{R} S_{mech}\right) \tag{7.216}$$

7.5 Definition of Material Properties

Material properties are vital for complete and accurate material constitutive models. Material properties must be defined over a large temperature and loading rate range and should be continuous and smooth over transition region. They should also account accurately for the rate dependency of the property. Expressions for material properties presented herein are indeed mathematical tools to describe influence of temperature and loading rate on material behavior in a continuous form. Formulations for these material properties representing temperature dependence are completely in a form that provides continuity in temperature domain and has continuous first derivative with respect to temperature. A significant part of material properties can be obtained by conducting isothermal tests at different temperatures (both above and below glass transition temperature) and at different loading rates such as $\{E, \nu, I_m, \nu_I^o, Q_I, n_I, B_g, X_B, V, \alpha_p, \gamma\}$. Some material parameters are difficult to measure directly such as $\{h_I, b, g, \nu_M^o, Q_M, h_M, n_M, \mu_M, \phi^*, S_M^*\}$, yet these properties/parameters can be obtained by statistical methods.

According to free volume theory of Williams et al (1955) plastic flow rule can be constructed for equivalent plastic shear strain rate at temperatures above θ_g using Williams-Landel-Ferry (WLF) equations. Similarly, rate dependence of glass transition temperature can be considered in terms of temperature-time equivalence of glass transition (Eq. (7.217)):

$$\theta_{\mathrm{g}} = \begin{cases} \theta_{\mathrm{g}}^{\mathrm{ref}} + \dfrac{c_2^{\mathrm{g}} \log\left(\nu/_{\nu^{\mathrm{ref}}} \right)}{c_1^{\mathrm{g}} - \log\left(\nu/_{\nu^{\mathrm{ref}}} \right)} & \nu > \nu^{\mathrm{ref}} \\ \theta_{\mathrm{g}}^{\mathrm{ref}} & \nu \leq \nu^{\mathrm{ref}} \end{cases} \tag{7.217}$$

where c_1^{g} and c_2^{g} are WLF parameters associated with θ_{g}, ν^{ref} is the reference stretch rate, and ν is the equivalent stretch rate which is defined by

$$\nu = \sqrt{2} \mid \mathbf{D} \mid \tag{7.218}$$

Temperature and rate dependence of elastic modulus (E) is given by

$$E = \left[\begin{array}{r} \dfrac{1}{2}\left(E_{\mathrm{g}} + E_{\mathrm{r}}\right) - \dfrac{1}{2}\left(E_{\mathrm{g}} - E_{\mathrm{r}}\right) \tanh\left(\dfrac{\theta - \left(\theta_{\mathrm{g}} + \theta_{\mathrm{E}}\right)}{\Delta_{\mathrm{E}}} \right) \\ + X_{\mathrm{E}}\left[\theta - \left(\theta_g + \theta_{\mathrm{E}}\right)\right] \end{array} \right] \times \left[1 + s_{\mathrm{E}} \log\left(\frac{\nu}{\nu^{\mathrm{ref}}} \right) \right] \tag{7.219}$$

$$X_{\mathrm{E}} = \begin{cases} X_{\mathrm{g}}^{\mathrm{E}} & \theta \leq \theta_{\mathrm{g}} + \theta_{\mathrm{E}} \\ X_{\mathrm{r}}^{\mathrm{E}} & \theta > \theta_{\mathrm{g}} + \theta_{\mathrm{E}} \end{cases} \tag{7.220}$$

where E_{g} and E_{r} are for glassy and rubbery elastic modulus corresponding to temperatures confining glass-rubber transition region; $X_{\mathrm{E}}^{\mathrm{g}}$ and $X_{\mathrm{E}}^{\mathrm{r}}$ represent rate of change of elastic modulus with respect to temperature in glassy and rubbery domains, respectively; s_E is the rate sensitivity of elastic modulus; while θ_{E} and Δ_{E} define origin temperature and width of glass-rubber transition, respectively. Experimental studies on temperature and rate dependence of storage modulus and elastic modulus of poly-methyl methacrylate (PMMA) indicates that PMMA is highly sensitive to rate and temperature. Modulus of PMMA continuously decreases with increasing temperature with a remarkable drop around θ_{g} over a 10–20 °C temperature domain depending on frequency of loading (Gunel 2010). Poisson's ratio ($\overline{\nu}$) is assumed to be only temperature dependent and defined as

$$\overline{\nu} = \frac{1}{2}\left(\overline{\nu}_{\mathrm{g}} + \overline{\nu}_{\mathrm{r}}\right) - \frac{1}{2}\left(\overline{\nu}_{\mathrm{g}} - \overline{\nu}_{\mathrm{r}}\right) \tanh\left(\frac{\theta - \left(\theta_{\mathrm{g}} + \theta_{\mathrm{E}}\right)}{\Delta_{\mathrm{E}}} \right) \tag{7.221}$$

where $\overline{\nu}_{\mathrm{g}}$ and $\overline{\nu}_{\mathrm{r}}$ are for glassy and rubbery Poisson's ratio, respectively. Shear modulus (G) and bulk modulus (K) are defined as follows:

$$\mathrm{G} = \frac{E}{2(1+\overline{\nu})} \tag{7.222}$$

$$K = \frac{E}{3(1-2\overline{\nu})} \tag{7.223}$$

Temperature and rate dependence of rubbery modulus (μ_{M}) is modeled similar to elastic modulus as follows:

$$\mu_{\mathrm{M}} = \left[\frac{1}{2}\left(\mu_{\mathrm{M}}^{\mathrm{g}} + \mu_{\mathrm{M}}^{\mathrm{r}}\right) - \frac{1}{2}\left(\mu_{\mathrm{M}}^{\mathrm{g}} - \mu_{\mathrm{M}}^{\mathrm{r}}\right) \tanh\left(\frac{\theta - (\theta_{\mathrm{g}} + \theta_\mu)}{\Delta_\mu}\right) + X_\mu\left[\theta - (\theta_{\mathrm{g}} + \theta_\mu)\right] \right] \times \left[1 + s_\mu \log\left(\frac{\nu}{\nu^{\mathrm{ref}}}\right)\right] \tag{7.224}$$

$$X_\mu = \begin{cases} X_\mu^{\mathrm{g}} & \theta \le \theta_{\mathrm{g}} + \theta_\mu \\ X_\mu^{r} & \theta > \theta_{\mathrm{g}} + \theta_\mu \end{cases} \tag{7.225}$$

Definitions of rubbery shear modulus parameters in Eq. (7.225) are identical to those for elastic modulus. Temperature dependence of critical value of order parameter (ϕ^*), limit of polymer chain extensibility (I_{M}), and saturation value of plastic flow resistance of molecular network $\left(S_{\mathrm{M}}^*\right)$ are defined below

$$\phi^* = \frac{1}{2}\left(\phi_{\mathrm{g}}^* + \phi_{\mathrm{r}}^*\right) - \frac{1}{2}\left(\phi_{\mathrm{g}}^* - \phi_{\mathrm{r}}^*\right) \tanh\left(\frac{\theta - (\theta_{\mathrm{g}} + \theta_\phi)}{\Delta_\phi}\right) + X_\phi\left[\theta - (\theta_{\mathrm{g}} + \theta_\phi)\right] \tag{7.226}$$

$$X_\phi = \begin{cases} X_\phi^{\mathrm{g}} & \theta \le \theta_{\mathrm{g}} + \theta_\phi \\ X_\phi^{\mathrm{r}} & \theta > \theta_{\mathrm{g}} + \theta_\phi \end{cases} \tag{7.227}$$

$$I_{\mathrm{M}} = \frac{1}{2}\left(I_{\mathrm{M}}^{\mathrm{g}} + I_{\mathrm{M}}^{\mathrm{r}}\right) - \frac{1}{2}\left(I_{\mathrm{M}}^{\mathrm{g}} - I_{\mathrm{M}}^{\mathrm{r}}\right) \tanh\left(\frac{\theta - (\theta_{\mathrm{g}} + \theta_{\mathrm{M}})}{\Delta_{\mathrm{M}}}\right) + X_{\mathrm{M}}\left[\theta - (\theta_{\mathrm{g}} + \theta_{\mathrm{M}})\right] \tag{7.228}$$

$$X_{\mathrm{M}} = \begin{cases} X_{\mathrm{M}}^{\mathrm{g}} & \theta \le \theta_{\mathrm{g}} + \theta_\mu \\ X_{\mathrm{M}}^{\mathrm{r}} & \theta > \theta_{\mathrm{g}} + \theta_\mu \end{cases} \tag{7.229}$$

$$S_{\mathrm{M}}^* = \frac{1}{2}\left(S_{\mathrm{M}}^{\mathrm{g}} + S_{\mathrm{M}}^{\mathrm{r}}\right) - \frac{1}{2}\left(S_{\mathrm{M}}^{\mathrm{g}} - S_{\mathrm{M}}^{\mathrm{r}}\right) \tanh\left(\frac{\theta - (\theta_{\mathrm{g}} + \theta_S)}{\Delta_S}\right) + X_S\left[\theta - (\theta_{\mathrm{g}} + \theta_S)\right] \tag{7.230}$$

$$X_S = \begin{cases} X_S^{\mathrm{g}} & \theta \le \theta_{\mathrm{g}} + \theta_S \\ X_S^{\mathrm{r}} & \theta > \theta_{\mathrm{g}} + \theta_S \end{cases} \tag{7.231}$$

Definitions of parameters used in above equations are identical to those for elastic modulus. Saturation value for network resistance $\left(S_{\mathrm{M}}^{*}\right)$ and critical value of disorder parameter (ϕ^{*}) were assumed to decrease with increasing temperature, while limited chain extensibility (I_{M}) increases with increasing temperature based on observations in experiments. Similar to models of Richeton et al (2005a, 2005b) and Anand et al (2009), back stress is assumed to vanish above θ_{g} . However, decrease in back stress modulus (B) with increasing temperature asymptotically approaches to zero at temperature around θ_{g}, can be defined as

$$B = B_{\mathrm{g}}\left(1 - \tanh\left(\frac{\theta - \theta_{\mathrm{g}}}{\Delta_{\mathrm{B}}}\right)\right) + X_{\mathrm{B}}\left(\theta_{\mathrm{g}} - \theta\right) \tag{7.232}$$

$$X_{\mathrm{B}} = \begin{cases} X_{\mathrm{B}}^{\mathrm{g}} & \theta \le \theta_{\mathrm{g}} \\ 0 & \theta > \theta_{\mathrm{g}} \end{cases} \tag{7.233}$$

Parameters b and g characterizing hardening-softening behavior in intermolecular structure can be defined as

$$\begin{aligned} \mathrm{g} = {} & \frac{1}{2}\left(\mathrm{g}_{\mathrm{g}} + \mathrm{g}_{\mathrm{r}}\right) - \frac{1}{2}\left(\mathrm{g}_{\mathrm{g}} - \mathrm{g}_{\mathrm{r}}\right)\tanh\left(\frac{\theta - \left(\theta_{\mathrm{g}} + \theta^{\mathrm{g}}\right)}{\Delta^{\mathrm{g}}}\right) \\ & + X_{\mathrm{g}}\left[\theta - \left(\theta_{\mathrm{g}} + \theta^{\mathrm{g}}\right)\right] \end{aligned} \tag{7.234}$$

$$X_{\mathrm{g}} = \begin{cases} X_{\mathrm{g}}^{\mathrm{g}} & \theta \le \theta_{\mathrm{g}} + \theta^{\mathrm{g}} \\ 0 & \theta > \theta_{\mathrm{g}} + \theta^{\mathrm{g}} \end{cases} \tag{7.235}$$

$$b = b_1 \exp\left(b_2\theta\right)\left(\frac{\nu_{\mathrm{I}}^{\mathrm{p}}}{\nu_{\mathrm{ref}}^{\mathrm{p}}}\right)^{b_3} \tag{7.236}$$

Other parameters involved in material model $\left\{\nu_{\mathrm{I}}^{\mathrm{o}}, Q_{\mathrm{I}}, V, \alpha_{\mathrm{p}}, n_{\mathrm{I}}, h_{\mathrm{I}}, \gamma, \nu_{\mathrm{M}}^{\mathrm{o}}, Q_{\mathrm{M}}, h_{\mathrm{M}}, n_{\mathrm{M}}\right\}$ are constants.

7.6 Applications of Finite Deformation Models

For verification of constitutive model, isothermal and non-isothermal stretching of PMMA was performed. Simulation results were compared with test results in terms of stress-strain curves for isothermal tests and temperature-displacement-force histories for non-isothermal tests.

7.6.1 Material Properties

In order to simulate material response under isothermal and non-isothermal conditions, it is essential to determine appropriate material parameters. Isothermal tests on PMMA are required to obtain material parameters for the dual-mechanism viscoplastic model presented in earlier sections. PMMA response is highly sensitive to loading rate and temperature. Therefore experiments must be conducted at different loading rates and different temperatures. In addition, thermal properties of PMMA are also necessary for a fully coupled temperature-displacement finite element analysis. A significant part of mechanical material properties of PMMA $\left(E, \nu, I_{\mathrm{m}}, \nu_{\mathrm{I}}^{\mathrm{o}}, Q_{\mathrm{I}}, n_{\mathrm{I}}, B_{\mathrm{g}}, X_{\mathrm{B}}, V, \alpha_{\mathrm{p}}, \gamma\right)$ can be obtained by conducting isothermal tests at different temperatures and different strain (displacement) rates. Since some material parameters such as $h_{\mathrm{I}}, b, \mathrm{g}, \nu_{\mathrm{M}}^{\mathrm{o}}, Q_{\mathrm{M}}, h_{\mathrm{M}}, n_{\mathrm{M}}, \mu_{\mathrm{M}}, \phi^{*}, S_{\mathrm{M}}^{*}$ cannot be directly observed in experiments or their influence on material response cannot be isolated from others, these parameters can be obtained by statistical methods, only.

Rate dependence of glass transition temperature of PMMA is based on free volume theory of Williams et al. (1955), while WLF parameters $\left(c_1^{\mathrm{g}}, c_2^{\mathrm{g}}\right)$ in Eq. (7.224) are provided by Richeton et al. (2005a, b), and reference glass transition temperature is provided by Nie (2005). Accordingly, variation of glass transition temperature with frequency (rate) can be obtained as shown in Fig. 7.5.

Figure 7.6 shows the elastic modulus as a function of temperature for different strain rates, *H* series (loading rate of 0.9 mm/s), *M* series (0.09 mm/s), and *L* (0.009 mm/s).

Figure 7.6 indicates that temperature sensitivity of elastic modulus becomes smaller at slower loading rates. Parameters $\left(E_{\mathrm{g}}, E_{\mathrm{r}}, \theta_{\mathrm{E}}, \Delta_{\mathrm{E}}, X_{\mathrm{E}}^{\mathrm{g}}, X_{\mathrm{E}}^{\mathrm{r}}, s_{\mathrm{E}}\right)$ as a function of temperature and rate can be obtained from test data by statistical means. Temperature dependence of limited chain extensibility (I_{M}) can be obtained from fracture strain values in isothermal tests. Figure 7.7 shows the limited chain extensibility as a

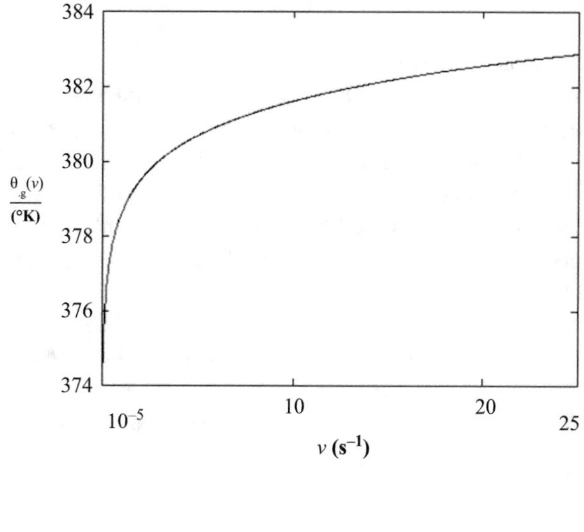

Fig. 7.5 Rate-dependent glass transition temperature of PMMA

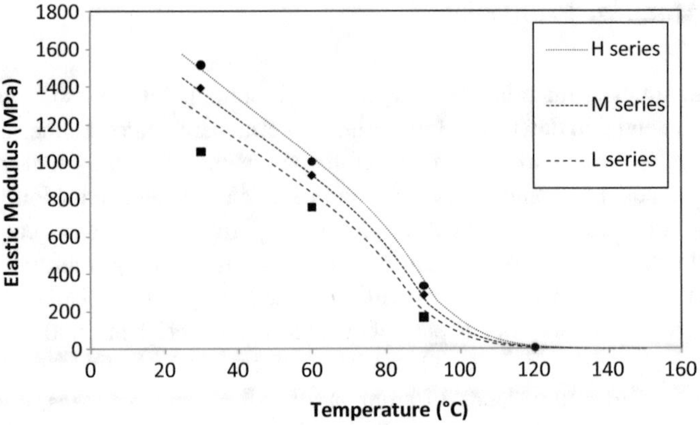

Fig. 7.6 Temperature and rate dependent elastic modulus of PMMA (Nie 2005)

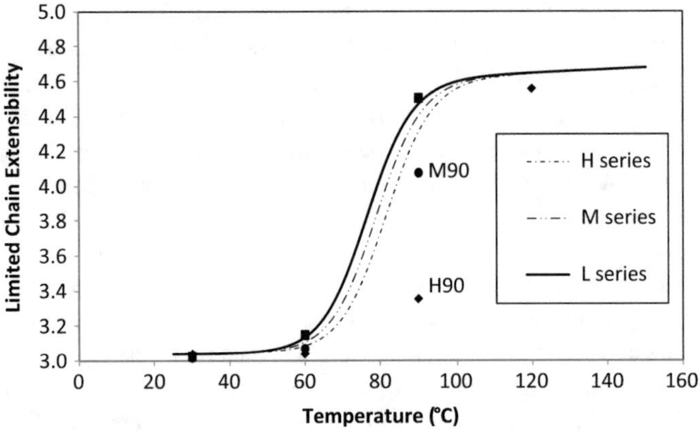

Fig. 7.7 Temperature- and rate-dependent limited chain extensibility of PMMA

function of temperature and strain rate. Clearly, two data points (H90 and M90 values) are outliers. Using test data, parameters $\left(I_{\mathrm{M}}^{\mathrm{g}}, I_{\mathrm{M}}^{\mathrm{r}}, X_{\mathrm{M}}^{\mathrm{g}}, X_{\mathrm{M}}^{\mathrm{r}}, \theta_{\mathrm{M}}, \Delta_{\mathrm{M}}\right)$ can be obtained as a function of temperature and loading rate.

Temperature variation of Poisson's ratio in glass-rubber transition region can be assumed to be identical to that of elastic modulus ($\theta_{\mathrm{E}}, \Delta_{\mathrm{E}}$), while glassy and rubbery values $\left(\overline{\nu}_{\mathrm{g}}, \overline{\nu}_{\mathrm{r}}\right)$ can be assumed to be constant. Poisson's ratio in glassy regime is given by Nie (2005), and Poisson's ratio in rubbery regime is assumed to be a value close to "0.5" to impose nearly incompressible conditions at high temperatures. Material parameters characterizing viscoplastic features of deformation associated with intermolecular structure $\left(\nu_{\mathrm{I}}^{\mathrm{o}}, Q_{\mathrm{I}}, n_{\mathrm{I}}, B_{\mathrm{g}}, X_{\mathrm{B}}, V, \alpha_{\mathrm{p}}, \gamma\right)$ can be determined by using ratio of yield stress to temperature ($\sigma_{\mathrm{y}}/\theta$) versus plastic strain rate (ν^{p}) plots (Eyring plots) obtained experimentally as presented in Fig. 7.8.

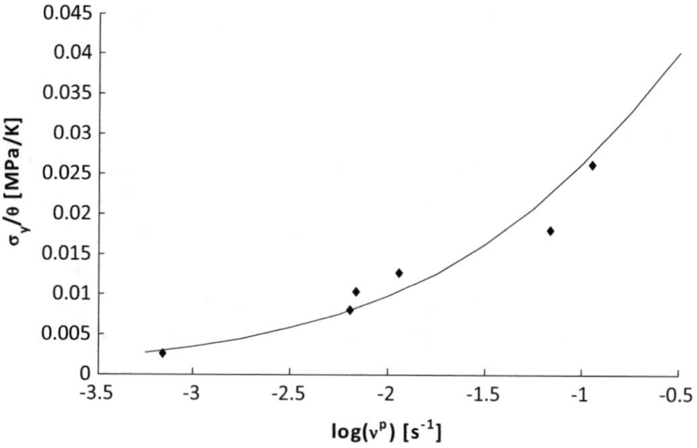

Fig. 7.8 Ratio of yield stress to temperature vs. plastic strain rate plot of PMMA (Nie 2005)

It is normally assumed that, when plastic flow occurs, plastic rate is equal to total strain rate. Since PMMA displays brittle failure under tension at low temperatures without pronounced yielding, data set for Eyring plots are limited. A master curve can be constructed by shifting data set horizontally and vertically to a reference temperature (θ^{ref}) as defined as

$$\Delta(\log(\nu_{\mathrm{I}}^{\mathrm{p}})) = H_{\mathrm{h}}\left(\frac{1}{\theta} - \frac{1}{\theta^{\mathrm{ref}}}\right) \tag{7.237}$$

$$\Delta\left(\frac{\sigma_{\mathrm{y}}}{\theta}\right) = H_{\mathrm{v}}\left(\frac{1}{\theta} - \frac{1}{\theta^{\mathrm{ref}}}\right) \tag{7.238}$$

According to Richeton et al. (2005a, b), if θ^{ref} is selected as θ_{g}, horizontal shift (H_{h}) and vertical shift (H_{v}) factors can be related to material properties through the following relation:

$$H_{\mathrm{h}} = \frac{Q_{\mathrm{I}}}{k_{\mathrm{B}}\ln(10)} \tag{7.239}$$

$$H_{\mathrm{v}} = B_{\mathrm{g}} - X_{\mathrm{B}}^{\mathrm{g}}\theta_{\mathrm{g}} = \frac{\gamma}{3} \tag{7.240}$$

where Q_{I} is the activation energy for plastic flow in intermolecular structure, k_{B} is Boltzmann's constant, B_{g} and $X_{\mathrm{B}}^{\mathrm{g}}$ are parameters for bulk modulus, and γ is the parameter characterizing dynamic recovery in hardening characteristics of intermolecular structure. Equation for master curve can be derived from plastic flow rule in intermolecular structure (Eq. (7.174)) as follows:

$$\frac{\sigma_y}{\theta_g} = \frac{2k_B}{V}\left(1 - \frac{\alpha_p}{3}\right)^{-1} \sinh^{-1}\left(\left(\frac{\nu_I^p}{\nu^*}\right)^{\frac{1}{n_I}}\right) \tag{7.241}$$

where n_I is the number of thermally activated transitions necessary for plastic flow, V is the activation volume, and α_p is the pressure sensitivity parameter. Temperature-dependent characteristic viscoplastic rate (ν^*) is defined in Eq. (7.175) which is derived from flow rule of the theory of plasticity defined in Eq. (7.174). Effective stress ($\overline{\tau}_I$) at θ_g can be approximated from Eq. (7.176) for one-dimensional case. In this example, it is assumed that back stress ($\mathbf{M}_{back}$) and plastic flow resistance in intermolecular structure (S_I) vanishes around glass transition temperature, whereas applied stress at yielding is equal to yield stress, and normal pressure is one third of applied stress:

$$\sigma_1 = \sigma_y \tag{7.242}$$

$$\mathbf{M}_{back} \cong \mathbf{0} \tag{7.243}$$

$$S_I \cong 0 \tag{7.244}$$

$$\tau_I \cong \sigma_1 = \sigma_y \tag{7.245}$$

$$p_I = \frac{1}{3}\sigma_1 = \frac{1}{3}\sigma_y \tag{7.246}$$

$$\overline{\tau}_I = \left(1 - \frac{\alpha_p}{3}\right)\sigma_y \tag{7.247}$$

Using regression analysis method for fitting master curve to experimental data with shift factors $H_h = 4900\,\mathrm{K}$ and $H_v = -40\mathrm{MPa}$ material parameters, ν_I^o, Q_I, n_I, B_g, X_B, V, α_p, and γ, can be calculated. Activation volume (V) and activation energy (Q_I) were assumed to be constant. Back stress modulus asymptotically approaches to zero around glass transition temperature. The remaining parameter in back stress modulus definition Δ_B which controls transition temperature range was selected as 5 °C to ensure a smooth change in hardening characteristics of material in non-isothermal simulations (Fig. 7.9).

Implementing 1-D version of the constitutive model in a program like MATLAB with isothermal conditions is an expedient way to determine the remaining parameters (h_I, b, g) in intermolecular structure and the parameters associated with molecular network resistance $\left(\nu_M^o, Q_M, h_M, n_M, \mu_M\right)$. These parameters cannot be directly observed in macroscale experiments. Hence, it is necessary to run simple 1-D simulations to estimate values for these parameters. Influence of parameters (h_I, b, g) which control hardening-softening characteristics of response on stress-strain curves and parameter $\left(\nu_M^o, Q_M, h_M, n_M, \mu_M\right)$ which control post-yield response are presented in Fig. 7.10a–h.

Arrows in Fig. 7.10a–h indicate increasing trend of corresponding parameter, while other parameters are held at constant value. Parameters from intermolecular network mechanism (b, g, h_I) clearly control strain hardening-softening behavior in

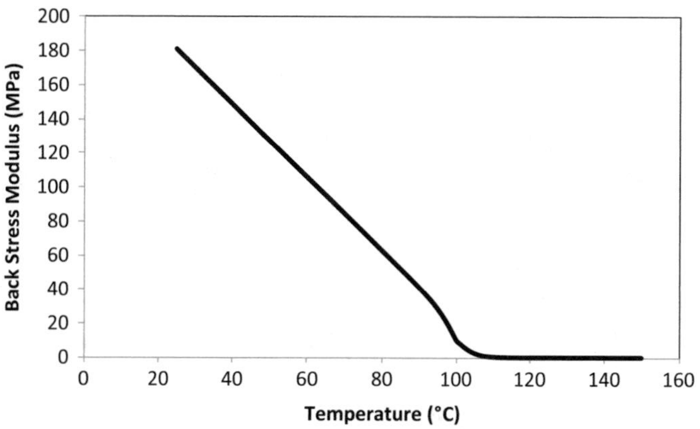

Fig. 7.9 Temperature-dependent back stress modulus of PMMA

yield region, whereas parameters from molecular network mechanism $(h_M, \mu_M, \nu_M^p, Q_M, n_M)$ control post-yield behavior at large deformations. 1-D MATLAB® simulations allow us to observe influence of parameters on different aspects of stress-strain curves and achieve an overall acceptable curve fitting to stress-strain curves from isothermal test data. When parameters (b, g, h_I) are determined, b is assumed to be both viscoplastic strain rate and temperature dependent, g is assumed to be temperature dependent, and h_I is assumed as a constant. Parameters for molecular network (h_M, ν_M^p, Q_M, n_M) are assumed to be constant except for rubbery modulus which is taken as both temperature and rate dependent. Critical value for parameter (ϕ^*) and saturation value for network resistance $\left(S_M^*\right)$ are obtained from test data of Ames et al. (2009) [33]. Initial values for parameter (ϕ) and intermolecular resistance to plastic flow (S_I) are usually assumed to be zero (while molecular network resistance to plastic flow (S_M) assumed to be 10% of saturation value $\left(S_M^*\right)$). Complete list of material parameters included in constitutive model are presented in Table 7.1.

$$\phi(\mathbf{r}, 0) = 0, S_I(\mathbf{r}, 0) = 0, S_M(\mathbf{r}, 0) = 0.1 S_M^*(\theta_o), \theta_o = \theta(\mathbf{r}, 0)$$

7.7 Numerical Implementation of Dual-Mechanism Viscoplastic Model

Dual-mechanism viscoplastic model is implemented numerically based on staggered method with isothermal split. In each load step, temperature value at the end of time increment is taken as constant, and after mechanical equilibrium satisfied, thermal equations is solved under fixed configuration to update temperature increment in the following time increment. Since this scheme is only conditionally stable, different

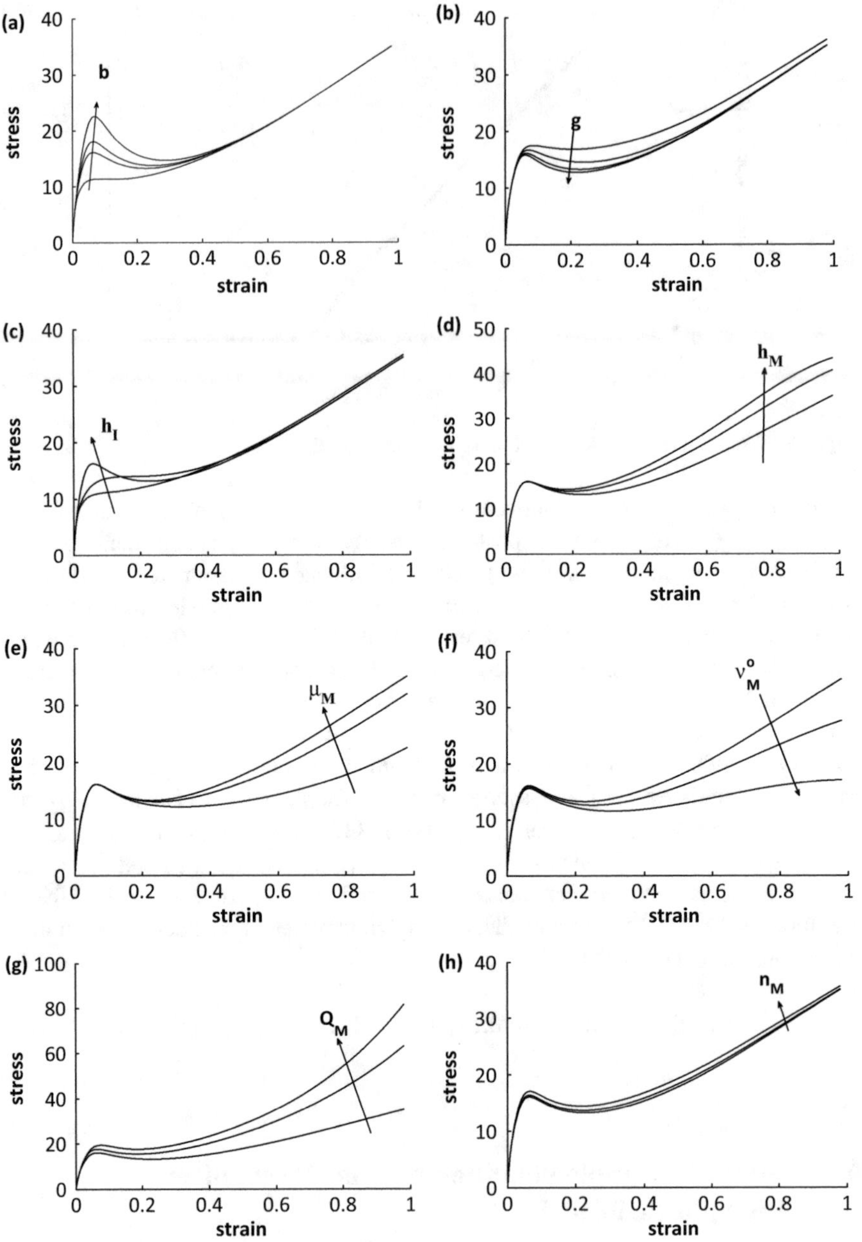

Fig. 7.10 Influence of material parameters $(b, g, h_I, h_M, \mu_M, \nu_M^o, Q_M, n_M)$ on stress-strain curves of PMMA (Gunel and Basaran 2010)

Table 7.1 Material parameters for PMMA constitutive model (Gunel 2010)

Parameters		Parameters		Parameters	
θ_g^{ref} (K)	**373**	ϕ_g^*	**0.001**	g_r	**9.55**
c_1^g(K)	**32.58**	ϕ_g^*	**0**	$\Delta\theta^g$ (K)	**5**
c_2^g	**87.5**	θ_ϕ(K)	**0**	θ^g (K)	**5**
c_1(K)	**9**	Δ_ϕ(K)	**10**	$X_g^g\ (K^{-1})$	**0.07**
c_2	**116**	$X_\phi^g(1/K)$	**-4.5×10^{-6}**	b_1(MPa)	**2.72×10^{10}**
$\nu^{ref}(s^{-1})$	**1.73×10^{-2}**	$X_\phi^r(1/K)$	**0**	b_2(1/K)	**-4.58×10^{-2}**
E_g(MPa)	**500**	I_M^g	**7.04**	b_3	**4.04×10^{-2}**
E_r (MPa)	**1.5**	I_M^r	**4.6**	$\nu_{ref}^p(s^{-1})$	**0.001**
θ_E(K)	**−8**	θ_M(K)	**−20**	$\nu_I^o(s^{-1})$	**2.43×10^{12}**
Δ_E(K)	**15**	Δ_M(K)	**12**	Q_I(J/K)	**1.56×10^{-19}**
s_E	**0.06**	$X_M^g(1/K)$	**0**	$V(m^3)$	**1.39×10^{-27}**
X_E^g(MPa/K)	**−15.05**	$X_M^r(1/K)$	**0.001**	α_p	**0.21**
X_E^r(MPa/K)	**0**	S_M^g(MPa)	**35**	n_I	**2.17**
$\bar{\nu}_g$	**0.31**	S_M^r(MPa)	**0.2**	h_I	**40.42**
$\bar{\nu}_r$	**0.49**	θ_S(K)	**5**	γ(MPa)	**60**
μ_M^g(MPa)	**9**	Δ_S(K)	**5**	$\nu_M^o(s^{-1})$	**4.33×10^9**
μ_M^r(MPa)	**0.5**	X_S^g(MPa/K)	**−0.01**	Q_M(J/K)	**1.30×10^{-19}**
θ_μ(K)	**5**	X_S^r(MPa/K)	**0**	h_M	**14.43**
Δ_μ(K)	**14**	B_g (MPa)	**10**	n_M	**5**
s_μ	**0.03**	X_B^g (MPa/K)	**2.15**	m_s(g/mol)	**100.13**
X_μ^g(MPa/K)	**−0.4**	$\Delta\theta_B$ (K)	**5**		
X_μ^r(MPa/K)	**0**	g_g	**7.97**		

time increments are used for isothermal cases at different temperatures, and a suitable time increment is chosen based on convergence study. Objectivity (frame indifference) was ensured by using stress components and their conjugate rate components in reference configuration (in material coordinates). Final forms of constitutive equations are obtained after a transformation of evolution equations to reference configuration. Evolution of elastic and plastic deformation gradients is approximated by exponential operator (Weber and Anand 1990). When exponential operator is used in combination with backward Euler scheme, plastic incompressibility and symmetry of state variable tensors are conserved with sufficient accuracy by only including the first two terms of series representation of exponential function. Rate dependency of some material properties (E, μ_M, θ_g) are implemented in numerical algorithm such that property values are calculated based on strain rates in previous increment and strain rates are updated based on initial and final deformation gradients at the end of time increment. Therefore, equilibrium is satisfied at discrete time increments assuring thermodynamic consistency.

Numerical integration of evolution equations of plastic deformation gradients $(\mathbf{F}_{\mathrm{I}}^{\mathrm{p}}, \mathbf{F}_{\mathrm{M}}^{\mathrm{p}})$ and stretch-like internal variable $(\mathbf{A})$ based on exponential mapping is obtained from Eqs. (7.172), (7.196), and (7.166):

$$(\mathbf{F}_{\mathrm{I}}^{\mathrm{p}})_{n+1} = \exp\left(\Delta t(\mathbf{D}_{\mathrm{I}}^{\mathrm{p}})_{n+1}\right)(\mathbf{F}_{\mathrm{I}}^{\mathrm{p}})_n \tag{7.248}$$

$$(\mathbf{F}_{\mathrm{M}}^{\mathrm{p}})_{n+1} = \exp\left(\Delta t(\mathbf{D}_{\mathrm{M}}^{\mathrm{p}})_{n+1}\right)(\mathbf{F}_{\mathrm{M}}^{\mathrm{p}})_n \tag{7.249}$$

$$\begin{aligned}\mathbf{A}_{n+1} = {} & \mathbf{A}_n \exp\left(\Delta t(\mathbf{D}_{\mathrm{I}}^{\mathrm{p}})_{n+1}\right) + \exp\left(\Delta t(\mathbf{D}_{\mathrm{I}}^{\mathrm{p}})_{n+1}\right)\mathbf{A}_n - \gamma \mathbf{A}_n \ln(\mathbf{A}_n) \\ & \times \frac{\left|\exp\left(\Delta t(\mathbf{D}_{\mathrm{I}}^{\mathrm{p}})_{n+1}\right)\right|}{\sqrt{2}}\end{aligned} \tag{7.250}$$

where $\Delta t = t_{n+1} - t_n$ is the time increment, γ is a constant representing dynamic recovery associated with kinematic hardening, $\nu_{\mathrm{I}}^{\mathrm{p}}$ and $\nu_{\mathrm{M}}^{\mathrm{p}}$ are the equivalent plastic stretch rates in intermolecular and molecular network structure, and $\mathbf{D}_{\mathrm{I}}^{\mathrm{p}}$ and $\mathbf{D}_{\mathrm{M}}^{\mathrm{p}}$ are the plastic stretch rate tensors in intermolecular structure and network structure, respectively. Owing to unconditional stability of implicit time integration schemes, classical backward Euler method is preferred for integration of evaluation equations of internal state variables, $\dot{\varsigma} = f(\varsigma, \theta)$.

For a generalized internal state variable (ς), numerical integration of evolution equation can be performed as

$$\varsigma_{n+1} = \Delta t f(\varsigma_{n+1}, \theta_{n+1}) + \varsigma_n \tag{7.251}$$

Further details on implementation of dual-mechanism viscoplastic model are given by Gunel and Basaran (2010).

7.7.1 *Simulating Isothermal Stretching of PMMA*

Fully coupled temperature-displacement analysis is performed to investigate adiabatic effect in stretching of PMMA. Eight-node linear brick elements (C3D8T) are used as element type, and convergence studies on different mesh sizes and time steps were conducted. In Fig. 7.11, influence of time increments on convergence of different aspects of material response is presented. In these simulations, a rectangular prism model was uniaxially stretched at a displacement rate of 1 mm/s for 50 s, while temperature was kept constant at 90 °C. In convergence studies for time increment, maximum axial stress (σ), equivalent viscoplastic strain rate $(\nu_{\mathrm{I}}^{\mathrm{p}}, \nu_{\mathrm{M}}^{\mathrm{p}})$, and equivalent stretch rate (d) must be monitored.

True stress-strain curves presented in Fig. 7.11.a indicate a fast convergence even for largest time increment of “0.1 s.” Convergence of equivalent plastic strain rates $(\nu_{\mathrm{I}}^{\mathrm{p}}, \nu_{\mathrm{M}}^{\mathrm{p}})$ requires a small time increment which also indicates a slow convergence of internal state variables $(S_{\mathrm{I}}, S_{\mathrm{M}}, \phi \ldots)$. Another interesting point that can be observed

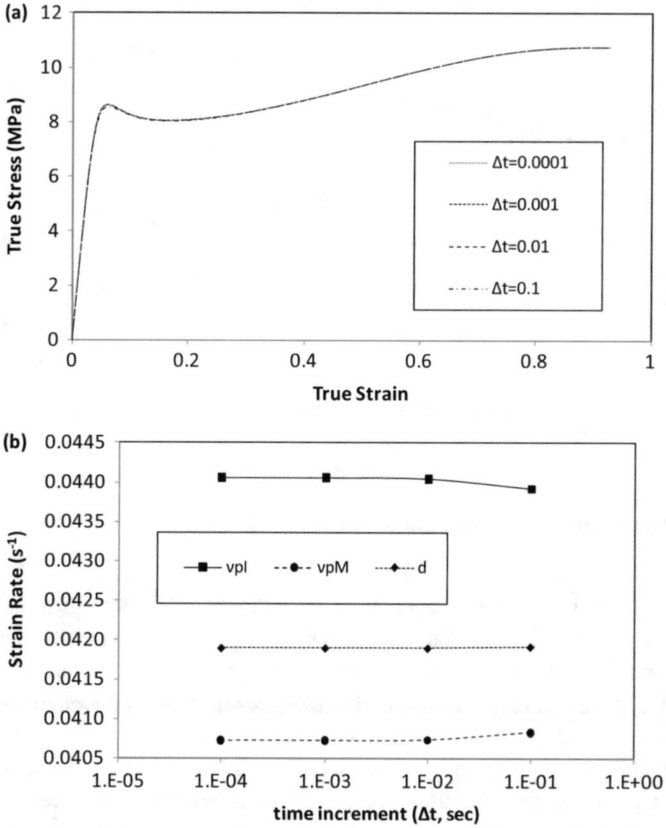

Fig. 7.11 Convergence study results for time increment in viscoplastic model in terms of (**a**) true stress-strain curves and (**b**) strain rates

from Fig. 7.11b is that equivalent viscoplastic strain rate in intermolecular network $(\nu_{\mathrm{I}}^{\mathrm{p}})$ is larger than equivalent stretch rate (d) which might seem like an error. However, this is due to the numerical method that is used to capture strain-softening characteristics and controlled by the parameters b and g in the constitutive model.

In Fig. 7.12, true stress, equivalent viscoplastic strain rates $(\nu_{\mathrm{I}}^{\mathrm{p}}, \nu_{\mathrm{M}}^{\mathrm{p}})$, and equivalent stretch rate (d) histories are presented. At the beginning of simulation, equivalent stretch rate (d) is larger than equivalent plastic strain rate in both mechanisms $(\nu_{\mathrm{I}}^{\mathrm{p}}, \nu_{\mathrm{M}}^{\mathrm{p}})$, and continuous increase in elastic strain in both mechanisms $\left(F_{\mathrm{I}}^{\mathrm{e}}, F_{\mathrm{M}}^{\mathrm{e}}\right)$ causes increase in stress level up to onset of yielding at which equivalent viscoplastic strain rate in intermolecular mechanism $(\nu_{\mathrm{I}}^{\mathrm{p}})$ becomes larger than equivalent stretch rate (d) and remains larger over some period which corresponds to yielding and strain-softening region. During this period, negative elastic strain rate results in decrease in elastic strain in intermolecular mechanism $\left(\mathbf{F}_{\mathrm{I}}^{\mathrm{e}}\right)$ and hence decrease in stress level $(\mathbf{T}_{\mathrm{I}})$. After strain softening, equivalent viscoplastic strain rate in intermolecular mechanism $(\nu_{\mathrm{I}}^{\mathrm{p}})$ remains equal to equivalent stretch rate (d). In the

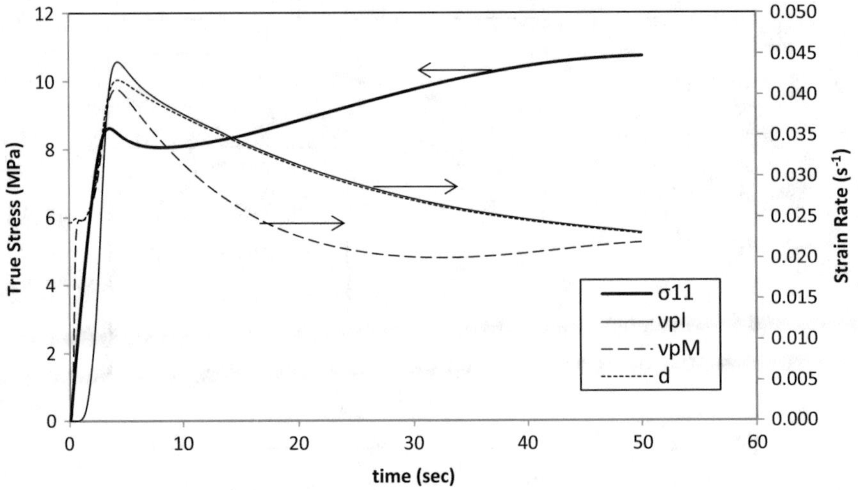

Fig. 7.12 Comparison of true stress and strain rate histories

meantime, difference between equivalent viscoplastic strain rate in molecular network ($\nu_{\mathrm{M}}^{\mathrm{p}}$) and equivalent stretch rate (d) starts to increase which leads to increase in elastic strain in molecular network $\left(\mathbf{F}_{\mathrm{M}}^{\mathrm{e}}\right)$. As a result, stress level in molecular network ($\mathbf{T}_{\mathrm{M}}$) also increases which describes strain-hardening characteristics of material due to chain locking in molecular network. However, further stretch in molecular network continuously increases stress level in molecular network ($\mathbf{T}_{\mathrm{M}}$) causing increase in driving stress for plastic flow (τ_{M}) which in return starts to increase equivalent viscoplastic strain rate in molecular network ($\nu_{\mathrm{M}}^{\mathrm{p}}$) once again. Therefore, change in stress level at large deformations gradually becomes smaller.

For structural models, mesh sensitivity of results is also very important in finite element analysis. In Fig. 7.13, the influence of size of mesh seeds (in ABAQUS) on convergence of different aspects of material response is presented. A rectangular prism sample was uniaxially stretched at a displacement rate of 1 mm/s for 50 s while temperature was kept constant at 100 °C. In mesh sensitivity studies, axial stress (σ) history, equivalent viscoplastic strain rate ($\nu_{\mathrm{I}}^{\mathrm{p}}, \nu_{\mathrm{M}}^{\mathrm{p}}$), and equivalent stretch rate (d) are monitored.

In Fig. 7.13a, number of elements for different mesh sizes is presented. Seed size encircled in Fig. 7.13a is used in actual simulations for PMMA since sufficient convergence is achieved at this mesh size for true stress value (σ) and equivalent plastic strain rate values ($\nu_{\mathrm{I}}^{\mathrm{p}}, \nu_{\mathrm{M}}^{\mathrm{p}}$) as shown in Fig. 7.13b, c, respectively. A coarser mesh than the selected one (seed size $=$ 3) would also be also reasonable for convergence of (true) stress values except for the coarsest one (seed size $=$ 10). However, a finer mesh is always desirable for convergence of equivalent strain rate values and associated state variables (Figs. 7.13c).

Results shown in Fig. 7.14 indicate that, simulations and experiment measurements are reasonably in good agreement. Constitutive model is especially successful

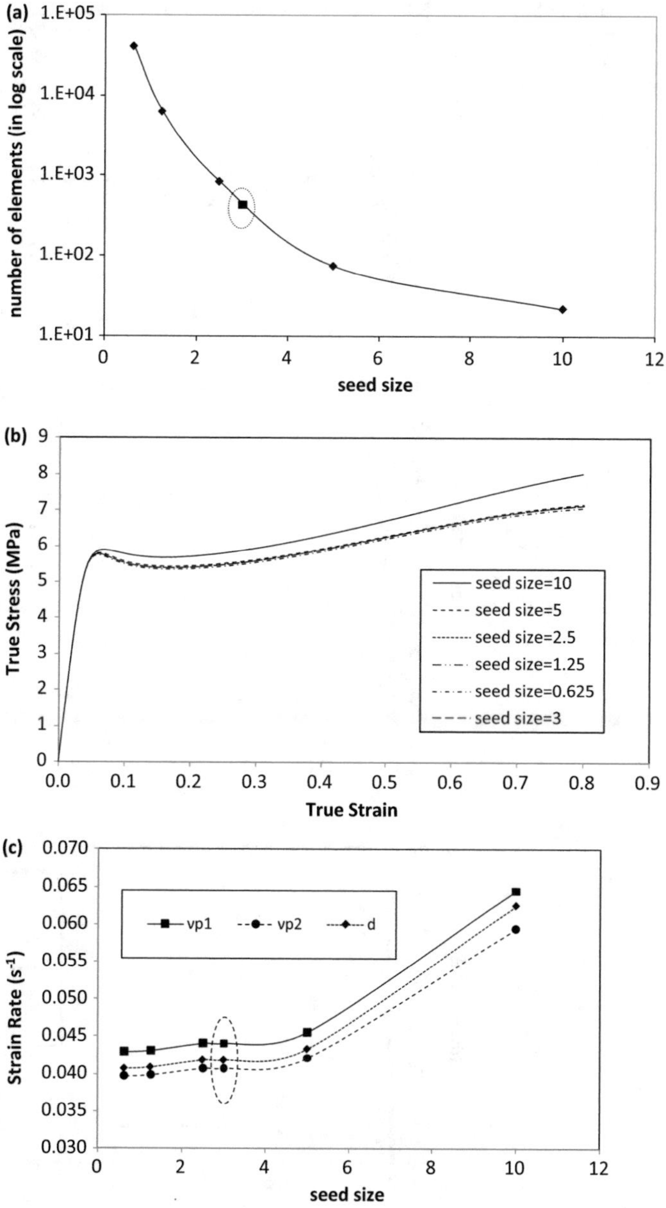

Fig. 7.13 Mesh sensitivity of ABAQUS model in terms of (**b**) true stress-strain curves and (**c**) strain rates (**a**) for different mesh densities

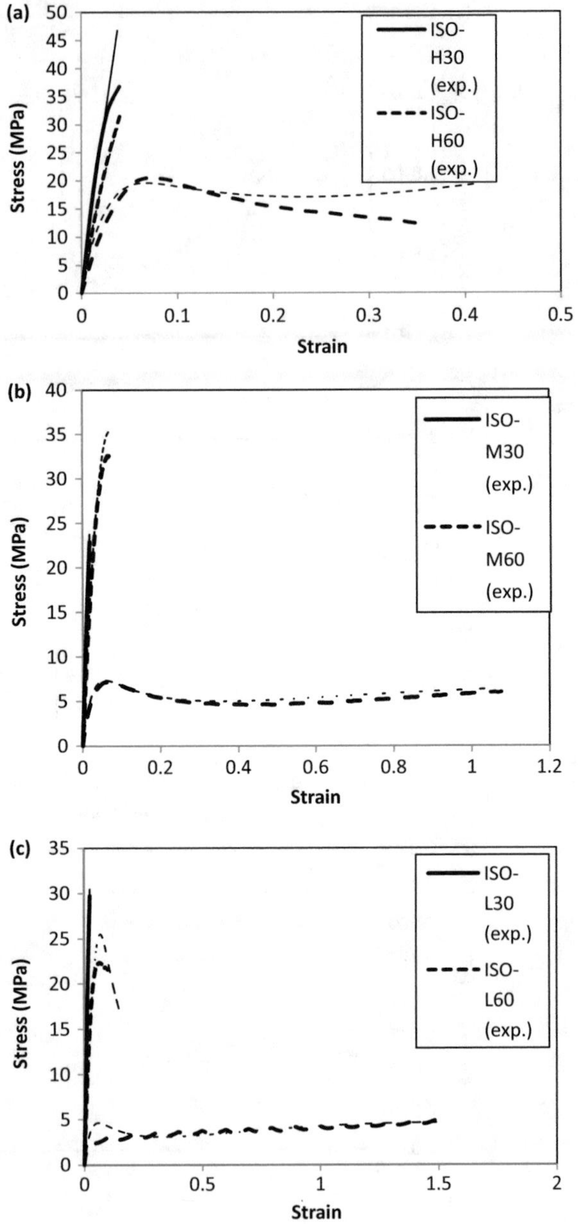

Fig. 7.14 Comparison of ABAQUS simulation and experimental results for isothermal testing of PMMA in (**a**) *H* series, (**b**) *M* series, and (**c**) *L* series

in describing temperature and rate dependence of material response as well as strain-softening and strain-hardening characteristics. Due to limited size of data set used for determination of viscoplastic flow rule for intermolecular mechanism, it is really difficult to determine material parameters involved in viscoplastic model which result in some minor differences in response.

Dual-mechanism viscoplastic model for PMMA is verified and validated for more challenging non-isothermal conditions as presented in the following section. Entropy production values are calculated as average of cumulative entropy production of all Gauss integration points at the cross section of center of narrow section of test sample. Difference between mean entropy production and maximum entropy production in the cross section is less than 5% for all isothermal test simulations which indicates that no damage localization is predicted at the cross section which is similar to case of brittle failure of materials. As brittle failure takes place in the form of sudden rupture of specimen with concurrent crack formation and propagation at many locations, damage localization is not observed contrary to ductile failure of materials for which damage localization is an essential part of crack initiation stage around microscopic defect regions while crack propagation takes place over a longer period of time compared to brittle failure. Therefore, damage quantification based on average entropy production in the cross section is an proper method for brittle failure, while local thermodynamic state index is essential for damage quantification in ductile failure. It is also important to note that damage evolution and material degradation due to mechanical load is assumed to result from only plastic energy dissipation excluding all other entropy generation mechanisms. Therefore, in order to fully account for evolution of microscopic defects that lead to stiffness and strength degradation in material, it is necessary to include all entropy generation terms associated with energy loss associated with new surface creation at microscopic defect regions, heat generation, and chemical reactions, if there are any. The latter approach would be more accurate. It is possible to account for difference between failure in tension and failure in compression or predict damage evolution in the case of ductile failure, by deriving a more comprehensive fundamental equation.

Total irreversible thermal entropy production in ISO-TEST simulations ($S_{\text{irr, ther, test}}$) is completely created by transfer of heat generated due to plastic dissipation during isothermal stretching. Temperature rise due to plastic dissipation in isothermal stretching is very small; hence, thermal gradients within specimen during testing is also small, as a result producing negligible amount of irreversible entropy production in comparison with $S_{\text{irr, ther, pre}-\text{test}}$ and total irreversible mechanical entropy production during isothermal tests ($\text{S}_{\text{irr, mech, test}}$).

Mechanical damage in materials can be described as formation of micro-cracks and voids at micro-level with corresponding degradation of stiffness at macro level. On the other hand, thermal damage in materials is deterioration of material properties by thermal gradients which weaken chemical bonds between molecules or change the length of the molecular bond. Evolution of thermal damage is much slower than evolution of mechanical damage. Remarkable thermal gradients over long periods of time are necessary to induce permanent cracks and voids in materials, e.g., freezing-thawing cycles, thermal shock, etc. Therefore, mechanical and thermal damage induced in a material have different sources, and damage evolution in both types must be defined separately in the case of thermo-mechanical loading. As failure in isothermal stretching of PMMA is completely induced by mechanical loading, damage evolution must be only based on evolution of irreversible mechanical entropy production. In experimental studies on metal fatigue fracture, it is found that cumulative entropy generation was constant at the time of failure (Fatigue Fracture Entropy) which was independent of structural geometry, load intensity, and loading frequency. It is observed by many researchers that rate dependence of critical entropy density is negligible while critical entropy density significantly increases with increasing temperature because of increasing plastic dissipation prior to failure. Mechanical damage evolution is completely based on irreversible mechanical entropy production, i.e., intrinsic dissipation; temperature variation of plastic dissipation is also observed in critical entropy production values, as well. Logarithm of critical entropy value $S_{\text{irr, cr}}$ as a function of temperature is given by Gunel and Basaran (2010): Crtical entropy is also referred to as Fattigue Fracture Entropy (FFE) in the literature. It is shown to be a unique material property.

$$\log\left(S_{\text{irr,cr}}\right) = 0.04050\theta - 2.5653 \tag{7.252}$$

7.7.2 *Simulating Non-isothermal Stretching of PMMA*

Dual-mechanism viscoplastic model is also applied to non-isothermal stretching of PMMA to large deformations with specific emphasis on transition of material response around glass transition. In non-isothermal mechanical tests on PMMA, temperature of samples spanned over $\theta_g \pm 50^\circ\mathrm{C}$ range during which a complete glass transition is bound to take place, and any change in material response due to this large temperature difference could be monitored. It is well-known that for some polymers viscoplastic behavior right around glass transition temperature exhibits significant change. This change is manifested as a significant increase in stiffness during cooling of specimen. Contrary to other polymer models here it is shown that PMMA does not experience any abrupt changes in response during stretching under non-isothermal conditions, and response is continuous and smooth.

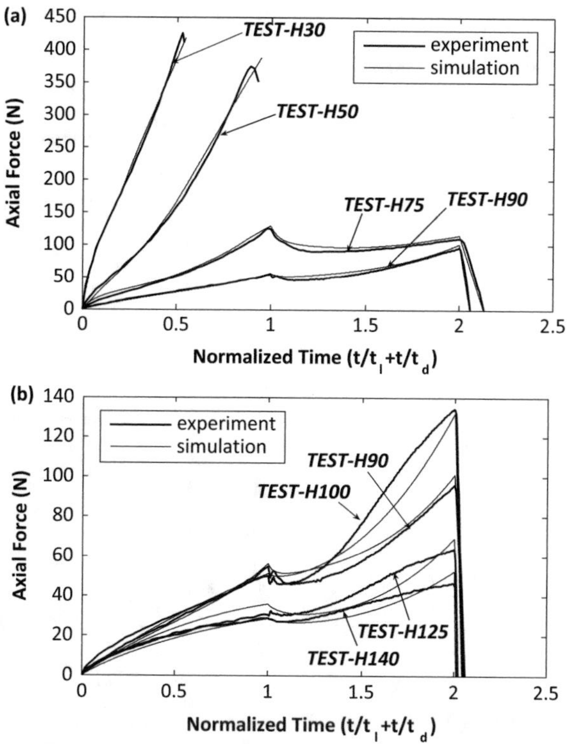

Fig. 7.15 Axial force histories in non-isothermal experiments and simulations of PMMA for H series (0.9 mm/s forming rate)

It is important to point out that material constitutive models for finite-strain behavior of amorphous polymers can only be verified in non-isothermal loading conditions.

Concurrent heat transfer with axial stretching of PMMA requires an absolutely fully coupled temperature-displacement analysis.

Axial force histories of H, M and L series with respect to normalized time (t_n) are presented in Figs. 7.15, 7.16 and 7.17 Simulation and experiment results in terms of axial force histories are reasonably in good agreement. Unified formulation of plastic flow rule for intermolecular structure and appropriate material property definitions ensure smooth transition of response around θ_g which is bound to take place at some stage of all non-isothermal simulations.

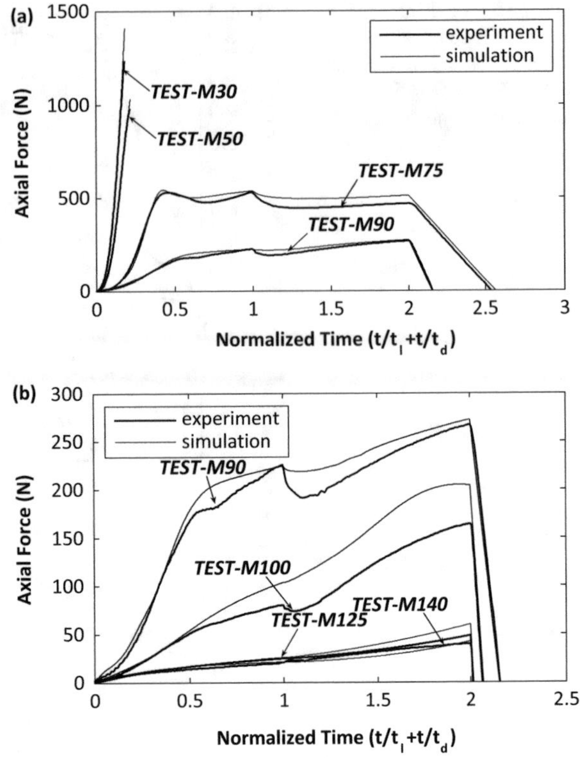

Fig. 7.16 Axial force histories in non-isothermal experiments and simulations of PMMA for M series (0.09 mm/s forming rate)

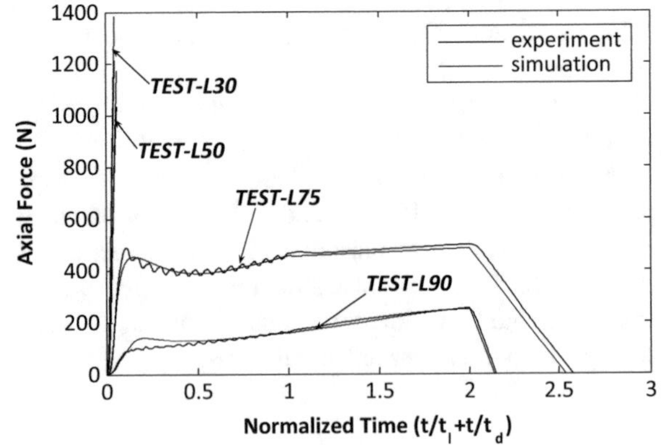

Fig. 7.17 Axial force histories in non-isothermal experiments and simulations of PMMA for L series (0.009 mm/s forming rate)

References

Ames, N. M., et al. (2009). A thermo-mechanically coupled theory for large deformations of amorphous polymers. Part II: Applications. *International Journal of Plasticity, 25*(8), 1495–1539.

Anand, L. (1986). Moderate deformations in extension-torsion of incompressible isotropic elastic materials. *Journal of the Mechanics and Physics of Solids, 34*, 293–304.

Anand, L., & On, H. (1979). Hencky's approximate strain-energy function for moderate deformations. *Journal of Applied Mechanics, 46*, 78–82.

Anand, L., et al. (2009). A thermo-mechanically coupled theory for large deformations of amorphous polymers. Part I: Formulation. *International Journal of Plasticity, 25*(8), 1474–1494.

Arruda, E. M., & Boyce, M. C. (1993). Evolution of plastic anisotropy in amorphous polymers during finite straining. *International Journal of Plasticity, 9*(6), 697–720.

Arruda, E. M., Boyce, M. C., & Jayachandran, R. (1995). Effects of strain rate, temperature and thermomechanical coupling on the finite strain deformation of glassy polymers. *Mechanics of Materials, 19*(2–3), 193–212.

Bergström, J. S., & Boyce, M. C. (1998). Constitutive modeling of the large strain time-dependent behavior of elastomers. *Journal of the Mechanics and Physics of Solids, 46*(5), 931–954.

Boyce, M. C., Socrate, S., & Llana, P. G. (2000). Constitutive model for the finite deformation stress-strain behavior of poly(ethylene terephthalate) above the glass transition. *Polymer, 41*(6), 2183–2201.

Fotheringham, D. G., & Cherry, B. W. (1978). The role of recovery forces in the deformation of linear polyethylene. *Journal of Materials Science, 13*(5), 951–964.

Fotheringham, D., Cherry, B. W., & Bauwens-Crowet, C. (1976). Comment on "the compression yield behaviour of polymethyl methacrylate over a wide range of temperatures and strain-rates". *Journal of Materials Science, 11*(7), 1368–1371.

Francisco, P., Gustavo, S., & Élida, B. H. (1996). Temperature and strain rate dependence of the tensile yield stress of PVC. *Journal of Applied Polymer Science, 61*(1), 109–117.

Gent, A. N. (1996). A new constitutive relation for rubber. *Rubber Chemistry and Technology, 69* (1), 59–61.

Gunel, M. E., (2010) *Large deformation micromechanics of particle filled acrylics at elevated temperatures*. PhD Dissertation submitted to University at Buffalo, SUNY.

Gunel, E. M., & Basaran, C. (2009). Micro-deformation mechanisms in thermoformed alumina trihydrate reinforced poly(methyl methacrylate). *Materials Science and Engineering: A, 523* (1–2), 160–172.

Gunel, E. M., & Basaran, C. (2010). Stress whitening quantification of thermoformed mineral filled acrylics. *Journal of Engineering Materials and Technology, 132*(3), 031002.

Gunel, E. M., & Basaran, C. (2011a). Damage characterization in non-isothermal stretching of acrylics. Part II: Experimental validation. *Mechanics of Materials, 43*(12), 992–1012.

Gunel, E. M., & Basaran, C. (2011b). Damage characterization in non-isothermal stretching of acrylics. Part I: Theory. *Mechanics of Materials, 43*(12), 979–991.

Mandel, J. (1972). *Plasticite classique et viscoplasticite (Lecture Notes)*. Udine: International Center for Mechanical Sciences.

Nie, S. (2005). *A micromechanical study of the damage mechanics of acrylic particulate composites under thermomechanical loading. (PhD Dissertation), in Civil, Structural and Environmental Engineering*. Buffalo: State University of New York at Buffalo.

Palm, G., Dupaix, R. B., & Castro, J. (2006). Large strain mechanical behavior of poly(methyl methacrylate) (PMMA) near the glass transition temperature. *Journal of Engineering Materials and Technology, 128*(4), 559–563.

Richeton, J., et al. (2005a). A formulation of the cooperative model for the yield stress of amorphous polymers for a wide range of strain rates and temperatures. *Polymer, 46*(16), 6035–6043.

Richeton, J., et al. (2005b). A unified model for stiffness modulus of amorphous polymers across transition temperatures and strain rates. *Polymer, 46*(19), 8194–8201.

Richeton, J., et al. (2006). Influence of temperature and strain rate on the mechanical behavior of three amorphous polymers: Characterization and modeling of the compressive yield stress. *International Journal of Solids and Structures, 43*(7–8), 2318–2335.

Richeton, J., et al. (2007). Modeling and validation of the large deformation inelastic response of amorphous polymers over a wide range of temperatures and strain rates. *International Journal of Solids and Structures, 44*(24), 7938–7954.

Srivastava, V., & Anand, L. (2010). A thermo-mechanically-coupled large-deformation theory for amorphous polymers in a temperature range which spans their glass transition. *International Journal of Plasticity, 26*(8), 1138–1182.

Weber, G. and Anand, L. (1990) Finite deformation constitutive equation and a time integration procedure for isotropic, Hyperelastic-Viscoplastic solids. *Computer Methods in Applied Mechanics and Engineering, 79*, 173–202.

Williams, M. L., Landel, R. F., & Ferry, J. D. (1955). The temperature dependence of relaxation mechanisms in amorphous polymers and other glass-forming liquids. *Journal of the American Chemical Society, 77*(14), 3701–3707.

Chapter 8
Unified Mechanics of Metals under High Electrical Current Density: Electromigration and Thermomigration

8.1 Introduction

Electromigration and thermomigration are in principal irreversible mass diffusion mechanisms under high current density and high temperature gradient, respectively. However, thermomigration can take place alone in the absence of electrical current, while electromigration cannot happen without thermomigration due to Joule heating, except for special circumstances.

8.2 Physics of Electromigration Process

When a solid conductor (metal) is subjected to an electrical potential gradient, the current enters from the anode side and travels to the cathode side, and the electrons travel from the cathode to the anode side. Electromigration is a mass diffusion-controlled phenomenon. When a conductor is subject to a high current density, the so-called electron wind transfer part of the momentum to the atoms (or ions) to make the atoms (or ions) moves in the direction of the current. As a result, the degradation of the conductor occurs mainly in two forms; in the anode side, the atoms will accumulate and finally form hillocks, and the vacancy concentration in the cathode side will form voids. Both hillocks and voids will cause the degradation of the material and eventual failure. The damage evolution due to electromigration can be modeled as an irreversible mass transport process. The purpose of this chapter is to present the formulation of modeling the electromigration- and thermomigration-induced material degradation process using the unified mechanics theory.

The physical mechanism of electromigration has been extensively investigated. Black (1969) established the relationship between the mean time to failure and current density for confined thin films on a substrate. The experiments by Blech (1976) revealed that the stress gradient could act as a counterforce to

C. Basaran, *Introduction to Unified Mechanics Theory with Applications*,
https://doi.org/10.1007/978-3-030-57772-8_8

electromigration under high current density. In addition, for thin films, Blech (1976) proposed a length scale called "Blech's critical length" below which mass diffusion due to electron wind force will be totally counterbalanced by stress gradient driving force in the opposite direction.

In order to solve an arbitrary boundary value problem with irregular boundaries or initial value problem with arbitrary conditions and composite material properties involving electromigration, there must be a rational mechanics-based current density- strain constitutive relation and a thermodynamics-based fundamental equation. Then this formulation can be implemented in a finite element method (or any other computational mechanics procedure) to solve any problem.

The classical definition of electromigration refers to the structural damage caused by ion transport in metal because of high current density. Electromigration is usually insignificant at low current density levels. Quantification of what is "high current density" is studied extensively by Ye et al. (2003a, b, c, d, e, f), Ye et al. (2004a, b, c, d), Ye et al. (2006) and Basaran et al. (2003). Electromigration is a mass transport in a diffusion-controlled process under certain driving forces. The driving force here is more complicated than what is involved in a pure diffusion process, in which the concentration gradient of the moving species is the only component. The electrical driving forces for electromigration consist of the electron wind force and the direct field force. The electron wind force refers to the effect of momentum exchange between the moving electrons and the ionic atoms when an electrical charge—a direct field force—is applied to a conductor with metallic bonds. When current density, which is proportional to the electron flux density, is high, enough, this momentum exchange effect becomes significant, resulting in a noticeable mass transport referred to as electromigration. Low melting point alloys when used at elevated homologous temperature ($>0.5T_{\mathrm{melt}}$ Kelvin) are prone to have considerable atomic diffusivity. For example, solder joints in microelectronics packaging are prime examples of this category.

8.2.1 *Driving Forces of Electromigration Process*

8.2.1.1 The Electric Field

Electrical current forces include two driving mechanisms: one is attributed to the electron wind forces, as originally suggested by Skaupy (1914), which refers to the effect of the momentum exchange between the moving electrons and the ionic atoms. This momentum exchange happens because of the scattering of free valence electrons. Scattered electrons collide with metal atoms and push them in the direction of electron flow (Seith 1955; Seith and Wever 1953; Wever and Seith 1955). On the other hand, the atoms move in the opposite direction of the applied electric field when they are ionized; the latter mechanism, the static force of the electric field, is considered the direct field force. The net effect of these two forces is the so-called electrical field driving force of the electromigration.

8.2.1.2 Stress Gradient

When atoms diffuse from the cathode side to the anode side, they leave behind vacancies on the cathode side and mass accumulation on the anode side. As a result there is tension on the cathode side and compressive stress on the anode side. This stress differential is responsible for the stress gradient.

Of course, mechanical stress gradient influences the electromigration process. There is an interaction between the stress gradient and the electromigration driving forces. Stress gradient can counteract or enhance the electromigration process, depending upon the interaction of all diffusion driving forces.

Essentially, stress gradient is another driving force of mass transport. There is a strong interaction between the stress gradient and the other driving forces.

However, in thermomigration in the absence of external stress gradients, the mass moves from the hot side to the cold side, and compression on the cold side and tension on the hot side develop, which counter the thermal gradient-induced force.

8.2.1.3 Thermal Gradient

The Joule heat generated under high current density is highly localized. Hence, there is a thermal gradient in the medium, which leads to thermomigration. Thermal gradient is one of the strongest driving forces of diffusion. Thermomigration cannot be ignored especially when the thermal gradient is large, in which case the thermomigration can be the dominant diffusion driving force.

8.2.1.4 Atomic Vacancy Concentration Gradient

It has been estimated that the mass flux due to vacancy concentration gradient is small compared to those induced by electrical field forces, stress gradient and thermal gradient. The atomic vacancy concentration gradient is usually small at the beginning compared to the former two forces. However, as the mass migration progresses, vacancy concentration gradient increases significantly. At a certain point in time, vacancy concentration gradient becomes large enough to slow the diffusion.

8.2.2 *Laws Governing Electromigration and Thermomigration Process*

Vacancy diffusion is governed by the vacancy conservation equation; mechanical deformation is governed by the Newtonian mechanics; heat transfer is governed by

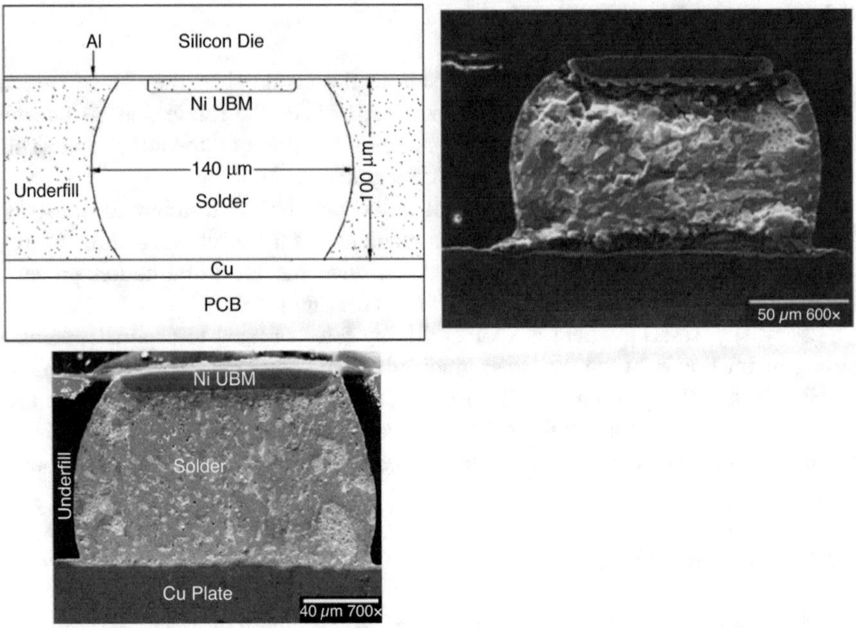

Fig. 8.1 Electronics solder joint profile after electromigration and thermomigration failure

the Fourier's law; and the electric field is governed by Maxwell's equation of conservation of charge. Degradation is governed by the laws of thermodynamics. Fig. 8.1 shows a microelectronics solder joint before and after electromigration/thermomigration failure.

In the following section, all these laws are used to model the unified mechanics of electromigration and thermomigration processes.

When the only acting load is an electrical potential field, the primary deformation mechanism is due to atomic diffusion, which is a mass transport by atomic motion process. Diffusion in solids occurs only at the atomic or molecular level. However, diffusion in solids may be observed in the macroscopic scale given enough time. Diffusion that results in the net transport of matter over macroscopic distances is considered a nonequilibrium process. It does not stop until the phase eventually achieves fully thermodynamic equilibrium or diffusion driving force is removed.

The laws of diffusion were first developed by Adolf Fick when studying the diffusion of the saltwater system (Fick 1855). Fick's diffusion laws establish the mathematical relationship between the rate of diffusion and the concentration gradients, which is the driving force of the mass transport.

8.2.2.1 Fick's First Law

Fick discovered by direct observation that the magnitude of the mass flux is proportional to the magnitude of the concentration gradient in isotropic continuum is given by.

$$\mathbf{J} = -D\nabla C \tag{8.1}$$

where:

J—is the mass flux, which is defined as a vector quantity, with units: mol m^{-2} s^{-1}.

D—is diffusivity, with unit's $m^2\,s^{-1}$; it is a function of $D = D_0 \exp\left(-\frac{Q_d}{RT}\right)$ where D_0 is temperature-independent constant, Q_d is activation energy for diffusion, *R* is the gas constant (8.31 J/mol K), and *T* is temperature in Kelvin.

C—is the concentration of the matter in question.

Combining the mass conservation equation with Fick's first law yields the Fick's second law;

8.2.2.2 Fick's Second Law: (for Nonsteady-State Diffusion)

Fick's second law can be given in its most elementary form as.

$$\frac{\partial C}{\partial t} = D\nabla^2 C \tag{8.2}$$

Until the twentieth century, diffusion in materials was addressed at microscopic scale only. Einstein showed the consistency of random motion of macroscopic particles in the presence of molecules. Einstein derived a relationship for the chaotic motion of small particles as.

$$v_D = \frac{DF}{k_B T} \tag{8.3}$$

where:

v_D—is the drift velocity, unit m/s.

F—is the driving force of the diffusion, unit *N*.

k_B—is Boltzmann constant.

T—is absolute temperature, *K*.

From Eq. (8.3) we can easily deduce that for self-diffusion, the driving force due to concentration gradient has the following form:

$$F = -k\,T\nabla C \tag{8.4}$$

In addition, Fick's second law has a more general form as follows:

$$\frac{\partial C}{\partial t} = \frac{D \nabla F}{kT} \tag{8.5}$$

where F is the driving force of the diffusion. In the presence of electrical current density, the driving force has four primary components:

$$F = \left(\sum_i F_i \right) = F_{em} + F_{tm} + F_\sigma + F_s \tag{8.6}$$

where F_{em} is the electron wind force, F_{tm} is the driving force due to temperature gradient, F_σ is the stress gradient driving force, and F_s is the driving force due to chemical potential gradient. Each driving force will be introduced separately in the following sections.

These forces also lead to grain coarsening and phase changes in the metals. However, here we are primarily interested in the diffusive implications.

8.2.3 Electromigration Electron Wind Force

The first ever observation of electromigration occurred long before it became an engineering problem. In 1861, a French scientist, M. Gerardin, first noticed the phenomenon of electromigration. Nevertheless, at that time, no real explanation could be provided for these effects. Although Michael Faraday's discovery of electrochemistry a few decades earlier may have suggested an explanation, it would have been incorrect because electrons had not been discovered yet. Without electrons, a proper explanation would not have been possible.

The first reasonable explanation of electromigration driving force came from Skaupy (Skaupy 1914; Landauer and Woo 1974). He recognized that in metals the moving free conduction band electrons could drag atoms along through a frictional force (scattering). After that a series of publications (Huntington and Grone 1961; Bosvieux and Friedel 1962; Genoni and Huntington 1977) proposed that at high current densities, conduction band electrons collide (scatter) with metal atoms/ions and push them in the direction of electron flow. In other words, the "electron wind" force drives positive ions (or atoms) in the direction of the electron flow, which is opposite to the direction of the electrostatic force of the electric field (Fig. 8.1).

In studies of the electromigration force term, it has been customary to distinguish two types of contribution: that from the scattering of the electrons by the ion in question (the electron wind term) and that from the direct interaction of an atomic charge with the applied field (the direct force) (Landauer and Woo 1974). Therefore, the electromigration driving force, $\vec{F}_{em}$, Fig. 8.2, exerted on an ion can be expressed as a summation of these two forces as follows;

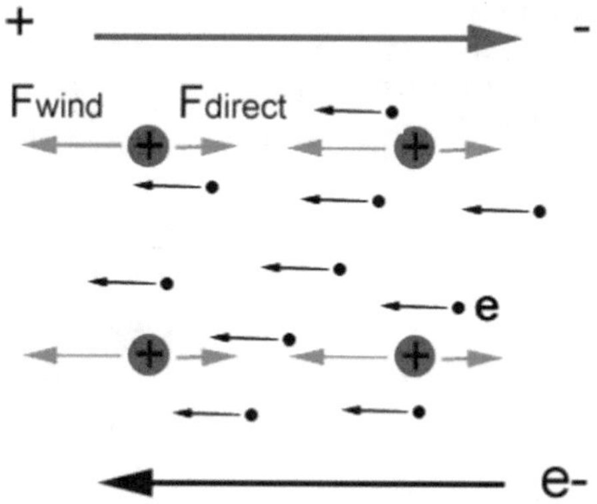

Fig. 8.2 Kinematics of electromigration

$$\begin{aligned}\vec{F}_{em} &= \vec{F}_{\text{wind}} + \vec{F}_{\text{direct}} \\ &= (-Z_{\text{wind}} + Z_{el})e\vec{E} \\ &= Z^* e\vec{E} \\ &= Z^* e\rho j\end{aligned} \tag{8.7}$$

where e is the electron charge, $\vec{E}$ is the electric field potential (V/cm), ρ is the resistance, j is the current density Amp/cm^2, and Z^* is a dimensionless quantity called the effective charge (also called the effective nuclear charge or the effective valence), which is the combined net attractive positive charge of nuclear protons acting on valence electrons. When a charged particle penetrates through condensed matter, it polarizes the medium, which in turn reacts, back slowing down the penetrating particle. In order to introduce a measure of the inertia of ions (Brandt and Kitagawa 1982) introduced the concept of effective charge. The effective charge, Z^*, relates how a diffusing atom interacts with the conducting electrons. The interaction with the conduction band electrons is a complex function of the electronic structure, a quantum mechanical effect. The effective charge is defined in terms of the stopping power of the charge *Z(S)* and power of the proton moving in the same medium at the same velocity (S_p) by (Barberan and Echenique 1986).

$$Z^* = \left(S/S_p\right)^{\frac{1}{2}} \tag{8.8}$$

Z^* is usually negative in metal conductors due to the influence of the "electron wind" (Ye et al. 2004a, b, c, d). For high conductivity metals, i.e., Al or Cu, the value of Z^* is about -1.0 (Lloyd et al. 2004). For pure tin, Z^* value was reported to range from -80 to -160 (Kuz'menko and Osirovskii 1962; Lodding 1965; Sun and Ohring 1976). For single-crystalline Sn, Z^* value was reported to range from -10 to -16, while for polycrystalline Sn is -12 at about 200 °C. These values seem to be favored by Sorbello's electromigration theory, which predicted the effective charge number for Sn at 185 °C as -10 (Sorbello 1973). His theory is based on pseudo

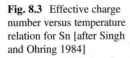

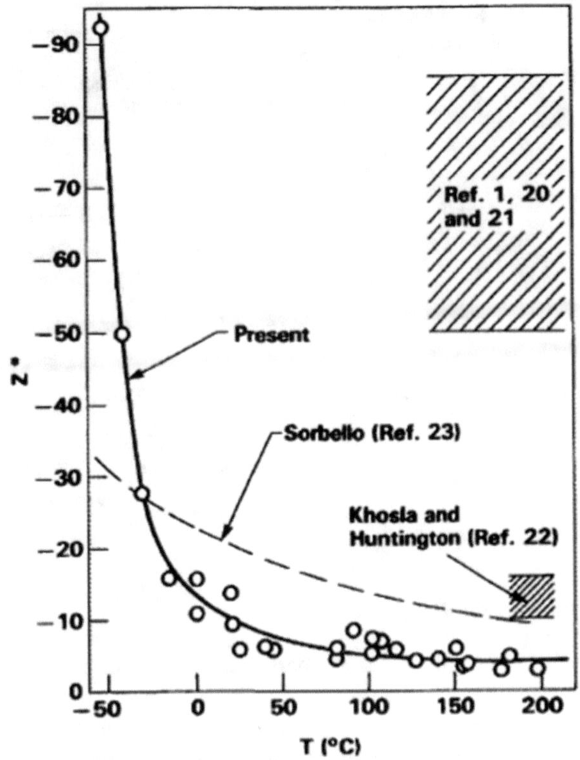

Fig. 8.3 Effective charge number versus temperature relation for Sn [after Singh and Ohring 1984]

potential calculation of driving force for atomic migration in metals in the presence of electric current. The fact that Z^* is temperature dependent is also reported by Singh and Ohring (Singh and Ohring 1984); see Fig. 8.3.

The effective charge, Z^*, is determined by the effective mass, m^*, of electrons available for interaction (scattering) in which the energy produced is not exceeding the Fermi energy, E_f, which is the total energy of uncharged metal at absolute zero temperature. The effective mass m^* is proportional to the second derivative in the energy-versus-momentum, *E-k* relationship, a quantum mechanical effect near Fermi level energy $E = E_f$ (Lloyd et al. 2004). Above the inflection point m^* *is* negative, and below it m^* is positive (Fig. 8.4).

8.2.4 Temperature Gradient Diffusion Driving Force

The role of thermomigration in the presence of high current density has been always ignored until recently. Recently, it has been shown by Ye et al. (2003a, b, c, d, e, f), Abdulhamid et al. (2008) and Basaran et al. (2008) that thermomigration plays a significant role in the presence of high current density. Diffusion driving force due to

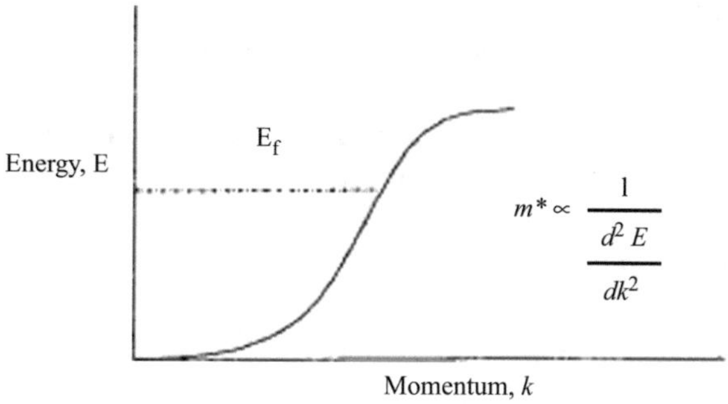

Fig. 8.4 Energy-momentum relationship that determines the value of effective mass (after Lloyd et al. 2004)

thermal gradient can be bigger than electron wind forces. If they are in opposite direction, it can negate the electron wind forces and if they are in the same direction can hasten the failure process significantly.

The temperature gradient-induced driving force is usually represented by (Huntington 1972).

$$F_{tm} = \frac{Q^*}{T} \nabla T \tag{8.9}$$

where Q^* is the heat of transport, which is the isothermal heat transmitted by the moving atom in the act of jumping a lattice site less the intrinsic enthalpy and T is temperature in Kelvin.

The most accepted thermomigration theory is very much based on the same concepts as electromigration. In electromigration, the force acting on the diffusing atom is the momentum exchange due to the collisions of diffusing atoms and scattered electrons in the conduction band. However, in thermomigration, in the absence of electrical current, the driving force in the direction of the temperature gradient results from the fact that the energy, therefore the momentum of the atoms (phonons) at higher temperatures, is greater than that of the lower ones. This gradient in the momentum produces a driving force for mass transport.

Based on this argument, Q^* for a metal with high Z^* value should also be high. Similar to Z^*, the sign of Q^* can either be positive or negative depending on the effective mass of the charge carriers.

According to Huntington (1972), thermomigration driving force can be divided into an intrinsic part and a part coupled to carriers, i.e. electrons, and phonons:

$$Q^* = Q^*_{in} + Q^*_{el} + Q^*_{ph} \tag{8.10}$$

The intrinsic considerations include the energy transported by the moving atom, the energy required to prepare a place [a lattice site] to receive it, and, for vacancy diffusion mechanism, the formation energy for the counter moving vacancy, which is given by.

$$Q^* = \beta E_m - E_f \tag{8.11}$$

where E_m and E_f are, respectively, the energies of motion and formation for the moving atom and β is a dimensionless fraction less than 1 which gives the fraction of the motion energy carried by the moving atom. From Eq. (8.11) we can see that Q^* can be either positive or negative depending on the magnitude of the energies of motion and formation. According to Huntington et al. (1970), β was reported to be about 0.81; hence Q^* is estimated to be about 1ev, which is close to the experimental result of *0.23ev*.

8.2.5 Stress Gradient Diffusion Driving Force

Due to mass transport from cathode to anode, or from the high-temperature side to low-temperature side, there are more vacancies on the one side and more mass on the other side. Consequently, on the mass accumulation side, atoms are subjected to compression, and on the vacancy accumulation side, atoms are subjected to tension. Therefore, there is a stress gradient between the two sides of the continuum. This stress gradient is also another diffusion driving force. Tensile stress makes diffusion easier, and compressive stress makes diffusion more difficult.

The role of mechanical stress gradients as a driving force in conjunction with electromigration was first explored in a series of experiments by Blech (1976) and Blech and Herring (1976). It was shown that a critical current density threshold exists below which Al mass transport is arrested in a thin film sample. This threshold was found to be inversely proportional to the thin film Al stripe length. It was suggested that electromigration in short segment tends to induce back-stress when they did an electromigration test on a set of short Al thin film strips depicted in Fig. 8.5.

Figure 8.5 shows the effect of stress gradient on Al thin film. The Al thin film strips are deposited on a baseline of TiN. It can be observed that the longer the length, the larger the depletion, and extrusion on the cathode and anode sides, respectively, in electromigration. When the thin film is below a "critical length," there was no observable depletion or extrusion. Stress gradient driving force was neutralizing the electron wind force driven diffusion.

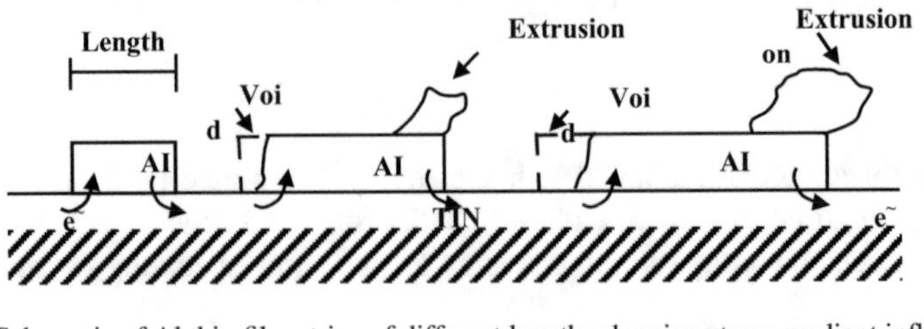

Fig. 8.5 Schematic of Al thin film strips of different lengths showing stress gradient influence due to electromigration

The dependence of mass transport in strip length was explained by the effect of stress gradient. When electromigration process transports Al atoms from cathode to anode, the anode will be in compression while the cathode will be in tension. Based on the Nabarro-Herring model of equilibrium vacancy concentration in a stressed solid, the tensile region has more and the compressive region has fewer vacancies. As a result, there exists a vacancy concentration gradient decreasing from cathode to anode side. The stress gradient induces an atomic flux of Al diffusing from anode to cathode, and it opposes the Al flux driven by electromigration from cathode to anode.

Blech length comes into effect when a thin film is subjected to current density with blocking boundary conditions. A stress gradient is produced which provides a mass transport force in opposition to the electromigration. If current density is sufficiently low, a stress gradient will be generated that can stop electron wind force driven diffusion.

The stress gradient depends on the length of the strip; the shorter the thin film strip, the greater the gradient. At a certain length defined as the critical length, the stress gradient is large enough to counterbalance electromigration, so no depletion at the cathode or extrusion at the anode occurs. The critical length can be calculated when the net force due to electromigration and stress gradient is nonexistence, as shown in equation Eq. (8.12). Force equilibrium equation for an atom in the x-direction can be rewritten as.

$$F = F_{ew} + F_{\sigma}$$

$$= -Z^{*} e\rho j + f\Omega \frac{\partial \sigma}{\partial x} = 0 \tag{8.12}$$

where f is the atomic vacancy relaxation ratio, the ratio of atomic volume to the volume of an atomic vacancy, Ω is the atomic volume, σ is the spherical stress.

Solving Eq. (8.12)

$$\frac{\partial \sigma}{\partial x} = \frac{Z^{*} e\rho j}{f\Omega}$$

$$j\partial x = \frac{f\Omega}{Z^{*} e\rho} \partial \sigma \tag{8.13}$$

The integration of Eq. (8.13) produces one of the most important characteristics of electromigration. Given a maximum stress that can be sustained by the conductor, the product of current density and the length of the conductor determine whether electromigration diffusion can occur or not. This product was discovered by I. A. Blech and is referred to as the "Blech product" (Blech 1976; Blech and Herring 1976):

$$jl_B = \frac{f\Omega}{Z^* e\rho}(\sigma_m - \sigma_0) \tag{8.14}$$

where l_B is known as the Blech length, σ_m is the maximum spherical stress that can be supported by the conductor, and σ_0 is the initial spherical stress in the conductor. If the product of the current density and the length is exceeded, electromigration mass transport can occur; if it is not, no electromigration can happen.

If all thin film conductors could be designed so that jl_B stays below the critical value and electromigration can be avoided since mass transport comes to a complete halt.

However, this scenario is only true for pure metal thin film on a substrate, because it assumes that metal is not an alloy and there is no temperature gradient acting on the system.

8.3 Laws of Conservation

Electromigration is an electron flow-assisted diffusion process. The process can be assumed to be controlled by a vacancy diffusion mechanism, in which the diffusion takes place by vacancies switching lattice sites with adjacent atoms. In isothermal condition, the process is driven by electrical current caused mass diffusion, stress gradient-induced diffusion, and diffusion due to atomic vacancy concentration. In the presence of electrical current, due to electrical resistance, there is always heat production that leads to thermomigration, which interacts with other diffusive forces. Under the presence of these four forces, the atomic vacancy flux equation can be given by combining Huntington (1972), Huntington et al. (1970) and Kirchheim (1992) flux definitions, and adding the influence of temperature gradient and vacancy concentration yields.

$$\frac{\partial C_v}{\partial t} = -\vec{\nabla} \cdot \vec{q} + \mathrm{G} \tag{8.15}$$

$$\vec{q} = -D_v \left[\vec{\nabla} C_v + \frac{C_v Z^* e}{kT}\left(-\rho \vec{j}\right) - \frac{C_v}{kT}(-f\Omega)\vec{\nabla}\sigma + \frac{C_v Q^*}{kT}\frac{\vec{\nabla} T}{T} \right] \tag{8.16}$$

Combining these two equations yield.

$$\frac{\partial C_v}{\partial t} = D_v \left[\nabla^2 C_v - \frac{Z^* e\rho}{kT}\vec{\nabla} \cdot \left(C_v \vec{j}\right) + \frac{f\Omega}{kT}\vec{\nabla} \cdot \left(C_v \vec{\nabla}\sigma\right) + \frac{Q^*}{kT^2}\vec{\nabla}(C_v \nabla T) \right] + G \tag{8.17}$$

8.3.1 Vacancy Conservation

Based on mass/vacancy conservation law, the following derivation can be given (Fig. 8.6):

$$
\begin{gathered}
\left[(q_x - q_{x+\Delta x})\Delta y \Delta z + G\Delta x \Delta y \Delta z\right]\Delta t = (c + \Delta c)(V + \Delta V) - cV \\
(-\nabla q + G)V\Delta t = \Delta(cV) \\
set \Delta t \rightarrow 0 \\
\Rightarrow (-\nabla q + G)V = \frac{\partial(ccV)}{\partial t} = V\frac{\partial c}{\partial t} + c\frac{\partial V}{\partial t} \\
\Rightarrow -\nabla q + G = \frac{\partial c}{\partial t} + c\frac{\partial V/V}{\partial t} = \frac{\partial c}{\partial t} + c\frac{\partial \varepsilon_V}{\partial t} \\
\Rightarrow \frac{\partial c}{\partial t} = -\nabla q + G - c\frac{\partial \varepsilon_V}{\partial t}
\end{gathered}
\tag{8.18}
$$

where:

C_{V0}—Thermodynamic equilibrium vacancy concentration in the absence of stress field.

c—Normalized vacancy concentration and $c = \frac{C_V}{C_{V0}}$.

C_V—Instantaneous vacancy concentration.

ε_V—Volumetric strain. It is assumed that a diffusing atom leaves behind a vacancy or takes an interstitial position. Both cases lead to volumetric strain only.

t—Time.

Substituting Eqs. (8.6), (8.8), and (8.11) into Eq. (8.5), we obtain the **q**,vacancy flux in the following form:

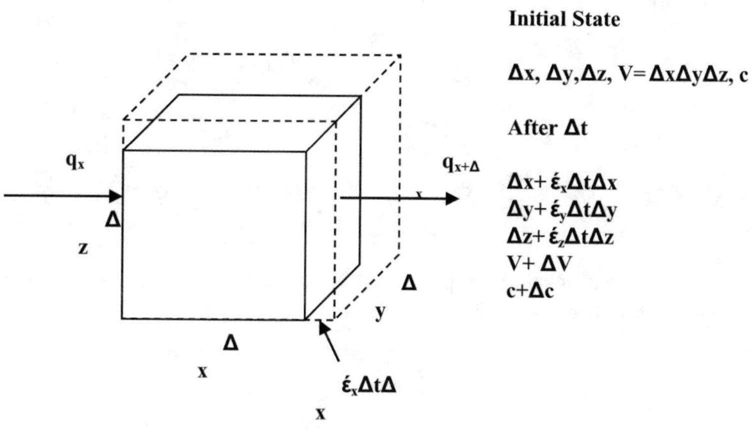

Fig. 8.6 Illustration of vacancy conservation law

$$\mathbf{q} = -D_v C_{v0}\left[\nabla c + \frac{Z^* e}{kT}(\nabla\phi)c + \frac{cf\Omega}{kT}\nabla\sigma^{sp} + \frac{cQ^*}{kT^2}\nabla T\right] \tag{8.19}$$

where:

D_V—Vacancy diffusivity, $D_V = D_o\left(e^{-\frac{Q^*}{kT}}\right)$

Z^*—effective charge number

e—Electron charge

ϕ—Electric potential

j—Current density (vector)

f—Vacancy relaxation ratio, ratio of atomic volume to the volume of a vacancy

Ω—Atomic volume

k—Boltzmann's constant

T—Absolute temperature in Kelvin

σ^{sp}—Spherical part of stress tensor, $\sigma^{sp} = trace(\sigma_{ij})/3$

Q^*—Heat of transport, the isothermal heat transmitted by the moving atom in the process of jumping a lattice site less the intrinsic enthalpy

G—Vacancy generation/annihilation rate expressed by (Sarychev et al. 1999)

$$G = -C_{v0}\frac{c - \exp\left(\frac{(1-f)\Omega\sigma^{sp}}{kT}\right)}{\tau_s} \tag{8.20}$$

τ_s—Characteristic vacancy generation/annihilation time

To solve Eq. (8.15) using the finite element method (method of weighted residuals), we can write the following relationship:

$$\int_V \delta c\left(C_{V0}\frac{\partial c}{\partial t} + \nabla\cdot q - G\right)dV = 0 \tag{8.21}$$

By expanding Eq. (8.21), we can write

$$\begin{aligned}&\int_V \delta c\cdot C_{V0}\frac{\partial c}{\partial t}dV + C_{V0}\int_V \delta c\cdot\frac{\exp\left(\frac{(1-f)\Omega\sigma^{sp}}{kT}\right) - c}{\tau_s}dV + \\ &\int_V \delta c\nabla\cdot D_V C_{V0}\left[\nabla c + \frac{Z^* e}{kT}\cdot\nabla\phi\cdot c + \frac{cf\Omega}{kT}\nabla\sigma^{sp} + + \frac{cQ^*}{kT^2}\nabla T\right]dV = 0\end{aligned} \tag{8.22}$$

Concentration vector can be represented in terms of nodal values and interpolation functions as follows

$$\delta c = N^T\cdot\delta c^i \tag{8.23}$$

Using Eq. (8.23) in (8.22), after simple organization we can derive the stiffness matrices. However, first we have to define the terms in Eq. (8.22).

If we define $C = C_v/C_{v0}$ as the normalized concentration, then the vacancy diffusion differential equation could be rewritten as

$$\frac{\partial C_v}{\partial t} = D_v\left[\nabla^2 C + \frac{Z^* e\rho}{kT}\vec{\nabla}\cdot\left(C\vec{j}\right) + \frac{f\Omega}{kT}\vec{\nabla}\cdot\left(C\vec{\nabla}\sigma\right) + \frac{Q^*}{kT^2}\vec{\nabla}\left(C\vec{\nabla}T\right)\right] + \frac{G}{C_{vo}} \tag{8.24}$$

where, initially, $C = 1$ $(or\ C_v = C_{v0})$.

The vacancy can be considered as a substitutional species at the lattice site with a smaller relaxed volume than the volume of an atom. When a vacancy switches, lattice site with an atom or a vacancy is generated/annihilated local volumetric strain occurs. As proposed by Sarychev et al. (1999), the vacancy causes volumetric strain in the lattice, because the volume of the vacancy is different from the volume of the atom. This volumetric strain is composed of two parts, ε_{ij}^{m}, the volumetric strain due to vacancy flux divergence, and ε_{ij}^{g}, the volumetric strain due to vacancy generation:

$$\dot{\varepsilon}_{ij}^{m} = \frac{1}{3}f\Omega\vec{\nabla}\cdot\vec{q}\delta_{ij} \tag{8.25}$$

$$\dot{\varepsilon}_{ij}^{g} = \frac{1}{3}(1-f)\,\Omega G\delta_{ij} \tag{8.26}$$

where $\delta_{ij}$is the Kronecker's delta.

Thus, the combined volumetric strain rate due to electrical current is

$$\dot{\varepsilon}_{ij}^{elec} = \dot{\varepsilon}_{ij}^{m} + \dot{\varepsilon}_{ij}^{g} = \frac{\Omega}{3}\left[f\vec{\nabla}\cdot\vec{q} + (1-f)G\right]\delta_{ij} \tag{8.27}$$

The total volumetric strain rate due to current stressing is then

$$\dot{\varepsilon}^{elec} = \Omega\left[f\vec{\nabla}\cdot\vec{q} + (1-f)G\right] \tag{8.28}$$

By analogy to thermal strain (which is the volumetric strain caused by temperature variation), the volumetric strain caused by the current stressing is superimposed onto the strains tensor with strains due to other loadings; thus total strain can be given by

$$\dot{\varepsilon}_{ij}^{\text{total}} = \dot{\varepsilon}_{ij}^{\text{mech}} + \dot{\varepsilon}_{ij}^{\text{therm}} + \dot{\varepsilon}_{ij}^{\text{elec}} \tag{8.29}$$

where $\dot{\varepsilon}_{ij}^{\text{total}}$ is the total strain rate tensor, $\dot{\varepsilon}_{ij}^{\text{mech}}$ is the strain rate due to mechanical loading, $\dot{\varepsilon}_{ij}^{\text{therm}}$ is the strain rate due to thermal load, and $\dot{\varepsilon}_{ij}^{\text{elec}}$ is the volumetric strain rate due to electromigration.

Following the standard procedure to obtain the finite element method stiffness matrices leads

$$\left[\boldsymbol{K}_{cc}^{1}\right] = \int_V N^T \cdot \frac{1}{\Delta t} \cdot NdV \tag{8.30}$$

$$\left[K_{cc}^{2}\right] = \int_V \frac{\partial N^T}{\partial x} \cdot D_V \cdot \begin{bmatrix} \frac{\partial}{\partial x} + \frac{f\Omega}{kT_{n+1}} \nabla \sigma_{n+1}^{sp} + \frac{Q^* \nabla T_{n+1}}{kT_{n+1}^2} + \\ + \frac{Z^* e}{kT_{n+1}} \cdot \nabla \phi_{n+1} + \frac{c_{n+1} f\Omega}{kT_{n+1}} \frac{\partial \nabla \sigma_{n+1}^{sp}}{\partial c_{n+1}} \end{bmatrix} \cdot NdV \tag{8.31}$$

$$\left[\boldsymbol{K}_{cc}^{3}\right] = \int_V -N^T \cdot \left[\frac{\frac{(1-f)\Omega}{kT_{n+1}} \exp\left(\frac{(1-f)\Omega\sigma_{n+1}^{sp}}{kT_{n+1}}\right) \frac{\partial \sigma_{n+1}^{sp}}{\partial c_{n+1}} - 1}{\tau_s} \right] \cdot NdV \tag{8.32}$$

Eventually we have

$$\left[\boldsymbol{K}_{n+1}^{cc}\right] = \left[\boldsymbol{K}_{cc}^{1}\right] + \left[\boldsymbol{K}_{cc}^{2}\right] + \left[\boldsymbol{K}_{cc}^{3}\right] \tag{8.33}$$

By taking derivative with respect to *temperature*, we can obtain the coupled stiffness matrix of vacancy concentration and temperature as follows:

$$\begin{aligned} [\boldsymbol{K}_{cT}] = & -\int_V N^T \cdot \frac{1}{\tau_s} \cdot \exp\left(\frac{(1-f)\Omega\sigma_{n+1}^{sp}}{kT_{n+1}}\right) \cdot \frac{(1-f)\Omega}{kT_{n+1}} \cdot \left(\frac{\partial \sigma_{n+1}^{sp}}{\partial T_{n+1}} - \frac{\sigma_{n+1}^{sp}}{T_{n+1}}\right) \cdot NdV \\ & + \int_V \frac{\partial N^T}{\partial x} \cdot D_V \cdot \left[\frac{cf\Omega}{kT_{n+1}} \cdot \left(\frac{\partial \nabla \sigma_{n+1}^{sp}}{\partial T_{n+1}} - \frac{\nabla \sigma_{n+1}^{sp}}{T_{n+1}}\right) + \frac{cQ^*}{kT_{n+1}^2} \left(\frac{\partial}{\partial x} - 2\frac{\nabla T_{n+1}}{T_{n+1}}\right) - \frac{Z^* e}{kT_{n+1}} \cdot \nabla \phi_{n+1} \cdot c_{n+1} \cdot \frac{1}{T_{n+1}} \right] \cdot NdV \end{aligned} \tag{8.34}$$

By taking derivative with respect to displacement u, we can write the following relation:

$$\left[\boldsymbol{K}_{cu}^{1}\right] = \int_V D_V \cdot \frac{\partial N^T}{\partial x} \cdot \frac{c_{n+1} f\Omega}{kT} \frac{\partial}{\partial x} \left(\frac{\partial \sigma_{n+1}^{sp}}{\partial \varepsilon_{n+1}}\right) \cdot BdV \tag{8.35}$$

$$\left[\boldsymbol{K}_{cu}^{2}\right] = -\int_V N^T \cdot \frac{(1-f)\Omega}{kT_{n+1}} \frac{e^{\frac{(1-f)\Omega\sigma_{n+1}^{sp}}{kT_{n+1}}}}{\tau_s} \frac{\partial \sigma_{n+1}^{sp}}{\partial \varepsilon_{n+1}} \cdot BdV \tag{8.36}$$

where $\boldsymbol{\varepsilon}_{n+1}^{M} = \boldsymbol{B}\,\mathbf{u}^{M}$

And we can obtain the coupled stiffness matrix of concentration and mechanical displacement like the following:

$$\left[\boldsymbol{K}_{n+1}^{cu}\right] = \left[\boldsymbol{K}_{cu}^{1}\right] + \left[\boldsymbol{K}_{cu}^{2}\right] \tag{8.37}$$

Proper approach would also take a derivative of the total potential with respect to entropy and find the related stiffness matrices. However, that will add entropy as a nodal unknown to our equations. For the sake of computational simplicity, we take the entropy generation rate values from the previous step and calculate them at the Gauss integration point. Assuming the increments are small, the error should be small. However, this simplification is not mathematically true, and does contribute to error.

8.4 Newtonian Mechanics Force Equilibrium

Force equilibrium equation must be written according to the laws of unified mechanics theory. However, we provide a simple implementation, which ignores the derivative of displacement with respect to entropy. Thermodynamic state index is introduced into the formulation later on. In the absence of body forces, the force equilibrium equation has the following form:

$$\sigma_{ij,j} = 0 \tag{8.38}$$

For small strains the elastic strain-stress constitutive model can be established as

$$\boldsymbol{\sigma} = \boldsymbol{C} \cdot \left(\boldsymbol{\varepsilon} - \boldsymbol{\varepsilon}_{vp} - \boldsymbol{\varepsilon}_{D} - \boldsymbol{\varepsilon}_{TE}\right) \tag{8.39}$$

where:

$\boldsymbol{\varepsilon}_{vp}$—is the viscoplastic strain vector.
$\boldsymbol{\varepsilon}_{D}$—is the strain vector due to diffusion.
$\boldsymbol{\varepsilon}_{TE}$—is the strain vector due to thermal expansion.
$\boldsymbol{C}$—is the constitutive matrix.
Using the principle of virtual work, we can write

$$\prod_{u} = \int_V \delta\boldsymbol{\varepsilon} \cdot \boldsymbol{\sigma}\, dV = \int_V \delta\boldsymbol{\varepsilon} \cdot \boldsymbol{C} \cdot \left(\boldsymbol{\varepsilon} - \boldsymbol{\varepsilon}_{vp} - \boldsymbol{\varepsilon}_{D} - \boldsymbol{\varepsilon}_{TE}\right) dV \tag{8.40}$$

After standard derivation process, we end up with the mechanical stress-induced deformation-related stiffness matrices as follows:

$$\left[K^{uu}_{n+1}\right] = \int_V \boldsymbol{B}^T \cdot \boldsymbol{C} \cdot \boldsymbol{B} dV \tag{8.41}$$

The stiffness matrix for unit vacancy concentration change-induced deformation can be obtained by taking the derivative of equation Eq. (8.40) with respect to c:

$$\left[\boldsymbol{K}_{n+1}^{uc}\right] = \int_V \boldsymbol{B}^T \cdot \frac{\partial \boldsymbol{\sigma}_{n+1}^{sp}}{\partial c_{n+1}} \cdot \boldsymbol{N} dV \tag{8.42}$$

With similar operation we can obtain the stiffness matrix components concerning the deformation induced by unit temperature change as follows:

$$\left[\boldsymbol{K}_{n+1}^{uT}\right] = \int_V \boldsymbol{B}^T \cdot \frac{\partial \boldsymbol{\sigma}_{n+1}^{sp}}{\partial T_{n+1}} \cdot \boldsymbol{N} dV \tag{8.43}$$

8.5 Heat Transfer

The transient heat transfer equation has the following form:

$$\rho C_p \frac{\partial T}{\partial t} - \nabla(k_h \nabla T) - \rho Q = 0 \tag{8.44}$$

where:

ρ—Density of the material
C_p—Specific heat
k_h—Coefficient of heat transfer
Q—Heat generated within the body, which can be expressed as

$$Q = Q_J + Q_P + Q_V \tag{8.45}$$

where:

Q_J—is heat due to Joule heating, which can be written as

$$Q_J = \frac{1}{\Delta t} \int_{\Delta t} Q_J(t) dt \tag{8.46}$$

Using linear assumption Eq. (8.46) can be expanded as

$$Q_J = E_{n+1} \cdot \frac{1}{R} \cdot E_{n+1} - E_{n+1} \cdot \frac{1}{R} \cdot \Delta E_{n+1} + \frac{1}{3} \Delta E_{n+1} \cdot \frac{1}{R} \cdot \Delta E_{n+1} \tag{8.47}$$

where E is the electrical field intensity defined as

$$E = -\frac{\partial \phi}{\partial x} \tag{8.48}$$

By Ohm's law the flow of electrical current is description as

$$j = \frac{E}{R} = -\frac{1}{R} \cdot \frac{\partial \phi}{\partial x} \tag{8.49}$$

Q_P is due to plastic deformation at step $n + 1$ which can be described as

$$Q_P = \boldsymbol{\sigma}_{n+1} : \dot{\boldsymbol{\varepsilon}}^{pl}_{n+1} \tag{8.50}$$

where $\dot{\boldsymbol{\varepsilon}}^{pl}$ is the plastic strain rate.

Q_V—is the heat due to vacancy flux which can be expressed by

$$Q_V = \mathbf{q} : F_k \tag{8.51}$$

where:

$\mathbf{q}$—is vacancy flux.

F_k—is the effective driving force which has the form of

$$F_k = -\left(Z^* e \nabla \phi + f \Omega \nabla \boldsymbol{\sigma}^{sp} + \frac{Q^*}{T} \nabla T + \frac{kT}{c} \nabla c \right) \tag{8.52}$$

Equation (8.51) can be expanded at step $n + 1$ as

$$Q_V = D_V C_V \frac{c_{n+1}}{kT_{n+1}} \left(\frac{kT_{n+1}}{c_{n+1}} \nabla c_{n+1} + Z^* e \nabla \phi_{n+1} + f \Omega \nabla \boldsymbol{\sigma}^{sp}_{n+1} + \frac{Q^* \nabla T_{n+1}}{T_{n+1}} \right)^2 \tag{8.53}$$

Using the method of weighted residuals, Eq. (8.44) can be rewritten as

$$\int_V \delta T \left(\rho C_p \frac{\partial T}{\partial t} - \nabla (k_h \nabla T) - \rho Q \right) dV = 0 \tag{8.54}$$

Using integration by parts and divergence theorem, we obtain

$$\int_V \delta T \cdot \frac{\partial T}{\partial t} dV - \frac{k_h}{\rho C_p} \int_V \nabla \delta T \cdot \nabla T dV - \frac{1}{C_p} \int_V \delta T \cdot Q dV = 0 \tag{8.55}$$

Substituting $\delta T = N^T \, \delta T$ into Eq. (8.54) after simple finite element method operations, we get

$$\left(\int_V N^T C_p \frac{\partial T}{\partial t} dV + \frac{k_h}{\rho} \int_V \frac{\partial N}{\partial x} \cdot \nabla T dV - \int_V N^T \cdot Q dV \right) \delta T = 0 \tag{8.56}$$

By taking first variation of Eq. (8.56) with respect to T, we can find the heat capacity matrix and heat transfer matrix which can be separately expressed as

$$\left[\boldsymbol{C}_{\boldsymbol{n+1}}^{\boldsymbol{TT}}\right] = \int_V \boldsymbol{N}^{\boldsymbol{T}} \cdot \frac{C_P}{\Delta t} \cdot \boldsymbol{N} dV \tag{8.57}$$

$$\left[\boldsymbol{K}_{TT}^{\boldsymbol{1}}\right] = \int_V \frac{\partial \boldsymbol{N}^{\boldsymbol{T}}}{\partial x} \cdot \frac{k_h}{\rho} \cdot \frac{\partial \boldsymbol{N}}{\partial x} dV \tag{8.58}$$

$$\left[\boldsymbol{K}_{TT}^{2}\right] = \int_V N^T \cdot \frac{\partial Q_{n+1}^J}{\partial T_{n+1}} \cdot NdV$$
$$= \int_V N^T \cdot \left(-\frac{1}{R^2}\frac{\partial R}{\partial T_{n+1}}\right)\left(\frac{\partial \phi_{n+1}}{\partial x}\cdot\frac{\partial \phi_{n+1}}{\partial x} - \frac{\partial \phi_{n+1}}{\partial x}\cdot\frac{\partial \Delta\phi_{n+1}}{\partial x} + \frac{1}{3}\frac{\partial \Delta\phi_{n+1}}{\partial x}\cdot\frac{\partial \Delta\phi_{n+1}}{\partial x}\right)\cdot NdV \tag{8.59}$$

$$\left[\boldsymbol{K}_{TT}^{3}\right] = \int_V N^T \cdot \frac{\partial Q_{n+1}^V}{\partial T_{n+1}} \cdot NdV$$
$$= \int_V N^T \cdot \frac{D_V C_V c_{n+1}}{k} \cdot \frac{F_{n+1}^k \cdot N - 2F_{n+1}^k\left[\frac{k\nabla c_{n+1}}{c_{n+1}}\cdot N + f\Omega\frac{\partial \sigma_{n+1}^{sp}}{\partial T_{n+1}}\cdot N + \frac{Q^*}{T_{n+1}}\cdot\frac{\partial N}{\partial x} - \frac{Q^*\nabla T_{n+1}}{T_{n+1}^2}\cdot N\right]}{T_{n+1}^2} dV \tag{8.60}$$

Eventually we can get the temperature stiffness matrix by adding them all together:

$$\left[\boldsymbol{K}_{n+1}^{TT}\right] = \left[\boldsymbol{C}_{n+1}^{TT}\right] + \left[\boldsymbol{K}_{TT}^{1}\right] + \left[\boldsymbol{K}_{TT}^{2}\right] + \left[\boldsymbol{K}_{TT}^{3}\right] \tag{8.61}$$

Ther term “stiffness” is normally used for mechanical behavior. Here we are using the “stiffness” term as an analogy. The coupled term concerning concentration field-induced temperature change can be obtained by taking derivative of the total potential equation with respect to c which yields

$$\left[\boldsymbol{K}_{\boldsymbol{n+1}}^{\boldsymbol{Tc}}\right] = -\int_V \boldsymbol{N}^{\boldsymbol{T}} \cdot \frac{\partial Q_{n+1}^V}{\partial c_{n+1}} \cdot \boldsymbol{N} dV$$
$$= -\int_V \frac{D_V C_V}{kT_{n+1}}\left[\left(F_{n+1}^k\right)^2 \cdot \boldsymbol{N} - 2F_{n+1}^k \boldsymbol{c}_{\boldsymbol{n+1}}\left(f\Omega\frac{\partial \boldsymbol{\sigma}_{\boldsymbol{n+1}}^{\boldsymbol{sp}}}{\partial c_{n+1}}\cdot\frac{\partial \boldsymbol{N}}{\partial x} + \frac{kT_{n+1}}{c_{n+1}^2}\left(\boldsymbol{c}_{n+1}\cdot\frac{\partial \boldsymbol{N}}{\partial x} - \nabla \boldsymbol{c}\cdot\boldsymbol{N}\right)\right)\right] \tag{8.62}$$

Similarly we can get the change of temperature induced by plastic deformation by taking the first variation of Eq. (8.56) with respect to u:

$$\left[\boldsymbol{K}_{\boldsymbol{n+1}}^{\boldsymbol{Tu}}\right] = -\int_V \boldsymbol{N}^{\boldsymbol{T}} \cdot \frac{\partial Q_{n+1}^P}{\partial u_{n+1}} \cdot \boldsymbol{N} dV = -\int_V \boldsymbol{N}^{\boldsymbol{T}} \cdot \left(\frac{\partial \boldsymbol{\sigma}_{\boldsymbol{n+1}}^{\boldsymbol{dev}}}{\partial \boldsymbol{\varepsilon}_{\boldsymbol{n+1}}} + \frac{1}{\Delta t}\frac{\partial \Delta \boldsymbol{\varepsilon}_{\boldsymbol{n+1}}^{\boldsymbol{Pl}}}{\partial \boldsymbol{\varepsilon}_{\boldsymbol{n+1}}}\right) \cdot \boldsymbol{B} dV \tag{8.63}$$

where $\boldsymbol{\sigma}^{\boldsymbol{dev}}$ represents deviatoric stress tensor .

And Joule heating can be considered by adding the following term into the general stiffness matrix:

$$\left[K_{n+1}^{T\phi}\right] = -\int_V N^T \cdot \frac{\partial Q_{n+1}^J}{\partial \phi_{n+1}} \cdot N dV == \int_V N^T \cdot \left(J - \frac{1}{3}\Delta J\right) \cdot \frac{\partial N}{\partial x} dV \tag{8.64}$$

8.6 Electrical Conduction Equations

The electrical field in a conducting material is governed by Maxwell's equation of conservation of charge. Assuming steady-state direct current and no internal volumetric current source, Maxwell equation reduces to

$$\int_{\Omega} \boldsymbol{j} \cdot \boldsymbol{n}\, d\boldsymbol{\Omega} = \boldsymbol{0} \tag{8.65}$$

where:

Ω—is the surface of a control volume.

n—is the outward normal to surface.

j—is the electrical current density defined in Eq. (8.49).

The divergence theorem can be used to convert the surface integral into a volume integral which yields to:

$$\int_V \frac{\partial}{\partial x} \cdot j dV = 0 \tag{8.66}$$

The equivalent weak form is obtained by introducing an arbitrary, variation, electrical potential field, $\delta\phi$, and integrating over the volume:

$$\int_V \delta\phi \cdot \frac{\partial}{\partial x} \cdot j dV = 0 \tag{8.67}$$

Using first the chain rule and then the divergence theorem, this statement can be rewritten as

$$\int_V \frac{\partial \delta\phi}{\partial x} \cdot j dV = 0 \tag{8.68}$$

By introducing the definition of Eqs. (8.48) and (8.49), the governing conservation of charge equation becomes

$$\int_V \frac{\partial \delta\phi}{\partial x} \cdot \frac{1}{R} \cdot \frac{\partial \phi}{\partial x} dV = 0 \tag{8.69}$$

Introducing $\delta\phi = N^T \cdot \delta\phi^N$ into the above equation yields

$$\left(\int_V \frac{\partial N^T}{\partial x} \cdot \frac{1}{R} \cdot \frac{\partial \phi}{\partial x} dV \right) \delta\phi^N = 0 \tag{8.70}$$

From Eq. (8.70) we can obtain the stiffness terms

$$\left[K_{n+1}^{\phi\phi} \right] = \int_V \frac{\partial N^T}{\partial x} \cdot \frac{1}{R} \cdot \frac{\partial N}{\partial x} dV \tag{8.71}$$

The influence of temperature change on the current field can be taken into consideration by adding the following term:

$$\left[K_{n+1}^{\phi T} \right] = \int_V \frac{\partial N^T}{\partial x} \cdot E \cdot \left(\frac{1}{R^2} \frac{\partial R}{\partial T_{n+1}} \right) \cdot N dV \tag{8.72}$$

Eventually we obtain the total stiffness matrix as

$$[\boldsymbol{K}_{n+1}]\{u\} = \begin{bmatrix} \left[\boldsymbol{K}_{n+1}^{uu}\right] & \left[\boldsymbol{K}_{n+1}^{uc}\right] & \left[\boldsymbol{K}_{n+1}^{uT}\right] & 0 \\ \left[\boldsymbol{K}_{n+1}^{cu}\right] & \left[\boldsymbol{K}_{n+1}^{cc}\right] & \left[\boldsymbol{K}_{n+1}^{cT}\right] & 0 \\ \left[\boldsymbol{K}_{n+1}^{Tu}\right] & \left[\boldsymbol{K}_{n+1}^{Tc}\right] & \left[\boldsymbol{K}_{n+1}^{TT}\right] & \left[\boldsymbol{K}_{n+1}^{T\phi}\right] \\ 0 & 0 & \left[\boldsymbol{K}_{n+1}^{\phi T}\right] & \left[\boldsymbol{K}_{n+1}^{\phi\phi}\right] \end{bmatrix} \begin{Bmatrix} \vec{\boldsymbol{u}} \\ \boldsymbol{c_v} \\ \boldsymbol{T} \\ \boldsymbol{\phi} \end{Bmatrix} = \begin{Bmatrix} \boldsymbol{f_u} \\ \boldsymbol{q_c} \\ \boldsymbol{q_T} \\ \boldsymbol{j} \end{Bmatrix} \tag{8.73}$$

It is easy notice that finite element nodal point has displacement, vacancy concentration, temperature, and electrical potential as nodal unknowns. An accurate analysis, according to unified mechanics theory, would also include entropy generation rate as an additional nodal unknown. Of course it does complicate the computation process:

Terms $\frac{\partial \sigma_{n+1}^{sp}}{\partial c_{n+1}}, \frac{\partial \sigma_{n+1}^{sp}}{\partial T_{n+1}}, \frac{\partial \sigma_{n+1}^{sp}}{\partial \varepsilon_{n+1}}, \frac{\partial \sigma_{n+1}^{dev}}{\partial \varepsilon_{n+1}}$, and $\frac{\partial \Delta \varepsilon_{n+1}^{P}}{\partial \varepsilon_{n+1}}$ in Eq. (8.73) can be derived individually once we introduce the viscoplastic model.

8.7 Fundamental Equation for Electromigration and Thermomigration

In this section the fundamental equation accounting for most but not all entropy production mechanisms is derived in a way appropriate for implementation in finite element method. Ignored entropy generation mechanisms such as grain coarsening and phase change are assumed to account for small portion of the total entropy generation. Of course including their contribution would be more accurate. In the derivation below, for the sake of completeness, and clarity, some obvious thermodynamics relations are re-stated.

According to the second law of thermodynamics for any macroscopic system, the entropy, S, of the system is a state function.

The differential of the entropy, dS, may be written as the sum of two terms:

$$dS = dS_e + dS_i \tag{8.74}$$

where dS_e is the entropy supplied to the system and dS_i is the entropy produced inside the system. The second law of thermodynamics states that dS_i must be equal to zero for reversible and positive for irreversible processes:

$$dS_i \geq 0 \tag{8.75}$$

We establish the local form of entropy as

$$S = \int_V \rho s dV \tag{8.76}$$

$$\frac{dS_e}{dt} = -\int_\Omega \mathbf{J}_{s,tot} \cdot d\Omega \tag{8.77}$$

$$\frac{dS_i}{dt} = \int_V \gamma dV \tag{8.78}$$

where s is the entropy per unit mass, $\mathbf{J}_{s,\ tot}$ is the total entropy flow per unit area and unit time, and γ is the entropy source strength or internal entropy production per unit volume per unit time.

Using the local form of entropy, we can get the local form of the variation of entropy as follows:

$$\frac{\partial \rho s}{\partial t} = -div\mathbf{J}_{s,tot} + \gamma \tag{8.79}$$

$$\gamma \geq 0 \tag{8.80}$$

With the help of the material time derivative of the volume integral, which is given by

$$\frac{d}{dt} = \frac{\partial}{\partial t} + \mathrm{V} \cdot \mathrm{grad} \tag{8.81}$$

Equation (8.79) can be rewritten in a slightly different form as

$$\rho \frac{ds}{dt} = -div\mathbf{J}_s + \gamma \tag{8.82}$$

where the entropy flux, J_s, is the difference between the total entropy flux and a convective term

$$\mathbf{J}_s = \mathbf{J}_{s,tot} - \rho s\mathrm{V} \tag{8.83}$$

In obtaining Eqs. (8.80) and (8.82), it is assumed that the definition given by Eqs. (8.74) and (8.75) also holds for infinitesimally small parts of the system. This assumption is in agreement with the Boltzmann (1877) relation where the entropy state is related to the disorder parameter via Boltzmann constant using statistical mechanics, discussed earlier in Chap. 3.

8.7.1 Entropy Balance Equations

Based on the above definitions, the degradation of the system can be related to the rate of change of the entropy (thermodynamic state index). This enables us to obtain more explicit expressions for the entropy flux and internal entropy production rate.

With the assumption of a caloric equation of state, Helmholtz free energy Ψ in differential form can be written as

$$d\Psi = du - Tds - sdT \tag{8.84}$$

where u is internal energy and T is temperature.

(Here we use upper case Ψ to distinguish from the electrical potential ϕ.)

Rearranging leads to

$$Tds = du - d\Psi - sdT \tag{8.85}$$

In order to get the expression for free energy, we need to specify a function with a scalar value concave with respect to temperature T and convex with respect to other variables. Here we use

$$\Psi = \Psi\left(\boldsymbol{\varepsilon}, \boldsymbol{\varepsilon}^e, \boldsymbol{\varepsilon}^{pl}, \boldsymbol{T}, \boldsymbol{V}_k\right) \tag{8.86}$$

where $\mathbf{V_k}$ can be any internal state variable determined by the specific application at hand and physical processes that is leading to the generating entropy for our definition of failure.

In small strain theory, the strain can be written in the form of their additive decomposition:

$$\boldsymbol{\varepsilon}_{\text{Total}} - \boldsymbol{\varepsilon}^{pl} = \boldsymbol{\varepsilon}^{e} \tag{8.87}$$

so that

$$\Psi = \Psi\big(\big(\boldsymbol{\varepsilon}_{\text{Total}} - \boldsymbol{\varepsilon}^{pl}\big), \boldsymbol{T}, \boldsymbol{V_k}\big) = \Psi(\boldsymbol{\varepsilon}^{e}, \boldsymbol{T}, \mathbf{V}_k) \tag{8.88}$$

Time derivative of the free energy is given by

$$\frac{d\Psi}{dt} = \frac{\partial\Psi}{\partial\boldsymbol{\varepsilon}^{e}} : \frac{d\boldsymbol{\varepsilon}^{e}}{dt} + \frac{\partial\Psi}{\partial T} : \frac{dT}{dt} + \frac{\partial\Psi}{\partial V_k} : \frac{dV_k}{dt} \tag{8.89}$$

From Eq. (8.85) we obtain

$$T\frac{ds}{dt} = \frac{du}{dt} - \frac{d\Psi}{dt} - s\frac{dT}{dt} \tag{8.90}$$

Using the first principle of thermodynamics, the conservation of energy equation can be given by

$$\rho\frac{du}{dt} = -div\mathbf{J}_q + \sigma : \mathbf{Grad}(\mathbf{v}) + \sum_k \mathbf{J}_k \cdot \mathbf{F}_k \tag{8.91}$$

where u is the total internal energy, $\mathbf{J_q}$ is the heat flux, $\boldsymbol{\sigma}$ is the stress tensor, $\mathbf{v}$ is the rate of deformation, $\mathbf{J_k}$ is the diffusional flux of component k, and $\mathbf{F}_k$ is the body force acting on the mass of component k.

Using the conservation of energy equation in the form given by Eq. (8.90) and Eqs. (8.82) and (8.83), we can write the rate of change of entropy density as follows:

$$\rho\frac{ds}{dt} = \frac{1}{T}\big(-div\mathbf{J}_q\big) + \boldsymbol{\sigma} : \mathbf{Grad}(\mathbf{v}) +$$

$$\sum_k \mathbf{J}_k \bullet \mathbf{F}_k - \rho\left(\frac{\partial\Psi}{\partial\boldsymbol{\varepsilon}^{e}} : \frac{d\boldsymbol{\varepsilon}^{e}}{dt} + \frac{\partial\Psi}{\partial T} : \frac{dT}{dt} + \frac{\partial\Psi}{\partial V_k} : \frac{dV_k}{dt} - s\frac{dT}{dt}\right) \tag{8.92}$$

where

$$\boldsymbol{\sigma} : \mathbf{Grad}(\mathbf{v}) = \boldsymbol{\sigma} : (\mathbf{D} + \mathbf{W}) \tag{8.93}$$

and **D** (symmetric) and **W** (skew-symmetric) are the rate of deformation tensor and spin tensor, respectively.

Due to the symmetry of $\boldsymbol{\sigma}$

$$\boldsymbol{\sigma} : (\boldsymbol{D} + \boldsymbol{W}) = \boldsymbol{\sigma} : \boldsymbol{D} \tag{8.94}$$

For small strain formulation, we can make the following assumption

$$\boldsymbol{\sigma} : \mathbf{D} = \boldsymbol{\sigma} : \frac{d\boldsymbol{\varepsilon}}{dt} = \boldsymbol{\sigma} : \left(\frac{d\boldsymbol{\varepsilon}^e}{dt} + \frac{d\boldsymbol{\varepsilon}^{pl}}{dt}\right) \tag{8.95}$$

Rearrange Eq. (8.92) and comparing with Eq. (8.82), we can get $\mathbf{J}_s = \frac{1}{T}\mathbf{J}_q$ and the following entropy production rate term:

$$\gamma = -\frac{1}{T^2}\mathbf{J}_q \cdot \mathrm{Grad}(T) + \frac{1}{T}\sum_k \mathbf{J}_k \cdot \mathbf{F}_k + \frac{1}{T}\boldsymbol{\sigma} : \frac{d\boldsymbol{\varepsilon}^p}{dt} + \frac{1}{T}\left(\boldsymbol{\sigma} : \frac{d\boldsymbol{\varepsilon}^e}{dt} - \rho\frac{\partial \Psi}{\partial \boldsymbol{\varepsilon}^e} : \frac{d\boldsymbol{\varepsilon}^e}{dt}\right) + \\ + \frac{\rho}{T}\left(s + \frac{\partial \Psi}{\partial T}\right)\frac{dT}{t} - \frac{\rho}{T}\frac{\partial \Psi}{\partial \mathbf{V}_k} : \frac{d\mathbf{V}_k}{dt} \tag{8.96}$$

In materials with internal friction, all deformations cause positive entropy production rate $\gamma \geq 0$. (Which is also referred as the Clausius-Duhem inequality, using the following relations.)

$$\boldsymbol{\sigma} = \rho(\partial\Psi/\partial\boldsymbol{\varepsilon}^e) \tag{8.97}$$

$$s = -\frac{\partial \Psi}{\partial T} \tag{8.98}$$

If we assume that entropy generation due to elastic strain rate is much smaller compared to other mechanisms, we can simplify Eq. (8.96) as follows:

$$\gamma = -\frac{1}{T^2}\mathbf{J}_q \cdot \mathrm{Grad}(T) + \frac{1}{T}\sum_k \mathbf{J}_k \cdot \mathbf{F}_k + \frac{1}{T}\boldsymbol{\sigma} : \dot{\boldsymbol{\varepsilon}}^p - \frac{\rho}{T}\frac{\partial \Psi}{\partial \mathbf{V}_k} : \frac{d\mathbf{V}_k}{dt} \tag{8.99}$$

or if the heat flux term J_q is replaced by

$$J_q = \frac{1}{T^2}\mathbf{C}|\mathrm{Grad}(T)|^2 \tag{8.100}$$

where is **C** is the thermal conductivity tensor, the simplified entropy production rate can be given by

$$\gamma = -\frac{1}{T^2}\mathbf{C}|\mathrm{Grad}(T)|^2 + \frac{1}{T}\sum_k \mathbf{J}_k \cdot \mathbf{F}_k + \frac{1}{T}\boldsymbol{\sigma} : \dot{\boldsymbol{\varepsilon}}^{pl} - \frac{\rho}{T}\frac{\partial \Psi}{\partial V_k} : \frac{dV_k}{dt} \tag{8.101}$$

In Eq. (8.101) as the effective driving force terms F_k as in the Onsager relation:

$$F_k = \left[Z^* ej\rho + (-f\Omega)\vec{\nabla}\sigma - \frac{Q}{T}\vec{\nabla}T - \frac{kT}{C}\vec{\nabla}C\right] \tag{8.102}$$

In Eq. (8.101) the irreversible dissipation includes two parts, the first term is called heat dissipation caused by conduction inside the system, while the second, third, and fourth terms account for other irreversible processes in the system; we will call it intrinsic dissipation.

It is important to point out that the simplification we did in entropy generation rate term is not necessarily accurate. However, it simplifies computation. An accurate computation should include all entropy generating micro-mechanisms in the fundamental equation. Ignoring the entropy generation due to elastic deformations as well, using Eqs. (8.101) and (8.102), we can write total entropy as

$$\Delta s = \int_{t_o}^{t_1} \gamma \, dt \tag{8.103a}$$

$$\Delta s = \int_{t_0}^{t} \left(\frac{1}{T^2} C|\mathrm{Grad}(T)|^2 + \frac{C_v D_{\mathrm{effective}}}{kT^2} \left(Z_l^* epj - f\Omega\nabla\sigma + \frac{Q\vec{\nabla}T}{T} + \frac{kT}{C}\vec{\nabla}C \right)^2 + \frac{1}{T}\sigma : \varepsilon^{pl} - \frac{\rho}{T}\frac{\partial\Psi}{\partial V_k} : \frac{dV_k}{dt} \right) dt \tag{8.103b}$$

Remember thermodynamic state index (TSI) is given by

$$\Phi = \left[1 - m_s \frac{\Delta s}{R}\right] \tag{8.104}$$

8.8 Example

The solder joint in the electronic package shown in Fig. 8.7 was studied using the formulation given in this chapter.

Ye et al. (2003a, b, c, d, e, f) published test data and finite element time to failure as shown in Table 8.1 for solder joint under different current density levels. Material properties used in the analysis are given in Table 8.2.

In these simulations, critical TSI is set for $\Phi_{cr} = 0.094$, because failure is defined as 10% change in resistance (1 ohm) (Figs. 8.8 and 8.9).

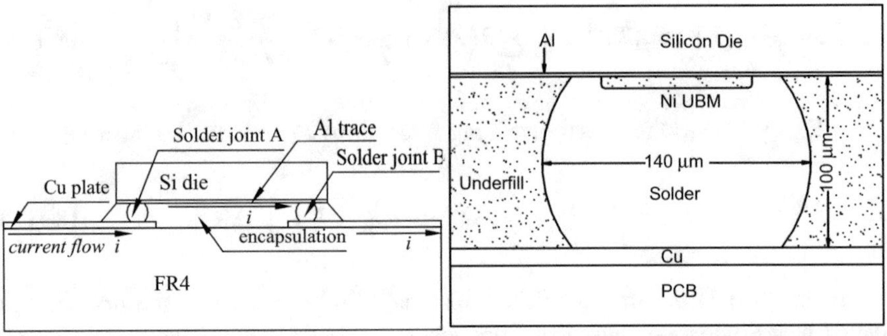

Fig. 8.7 Cross section of the test module and solder joints (Ye et al. 2003a)

Table 8.1 Time to failure

Current density (Amp/cm^2)	Experiment Time to failure (h)	Finite element simulations Time to failure (h)
0.6×10^4	1058.7	1098.2
0.8×10^4	446.6	435.33
1.0×10^4	228.7	222.41

Table 8.2 Material properties for SAC405

Property	Unit	SAC405
Vacancy relaxation time (T_s)	seconds	1.80×10^{-03}
Effective charge number (Z^*)	N/A	10
Resistivity (R)	um^{3}s^{-3}A^{-2} kg	$R(T) = 1.52 \times 10^{11}$ $+3.50 \times 10^8 T(K)$
Atomic volume (Ω)	um^{-3}	2.71×10^{-11}
Vacancy concentration at stress-free state (C_{v0})	um^{-3}	1.107E+06
Young's modulus (E)	GPa	$E(T) = 57.7 - 0.056T(K)$
Poisson's ratio	N/A	0.33
Initial yield stress (σ_{y0})	MPa	$\sigma_y(T) = 79.98 + 95/\sqrt{d}$ $-0.2133T(K)$
Coefficient of thermal expansion (CTE)	K^{-1}	18.9×10^{-06}
Linear kinematic hardening cons (c_1)	kg s um^{-1}	9.63×10^3
Nonlinear kinematic hardening cons (c_2)	kg s K^{-1} um^{-1}	7.25×10^2
Saturation isotropic hardening (R_∞)	kg s um^{-1}	0
Isotropic hardening rate (c)	N/A	383.3
Dimensionless constant (A)	N/A	7.60×10^9
Burger vector (B)	um	3.18×10^{-04}
Initial average grain size	um	2.45
Grain size exponent	N/A	3.34
Stress exponent	N/A	6.65

(continued)

Table 8.2 (continued)

Property	Unit	SAC405
Creep activation energy	$um^2\ s^{-2}\ kg\ mol^{-1}$	7.95×10^{16}
Heat of transport(Q^*)	$kg\ um^2\ s^{-2}$	-3.68×10^{-08}
Density	$kg\ um^{-3}$	7.39×10^{-15}
Specific heat	$um^2\ s^{-2}\ K^{-1}$	2.19×10^{14}
Heat conductivity	$kg\ um\ s^{-3}\ K^{-1}$	5.73×10^{7}

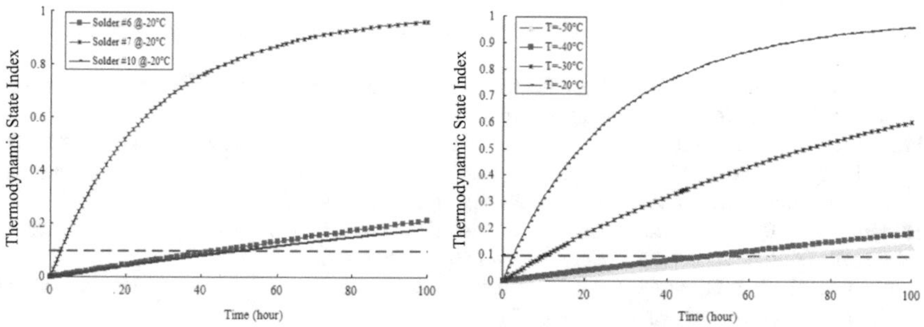

Fig. 8.8 TSI evolution (**a**) in three different solder joints at −20 °C (**b**) in solder #7 at different temperatures

Fig. 8.9 Comparison of experiment and simulation result

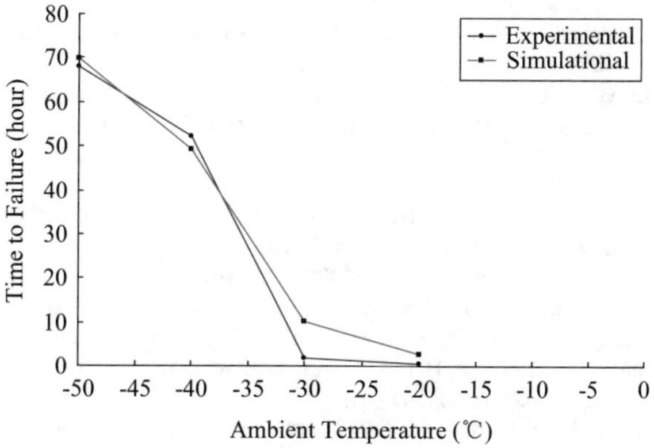

References

Abdulhamid, M., Li, S., & Basaran, C. (2008). Thermomigration in lead-free solder joints. *International Journal of Materials and Structural Integrity, 2*(1/2), 11–34.

Barberan, N., & Echenique, P. M. (1986). Effective charge of slow ions in solids. *Journal of Physics B: Atomic and Molecular Physics, 19*, 3.

Basaran, C., Li, S., & Abdulhamid, M. (2008). Thermomigration induced degradation in solder alloys. *Journal of Applied Physics, 103*, 123520.

Basaran, C., Lin, M., & Ye, H. (2003). A thermodynamic model for electrical current induced damage. *International Journal of Solids and Structures, 40*(26), 7315–7327.

Black, J. R. (1969). Electromigration failure modes in aluminum metallization for semiconductor devices. *Proceedings of the IEEE, 57*(9), 1587–1594.

Blech, I. A. (1976). Electromigration in thin aluminum films on titanium nitride. *Journal of Applied Physics, 47*(4), 1203–1208.

Blech, I. A., & Herring, C. (1976). Stress generation by electromigration. *Applied Physics Letters, 29*(3), 131–133.

Boltzmann, L. (1877). Sitzungberichte der Kaiserlichen Akademie der Wissenschaften. Mathematisch-Naturwissen Classe. Abt. II, LXXVI, pp 373–435 (Wien. Ber. 1877, 76:373–435). Reprinted in Wiss. Abhandlungen, Vol. II, reprint 42, p. 164–223, Barth, Leipzig, 1909.

Bosvieux, C., & Friedel, J. (1962). Electrolysis of metallic alloys. *Physics Chemistry Solids, 23*, 123–136.

Brandt, W., & Kitagawa, M. (1982). Effective stopping-power charges of swift ions in condensed matter. *Physical Review B, 25*, 5631. Erratum Phys. Rev. B 26, 3968 (1982).

Fick, A. (1855). On liquid diffusion. *Philosophical Magazine and Journal of Science, 10*, 31–39.

Genoni, T. C., & Huntington, H. B. (1977). Transport in nearly-free-electron metals. IV. Electromigration in zinc. *Physical Review B, 16*, 1344–1315.

Huntington, H. B. (1972). Electro-and thermomigration in metals. In *Diffusion, papers presented at a seminar of the American Society of Metals* (pp. 155–184).

Huntington, H. B., & Grone, A. R. (1961). Current-induced marker motion in gold wires. *Journal of Physics and Chemistry of Solids, 20*(1-2), 76–87.

Huntington, H. B., Feit, M. D., et al. (1970). Dynamic studies of vacancy motion, Crystal Lattice Defects *1*(3), 8.

Kirchheim, R. (1992). Stress and electromigration in Al-lines of integrated circuits. *Acta Metallurgica et Materialia, 40*(2), 309–323.

Kuz'menko, P. F., & Osirovskii, L. F. (1962). Electric transfer of silver in copper. *Izv. Vyssh. Uchebn. Zaved. Chern Met, 11*, 146–149.

Landauer, R., & Woo, J. W. F. (1974). Driving force in electromigration. *Physical Review B, 10*, 1266–1215.

Lloyd, J. R., Tu, K. N.. and Jasvir Jaspal, J. The physics and materials science of electromigration and thermomigration in solders, Chapter 20 in Handbook of lead-free solder technology for microelectronic assemblies, Editors Karl J. Puttlitz, Kathleen A. Stalter, (2004) Taylor & Francis.

Lodding, A. (1965). Current induced motion of lattice defects in indium metal. *Journal of Physics and Chemistry of Solids, 26*(1), 143–151.

Sarychev, M. E., et al. (1999). General model for mechanical stress evolution during electromigration. *Journal of Applied Physics, 86*(6), 3068–3075.

Seith, W. (1955). *Diffusion in Metallen, Platzwechselreaktionen*. Heidelberg Berlin: Springer.

Seith, W., & Wever, H. (1953). A new effect in the electrolytic transfer in solid alloys. *Zeitscrift für Elektrochemie, 57*(891), 61.

Singh, P., & Ohring, M. (1984). Tracer study of diffusion and electromigration in thin tin films. *Journal of Applied Physics, 56*(4), 899–907.

Skaupy, F. (1914). Electrical conduction in metals. *Verband Deutscher Physikalischer Gesellschaften, 16*, 156–157.

Sorbello, R. S. (1973). A pseudopotential based theory of the driving forces for electromigration in metals. *Journal of Physics and Chemistry of Solids, 34*(6), 937–950.

Sun, P. H., & Ohring, M. (1976). Tracer self-diffusion and electromigration in thin tin films. *Journal of Applied Physics, 47*, 478.

Wever, H., & Seith, W. (1955). Zeitschrift fuer Elektrochemie und Angewandte Physikalische Chemie, 59, 942–946.

Ye, H., Basaran, C., & Hopkins, D. (2003a). Thermomigration in Pb-Sn solder joints under joule heating during electric current stressing. *Applied Physics Letters, 82*(8).

Ye, H., Basaran, C., & Hopkins, D. (2003c). Damage mechanics of microelectronics solder joints under high current densities. *International Journal of Solids and Structures, 40*(15), 4021–4032. #34.

Ye, H., Basaran, C., & Hopkins, D. (2003e). Mechanical degradation of microelectronics solder joints under current stressing. *International Journal of Solids and Structures, 40*(26), 7269–7284.

Ye, H., Basaran, C., & Hopkins, D. (2003f). Measurement of high electrical current density effects in solder joints. *Microelectronics Reliability, 43*(12), 2021–2029.

Ye, H., Basaran, C., & Hopkins, D. (2004a). Pb phase coarsening in eutectic Pb/Sn flip chip solder joint under electrical current stressing. *International Journal of Solids and Structures, 41*, 2743–2755.

Ye, H., Basaran, C., & Hopkins, D. (2004b). Deformation of microelectronic solder joints under current stressing and numerical simulation I. *International Journal of Solids and Structures, 41*, 4939–4958.

Ye, H., Basaran, C., & Hopkins, D. (2004c). Deformation of microelectronic solder joints under current stressing and numerical simulation II. *International Journal of Solids and Structures, 41*, 4959–4973.

Ye, H., Basaran, C., & Hopkins, D. (2004d). Mechanical implications of high current densities in flip chip solder joints. *International Journal of Damage Mechanics, 13*(4), 335–346.

Ye, H., Basaran, C., & Hopkins, D. (2006). Experimental damage mechanics of micro/power electronics solder joints under electrical current stresses. *International Journal of Damage Mechanics, 15*, 41–68.

Ye, H., Basaran, C., & Hopkins, D. (2003d). Numerical simulation of stress evolution during Electromigration in IC interconnect lines. *IEEE Transactions on Components and Packaging Technologies, 26*(3), 673–681.

Ye, H., Hopkins, D., & Basaran, C. (2003b). Measuring joint reliability: applying the Moire interferometry technique. *Advanced Packaging, 1*, 17–20. Review article.

Index

A
ABAQUS model, 387
ABAQUS simulation, 388
Action-reaction law, 10
Activation energy, 366, 379, 380
Activation volume, 380
Almansi tensor, 348, 349
Al thin film strips, 404
Alumina trihydrate (ATH), 278, 279, 312
Amorphous polymers, 363, 373
Anand model, 366, 370
Anand/Richeton viscoplastic models, 366
Applied stress, 380
Arrhenius-type temperature dependence, 363
Atomic stresses, 1
Atomic vacancy
    concentration gradient, 397
    flux equation, 406
Atomic vibrations, 81, 100
Average stress, 13
Average stress norm, matrix
    alternative expression, 289
    Eigen-strain, 285
    ensemble average, 286
    ensemble-averaged loading function, 290
    Eshelby theory, 285
    far-field stress, 285
    forth rank identity tensor, 286
    fourth-rank tensors, 287
    homogeneous matrix, 288
    macroscopic stress, 288
    particle domain, 285
    phase configuration, 286
    probability density functions, 286
    single ellipsoidal inclusion, 285
    single inclusion problem, 286
    statistical isotropy and uniformity, 287
    straightforward derivations, 287
    viscoplastic strain, 290

B
Back stress modulus, 376
Backward Euler scheme, 383
Balanced equation, 217, 219
Bénard cells, 116
Blech Length, 405, 406
Blech Product, 405
Blech's critical length, 396
Boltzmann constant, 97, 98, 207, 379, 418
Boltzmann's equation, 98
Boltzmann's mathematical derivation, 180, 181
Boltzmann-Planck formulation, 191
Bulk modulus, 284

C
Caloric equation of state, 95, 96
Cantilever beam, 90
Cartesian components, 219
Cartesian coordinate system, 13, 18, 21, 44, 77
Cartesian tensor, 46
Cauchy and Almansi tensors, 345
Cauchy deformation tensor, 50, 51
Cauchy stress, 356, 361, 368
Cauchy stress tensor, 18, 60, 67, 240
Cauchy tensor, 347–349
Cauchy's equilibrium equation, 77, 78
Chen's analytical micromechanical approach, 279

C. Basaran, *Introduction to Unified Mechanics Theory with Applications*,
https://doi.org/10.1007/978-3-030-57772-8

Classical irreversible thermodynamics, 116
Clausius –Duhem inequality, 80, 91, 93, 94, 101, 103, 107, 110, 224, 226, 351, 420
Coefficient of thermal expansion (CTE), 279, 334
Composite materials, 277
Composite sphere assemblage (CSA), 309, 316
Computational mechanics, 2
Concentration gradient
- atomic vacancy, 397
- component, 396
- driving force, 399
- magnitude, 399

Conduction band electrons collide (scatter), 400
Conservation laws, 216
- energy, 221–223
- entropy (*see* Entropy law and balance)
- fully coupled thermo-mechanical equations, 229
- local momentum, 216
- mass, 217, 218
- momentum principle (*see* Momentum principle)
- systematic macroscopic scheme, 216

Conservation of energy law, 351
Conservation of mass principle, 63, 64
Conservation of moment of momentum principle, 68–70
Conservation of momentum principle
- Cartesian coordinate system, 67, 68
- Cauchy stress tensor, 67
- equilibrium of forces, 65
- force, 66
- Newtonian continuum mechanics, 66
- normal stresses, 65
- shear stresses, 65
- tensile stress, 65
- time rate of change, 66, 67

Consistency parameter-Classical theory, 215
Constitutive model approaches, 206
Constitutive model framework, 370
Constitutive relations
- dual- and trial-mechanism models, 360
- dual-mechanism constitutive models, 360
- intermolecular resistance, 361–367
- inter-molecular structure, 360
- material model, 360
- molecular network resistance, 367–371
- molecular network structure, 360

Continuous medium
- continuity, 2
- homogeneity, 2
- isotropy, 2
- stress, 1

Continuous one-to-one mapping, 343
Continuum mechanics
- atomic stresses, 1
- continuous functions, 1
- continuous medium, 1
- definitions, 1
- displacement vectors, 57
- laws of Newton, 115
- molecular structure, 1
- Newton's universal laws of motion (*see* Newton's universal laws of motion)
- shear strain, 59
- small strain, 57, 58
- St Venant's compatibility equations, 59, 60
- strain, 2
- strain-displacement relations, 58
- stress, 2
- thermodynamics (*see* Thermodynamics)

Continuum plasticity, 277
Continuum stresses, 1
Contribution
- types, 400

Cosserat continuum, 3
Cosserat continuum thermomechanical analysis
- couple stress theory (*see* Cosserat's couple stress theory)
- equilibrium equations/problem formulation
  - deformation state, 245
  - differential element, 245
  - general solid, 245
  - general strain gradient plasticity theory, 248
  - interpretations, 245
  - kinematic degrees of freedom, 245
  - rotation and micro rotation, 245
  - stiffness, 245
  - surface traction vector, 246
  - symmetric/anti-symmetric components, 246
  - theories, 245
- function/reduced integration scheme, 237
- geometrical factor, 235, 236
- gradient plasticity theories, 235
- higher order stress theory (*see* Toupin-Mindlin higher order stress theory)
- mechanics formulation, 234
- translational/rotational degrees, 236

Cosserat's couple stress theory
- classical continuum mechanics, 237
- elastic constitutive tensors, 241

general couple stress theory, 238
linear elastic behavior, 238
micro rotation, 239
rotation vector, 238
strain and curvatures, 239
symmetric/anti-symmetric components, 240
traction vector, 240
Couple stress, 20, 21
Coupled thermo-mechanical loading, 371
Critical entropy level, 372
Current density
and temperature gradient, 395
blocking boundary conditions, 405
conductor, 405
definition, 415
electrical, 400
electron flux density, 396
electron-wind transfer, 395
mechanics based, 396
product, 406
thermomigration, 402
Cyclic loading
stress-strain diagram, 195–197
TSI, 196
Cyclic stress-strain response, 337, 338, 340

D

Defect energy, 352
Deformable bodies, 11
Deformation gradient, 349
Deformation gradient tensor, 220
Deformation tensor, 420
Degradation evolution, 322
Deviatoric stress tensor, 26–28
Diffusion
activation energy, 399
atomic, 398
electron flow assisted diffusion process, 406
irreversible mass diffusion mechanisms, 395
mass transport, 396
non-steady state, 399, 400
salt-water system, 398
stress gradient, 404–406
temperature gradient, 402–404
vacancy, 397, 409
Direction cosines, 16–18, 23, 24
Disorder, 96–99
Disordered state, 96
Dissipated energy, 353
Dissipation function, 106, 111
Dissipation potential, 106
Dissipation power, 109–112
Dissipation variables, 107
Dissipative resistance, 370
Divergence theorem, 63, 79, 413, 415
Driving forces of electromigration
atomic vacancy, 397
electric field, 396
stress gradient, 397
thermal gradient, 397
Driving stress, 362
Dual- and trial-mechanism models, 360
Dual-mechanism constitutive models, 360
Dual-mechanism model, 365, 366
Dual-mechanism viscoplastic model
elastic and plastic deformation gradients, 383
frame-indifference, 383
numerical integration, 384
plastic stretch rate tensors, 384
PMMA
isothermal stretching, 384–390
non-isothermal stretching, 390–391
rate dependency, 383
staggered method, 381

E

Effective bulk modulus, 310
Effective charge, 401
definition, 401
vs. temperature relation, 402
Effective coefficient of thermal expansion (ECTE), 310
Effective nuclear charge, 401
Effective shear modulus, 310–312
Effective stress concept
back-stress, 209
continuous state variables, 208
incremental procedure, 208
micro-cracks/micro-cavities, 207
nominal uniaxial, 207
RVE, 207, 208
strain equivalence principle, 208
tangential constitutive tensor, 208
thermodynamics state index space, 209
UMT strain energy density, 208
Effective thermo-mechanical properties
analytical and numerical evaluation, 313
bulk modulus, 310, 315
CSA, 309
ECTE, 310
filler and matrix, 309
imperfect interface bond, 309

Effective thermo-mechanical properties (*cont.*)
interphase layer, 309
numerical examples, 312, 313
Poisson's ratio, 315
shear modulus, 310–312
theoretical solution, 309
Young modulus, 312
Effective valence, 401
Eigen-strain, 284, 285
Eight-node linear brick elements (C3D8T), 384
Elastic and plastic deformation gradients, 346
Elastic and plastic rotation tensors, 349
Elastic distortional Almansi tensor, 368
Elastic Eigen-strain, 285
Elastic logarithmic strain, 361, 362
Elastic Mandel stress, 358, 361, 362, 369
Elastic modulus, 334, 335, 337, 374
Elastic recovery, 370
Elastic response, 106
Elastic right Cauchy tensor, 361
Elastic shear modulus, 284
Elastic strain energy, 353
Elastic strain rate, 209
Elastic strain-stress constitutive model, 411
Elastic stretch rate tensor, 350
Elastic symmetric Mandel stress, 357
Elastomers, 369
Electric field, 396
Electrical conduction equations, 415, 416
Electrical current, 403
Electrical current density, 400
Electrical field intensity, 412
Electro-magnetic and chemical loads, 80
Electromagnetic radiation theory, 182
Electromigration
Blech's critical length, 396
Boltzmann constant, 418
characteristics, 405
damage evolution, 395
definition, 396
degradation, conductor, 395
driving forces (*see* Driving forces of electromigration)
electron wind forces, 396, 400–405
electron-wind transfer, 395
experiment and simulation result, 423
finite element method, 396, 416
heat transfer, 412–415
material properties, 421, 422
physical mechanism, 395
properties, 396
quantification, 396
second law of thermodynamics, 417
solder joint, 421, 422
solid conductor, 395
stress gradient diffusion driving force, 404–406
temperature gradient diffusion driving force, 402, 403
and thermomigration
diffusion (*see* Diffusion)
Fick's first law, 399
Fick's second law, 399
mechanical deformation, 397
mechanics, 398
mechanism, 398
microelectronics solder joint, 398
time to failure, 421, 422
TSI, 421, 423
Electron flux density, 396
Electron wind, 401
Electron wind forces, 396, 400–405
Electronics
solder joint, 398
Electron-wind forces, 396, 400
Electron-wind transfer, 395
Electro-thermo-chemical-radiation mechanisms, 95
Electro-thermo-mechanical loading application, 3
Endochronic plasticity theory, 118
Endochronic theory, 118
Energetic fundamental relation, 87
Energy conservation, 221–223
Energy conservation law, 79
Energy-versus-momentum, 402
Engineering mechanics, 116
Engineering strain, 45
Ensemble-average stress norm, filler particles
averaged internal stresses, 299
non-interacting solution, 298
positive definite fourth-rank tensor, 300
volume-averaged stress tensor, 299
Ensemble-average stress norm, matrix
alternative expression, 297
components, 297
CTE mismatch, 298
Eigen-strain, 295
far-field stress, 295
macroscopic stress, 296
non-interacting solution, 296
positive definite fourth-rank tensor, 297
Ensemble-averaged loading function, 290
Ensemble-volume averaging process
effective elastic properties, 282
elastic stiffness tensor, 280

ellipsoidal shape and orientation, 283
Eshelby's equivalence principle, 280
fictitious equivalent eigenstrain, 281
Green's function, 281
inhomogeneities, 280
integral equation, 282–284
inter-particle interaction effects, 283
linear elastic matrix material, 282
micromechanical, 283
perturbed strain field, 281
phase particles, 280, 283
recast, 282
renormalization procedure, 282
RVE, 280, 281
Enthalpy, 101
Entropic fundamental relation, 87
Entropy, 81, 82, 98
balance equations, 418–421
Clausius-Duhem Inequality, 91–94
concept, 80
definition, 80, 85, 118, 126
degradation metric, 129
density, 419
disorder, 96–99, 183, 192
Euler equation, 87–90
extensive parameter, 82
flat surface, 82
flux to heat flux, 359
fundamental governing equations, 84
fundamental relation, 83
generation mechanisms, 191, 389
generation rate, 132, 133, 135, 416
Gibbs-Duhem Relation, 85–87
hypothetical reversible engine, 85
inductive-reasoning, 81
intensive variable, 82
internal energy, 83
irreversible, 359
ISO-TEST simulations, 389
mathematical quantity, 81
measurements, 131
Newtonian (continuum) mechanics, 94–96
Newtonian mechanics, 120
Newtonian mechanics formulation, 118
non-equilibrium states, 83
nonnegative, 373
permutability measure, 137, 176, 180
plastic deformation energy, 125
production, 82, 116, 127, 351, 352, 371, 390
production in irreversible process, 90–91
production rate, 126, 420
production values, 389
property, 83
resonator, 185–188
source strength or internal, 417
stable equilibrium, 84
state functions, 85
states of energy, 82
statistical mechanics, 136, 367
and stiffness matrices, 411
temperature means, 81
thermodynamic potential
(*see* Thermodynamic potential)
third law of thermodynamics, 84
TSI, 194
zero Kelvin temperature, 84
Entropy balance equation, 216
Entropy generation mechanisms, 324
Entropy generation rate, 216
Entropy law and balance
Clausius-Duhem inequality, 223, 224, 226, 228
conservation of energy, 226
continuum mechanics, 225
direction, 225
elastic deformation, 229
elastic energy, 226
entropy, 228
equation, 227
flux quantity, 227
Gibbs relation, 226
heat conduction, 228
heat transfer, 223
Helmholtz free energy, 226
imaginary reversible isothermal path, 223
irreversible process, 224
irreversible thermodynamics, 224
local equilibrium assumption, 224
macroscopic system, 223
mathematical expressions, 225
production, 227
RVE, 224
temperatures, 228
thermal boundary conditions, 228
thermo-mechanical problems, 228
Entropy production, 351
Equivalent stretch rate, 386
Eshelby tensor, 283
Eshelby theory, 285, 302
Eshelby's equivalence principle, 280
Euler equation, 87–90
Euler form, 87
Euler relations, 88
Eulerian description, 32, 42
Eulerian strain tensor, 52, 54
Exponential-mapping, 384

External forces, 161
Eyring cooperative model, 367

F

Failure in electronics, 395, 398, 403, 418, 421
Far-field stress, 285
Fatigue fracture entropy (FFE), 128, 129, 371
Fick's first law, 399
Fick's second law, 399, 400
Finite deformations
  Cauchy and Almansi tensors, 345
  continuous one-to-one mapping, 343, 344
  dual-mechanism viscoplastic model (*see* Dual-mechanism viscoplastic model)
  elastic and plastic, 345, 346
  elastic part, 344
  finite strain (*see* Finite strain)
  formulation, 343
  frame indifferent model, 346
  frame of reference indifference, 347–351
  gradient, 343
  incompressible plastic flow, 346
  irrotational plastic flow, 346
  material properties, 343, 377–381
  mechanics, 343
  multiplicative split, 346
  plastic, 344
  polymer, 343
  rotation tensor, 345
  spin tensor, 345
  stretch tensors, 345
  stretching rate tensor, 344, 345
  TSI, 371–373
  velocity gradient, 343, 344
Finite element analysis, 386
Finite element method, 396, 408
Finite element method implementation
  anti-symmetric Cauchy stress tensor, 248
  general couple stress theory, 250, 252
  prime superscript notation, 249
  reduced couple stress theory (*see* Reduced couple stress theory)
  reduced theory continuum, 249
  symmetric bilinear forms, 249
  symmetric Cauchy stress tensor, 248
  translational degrees of freedom, 249
  translational/rotational degrees, 248
Finite strain, 29
  Clausius-Duhem inequality, 351
  conservation of energy law, 351
  constitutive relations (*see* Constitutive relations)
  definitions, 46
  Eulerian description, 48
  green deformation tensor, 49, 50
  Helmholtz free energy, 351
  Lagrangian formulation, 49
  local coordinate system, 46, 47
  Malvern formulation, 47
  rate of deformation, 54, 55
  referential coordinate system, 47
  rotation tensors, 57
  second law of thermodynamics, 351
  spatial description, 49
  spatial gradient, velocity *vs.* deformation gradient tensor, 56
  strain rate, 54
  stretch tensors, 57
  thermodynamic restrictions, 352–359
  thermo-mechanical effects, 351
Finite strain formulation, 3
First law of thermodynamics, 351
  continuum mechanics, 75
  heat input, 78–80
  potential function, 76
  work done on the system (power input), 76–78
First Piola-Kirchhoff stress tensor, 60, 61
Flow theory
  elastic constitutive relationship, 203, 204
  flow rule, 204
  isotropic hardening, 205
  NLK rule, 205, 206
  viscoplastic creep rate law, 207
Fluctuation theorem, 127
Flux quantity, 227
Force equilibrium, 411, 412
Fourier heat conduction equation, 228
Fourier's law, 398
Frame indifferent model, 346
Free-volume theory, 373
Fully coupled thermo-mechanical equations
  dissipative phenomena, 229
  entropy generation rate, 232
  Fourier's Law, 231
  fundamental equations, 232
  gradient functions, 230
  Helmholtz free energy, 230
  inter-convertibility, 232
  internal state, 229
  isotropic thermal expansion coefficient, 231
  Laplacian operator, 231

Newtonian mechanics, 229, 231
non-recoverable energy, 231
plastic work dissipation, 232
small strain formulation, 229
straining process, 233
stress and strain tensor, 229
temperature distribution, 232
temperature thermomigration, 233
Function of the intrinsic parameters of the system, 83
Fundamental equation, 89
Fundamental relation, 83, 84, 87, 90, 99, 100

G

Gauss integration point, 411
Gauss's theorem, 217, 221
General couple stress theory
constitutive relationships, 253
finite element discretization, 253
independent degrees of freedom, 252
Lagrange multiplier formulation, 254, 255
pure displacement formulation, 253, 254
General return-mapping algorithm
elastic consistent tangent moduli, 325
finite element formulation, 329
internal heat source, 325
irreversible entropy generation, 325
isotropic hardening modulus, 326
iterations, 324
residual functions, 327
systematic application, 329
unconditionally stable, 329
viscoplastic strain, 329
General systems theory, 117, 130
Generalized coordinates, 161
Generalized couple stress theory, 238
Generalized irreversible forces X and fluxes J, 110
Generalized self-consistent scheme (GSCS), 310
Generalized standard materials, 107
Gent free energy, 368
Gibbs-Duhem relation, 85–89
Gibbs relation, 226
Glass-rubber transition, 374, 378
Glass transition temperature, 366
Gravity, 12
Green deformation tensor, 49–52
Green's function, 281, 285

H

Hardening-softening behavior, 367, 376
Heat conduction, 371
Heat exchange, 91, 371
Heat flux, 352
Heat transfer, 397, 412–415
Helmholtz free energy, 103, 226, 230, 266, 351–353, 359, 370, 418
Helmholtz free energy density, 361
Helmholtz free energy function, 362
Helmholtz free specific energy, 101, 104
High temperature mechanics, 365, 373, 378
Hofstadter lattice, 127
Homogeneity, 2
Homogenization, 277
Hooke's law, 203
Hydrostatic stress, 26, 27
Hydrostatic tensor, 26

I

Imaginary point, 16, 17
Incompressible materials, 65
Independent plasticity, 206
Independent variables, 87, 99
Indeterminate couple stress theory, 238
Integral equation, 282
Intensive parameters, 86, 88, 89
Intensive variable
definition, 82
Intermediate (relaxed) configurations, 346
Intermediate configuration, 354
Intermolecular mechanism, 352
Intermolecular network mechanism, 380
Intermolecular resistance
amorphous polymers, 363
Arrhenius-type temperature dependence, 363
Cauchy stress tensor, 361
characteristic viscoplastic shear strain rates, 365
dislocation and dislocation motion, 363
dual-mechanism model, 366
elastic logarithmic strain, 361, 362
elastic Mandel stress, 362
Helmholtz free energy, 361
Helmholtz free energy density, 361
Kinematic hardening, 362
material properties, 365
non-isothermal loading, 365
order parameter, 365
Piola-Kirchhoff stress tensor, 361
plastic deformation gradient, 363, 365
plastic flow, 364
plastic flow resistance, 364
plastic stretch rate, 362
Richeton's model, 366
stretch-like internal variable, 362

Intermolecular resistance (*cont.*)
temperature, 361
temperature and rate dependent parameter, 365
tensorial variables, 362
viscoplastic models, 366, 367
viscous flow, 363
WLF parameters, 364, 365
yield behavior, 363
Intermolecular structure, 360, 362, 376
Internal (strain) energy, 78
Internal energy, 352
Internal friction, 160
Inter-particle interaction effects, 283
Interphase bond modulus, 313
Intrinsic and thermal dissipation, 108
Intrinsic dissipation, 390
Invariants, 25
Irreversible dissipation, 421
Irreversible entropy, 359
Irreversible entropy generation mechanisms, 325
Irreversible process, 90–91
Irreversible thermodynamics, 216, 224
Irrotational plastic flow, 355
ISO-TEST simulations, 389
Isothermal stretching
PMMA, 384–390
Isotropic hardening function, 284
Isotropic hardening modulus, 326
Isotropy, 2

J

Jacobian coupling, 213
Joule heating, 395, 415
Ju and Chen model, 278

K

Kelvin-Planck's statement, 80
Kinematic hardening behavior, 322
Kinematic hardening function, 209
Kinematic hardening modulus, 258, 324
Kinematic of continuous medium
continuum, 41
Eulerian description, 42
Lagrangian description, 42
material description, 42
referential coordinate system, 42
Kinetic energies exchange, continuous manner
additive constant, 156, 157
continuum, 154
external forces, 161–163, 165
integral, 154
internal friction, 160
molecules, 152, 155, 156
omitting, 156
polyatomic gas molecules, 161–163, 165
power-exponent free product (*see* Power-exponent free product)
probability distribution (*see* Probability distribution)
state distribution, 153, 157, 159, 160
thermal equilibrium, 153, 155, 157
velocity components, 155, 158–160
Kinetic energy, 78, 79
colliding molecules, 140
complexions, 140, 142, 151
Descartes' theorem, 146
exchange (*see* Kinetic energies exchange, continuous manner)
exchange of energy, 139
gamma function integral, 144
molecules, 140, 141, 143, 147–149
Newton iteration method, 149
number of permutations, 151
polynomial equation, 147
state distribution, 140–143, 152
velocities, 139
Kirchoff-Clausius law, 186
Kronecker's delta, 409

L

Lagrangian description, 42, 44
Lagrangian formulation, 46, 49
Lagrangian mechanics, 115
Lagrangian strain tensor, 52, 54
Lagrangian stress tensor, 60, 61
Law of inertia, 6, 7
Laws of conservation
mass/vacancy, 407–411
temperature gradient, 406
vacancy diffusion mechanism, 406
Laws of thermodynamics, 3
energy, 135
first law, 75–80
forces, 132
Newtonian mechanics, 115
Pseudo-thermodynamics forces, 120
second law (*see* Second law of thermodynamics)
thermo-elastic laws, 105
third law, 84
TSI, 191, 193

Laws of unified mechanics theory
second law, 133, 134
third law, 134
Legendre-Fenchel transformation, 107
Light emitting diodes (LEDs), 131
Linearization (Consistent Jacobian)
coupling, 213
identity tensors, 212
Newton-Raphson algorithm, 213
thermodynamic state index rate, 212
yield surface, 212
Local coordinates, 26, 32, 33, 36–38
Local equilibrium assumption, 224
Local Newton iteration, 274

M

Macroscopic conservation laws, 215
Macroscopic stress, 288
Mass conservation, 217, 218
Mass transport, 403
Mass/vacancy conservation law, 407–411
Material behavior, 365, 370
Material behavior modeling, 2
Material coordinates, 32–35, 38
Material properties
definition, 373–376
isothermal and non-isothermal conditions, 377
PMMA, 377–379, 381–383
MatLAB, 380
Matrix-filler inter-phase properties
cyclic stress-strain response, 337
finite element simulation, 336, 337
particulate composites, 333, 334, 336
PMMA, 333
Poisson ratio and CTE, 333
Maximum normal stress, 22
Maxwell's equation, 398, 415
Mechanical damage, 389
Mechanical theory of heat
entropy, 138
equilibrium, 138
formulation, 138
kinetic energy (*see* Kinetic energy)
thermal equilibrium, 139
Mechanical work dissipation, 109
Mechano-Thermodynamics theory, 129
Memory, 74
Mesh sensitivity, 386, 387
Metal fatigue fracture, 390
Micro-cracks/micro-cavities, 207
Microelectronics, 396, 398
Micromechanical constitutive model, particulate composite
analytical and numerical models, 316
consistent elastic-viscoplastic tangent modulus
combining equation, 331
differentiating equation, 329, 330
increment stress-strain relationship, 329
notion, 329
elastic properties, 317–320
imperfect interface, 316
interfacial transition zone, 316
matrix material, 316
matrix-filler inter-phase (*see* Matrix-filler inter-phase properties)
modeling process
equivalence, 317, 318
interface properties, 316
micromechanics, 319
simplification, 316
thermomechanical properties, 317, 318
solution algorithm (*see* General return-mapping algorithm)
TSI, 324
verification examples
ATH material properties, 332
PMMA properties, 332
viscoplasticity (*see* Viscoplasticity model)
Microstates, 97
Minimum normal stress, 22
Mises rate-independent plasticity, 203
Molecular network mechanism, 352
Molecular network resistance, 367–371, 380
Molecular network structure, 360
Molecular orientation, 367
Molecular relaxation, 369
Molecular structure, 1
Momentum principle, 220
balanced equation, 219
Cartesian coordinates, 219
conservative body forces, 220
deformation tensor rate, 220
displacement gradient components, 220
expression, 218
kinetic energy, 221
long/short-range interactions, 219
Newton's Third Law, 218
potential energy density, 221
rectangular coordinates, 218
surface integral, 219
time rate, 218

Multiphase composites, 277
Multi-phase composites elastic moduli, 301

N
Nabarro–Herring model, 405
Natural radiation, 182
Natural sciences, 117
Newton's universal laws of motion
  classical mechanics, 5
  conservation laws, 12
  first law, 6, 7
  range of validity, 11
  second law, 8, 9
  thermodynamics, 12
  third law, 9–11
Newtonian (continuum) mechanics, 94–96
Newtonian mechanic formulation, 281
Newtonian mechanics, 73, 80, 94, 100, 106, 107, 124, 336
  force equilibrium, 411, 412
Newton-Raphson integration algorithm, 274
Nodal values, 408
Non-associative plasticity, 109
Non-conservative forces, 371
Non-dimensional interface parameter, 313
Non-equilibrium states, 84
Non-interacting solution
  average stress norm (*see* Average stress norm, matrix)
  elastoplastic response, 284
  isotropic hardening function, 284
  particles average stress, 290, 291
  statistical/microstructural information, 284
  yield function, 284
Non-isothermal conditions, 360
Non-isothermal loading, 365
Non-isothermal simulations, 380
Non-isothermal stretching
  PMMA, 390–391
Non-isothermal tests, 376
Nonlinearities, 205
Non-linear-kinematic hardening (NLK) rule, 205, 206
Non-negative consistency parameter, 206
Non-steady state diffusion, 399, 400
Normal stresses, 65

O
Observable thermodynamics variables, 100
Ohm's law, 412
Onsager reciprocal relations, 109–112
Ordered state, 96

P
Pairwise interacting solution
  approximate solutions (*see* Two-phase interaction approximate solutions)
  ensemble-average stress norm, filler particles, 298
  ensemble-average stress norm, matrix, 295
Particle filled composites, 279
Particle-matrix interactions and microstructure, 282
Particulate composites
  analytical predictions, 278
  constituent phases, 277
  interfacial layer, 309
  matrix and particles, 280
  microstructure, 279
  plastic strain, 286
  relationships, 278
  shear modulus, 311
Permutability measure, 136, 137, 174–176
Phenomenological curve fitting methods, 115
Phenomenological equations, 111
Phonons, 81, 93
Piola-Kirchhoff stress tensor, 103, 358, 361, 362
  Cauchy stress tensor, 60
  Lagrangian stress tensor, 60, 61
  material formulation, 60
  referential coordinate system, 60
  second Piola-Kirchhoff, 61, 62
Planck, 136
Plastic deformation, 194
Plastic deformation gradient, 344, 349, 363, 369
Plastic flow, 364
Plastic flow behavior, 367
Plastic flow resistance, 364
  molecular network, 375
Plastic strain rate, 364
Plastic stretch, 350, 357, 362, 369, 370
Plastic stretch rate tensors, 384
Plasticity, 106, 107
  theory, 380
Ply-methyl-meth-acrylate (PMMA)
  glass transition temperature, 377
  isothermal and non-isothermal stretching, 376
  isothermal stretching, 384–390
  isothermal tests, 377
  material parameters, 383
  material properties, 377
  non-isothermal stretching, 390–392
  rate dependent glass transition temperature, 377

ratio of yield stress to temperature *vs*. plastic strain rate plot, 379
reference glass transition temperature, 365
storage modulus and elastic modulus, 374
stress-strain curves, 382
temperature and rate dependent elastic modulus, 377, 378
temperature and rate dependent limited chain extensibility, 378
temperature dependent back stress modulus, 381
thermal properties, 377
Poisson's ratio, 278, 281, 312, 313, 333, 374, 378
Polyatomic gas molecules
external forces, 161
generalized coordinates, 161
gravity, 163
integration, 165
molecules, 162, 164
state distributions, 163
Polymer, 343
Poly-methyl methacrylate (PMMA), 278–280
Polynomial theorem, 173
Potential
thermodynamic (*see* Thermodynamic potential)
Potential function, 76, 101, 109
Power-exponent free product, 165
complexions, 171
kinetic energy, 167–169, 171
mono-atomic gases, 165
polynomial coefficient, 173
polynomial theorem, 173
probability determination, 172
quantity, 165
Regula falsi, 166
state distribution, 171–174
values, 167
velocity components, 169
Prager-Ziegler rule, 258
Prigogine's dissipative structure theory, 117
Prigogine's thermodynamics, 116
Probability density functions, 286
Probability distribution
permutability measure, 174, 176–178
potential energy, 179
reversible change of state, 175
second law of thermodynamics, 179
theory of gases, 180
thermal equilibrium, 174, 176
velocity components, 174
Prolonged ductile failure, 373
Pseudo potential, 106
Pseudo-static loading, 105
Pseudo-thermodynamics forces, 120
Pure rigid body motion, 31

Q
Quantum mechanics, 10, 13
Quasi-static electro-chemical work, 87, 88
Quasi-static processes, 111
Quasi-static work, 86

R
Rate of deformation tensor
definition, 43
small strain tensor, 44, 45
spin matrix, 44
stretching tensor, 43
true strain, 45, 46
Rayleigh-Bénard instability, 116
Reaction force, 11
Recoverable work, 103
Reduced Cosserat, 242
Reduced couple stress theory
asymmetric component, 252
displacement functions, 251
finite element method implementation, 252
virtual displacements, 251
virtual work, 252
Reference coordinate system, 343
Reference frame
Almansi tensor, 348
Cauchy tensor, 347, 348
elastic and plastic velocity gradient, 350
elastic stretch rate tensor, 350
plastic deformation gradient, 349
quantity/equation, 347
rotation tensor, 347
spin tensor, 348, 350
stress tensor, 347
stretch rate tensor, 348, 349
transformation rules, 347
velocity gradient, 348
Reference temperature, 379
Referential coordinate system, 42, 63
Regression analysis method, 380
Regula falsi, 166
Relaxation process, 367
Relaxation time, 323, 335
Renormalization procedure, 282
Representative volume element (RVE), 2, 207, 224, 277, 280

Resistors, 131
Return mapping algorithm
  classical theory rate dependent model, 214
  deviatoric strain increment vector, 209
  explicit/implicit method, 209
  flow rule, 210
  generalized midpoint rule, 209
  integration parameter, 210
  linearization (*see* Linearization (Consistent Jacobian))
  Newton method, 211
  rate independent case, 211
  relative stress tensor, 210
  yield surface, 210
Rice, 118, 119, 121, 122
Richeton's model, 366
Rotation tensor, 345, 347
Rubber-elasticity model, 367
Rubbery modulus, 375
Rubbery shear modulus, 375

S
Saturation value, 376
Second law of thermodynamics, 119, 351, 372, 417
  Clausius-Duhem inequality, 80
  definition, 80, 81
  entropy (*see* Entropy)
  pendulum, 80
Second law of unified mechanics theory, 133, 134
Second Piola-Kirchhoff stress, 61–63, 368
Second-order tensor, 14, 18
Self-diffusion, 399
Self-organizing systems, 117
Shear strain, 30, 59
Shear stresses, 19, 20, 28, 65
Single ellipsoidal inclusion, 285
Skew symmetric tensor, 220
Small rotation formulation, 31, 32
Small strain
  definition, 29
  local coordinates, 33, 36–38
  material coordinates, 32–35
  pure rigid body motion, 31
  pure uniaxial strain, 29
  shear strain, 30, 31
    local coordinates, 39
    spatial *vs.* local coordinates, 40
  small rotation formulation, 31, 32
Small strain theory, 419
Sorbolle's electromigration theory, 401
Spatial description, 32, 49
Spherical stress tensor, 26
Spin tensor, 43, 344, 345, 348, 350, 420
State distribution, 140, 141
State functions, 85, 94, 95
State variables, 85, 94, 95, 98, 99, 101, 102, 105–108
Statistical mechanics, 135, 136, 191
Stiffness matrix, 416
Strain
  definition, 28
  formulation, 28
  samll (*see* Small strain)
Strain equivalence principle, 208, 322
Strain rate, 363–367, 369, 371, 373, 377–379, 381, 383–386
Strain rate histories, 385, 386
Strain rate measures, 371
Stress
  arbitrary plane, 15–18
  definition, 12–14
Stress components, 371
Stress gradient, 395–397, 400
  diffusion driving force, 404–406
  induced diffusion, 406
Stress invariants
  characteristic equation, 24
  deviatoric stresses, 27, 28
  hydrostatic stress, 26
  principal axes, 25
  scalar quantity, 25
  spherical stress tensor, 26
Stress measures, 354, 357
Stress modulus, 380
Stress-strain curves, 376, 380, 382
Stress-strain relationship, 231, 336
Stress tensor, 19, 347, 355
  arbitrary tetrahedron, 22
  Cartesian coordinate system, 21
  direction cosines, 23, 24
  principal axes, 22
  principal planes, 22
  principal stresses, 24
  symmetric, 22, 24
  3-dimensional stress-tensor, 22
Stretching rate tensor, 344, 345
Stretching tensor, 43
Stretch-like internal variable, 362
Sudden brittle failure, 373
Surface stress vector, 123
Surface tractions, 355

Symmetric Cauchy stress tensor, 244
Symmetric tensors, 345
Symmetric unimodular tensor, 362

T
Temperature dependent, 373
Temperature gradient diffusion driving force, 402, 403
Tensile stress, 65, 404
Tension-compression cyclic loading, 195
Tensorial variables, 362
Tetrahedron, 15, 16
Theory of gases, 180
Thermal conductivity tensor, 420
Thermal damage, 390
Thermal dissipation, 372
Thermal gradient, 397
Thermodynamic chambers, 75
Thermodynamic forces, 105, 106, 216, 227
Thermodynamic fundamental equation
  balance equation, 213
  conservation laws (*see* Conservation laws)
  entropy, 215
  entropy source, 214
  force, 216
  Gibbs relation, 215
  irreversible fluxes, 216
  isolated/closed system, 214
  probability and statistics, 215
  second law, 215
  thermal equilibrium, 215
Thermodynamic potential
  dissipation potential, 106
  dissipation power and onsager reciprocal relations, 109–112
  forces, 105
  formulation, 101–105
  intrinsic and thermal dissipation decoupling, 108
  time independent dissipation, 108
  variables (*see* Thermodynamic variables)
Thermodynamic restrictions
  Cauchy's stress, 356
  defect energy, 352
  deformed configuration, 353
  dual-mechanism elastic-viscoplastic constitutive model, 358
  elastic symmetric Mandel stress, 357
  Helmholtz free energy, 352, 353, 359
  intermediate configuration, 354
  intermolecular mechanism, 352
  irreversible entropy, 359
  molecular network mechanism, 352
  multi-mechanism generalization, 352
  Piola-Kirchhoff stress tensor, 358
  plastic stretch rate, 357
  reference configuration, 353
  specific entropy, 359
  stress measures, 354, 357
  stress tensor, 355
  surface tractions, 355
  transformation rules, 354
  velocity gradient, 357
  virtual work rate, 353
Thermodynamic state index (TSI), 232, 324, 329, 335, 336, 411, 418, 421, 423
  Boltzmann equation, 192
  computational mechanics, 194
  definition, 193
  degradation, energy storage capacity, 193, 194
  disorder parameter, 191
  entropy, 133, 191, 194
  fundamental equation, 191
  gas constant, 192
  laws of thermodynamics, 193
  material, 192
  monotonic loading test, 195–197
  one-dimensional spring, 193
  plastic deformation, 194
  properties, 195
  ramification, 194
  second law of thermodynamics, 191
  specimen dimensions, 195
  statistical mechanics, 191
  strain energy, 193
  tension-compression cyclic loading, 195
  TSI axis, 191
Thermodynamic variables
  appropriate choice, 100
  microstate, 100
  Newton's laws of motion, 100
  observable, 100
Thermodynamics
  Boltzmann distribution, 132
  classical irreversible, 116
  continuum mechanics, 2
  continuum system, 73
  corrosion-fatigue volumetric entropy, 129
  elastic, 122
  endochronic plasticity theory, 118
  engineering mechanics, 116
  entropy (*see* Entropy)
  equilibrium, 74–75, 127
  equilibrium vacancy, 407

Thermodynamics (*cont.*)
- equilibrium/diffusion-driving force, 398
- fatigue, 119
- fatigue crack nucleation, 128
- FFE, 128, 129
- first law, 75–80, 351
- flow potential, 122
- fluctuation theorem, 127
- fluxes, 116
- forces, 119–121, 126
- fracture, 128
- fundamental equation, 396
- general systems theory, 117, 130
- internal energy, 125
- irreversible entropy generation rate, 131
- irreversible processes, classical theory, 122
- Lagrangian mechanics, 115
- LEDs, 131
- macroscopic scale, 73
- Mechano-Thermodynamics theory, 129, 130
- modeling mechanics, 124
- Newtonian mechanics, 73, 115, 124
- number of cycles to failure *vs.* FFE, 128
- phenomenological curve fitting methods, 115
- Prigogine, 116
- Prigogine's dissipative structure theory, 117
- principle, 419
- probabilistic model, 124
- pseudo-thermodynamics forces, 120
- Rayleigh-Bénard instability, 116
- rice, 118, 119, 121–124
- second law, 119, 351
- self-organization, 117
- stability, 351
- surface stress vector, 123
- thermal conductivity, 125
- tribo-fatigue entropy, 129, 130
- TSI, 411, 418, 421
- variables, 127
- yield surface, 119

Thermodynamics equilibrium, 74–75

Thermodynamics Gibbs relation, 215

Thermodynamics state index (TSI), 94, 371–373
- Boltzmann equation, 135
- entropy, 136
- materials, 135
- Mechanical theory of heat (*see* Mechanical theory of heat)
- non-equilibrium states, 137
- permutability measure, 136, 137
- space-time coordinate system, 134
- statistical mechanics, 135, 136

Thermo-elastic effect, 371

Thermo-elastic laws, 105

Thermo-elastic materials, 105

Thermo-mechanical analysis, 3

Thermo-mechanical constitutive model, 233

Thermo-mechanical loading, 106, 373, 390

Thermomigration, 233
- and electromigration (*see* Electromigration)
- diffusion driving force, 397
- electrical current, 395
- external stress gradients, 397
- intrinsic part, 403
- Joule heating, 395
- material degradation process, 395
- principal, 395

Thermos-electro-mechanical-chemical-radiation processes, 95

Thermos-mechanical loading, 233

Third law of thermodynamics, 84

Third law of unified mechanics theory, 134

Three-phase composites, non-interacting solution
- elastic moduli, 301
- ensemble-average stress norm, matrix
    - algebra, 303
    - alternative expression, 305
    - CTE mismatch, 306
    - Eigen-strain, 302
    - Eshelby theory, 302
    - loading function, 304
    - macroscopic stress, 304
    - perturbations, 302
    - plastic strain, 302
    - positive definite fourth-rank tensor, 303
    - probability density functions, 303
    - q-phase inhomogeneity, 302
- ensemble-average stress, filler particles
    - average internal stress, 307
    - definite fourth-rank tensor, 308
    - eigen-strain, 306
    - forms, 306
    - volume-averaged stress tensor, 307
- spherical voids, 300

Time independent dissipation (instantaneous dissipation), 108

Toupin-Mindlin higher order stress theory
- displacement gradients, 242
- elastic strain gradients, 244
- Lamè constants, 244
- strain-displacement kinematic relations, 244
- symmetric Cauchy stress tensor, 244

Traction, 12
Transformation rules, 347
Tribo-fatigue entropy, 129, 130
True stress, 385, 386
True stress-strain curves, 384, 385
Two-phase composites
  dispersed elastic spherical inhomogeneity, 277
  effective elastic moduli, 279
  elastic matrix, 277
  inter-particle interactions, 278
  non-interacting solution (*see* Non-interacting solution)
  pairwise interacting solution (*see* Pairwise interacting solution)
  particle interaction, 290
  plastic deformation, 290
  viscoplastic matrix, 284
Two-phase interaction approximate solutions
  conditional probability function, 294
  ensemble integration, 294
  expression, 294
  mathematical manipulation, 295
  non-interacting, 292, 293
  particle configuration, 292
  procedure, 293
  spherical particle, 293
Two-point conditional probability function, 294

U
UMT Cosserat continuum implementation
  coupling, 256
  entropy generation rate, 266
  radial return algorithm- rate dependent model, 269–273
  radial return algorithm- rate independent model, 266–269
  rate dependent material without degradation
    constitutive equation, 262
    TSI, 263, 264
    yield surface, 262
  rate independent material without degradation
    consistency parameter, 258, 260
    constitutive model, 257, 261
    constitutive tensor, 262
    couple back-stress, 257
    deviatoric component, 257
    evolution equations, 259
    flow rules, 258
    flow theory, 261
    generalized strain tensor, 257
    hardening parameter, 257
    Hooke's law, 259–262
    kinematic hardening modulus, 258
    rate form, 256
    yield surface, 257
  size effects, 256
  viscoplastic creep law, 265
UMT rate dependent model-strain gradient-formulation, 267
Uniaxial stress-strain hysteresis loop, 339, 340
Uniaxial tension, 205
Unified mechanics theory, 100, 395, 411, 416
  finite deformations (*see* Finite deformations)
  TSI (*see* Thermodynamic state index (TSI))
Uniform motion, 7
Universal law of degradation
  additive constant, 185
  electromagnetic radiation theory, 182
  entropy, 183
  molecular-kinetic considerations, 182
  probability, 184
  Stirling's theorem, 185
  thermodynamics, 182

V
Vacancy diffusion mechanism, 406
Vanishing internal resistance, 365
Vector quantity, 399
Velocity gradient, 343, 344, 348, 357
Virtual work
  principal, 355–357
Virtual work rate, 353
Viscoplastic creep rate law, 207
Viscoplastic flow rule, 360, 389
Viscoplastic model, 365–367, 416
Viscoplastic strain, 290
Viscoplastic strain rate, 366
Viscoplasticity, 107
Viscoplasticity model
  back stress evolution, 323
  ensemble-volume averaged stress norm, 322
  filler particles, 322
  kinematic hardening behavior, 322
  kinematic hardening function, 324
  plastic strain increment equations, 322
  UMT framework, 322
  viscoplastic behavior, 322
  viscoplastic strain rate, 323
  von-Mises type, 322
  yield function, 323
Viscous flow, 363

Void, 395
Volume-averaged strain tensor, 280
Volume-averaged stress tensor, 280, 290
von Mises yield criterion, 284
Von Misses equivalent stress, 264
Vorticity tensor, 43

W
Walpole's formulae, 310
Wien's displacement law
  energy density, 186, 187
  entropy, 186–188
  Kirchoff's theorem, 186
  Kirchoff-Clausius law, 186
  numerical values, 189, 190
  volume density, 186
  wavelength, 188
Williams-Landel-Ferry (WLF), 364, 365, 373, 374

Y
Yield behavior, 363
Yield stress, 363
Yield surface, 122, 204
Young's modulus
  bulk modulus, 313
  effective variations, 314
  experimental mean value, 279
  interphase, 333
  interphase properties, 320
  particle volume fraction, 279
  PMMA, 332
  Poisson's ratio, 278, 312

Z
Zero Kelvin temperature, 84

Printed in the United States
by Baker & Taylor Publisher Services